INTELLIGENT DECISION SUPPORT

[illegible]ON LIB[illegible]

General Editors: W. [illegible] and G. Eberlein (Munich)

Series A: Philosophy and Methodology of the Social Sciences

Series B: Mathematical and Statistical Methods

Series C: Game Theory, Mathematical Programming and Operations Research

Series D: System Theory, Knowledge Engineering and Problem Solving

SERIES D: SYSTEM THEORY, KNOWLEDGE ENGINEERING AND PROBLEM SOLVING

VOLUME 11

Scope: Design, study and development of structures, organizations and systems aimed at formal applications mainly in the social and human sciences but also relevant to the information sciences. Within these bounds three types of study are of particular interest. First, formal definition and development of fundamental theory and/or methodology, second, computational and/or algorithmic implementations and third, comprehensive empirical studies, observation or case studies. Although submissions of edited collections will appear occasionally, primarily monographs will be considered for publication in the series. To emphasize the changing nature of the fields of interest we refrain from giving a clear delineation and exhaustive list of topics. However, certainly included are: artificial intelligence (including machine learning, expert and knowledge based systems approaches), information systems (particularly decision support systems), approximate reasoning (including fuzzy approaches and reasoning under uncertainty), knowledge acquisition and representation, modeling, diagnosis, and control.

(Volumes published in this series are listed at the end of this volume.)

THEORY AND DECISION LIBRARY

General Editors: W. Leinfellner (*Vienna*) and G. Eberlein (*Munich*)

Series A: Philosophy and Methodology of the Social Sciences

Series B: Mathematical and Statistical Methods

Series C: Game Theory, Mathematical Programming and Operations Research

Series D: System Theory, Knowledge Engineering and Problem Solving

SERIES D: SYSTEM THEORY, KNOWLEDGE ENGINEERING AND PROBLEM SOLVING

VOLUME 11

Scope: Design, study and development of structures, organizations and systems aimed at formal applications mainly in the social and human sciences but also relevant to the information sciences. Within these bounds three types of study are of particular interest. First, formal definition and development of fundamental theory and/or methodology, second, computational and/or algorithmic implementations and third, comprehensive empirical studies, observation or case studies. Although submissions of edited collections will appear occasionally, primarily monographs will be considered for publication in the series. To emphasize the changing nature of the fields of interest we refrain from giving a clear delineation and exhaustive list of topics. However, certainly included are: artificial intelligence (including machine learning, expert and knowledge based systems approaches), information systems (particularly decision support systems), approximate reasoning (including fuzzy approaches and reasoning under uncertainty), knowledge acquisition and representation, modeling, diagnosis, and control.

The titles published in this series are listed at the end of this volume.

INTELLIGENT DECISION SUPPORT

Handbook of Applications and Advances of the Rough Sets Theory

Edited by

ROMAN SŁOWIŃSKI

Institute of Computing Science,
Technical University of Poznań, Poland

KLUWER ACADEMIC PUBLISHERS

DORDRECHT / BOSTON / LONDON

Library of Congress Cataloging-in-Publication Data

ISBN 978-90-481-4194-4

Published by Kluwer Academic Publishers,
P.O. Box 17, 3300 AA Dordrecht, The Netherlands.

Kluwer Academic Publishers incorporates
the publishing programmes of
D. Reidel, Martinus Nijhoff, Dr W. Junk and MTP Press.

Sold and distributed in the U.S.A. and Canada
by Kluwer Academic Publishers,
101 Philip Drive, Norwell, MA 02061, U.S.A.

In all other countries, sold and distributed
by Kluwer Academic Publishers Group,
P.O. Box 322, 3300 AH Dordrecht, The Netherlands.

Printed on acid-free paper

CONTENTS

PREFACE

Making decisions under uncertainty and imprecision is one of the most challenging problems of our age, which for a long time have been tackled by philosophers, logicians and others. Recently AI researchers have given new momentum and flavor to this area.

Expert systems, decision support systems, machine learning, inductive reasoning, pattern recognition, are areas where decision making under uncertainty is of primary importance.

There are known several mathematical models of uncertainty (e.g. fuzzy sets, theory of evidence), however, there is widely shared view that the problem is far from being fully understood.

The concept of a rough set has been proposed as a new mathematical tool to deal with uncertain and imprecise data, and it seems to be of significant importance to AI and cognitive sciences both from theoretical and practical points of view.

A special attention should be paid to the decision support systems, basic topic of this book – where the rough sets approach offers a new insight and efficient algorithms.

The rough sets philosophy means a specific view on representation, analysis and manipulation of knowledge as well as a new approach to uncertainty and imprecision.

Knowledge is understood here as an ability to classify objects (states, events, processes etc.), i.e. we assume that knowledge is identified with a family of various classification patterns. Objects being in the same class are *indiscernible* by means of knowledge provided by the classification and form elementary building blocks (*granules, atoms*) which are employed to define all basic concepts used in the rough sets philosophy.

In particular, the granularity of knowledge causes that some notions cannot be expressed precisely within available knowledge and can be defined vaguely only. This leads to the so called "boundary–line" view on imprecision, due to Frege who writes (cf. Frege (1903)):

The concept must have a sharp boundary. To the concept without a sharp boundary there would correspond an area that had not a sharp boundary–line all around.

In the rough sets theory each imprecise concept is replaced by a pair of precise concepts called its *lower* and *upper approximation*; the lower approximation of a concept consists of all objects which *surely* belong to the concept whereas the upper approximation of the concept consists of all objects which *possibly* belong to the concept in question. Difference between the lower and the upper approximation is a *boundary region* of the concept, and it consists of all objects which cannot be classified with certainty to the concept or its complement. These approximations are fundamental tools of reasoning about knowledge.

For algorithmic reasons, i.e. in order to provide easy processing and manipulation of knowledge, suitable representation of knowledge is needed. To this end the tabular form, known as an *information system, attribute-value system* or *knowledge representation system* – is used. Attributes in the information system represent various classification patterns. In this way knowledge can be replaced by data and knowledge processing can be replaced by data manipulation. In particular, concepts (subsets of objects) can now be defined (exactly or approximately) in terms of attribute-values, and a variety of other concepts needed for reasoning about knowledge can by expressed in attribute-value terms. Mostly, we are interested in discovering various relations between attributes, like exact or approximate dependency of attributes (cause-effect relations), redundancy of attributes and significance of attributes, and in generation of decision rules from data.

The rough sets philosophy turned out to be a very effective, new tool with many successful real-life applications to its credit. It is worthwhile to stress that no auxiliary assumptions about data, like probability or membership function values, are needed, which is its great advantage.

The rough set concept has an overlap with other ideas developed to deal with uncertainty and imprecision, in particular with fuzzy sets (cf. Dubois and Prade (1990)), evidence theory (cf. Skowron and Grzymala-Busse (1992)), statistics (cf. Krusińska, Słowiński and Stefanowski (1992)) albeit it can be viewed in its own rights.

The book edited by Prof. Roman Słowiński shows a wide spectrum of applications of the rough set concept, giving the reader the flavor of, and the insight in, the methodology of the newly developed discipline.

Although the book emphasizes applications, comparison to other related methods and further developments receive due attention. In this sense, the book can be seen as a continuation of the book on theoretical foundations of rough sets (cf. Pawlak (1991)).

I am sure that the reader will benefit from studying the book by gaining a new tool to solve his or her problems as well as a new exciting area of research.

Zdzisław PAWLAK
Warsaw, April 1992

References

Dubois, D. and Prade, H. (1990). Rough Fuzzy Sets and Fuzzy Rough Sets. *International Journal of General Systems* **17**, pp.191-209

Frege, G., (1903). Grundgesetze der Arithmetik, 2. In Geach and Black (Eds.), *Selections from the Philosophical Writings of Gotlob Frege*, Blackwell, Oxford, 1970.

Krusińska, E., Słowiński, R. and Stefanowski, J. (1992). Discriminant Versus Rough Sets Approach to Vague Data Analysis. *Journal of Applied Stochastic Models and Data Analysis* **8**, (to appear).

Pawlak, Z. (1991). *Rough Sets. Theoretical Aspects of Reasoning about Data.* Kluwer Academic Publishers, Dordrecht/Boston/London.

Skowron, A. and Grzymala-Busse, J. (1992). From Rough Set Theory to Evidence Theory. In M. Fedrizzi, J. Kacprzyk and R.R. Yager (Eds.), *Advancess in Dempster-Shafer Theory*, Wiley, New York (to appear).

SCOPE AND GOALS OF THE BOOK

Roman SŁOWIŃSKI
Institute of Computing Science
Technical University of Poznań
60-965 Poznań, Poland

Intelligent decision support is based on human knowledge understood as a family of classification patterns related to a specific part of a real or abstract world. When the knowledge is gained in the process of learning by experience, it is induced from empirical data. The data are often presented as a record of objects (events, observations, states, patients, ...) described by a set of multi-valued attributes (features, variables, characteristics, conditions, ...). The objects are associated with some decisions (actions, opinions, classes, diagnoses, ...) taken by an expert (decision-maker, operator, doctor, diagnostician, ...). Such a record is called information system.

A natural problem of knowledge analysis consists then in discovering relationships between objects and decisions, so as to get a minimum representation of the information system in terms of decision rules.

The rough sets theory created by Z. Pawlak in early 80's provides a tool for such an analysis. The observation that one cannot distinguish objects on the basis of given information about them is the starting point of the rough sets philosophy. In other words, imperfect information causes indiscernibility of objects. The indiscernibility relation induces an approximation space made of equivalence classes of indiscernible objects. A rough sets is a pair of a lower and an upper approximation of a set in terms of these equivalence classes.

Using the rough sets approach, one can obtain among others the following results:

- evaluate importance of particular attributes in relationships between objects and decisions;
- reduce all redundant objects and attributes, so as to get minimal subsets of attributes (reducts) ensuring a satisfactory approximation of the classification made by decisions;
- create models of the most representative objects in particular classes of decisions;
- represent the relationships between objects, described by a reduct, and decisions, in the form of a set of decision rules (if ... then ...) called decision algorithm.

The decision algorithm represents knowledge gained by the expert on all the set of objects. This representation is free of redundancies, so typical for real data bases, which cover the most important factors of experience. The algorithm and the models of objects can be used to support decisions concerning new objects.

Since its foundation, the rough sets theory has been applied with success in many domains. The success is mainly due to the following features:

- it analyses only facts hidden in data,
- it does not need any additional information about data,
- it does not correct inconsistencies manifested in data; instead, produced rules are categorized into certain and possible,
- it finds a minimal knowledge representation,
- it is conceptually simple and needs simple algorithms.

The theory has gained the degree of maturity and acceptance which urges to make a synthesis of experience with rough sets. After the monograph on foundations of the rough sets theory (cf. Z. Pawlak, *Rough Sets – Theoretical Aspects of Reasoning about Data*, Kluwer Academic Publishers, TDL.D, Dordrecht, 1991), the present volume is a first tentative of such a synthesis. It is thought of as a handbook, i.e. a source book of applications, new developments and comparisons with related methodologies.

The book is composed of three parts:

Part I: Applications of the rough sets approach to intelligent decision support
Part II: Comparison with related methodologies
Part III: Further developments

In Part I, GRZYMALA-BUSSE presents the system LERS for rule induction. The system handles inconsistencies in the input data due to its usage of rough set theory principle. It induces certain rules and posible rules which may be practically used by putting them into the knowledge base of rule–based expert system.

MRÓZEK discusses the use of rough sets in computer implementation of a rule–based control of industrial processes. Decision rules are induced from control protocols describing a human operator control. The control process of a rotary clinker kiln is chosen for illustration.

NOWICKI, SŁOWIŃSKI and STEFANOWSKI apply the rough sets approach to diagnostic classification of mechanical objects on the basis of vibroacoustic symptoms. The rough sets analysis concerns: evaluation of diagnostic capacity of symptoms, comparison of different methods of defining symptom limit values, reduction of the set of symptoms, construction of a classifier of the technical state. An example of rolling bearings is considered.

SZLADOW and ZIARKO develop a process model of heavy oil upgrading using a machine learning system based on rough sets. The key advantages of using the rough sets approach are: it allows the use of qualitative and quantitative process information in the model; it provides a unified description of temporal events and patterns; it permits the use of "raw" sensor data without preprocessing.

ZIARKO presents one more application of the rough sets approach to acquisition of control algorithms from process or device operation data. The presentation is illustrated with a comprehensive example showing the generation of the control algorithms from simulated data representing movements of a robot arm.

SŁOWIŃSKI describes his experience with a medical application of the rough set concept. In particular, he analyses an information system containing patients with duodenal ulcer treated by highly selective vagotomy (HSV). Qualitative and quantitative attributes are used to describe the patients. Using the rough sets approach, all superflous objects and attributes have been reduced from the information system and a decision algorithm has been derived in order to support decisions about the treatment of new duodenal ulcer patients by HSV.

KANDULSKI, MARCINIEC and TUKAŁŁO apply the rough sets analysis of data concerning some commonly considered factors of surgical wound infection. The aim is to establish a hierarchy of these factors with reference to a particular surgical clinic or department.

TANAKA, ISHIBUCHI and SHIGENAGA present another medical application. They describe a method of constructing a fuzzy inference system by introducing fuzzy intervals representing fuzzified attribute values. The reason for using fuzzy intervals is that fuzzy if–then rules can cover the whole space of attribute values while it is not the case of rules derived from only lower approximations. The classification power of the proposed inference model is demonstrated on a diagnostic problem.

KRYSIŃSKI applies the rough sets approach to analysis of relationships between chemical structure and antimicrobial activity of quaternary ammonium compounds. The decision algorithm derived from the information system can be used to support decisions concerning synthesis of new antimicrobial compounds.

HADJIMICHAEL and WASILEWSKA present an interactive probabilistic inductive learning system based on the rough set concept. The system is used to study voter preferences in the 1988 U.S.A. presidential election. Results include an analysis of the predictive accuracy of the generated rules, and an analysis of the semantic content of the rules.

REINHARD, STAWSKI, WEBER and WYBRANIEC-SKARDOWSKA report their experience with the use of rough sets to decision support concerning control of water conditions on a polder. Two information systems composed of objects corresponding to periodic measurements of weather conditions, water surface levels and soil humidity are analysed.

TEGHEM and CHARLET apply the rough sets approach to draw premonitory factors for earthquakes on the basis of radon emanation. The field of application concerns the Mons Basin (Belgium) with various geological environment, a geothermal system and a rather low seismic activity. The analysis has allowed to discriminate the sites with a particularly high sensitivity to a seismic event.

LUBA and RYBNIK approach two seemingly different problems. One concerns information systems theory and the other is connected to logic synthesis. An efficient algorithm for reduction of attributes (arguments) as well as functional decomposition of decision (truth) tables is presented. Using manipulations based on both rough sets and Boolean algebra, the decision table is reduced and decomposed so as to get an efficient implementation.

In Part II, DUBOIS and PRADE demonstrate that fuzzy sets and rough sets aim to different purposes and that it is natural to try to combine the two models of uncertainty (vagueness for fuzzy sets and coarseness for rough sets) in order to get a more accurate account of imperfect information. In result of the combination, a fuzzy rough set is defined whose properties and potential usefulness are indicated.

NAKAMURA considers a concept of fuzzy–rough classification. Two logics based on fuzzy–rough classifications are proposed and compared with the indiscernibility relations. Decision procedures for the proposed logics are given in a tableau style, useful for an automated reasoning.

KRUSIŃSKA, BABIC, SŁOWIŃSKI and STEFANOWSKI make a comparison study of the rough sets approach and probabilistic techniques, in particular, discriminant analysis and probabilistic inductive learning, on a common set of medical data. A general discussion on similarities and differences among compared methods is given. In the comparative computational experiment, particular attention is paid to data reduction and construction of decision rules.

TEGHEM and BENJELLOUN describe some experiments made to compare the rough sets method of data analysis with the Quinlan's method using the notion of entropy and an ordinal discriminant analysis method using the Benzekri's distance. Various pros and cons are derived from this comparison.

LIN studies approximation theory via rough sets, fuzzy sets and topological spaces. Rough sets theory is extended to fuzzy sets and Frechet topological spaces which are shown to be equivalent. It is shown, moreover, that the three theories allow one to draw an exact solution whenever there are adequate approximations.

POLKOWSKI is also dealing with topological properties of rough sets. He proposes metrics for rough sets that make them into a complete metric space and shows that the metric convergence encompasses and generalizes approximative convergence.

In Part III, ORLOWSKA and ORLOWSKI show the role of reducts in the definitions and other fundamental notions of the information systems theory, such as functional dependencies and redundancy. They demonstrate different ways of computing reducts and propose a new algorithm which improves the execution time and avoids repetition of the whole procedure for dynamically changing data.

SKOWRON and RAUSZER introduce the notions of discernibility matrix and discernibility function related to any information system. Several algorithms for solving problems related to the rough definability, reducts, core and dependencies follow from demonstrated properties of these notions.

SŁOWIŃSKI and SŁOWIŃSKI analyse rough classification of patients after highly selective vagotomy (HSV) for duodenal ulcer from the viewpoint of sensitivity of previously obtained results to minor changes in the norms of attributes. An extensive computational experiment leads to the general conclusion that original norms following from medical experience were well defined.

LENARCIK and PIASTA consider the problem of norms for discretization of continuous attributes from a probabilistic perspective. They introduce the concepts of a random information system and of an expected value of the quality of classification. The method of finding suboptimal norms based on these concepts is presented and illustrated with data from concretes' frost resistance investigations.

VAKARELOV develops a representation theory for Scott consequence systems in Pawlak's information systems.

WÓJCIK and WÓJCIK present a novel methodology for high performance allocation of processors to tasks based on an extension of the rough sets to the rough grammar. It combines a global load balancing with a dynamic task scheduling on a multiprocessor machine.

YASDI describes a method for learning classification rules from a data base. The rough sets theory is used in designing the learning algorithm. Knowledge representation is considered in the framework of a conceptual schema consisting of a semantic model and an event model.

SŁOWIŃSKI and STEFANOWSKI characterize two software implementations of the rough sets approach. The first one, 'RoughDAS', performs main steps of the analysis of data in an information system. The second, 'RoughClass', is intended to support classification of new objects. Some problems of sensitivity analysis of rough classification are also discussed.

A glossary of basic concepts of the rough sets theory is appended to the book for reader's convenience.

The book ends in a subject index.

If the book can be considered as representative for the state–of–the–art in the rough sets theory and applications, this is due to a large participation of outstanding specialists in this field. I wish therefore to thank very much all the Authors for their valuable contributions and active cooperation in this collective work. I wish also to extend my thanks to Prof. Zdzisław PAWLAK who kindly accepted to write the Preface to the book.

Part I

APPLICATIONS OF THE ROUGH SETS APPROACH TO INTELLIGENT DECISION SUPPORT

Part I
Chapter 1

LERS–A SYSTEM FOR LEARNING FROM EXAMPLES BASED ON ROUGH SETS

Jerzy W. GRZYMALA-BUSSE
Department of Computer Science
University of Kansas
Lawrence, KS 66045, U.S.A.

Abstract. The paper presents the system LERS for rule induction. The system handles inconsistencies in the input data due to its usage of rough set theory principle. Rough set theory is especially well suited to deal with inconsistencies. In this approach, inconsistencies are not corrected. Instead, system LERS computes lower and upper approximations of each concept. Then it induces certain rules and possible rules. The user has the choice to use the machine learning approach or the knowledge acquisition approach. In the first case, the system induces a single minimal discriminant description for each concept. In the second case, the system induces all rules, each in the minimal form, that can be induced from the input data. In both cases, the user has a choice between the local or global approach.

1. Introduction

Research in machine learning, an area of artificial intelligence, has been much intensified recently. Development of machine learning is so intensive that entire subareas have emerged, such as similarity-based learning, explanation-based learning, computational learning theory, learning through genetic algorithms, and learning in neural nets.

The most research effort has been done in similarity-based learning, more specifically, in one of the sub-areas, machine learning from examples. Here, learning is based on establishing similarities between positive examples, representing the same concept, and dissimilarities between positive examples and negative examples, representing other concepts. Similarity-based learning is also called empirical learning, to emphasize the fact that it is based on inducing the underlying knowledge from empirical data. Thus, a priori knowledge is not required for the similarity-based learning method, although it is biased by a corresponding, specific method of learning (e.g., a form in which the knowledge is finally expressed).

Frequently, the goal of machine learning from examples is finding a discriminant description of the concept [40]. Thus, the task is to include in the description of the concept all positive examples from the concept and to exclude from the description the complementary set, containing all negative examples. Some of the systems of machine learning from examples induce a minimal discriminant description.

Knowledge in the form of rules, decision trees, etc, induced by learning from training examples, is easy for humans to comprehend. Besides, such rules may be practically used by putting them into the knowledge base of rule-based expert systems and used for inferences made by the systems. However, after learning from training data, knowledge must be validated. An example which was not used in the training input data for learning may be incorrectly classified by rules. Therefore, knowledge should be modified.

Moreover, the learning system is frequently forced to deal with uncertainty, or work with presence of noise [32, 39, 40, 42, 55, 64]. There are two main reasons for uncertainty: incomplete evidence or conflicting evidence [20]. The problem of inducing rules from attributes with incomplete values was discussed in [25, 32, 39, 54]. The main focus of this paper is rule induction from conflicting evidence, i.e., from inconsistent examples. Practically, it means that two examples, classified by the same values of attributes, belong to two different concepts.

Most existing learning systems handling uncertainty are based on probability theory. Some of them use the subjective Bayesian approach [8, 15–17, 21, 22, 45, 57], while some add statistical tools [4, 5, 32, 39]. A popular method is to induce decision trees [7, 10, 35, 62]. The most popular system from this class, ID3, uses the principle of information gain for the best choice of an attribute in constructing a decision tree. Although the original system ID3 was unable to deal with inconsistencies, its modified version has an incorporated chi-square test and can deal with uncertain data [33, 34, 43, 44, 46, 50–56]. The genuine system was enhanced in many ways by many authors and influenced many other systems. For example, incremental versions of ID3 were developed under names ID4 [60] and ID5 [65], system pruning trees, like C4 and C4.5 were introduced [14, 55, 56], and some systems, e.g., CN2 [18] that induce decision lists [58] or system PRG [37], were influenced by ID3. The performance of many systems is compared with that of ID3, see e.g., [1, 13, 60, 63, 65, 66]. The main advantage of all of these systems is that their algorithm is relatively simple and efficient. The main disadvantage is that they produce decision trees, which must be transformed into rules in order to be utilized in production systems [53]. As a result, rules are not in their minimal discriminant form. The learning system AQ15 [41] induces rules from not necessarily consistent data; if probabilities of inconsistent examples are given, they are classified according to the maximum likelihood. Another system inducing rules from inconsistent data is presented in [31]. Among other approaches to inductive learning under uncertainty, human categorization was used in [28]. Fuzzy set theory, used in machine learning from examples [2, 38, 49], is another approach to deal with uncertainty.

Recently, a lot of attention has been paid to a study of machine learning algorithms using an approach called probably approximately correct learning. The idea was introduced in [68] and developed in many papers, see e.g. [29, 30, 69]. Using this approach authors expect to get some answers for learnability in the presence of noise [3, 6, 9, 36, 59, 61].

In the early eighties Z. Pawlak introduced a new tool to deal with uncertainty, rough set theory [47, 48]. The main advantage of rough set theory is that it does not need any preliminary or additional information about data (like prior probability in probability theory, grade of membership in fuzzy set theory, etc.). Rough set theory is especially well suited to deal with inconsistencies. In this approach, inconsistencies are not corrected. Instead, lower and upper approximations of the concept are computed. On the basis of these approximations, two corresponding sets of rules are induced: certain and possible. This idea was presented for the first time in [23]. Certain rules are categorical and may be further employed using classical logic. Possible rules may be further processed using either classical logic or any theory to deal with uncertainty. The rule induction system, producing certain and possible rule, is called LERS (Learning from Examples based on Rough Sets). The first implementation of the system was done in 1988 [19, 27]; further improvements of the system and its performance were reported, e.g., in [11, 12, 13, 24, 26]. Another approach to rule induction, also based on rough set theory, but not using the idea of certain and possible rules, was developed in [67, 70, 71].

System LERS tests the input data for consistency. In the case of inconsistent data the system computes lower and upper approximations of each concept. The user has a choice to use the machine learning approach or the knowledge acquisition approach. In the first case, the system induces a single minimal discriminant description for each concept. In the second case, the system induces all rules, in the minimal form each, that can be induced from the input data. In both cases, the user has a choice between local or global approach. In the local approach, a subsystem LEM2 (Learning from Examples Module, version 2) computes local coverings of attribute-value pairs [13]. Local coverings are constructed from minimal complexes. The minimal complex contains attribute-value pairs, selected on the basis of their relevancy to the concept. In the case of a tie, the next criterion is the maximum of conditional probability of the concept given the attribute-value pair. In the global approach, each concept is represented by a substitutional partition with two blocks: the first block is the concept, the second block is its complement. A minimal subset of the set of all attributes, such that the substitutional partition depends on it, is called a global covering. A global covering may be selected on the basis of lower boundaries [12]. The learning system LERS, based on rough set theory, and producing certain and possible rules, is installed at the University of Kansas. It is implemented in Common Lisp and runs on VAX 9000.

2. LERS Preliminary Processing

The input data file for the system LERS must follow the following format. The first line of the input data file consists of a list, starting from '<', then a sequence of the following symbols: 'a', 'x', or 'd'; the last symbol of the list is '>'. Symbol 'a' stands for an attribute, 'x' means ignore the value, and 'd' denotes a decision. The user may assign priorities to attributes by placing integers immediately following any character 'a'. Bigger integers correspond to higher priorities. The assigned priorities may be changed after entering the program.

Table 1. Hypothermic Post Anesthesia Patients

< a	a	a	a	d >
[Temperature	Hemoglobin	Blood_Pressure	Oxygen_Saturation	Comfort]
low	fair	low	fair	low
low	fair	normal	poor	low
normal	good	low	good	low
normal	good	low	good	medium
low	good	normal	good	medium
low	good	normal	fair	medium
normal	fair	normal	good	medium
normal	poor	high	good	very_low
high	good	high	fair	very_low

System LERS will attempt to form rules primarily from attributes with highest priorities. Symbol 'x' means that the corresponding variable will be ignored in all following computations. The total number of symbols *a*, *d*, and *x* is equal to the total number of attributes and decisions in the input data file or decision table. All symbols are separated by white space characters.

The second line of the input data file starts from '['; then comes a list of attribute and decision names, and then ']'. The following lines contain values of the attributes and decisions. Table 1 depicts a typical input data file for system LERS.

In computations of LERS the examples from the decision table are numbered. Thus, Table 1 is represented by the decision table presented as Table 2.

Table 2. Hypothermic Post Anesthesia Patients

	Attributes				Decision
	Temperature	Hemoglobin	Blood_Pressure	Oxygen_Saturation	Comfort
1	low	fair	low	fair	low
2	low	fair	normal	poor	low
3	normal	good	low	good	low
4	normal	good	low	good	medium
5	low	good	normal	good	medium
6	low	good	normal	fair	medium
7	normal	fair	normal	good	medium
8	normal	poor	high	good	very_low
9	high	good	high	fair	very_low

In Table 2, decision *Comfort* has three values: *low*, *medium*, and *very_low*. Each such value represents a concept, a subset of the set *U* of all examples {1, 2, 3, 4, 5, 6, 7, 8, 9}. Table 2 represents three concepts: {1, 2, 3}, {4, 5, 6, 7}, and {8, 9}. Let *d* denote a decision and let *w* denote its value. A concept is a set [(*d*, *w*)] of all examples that have value *w* for decision *d*. In other words, [(*d*, *w*)] is a {d}-elementary set of *U* [47, 48]. Similarly, the ***block of an attribute-value pair*** *t* = (*q*, *v*), denoted [*t*], is the set of all examples that for attribute *q* have value *v* [13].

The system LERS first checks input data for consistency. This operation is done immediately after reading the input data file. If the input data are inconsistent lower and upper approximations are computed for each concept. In local options of LERS the system induces certain and possible rules from lower and upper approximations for each concept, respectively. In the case of global options of LERS new partitions on the set *U* are computed.

Say that the original decision table described k concepts, i.e., decision has k values. Then $2k$ new partitions on U, called ***substitutional partitions*** [12] are created. Each substitutional partition has exactly two blocks, the first block is either lower or upper approximation of the concept, the second block is the complement of the first block. Substitutional partitions computed from lower approximations are called *lower substitutional partitions*; substitutional partitions computed from upper approximations are called *upper substitutional partitions.* Decisions, corresponding to lower and upper partitions, are called *lower* and *upper substitutional decisions* [12].

In the example from Table 2, concept [(Comfort, low)] = {1, 2, 3} has the following lower approximation: {1, 2} and the following upper approximation: {1, 2, 3, 4}. Thus, its lower substitutional partition is {{1, 2}, {3, 4, 5, 6, 7, 8, 9}} and its upper substitutional partition is {{1, 2, 3, ,4}, {5, 6, 7, 8, 9}}.

3. Single Local Covering Option

The single local covering approach algorithm described here is called LEM2. This algorithm represents the machine learning approach, i.e., it produces minimal discriminant description. Algorithm LEM2 is based on the following ideas [13]. Let B be a nonempty lower or upper approximation of a concept represented by a decision-value pair (d, w). Set B *depends on a set* T of attribute-value pairs if and only if

$$\emptyset \neq [T] = \bigcap_{t \in T} [t] \subseteq B.$$

Set T is a *minimal complex* of B if and only if B depends on T and no proper subset T' of T exists such that B depends on T'. Let $\mathbb{T}$ be a nonempty collection of nonempty sets of attribute-value pairs. Then $\mathbb{T}$ is a *local covering of* B if and only if the following conditions are satisfied:

(1) each member T of $\mathbb{T}$ is a minimal complex of B,

(2) $\bigcup_{T \in \mathbb{T}} [T] = B$, and

(3) $\mathbb{T}$ is minimal, i.e., $\mathbb{T}$ has the smallest possible number of members.

The algorithm LEM2 is based on computing a single local covering for each of the concepts from the decision table. The user may select an option of LEM2 with or without taking into account attribute priorities. The procedure LEM2 with attribute priorities is presented below. The other option differs from the one presented below in the selection of a pair $t \in T(G)$ in the inner loop WHILE. In the option of LEM2 without taking attribute priorities into account the first criterion is ignored.

Procedure LEM2
(input: a set B,
output: a single local covering $\mathbb{T}$ of set B);
begin

```
G := B;
𝕋 := ∅;
while G ≠ ∅
    begin
    T := ∅;
    T(G) := {t | [t] ∩ G ≠ ∅};
    while T = ∅ or [T] ⊄ B
        begin
            select a pair t ∈ T(G) with the highest attribute priority, if a tie occurs, select a
              pair t ∈ T(G) such that |[t] ∩ G| is maximum; if another tie occurs, select
              a pair t ∈ T(G) with the smallest cardinality of [t]; if a further tie occurs, select
              first pair;
            T := T ∪ {t};
            G := [t] ∩ G;
            T(G) := {t | [t] ∩ G ≠ ∅};
            T(G) := T(G) – T;
        end {while}
    for each t in T do
        if [T – {t}] ⊆ B then T := T – {t};
    𝕋 := 𝕋 ∪ {T};
    G := B – ⋃_{T∈𝕋} [T];
end {while};
for each T in 𝕋 do

    if ⋃_{S∈𝕋–{T}} [S] = B then 𝕋 := 𝕋 – {T};
end {procedure}.
```

For the decision table presented in Table 2, let a concept be defined by the value *low* of the decision *Comfort*. Also, we will assume that the user has chosen an option of LEM2 without taking into account attribute priorities. Thus, [(Comfort, low)] = {1, 2, 3}. Let B be the lower approximation of the concept, i.e., $B = \{1, 2\}$. Then the set $T(G)$ from the procedure LEM2 is equal to {(Temperature, low), (Hemoglobin, fair), (Blood_Pressure, low), (Blood_Pressure, normal), (Oxygen_Saturation, fair), (Oxygen_Saturation, poor)}. Initially, $B = G = \{1, 2\}$. Moreover,

[(Temperature, low)] = {1, 2, 5, 6},
[(Hemoglobin, fair)] = {1, 2, 7},
[(Blood_Pressure, low)] = {1, 3, 4},
[(Blood_Pressure, normal)] = {2, 5, 6, 7},
[(Oxygen_Saturation, fair)] = {1, 6, 9},
[(Oxygen_Saturation, poor)] = {2}.

Thus, (Temperature, low) and (Hemoglobin, fair) are selected because their blocks have the maximum intersection with set B. Since the tie occurred, the next criterion is used, and (Hemoglobin, fair) is selected because its block is of smaller cardinality. Set G is still equal to B, because [(Hemoglobin, fair)] $\cap G = G$. Set $T(G)$ becomes {(Temperature, low), (Blood_Pressure, low), (Blood_Pressure, normal), (Oxygen_Saturation, fair), (Oxygen_Saturation, poor)}. Since [(Hemoglobin, fair)] $\not\subseteq B$, the second iteration is needed.

This time pair (Temperature, low) is selected, T = {(Hemoglobin, fair), (Temperature, low)}, and [*T*] = [*B*]. Therefore, the only induced certain rule is

(Hemoglobin, fair) ∧ (Temperature, low) → (Comfort, low).

Remaining certain rules, induced by LEM2, are

(Hemoglobin, good) ∧ (Blood_Pressure, normal) → (Comfort, medium),
(Temperature, normal) ∧ (Hemoglobin, fair) → (Comfort, medium),
(Blood_Pressure, high) → (Comfort, very_low),

and all possible rules, induced by LEM2, are

(Blood_Pressure, low) → (Comfort, low)
(Oxygen_Saturation, por) → (Comfort, low)
(Oxygen_Saturation, good) ∧ (Hemoglobin, good) → (Comfort, medium)
(Oxygen_Saturation, fair) ∧ (Blood_Pressure, normal) → (Comfort, medium),
(Temperature, normal) ∧ (Hemoglobin, fair) → (Comfort, medium),
(Blood_Pressure, high) → (Comfort, very_low),

4. Single Global Covering Option

System LERS computes minimal discriminant description of the concept on the basis of a single global covering using procedure LEM. This procedure employs the following ideas.

Let *C* denote the set of all attributes and let *d* denote a lower or upper substitutional decision. Let *P* be a subset of *C*. The family of all *P*-elementary sets will be denoted P^*. We say that $\{d\}$ *depends on* P if and only if $P^* \leq \{d\}^*$. A *global covering* of $\{d\}$ is a subset *P* of *C* such that $\{d\}$ depends on *P* and *P* is minimal in *C*. Thus, coverings of $\{d\}$ are computed by comparing partitions P^* with $\{d\}^*$. The procedure LEM, option SINGLE_COVERING, used in LERS to compute a single global covering is presented below.

```
Procedure LEM_SINGLE_COVERING
(input: the set C of all attributes, partition {d}* on U;
output: a single global covering R);
begin
   compute partition C*;
   P: = C;
   R:= Ø;
   if C* ≤ {d}*
      then
         begin
            for each attribute q in C do
               begin
                  Q:= P – {q};
                  compute partition Q*;
                  if Q* ≤ {d}* then P:= Q
               end {for}
            R:= P
```

end {then}
end {procedure}.

On the basis of global covering certain and possible rules may be computed. However, further processing in the form of *dropping conditions* [40] is necessary. System LERS uses two different forms of dropping conditions. The first one is called *linear* because its time complexity is linear. For a rule of the form

$$C_1 \wedge C_2 \wedge \cdots \wedge C_l \rightarrow A,$$

linear dropping conditions means scanning the list of all conditions, from the left to the right, with an attempt to drop any of l condition, checking against the decision table where the simplified rule does not violate consistency of the discriminant description, where $C_1, C_2, \ldots, C_l$ are conditions and A is an action. Another possibility of LERS is *exponential dropping conditions*, in which any subset of the set of all l conditions of the above rule is checked for dropping conditions. Single global covering option of system LERS uses linear dropping conditions.

For the decision table presented in Table 2, let d be a lower substitutional decision, corresponding to the lower substitutional partition {{1, 2}, {3, 4, 5, 6, 7, 8, 9}}. All global coverings of {d} are the following sets {Temperature, Hemoglobin}, {Hemoglobin, Oxygen_Saturation}, and {Blood_Pressure, Oxygen_Saturation}. SINGLE_COVERING option of LEM will select one of these three global coverings. The choice is affected by attribute priorities, assigned by the user. Default attribute priorities are in ascending order. Therefore, the selected global covering will be {Blood_Pressure, Oxygen_Saturation}. After linear dropping conditions, induced certain rules are

(Oxygen_Saturation, fair) ∧ (Blood_Pressure, low) → (Comfort, low),
(Oxygen_Saturation, poor) → (Comfort, low),
(Oxygen_Saturation, good) ∧ (Blood_Pressure, normal) → (Comfort, medium),
(Oxygen_Saturation, fair) ∧ (Blood_Pressure, normal) → (Comfort, medium),
(Blood_Pressure, high) → (Comfort, very_low),

and possible rules are

(Blood_Pressure, low) → (Comfort, low),
(Oxygen_Saturation, poor) → (Comfort, low),
(Oxygen_Saturation, good) ∧ (Blood_Pressure, low) → (Comfort, medium),
(Oxygen_Saturation, good) ∧ (Blood_Pressure, normal) → (Comfort, medium),
(Oxygen_Saturation, fair) ∧ (Blood_Pressure, normal) → (Comfort, medium),
(Blood_Pressure, high) → (Comfort, very_low).

5. All Global Coverings Option

The algorithm presented here is called LEM, option ALL_COVERINGS. This option of LEM represents the knowledge acquisition approach to rule induction. That means that the algorithm discovers the set ℝ of all global coverings for every lower and upper substitutional partition of {d}* and then it induces rules using the exponential dropping condition algorithm. The time complexity of the algorithm LEM, option ALL_COVERINGS is exponential. Hence, the user is asked to enter the maximum size of the global covering (i.e., the number of attributes in a global covering). Only global coverings whose size does not exceed the maximum size will be generated and used to compute rules.

Algorithm LEM_ALL_COVERINGS
(**input:** the set C of all attributes, partition {d}* on U;
output: the set ℝ of all coverings of {d});
begin
ℝ:= Ø;
for each attribute q in C **do**
compute partition {q}*;
k:= 1;
while k ≤ |C| **do**
begin
for each subset P of the set C with |P| = k **do**
if (P is not a superset of any member of ℝ) **and** ($\prod_{x \in P} \{q\}^* \leq \{d\}^*$)
then add P to ℝ,
k:= k+1
end {while}
end {procedure}.

After execution of LEM, ALL_COVERINGS option, system LERS executes the exponential dropping conditions algorithm. For the decision table presented in Table 2, the set of all induced rules is presented below. Certain rules, induced by all global coverings option of the system LERS are

(Temperature, low) ∧ (Hemoglobin, fair) → (Comfort, low),
(Oxygen_Saturation, poor) → (Comfort, low),
(Hemoglobin, fair) ∧ (Oxygen_Saturation, fair) → (Comfort, low),
(Blood_Pressure, low) ∧ (Oxygen_Saturation, fair) → (Comfort, low),
(Temperature, low) ∧ (Hemoglobin, good) → (Comfort, medium),
(Temperature, normal) ∧ (Hemoglobin, fair) → (Comfort, medium),
(Blood_Pressure, normal) ∧ (Oxygen_Saturation, good) → (Comfort, medium),
(Blood_Pressure, normal) ∧ (Oxygen_Saturation, fair) → (Comfort, medium),
(Blood_Pressure, high) → (Comfort, very_low),
(Temperature, high) → (Comfort, very_low),
(Hemoglobin, poor) → (Comfort, very_low),

and possible rules are

(Temperature, low) ∧ (Hemoglobin, fair) → (Comfort, low),
(Temperature, normal) ∧ (Hemoglobin, good) → (Comfort, low),
(Blood_Pressure, low) → (Comfort, low),
(Oxygen_Saturation, poor) → (Comfort, low),
(Temperature, normal) ∧ (Hemoglobin, good) → (Comfort, medium),
(Temperature, low) ∧ (Hemoglobin, good) → (Comfort, medium),
(Temperature, normal) ∧ (Hemoglobin, fair) → (Comfort, medium),

(Blood_Pressure, low) ∧ (Oxygen_Saturation, good) → (Comfort, medium),
(Blood_Pressure, normal) ∧ (Oxygen_Saturation, good) → (Comfort, medium),
(Blood_Pressure, normal) ∧ (Oxygen_Saturation, fair) → (Comfort, medium),
(Blood_Pressure, high) → (Comfort, very_low),
(Temperature, high) → (Comfort, very_low),
(Hemoglobin, poor) → (Comfort, very_low)

6. All Rules Option

This option of the LERS algorithm induces all rules that can be induced from the input data file; each rule is produced in the simplest form. Thus, that option represents the knowledge acquisition approach to rule induction. Although this option is the most demanding from the viewpoint of time complexity, the description of the algorithm is very simple. For every concept, represented by a decision-value pair, the lower and upper approximations are computed. Then rules are induced directly from the decision table. Thus, for each rule, the number of conditions is equal to the number of attributes. Finally, exponential dropping condition algorithm is used. Hence, all certain and possible rules are induced in their minimal form.

For the decision table presented in Table 2, the set of all induced rules is presented below. Certain rules, induced by all rules option of the system LERS are

(Oxygen_Saturation, poor) → (Comfort, low),
(Temperature, low) ∧ (Hemoglobin, fair) → (Comfort, low),
(Temperature, low) ∧ (Blood_Pressure, low) → (Comfort, low),
(Hemoglobin, fair) ∧ (Blood_Pressure, low) → (Comfort, low),
(Hemoglobin, fair) ∧ (Oxygen_Saturation, fair) → (Comfort, low),
(Blood_Pressure, low) ∧ (Oxygen_Saturation, fair) → (Comfort, low),
(Temperature, low) ∧ (Hemoglobin, good) → (Comfort, medium),
(Temperature, normal) ∧ (Hemoglobin, fair) → (Comfort, medium),
(Temperature, normal) ∧ (Blood_Pressure, normal) → (Comfort, medium),
(Temperature, low) ∧ (Oxygen_Saturation, good) → (Comfort, medium),
(Hemoglobin, good) ∧ (Blood_Pressure, normal) → (Comfort, medium),
(Hemoglobin, fair) ∧ (Oxygen_Saturation, good) → (Comfort, medium),
(Blood_Pressure, normal) ∧ (Oxygen_Saturation, good) → (Comfort, medium),
(Blood_Pressure, normal) ∧ (Oxygen_Saturation, fair) → (Comfort, medium),
(Temperature, high) → (Comfort, very_low),
(Hemoglobin, poor) → (Comfort, very_low),
(Blood_Pressure, high) → (Comfort, very_low),

and possible rules are

(Blood_Pressure, low) → (Comfort, low),
(Oxygen_Saturation, poor) → (Comfort, low),
(Temperature, low) ∧ (Hemoglobin, fair) → (Comfort, low),
(Temperature, normal) ∧ (Hemoglobin, good) → (Comfort, low),
(Hemoglobin, fair) ∧ (Oxygen_Saturation, fair) → (Comfort, low),

(Temperature, normal) ∧ (Hemoglobin, good) → (Comfort, medium),
(Temperature, low) ∧ (Hemoglobin, good) → (Comfort, medium),
(Temperature, normal) ∧ (Hemoglobin, fair) → (Comfort, medium),
(Temperature, normal) ∧ (Blood_Pressure, low) → (Comfort, medium),
(Temperature, normal) ∧ (Blood_Pressure, normal) → (Comfort, medium),
(Temperature, low) ∧ (Oxygen_Saturation, good) → (Comfort, medium),
(Hemoglobin, good) ∧ (Blood_Pressure, low) → (Comfort, medium),
(Hemoglobin, good) ∧ (Blood_Pressure, normal) → (Comfort, medium),
(Hemoglobin, good) ∧ (Oxygen_Saturation, good) → (Comfort, medium),
(Hemoglobin, fair) ∧ (Oxygen_Saturation, good) → (Comfort, medium),
(Blood_Pressure, low) ∧ (Oxygen_Saturation, good) → (Comfort, medium),
(Blood_Pressure, normal) ∧ (Oxygen_Saturation, good) → (Comfort, medium),
(Blood_Pressure, normal) ∧ (Oxygen_Saturation, fair) → (Comfort, medium),
(Temperature, high) → (Comfort, very_low),
(Hemoglobin, poor) → (Comfort, very_low),
(Blood_Pressure, high) → (Comfort, very_low).

7. Conclusions

The paper presents the system LERS for rule induction. The system handles inconsistencies in the input data due to its usage of rough set principle. Namely, the system induces certain rules and possible rules, on the basis of lower and upper approximations of each concept.

The system gives many options to the user. The first choice is between using the machine learning approach or the knowledge acquisition approach. In the first case, the system induces a single minimal discriminant description for each concept. Practically that means that the system induces a set of sufficient rules, completely describing every concept, although only some attribute-value pairs are involved in rules. In the input data much more knowledge, not discovered by the machine learning approach, may still exist. If the user wants to discover more rules, the knowledge acquisition approach should be used. The latter approach is used, e.g., for expert systems, when it is crucial to know as much about the problem as possible. However, time complexity for the first choice is polynomial, while it is exponential for the second choice. Thus, for big input data files the second choice may be not practical.

Within the machine learning approach, the user has another choice, between local and global options of computing a single covering. The global option is less demanding computationally; however, the local option induces simpler rules. Again, the size of the input data file may dictate the choice.

When using the knowledge acquisition approach, the user's choice is between inducing rules from all global coverings or directly inducing all rules using local computations. Although time complexity of both options is exponential, the first one is less demanding. That may force the user to use it, in spite of the fact that only the all rules option of system LERS truly induces all rules.

References

[1] Aha, D. W. and D. Kibler. Noise-tolerant instance-based learning algorithms. *Proc. of the IJCAI 89, 11th Int. Joint Conf. on AI, 794–799.*

[2] Bergadano, F., A. Giordana, and L. Saitta. Automated concept acquisition in noisy environment. *IEEE Trans. PAMI* 10, 1988, 555–578.

[3] Bergadano, F. and L. Saitta. On the error probability of Boolean concept description. *Proc. of the 4th European Working Session on Learning*, 1989, 25–36.

[4] Berzuini, C. Combining symbolic learning techniques and statistical regression analysis. *Proc. of the AAAI-88, 7th Nat. Conf. on AI*, 612–617.

[5] Berzuini, C. Partially supervised learning from examples with the aid of statistical regression analysis. *Proc of the ISMIS-88, 3rd Int. Symp. on Methodologies for Intelligent Systems*, 281-292.

[6] Boucheron, S. and J. Sallantin. Learnability in the presence of noise. *Proc. of the 3rd European Working Session on Learning*, 1988, 25–35.

[7] Buntine, W. Decision tree induction systems: A Bayesian analysis. In *Uncertainty in AI 3, L. N. Kanal, T. S. Lewitt, and J. L. Lemmer (eds.),* Elsevier, 1989, 109–127.

[8] Buntine, W. Learning classification rules using Bayes. *Proc. of the 6th Int. Workshop on Machine Learning*, 1989, 94–98.

[9] Buntine, W. A critique of the Valiant model. *Proc. of the IJCAI-89, 11th Int. Joint Conf. on AI*, 837–842.

[10] Breiman, L., J. H. Friedman, R. A. Olshen, and C. J. Stone. *Classification and Regression Trees.* Wadsworth & Brooks, 1984.

[11] Budihardjo, A., J. W. Grzymala-Busse, and L. Woolery. Program LERS_LB 2.5 as a tool for knowledge acquisition in nursing. *Proc. of the 4th Int. Conf. on Industrial & Engineering Applications of Artificial Intelligence & Expert Systems*, 1991, 735–740.

[12] Chan, C. C. and J. W. Grzymala-Busse. Rough-set boundaries as a tool for learning rules from examples. *Proc. of the ISMIS–89, 4th Int. Symp. on Methodologies for Intelligent Systems*, 281–288.

[13] Chan, C. C. and J. W. Grzymala-Busse. On the attribute redundancy and the learning programs ID3, PRISM, and LEM. *Submitted for publication.*

[14] Chan, P. K. Inductive learning with BCT. *Proc. of the 6th Int. Workshop on Machine Learning*, 1989, 104–108.

[15] Chang, K. C. C. and A. K. C. Wong. Performance analysis of a probabilistic inductive learning system. *Proc. of the 7th Conf. on Machine Learning*, 1990, 16–23.

[16] Cheeseman, P., J. Kelly, M. Self, J. Stutz, W. Taylor, and D. Freeman. AutoClass: A Bayesian classification system. *Proc. of the 5th Int. Conf. on Machine Learning*, 1988, 54–64.

[17] Cheeseman, P., M. Self, J. Kelly, J. Stutz, W. Taylor, and D. Freeman. Bayesian classification. *Proc. of the AAAI-88, 7th Nat. Conf. on AI*, 607–610.

[18] Clark, P. and T. Niblett. The CN2 induction algorithm. *Machine Learning* 3, 1989, 261–283.

[19] Dean, J. S. and J. W. Grzymala-Busse. An overview of the learning from examples module LEM1. *Report TR-88-2, Department of Computer Science, University of Kansas*, 1988.

[20] Dubois, D. and H. Prade. Epistemic entrenchment and possibilistic logic. *Artificial Intelligence* 50, 1991, 223–239.

[21] Fung, R. M. and S. L. Crawford. CONSTRUCTOR: A system for the induction of probabilistic models. *Proc. of the AAAI-90, 8th Nat. Conf. on AI*, 762–769.

[22] Goodman, R. M. and P. Smyth. The induction of probabilistic rule sets—the ITRULE algorithm. *Proc. of the 6th Int. Workshop on Machine Learning*, 1989, 129–132.

[23] Grzymala-Busse, J. W. Knowledge acquisition under uncertainty—A rough set approach. *J. Intelligent & Robotic Systems* 1, 1988, 3–16.

[24] Grzymala-Busse, J. W. An overview of the LERS1 learning system. *Proc. of the 2nd Int. Conf. on Industrial and Engineering Applications of Artificial Intelligence and Expert Systems*, 1989, 838–844.

[25] Grzymala-Busse, J. W. On the unknown attribute values in learning from examples. *Proc. of the ISMIS'91, 6th Int. Symp. on Methodologies for Intelligent Systems*, 368–377.

[26] Grzymala-Busse, J. W. *Managing Uncertainty in Expert Systems*. Kluwer Academic Publishers, 1991.

[27] Grzymala-Busse, J. W. and D. J. Sikora. LERS1—A system for learning from examples based on rough sets. *Report TR-88-5, Department of Computer Science, University of Kansas*, 1988.

[28] Hanson, S. J. and M. Bauer. Machine learning, clustering and polymorphy. In *Uncertainty in Artificial Intelligence, L. N. Kanal and J. F. Lemmer (eds.)* Elsevier Science Publ., 1986, 415–428.

[29] Haussler, D. Bias, version spaces and Valiant's learning framework. *Proc. of the 4th Int. Workshop on Machine Learning*, 1987, 324–336.

[30] Haussler, D. Probably approximately correct learning. *Proc. of the AAAI-90, 8th Nat. Conf. on AI*, 1101–1108.

[31] Hirsch, H. Learning from data with bounded inconsistency. *Proc. of the 7th Conf. on Machine Learning*, 1990, 32–39.

[32] Kodratoff, Y., M. Manago, and J. Blythe. Generalization and noise. *Int. J. Man-Machine Studies* 27, 1987, 181–204.

[33] Kononenko, I. ID3, sequential Bayes, naive Bayes and Bayesian neural networks. *Proc. of the 4th European Working Session on Learning*, 1989, 91–98.

[34] Kononenko, I. and I. Bratko. Information-based evaluation criterion for classifiers performance. *Machine Learning* 6, 1991, 67–80.

[35] Kwok, S. W. and C. Carter. Multiple decision trees. In *Uncertainty in AI 4, R. D. Shachter, T. S. Levitt, L. N. Kanal, J. F. Lemmer (eds.)*, Elsevier, 1990, 327–335.

[36] Laird, P. D. *Learning from Good and Bad Data*. Kluwer Academic Publishers, 1988.

[37] Lee, W. D. and S. R. Ray. Rule refinement using the probabilistic rule generator. *Proc. of the AAAI–86, 5th Nat. Conf. on AI*, 442–447.

[38] Lesmo, L., L. Saitta, and P. Torasso. Learning of fuzzy production rules for medical diagnosis. In *Approximate reasoning in Decision Analysis, M. M. Gupta and E. Sanchez (eds.)*, North Holland Publ. Co., 1982, 249–260.

[39] Manago, M. V. and Y. Kodratoff. Noise and knowledge acquisition. *Proc. of the IJCAI 87, Int. Joint Conf. on AI*, 348–354.

[40] Michalski, R. S. A theory and methodology of inductive learning. In *Machine Learning, R. S. Michalski, J. G. Carbonell, T. M. Mitchell (eds.)*, Morgan Kaufmann 1983, 83–134.

[41] Michalski, R. S., I. Mozetic, J. Hong, and N. Lavrac. The AQ15 inductive learning system: An overview and experiments. *Report 1260, Department of Computer Science, University of Illinois at Urbana-Champaign*, 1986.

[42] Michalski, R. S. How to learn imprecise concepts: A method for employing a two-tiered knowledge representation in learning. *Proc of the 4th Int. Workshop on Machine Learning*, 1987, 50–58.

[43] Mingers, J. An empirical comparison of selection measures for decision-tree induction. *Machine Learning* 3, 1989, 319–342.

[44] Mingers, J. An empirical comparison of pruning methods for decision tree induction. *Machine Learning* 4, 1989, 227–243.

[45] Nakakuki, Y., Y. Koseki, and M. Tanaka. Inductive learning in probabilistic domain. *Proc. of the AAAI-90, 8th Nat. Conf. on AI*, 809–814.

[46] Niblett, T. and I. Bratko. Learning decision rules in noisy domains. *Proc. of Expert Systems '86, the 6th Annual Techn. Conference of the British Computer Society, Specialist Group on Expert Systems*, 1986, 25–34.

[47] Pawlak, Z. Rough sets. *Int. J. Computer and Information Sci.*, 11, 1982, 341–356.

[48] Pawlak, Z. Rough Classification. *Int. J. Man-Machine Studies* 20, 1984, 469–483.

[49] Plaza, E. and R. Lopez de Mantaras. A case-based apprentice that learns from fuzzy examples. *Proc. of the 5th Int. Symp. on Methodologies for Intelligent Systems*, 1990, 420–427.

[50] Quinlan, J. R. Learning efficient classification procedures and their application to chess end games. In *Machine Learning, R. S. Michalski, J. G. Carbonell, T. M. Mitchell (eds.)*, Morgan Kaufmann Publishers, Inc., 1983, 461–482.

[51] Quinlan, J. R. Induction of decision trees. *Machine Learning* 1, 1986, 81–106.

[52] Quinlan, J. R. Decision trees as probabilistic classifiers. *Proc of the 4th Int. Workshop on Machine Learning*, 1987, 31–37.

[53] Quinlan, J. R. Generating production rules from decision trees. *Proc. of the 10th Int. Joint Conf. on AI*, 1987, 304–307.

[54] Quinlan, J. R. Unknown attribute values in induction. *Proc. of the 6th Int. Workshop on Machine Learning*, 1989, 164–168.

[55] Quinlan, J. R. The effect of noise on concept learning. In *Machine Learning. An Artificial Intelligence Approach. Vol. II, R. S. Michalski, J. G. Carbonell, T. M. Mitchell, (eds)*. Morgan Kaufmann Publishers, Inc., 1986, 149–166.

[56] Quinlan, J. R. Probabilistic decision trees. In *Machine Learning. An Artificial Intelligence Approach. Vol. III*, Morgan Kaufmann Publishers, Inc., 1990, 140–152.

[57] Rendell, L. Induction, of and by probability. In *Uncertainty in Artificial Intelligence, L. N. Kanal and J. F. Lemmer (eds.)*, Elsevier Science Publ., 1986, 415–443.

[58] Rivest, R. L. Learning decision lists. *Machine Learning* 2, 1987, 229–246.

[59] Shackelford, G. and D. Volper. Learning k-DNF with noise in the attributes. *Proc. of the 1988 Workshop on Computational Learning Theory*, 97–101.

[60] Schlimmer, J. C. and D. Fisher. A case study of incremental concept induction. *Proc of the AAAI-86, 5th Nat. Conf. on AI*, 496–501.

[61] Sloan, R. Types of noise in data for concept learning. *Proc. of the 1988 Workshop on Computational Learning Theory*, 91–96.

[62] Spangler, S., U. M. Fayyad, and R. Uthurusamy. Induction of decision trees from inconclusive data. *Proc. of the 6th Int. Workshop on Machine Learning*, 1989, 146–150.

[63] Tan, M. and L. Eshelman. Using weighted networks to represent classification knowledge in noisy domains. *Proc. of the 5th Int. Conf. on Machine Learning*, 1988, 121–134.

[64] Tan, M. and L. Eshelman. The impact of noise on learning. *Rep. CMU-CS-88-144, Department of Computer Science, Carnegie Mellon*, 1988.

[65] Utgoff, P. E. ID5: An incremental ID3. *Proc. of the 5th Int. Conf. on Machine Learning*, 1988, 107–120.

[66] Utgoff, P. E. and C. E. Brodley. An incremental method for finding multivariate splits for decision trees. *Proc. of the 7th Conf. on Machine Learning*, 1990, 58–65.

[67] Wong, S. K. M. and W. Ziarko. On learning and evaluation of decision rules in the context of rough sets. *Proc. of the ACM SIGART Int. Symp. on Methodologies for Intelligent Systems*, 1986, 308–324.

[68] Valiant, L. G. A theory of learnable. *Com. ACM* 27, 1984, 1134–1142.

[69] Valiant, L. G. Learning disjunctions of conjunctions. *Proc. of the 9th Int. Joint Conf. on AI*, 1985, 560–566.

[70] Yasdi, R. and W. Ziarko. An expert system for conceptual schema design: A machine learning approach. *Int. J. Man-Machine Studies* 29, 1988, 351–376.

[71] Ziarko, W. and J. D. Katzberg. Control algorithm acquisition, analysis, and reduction. In *Knowledge-Based System Diagnosis, Supervision and Control, S. T. Tzafestas (ed.), Plenum Press*, 1989, 167–178.

Part I
Chapter 2

ROUGH SETS IN COMPUTER IMPLEMENTATION OF RULE-BASED CONTROL OF INDUSTRIAL PROCESSES

Adam MRÓZEK
Institute of Theoretical and Applied Computer Science
Polish Academy of Sciences
44-100 Gliwice, Poland

Abstract. We discuss the use of elements of rough set theory in computer implementation of rule-based control of industrial processes. The notions of expert's inference model for industrial process control and of control protocol which registers the expert's decisions are introduced. The process of extraction of decision rules contained in the control protocol based on the rough set theory formalism is presented. An example of the analysis of control protocol concerning the control process of a rotary clinker kiln is chosen as an illustration.

1. Introduction

Elements of rough set theory can be constructively used in computer implementations of the rule-based control of industrial processes [2,3,5]. The essence of such control lies in intercepting and imitating by computer the way how the expert controls a chosen industrial process.

It is also a way of implementation of real-time computer control in industry. This approach is of special importance when mathematical models of the industrial process in question are not obtainable or control algorithms related to them are numerically difficult. Naturally, the human operator control which is a basis for our approach should be satisfactory. Practical realization of such computer control needs some formal description of the expert's decision process.

A proposition of such formalization is presented below. Let us introduce the following notions:

1. **expert's inference model,**
2. **control protocol** as an ordered record of expert's decisions,
3. elements of rough set theory are needed to select **decision rules** contained in the control protocol.

A set of decision rules, properly selected and describing expert's performance during the control may be then implemented in computer.

An example of control of a rotary clinker kiln in a cement plant illustrates the above proposition.

2. The notion of expert's inference model

The primary notion of every control is the one of **control goal.** This notion is difficult to formalize when expert's control process is considered. However, the expert can subjectively judge whether the control goal is achieved or not. The expert, when he controls an industrial process, first of all focuses his attention on parameters of the process which are directly related to the imposed control goal. Following their values he can judge upon the states of the process and thus, indirectly, he knows whether he attains the control goal. Hence, from the expert's viewpoint, an industrial process is characterized by:

1. the **space of observations** determined by **measurable** and **observable variables;**

2. the **space of control** determined by **control variables.**

The space of observations is explicitly determined by a finite set of coordinates that we call **condition attributes** and denote $c_1, c_2, \cdots, c_n$.

The space of control is explicitly defined by a finite set of coordinates that we call **decision attributes** and denote $d_1, d_2, \cdots, d_k$.

Experts's inference model is composed of the following steps:

1. decomposition of the space of observations into areas called **characteristic states** of the industrial process;

2. decomposition of the space of control into areas called **characteristic controls;**

3. **assignement** of a proper characteristic control to each characteristic state.

In the context of the inference model presented above, the control performed by the expert consists in repetitive execution of the following steps:

1. evaluation of the current situation within the space of observations;

2. assignement of this situation to the proper characteristic state of the object;

3. choice and realization of the proper characteristic control within the space of control;

A natural way of expert's knowledge representation is thus a construction of a set of **decision rules** having the following form: **IF** {set of conditions} **THEN** {set of decisions}. In such notation the decision rules represent dependencies between the set of conditions and the set of corresponding decisions.

The selection of the set of decision rules needs a proper approach to the representation and analysis of data describing the expert's behaviour during the control process.

3. The control protocol

The selection of decision rules needs the proper characterization of the conditions and decisions sets. Each condition attribute $c_i, i = 1, 2, \cdots, n$, has a finite domain of its values V_{c_i} and each decision attribute $d_j, j = 1, 2, \cdots, k$, has its domain V_{d_j}.

In a finite time-horizon the expert's behaviour during process control can be written as a **control protocol** having structure presented in Table 1.

Time moments (t)	Values of condition attributes					Values of decision attributes				
	c_1	$\cdots$	c_2	$\cdots$	c_n	d_1	$\cdots$	d_j	$\cdots$	d_k
t_1	$v_{c_1}^1$	$\ldots$	$v_{c_i}^1$	$\ldots$	$v_{l_n}^1$	$u_{d_1}^1$	$\ldots$	$u_{d_j}^1$	$\cdots$	$u_{d_k}^1$
$\cdots$	$\cdots$		$\cdots$		$\cdots$	$\cdots$		$\cdots$		$\cdots$
t_l	$v_{c_1}^l$	$\ldots$	$v_{c_i}^l$	$\ldots$	$v_{c_n}^l$	$u_{d_1}^l$	$\ldots$	$u_{d_j}^l$	$\cdots$	$u_{d_k}^l$
$\cdots$	$\cdots$		$\cdots$		$\cdots$	$\cdots$		$\cdots$		$\cdots$
t_N	$v_{c_1}^N$	$\ldots$	$v_{c_i}^N$	$\ldots$	$v_{c_n}^N$	$u_{d_1}^N$	$\ldots$	$u_{d_j}^N$	$\cdots$	$u_{d_k}^N$

Table 1. Structure of the control protocol.

Remark: If the set of decision attributes contains only one element, e.g. the number (name) of characteristic state then the control protocol represents the classification of chracteristic states of industrial process performed by the expert with the use of the values of condition attributes.

The control protocol can be analysed in order to obtain:

1. selection of all different decision rules;
2. checking whether there is no contradictory decision rules, i.e. rules in which different values of decision attributes correspond to the same values of condition attributes;
3. determination of the relevence between the set of values of condition attributes and decision attributes;
4. elimination of these condition attributes which do not influence the relations between the set of values of condition attributes and decision attributes.

In the case when the control protocol represents classification of characteristic states done by the expert, the analysis may have also the following goals:

1. evaluation of quality and accuracy of the classification;
2. determination of the influence of the values of particular condition attributes on this classification;
3. selection of the set of condition productions as a basis of the recognition of characteristic states.

4. Rough sets in analysis of control protocols

A control protocol presented in Table 1 is formally equivalent to a decision table $DT =< U, C \cup D, V, f >$, where:

- U corresponds to a finite time-horizon of the control performed by the expert (in a particular case the time moments can be interpreted as situations of the industrial process described by a set of condition attributes);

- C is a finite set of condition attributes;
- D is a finite set of decision attributes;
- $V = \bigcup_{q \in C \cup D} V_q$, where V_q is the domain of the attribute $q \in C$.

Hence, the problems of control protocol analysis can be reduced to the investigation of properties of the equivalent decision table DT. Depending on the properties of semantics of decision attributes in the set D, one can distinguish two tasks of the analysis:

1. evaluation of the classification of current situation to corresponding characteristic states performed by the expert;

2. discovering decision rules from control protocol and investigation of their properties.

Let, in the case of the first task, $D = \{d\}$ and d is a class number. The family $D = \{D_1, D_2, \cdots D_n\}$ of subsets U represents then the expert's classification of situations to n characteristic states of the considered industrial process distinguished by the expert.

In this case, the analysis of the proper decision table is reduced to the investigation of the quality and accuracy of the expert's classification and of the influence of the condition attributes contained in the set C on this classification. It is the basis of the synthesis of the set of corresponding decision rules for characteristic states recognition.

From the formal viewpoint, it corresponds to the determination for the family D^* : $\underline{C}D^*$, $\overline{C}D^*$, $Pos_C(D^*)$, $Bn_C(D^*)$ $Neg_C(D^*)$, $CORE(C)$, $RED(C)$ and the calculation of $\gamma_C(D^*)$, $\beta_C(D^*)$.

One can determine the influence of each condition attribute from the set C on subset of the family D^*, what has a practical importance. It can be done by introduction of the notion of **decidability of sets** by condition attributes.

Let $D_i \in D^*$ and $c_j \in C$.

1. The set $D_i^+(c_j) = \underline{C}D_i \setminus \underline{C - \{c_j\}}D_i$ is called **positively decidable** by condition attribute $c_j \in C$;

2. The set $D_i^-(c_j) = \overline{C - \{c_j\}}D_i \setminus \overline{C}D_i$ is called **negatively decidable** by a condition attribute $c_j \in C$.

In the same way the idea of decidability of sets by any subset of condition attributes in C can be defined. We also introduce two measures connected with the above notions:

1. for sets positively decidable by condition attribute $c_j \in C$,

$$m_i^+ = 1 - \frac{card(\underline{C - \{c_j\}}D_i)}{card(\underline{C}D_i)},$$

2. for the sets negatively decidable by a condition attribute $c_j \in C$,

$$m_i^- = \frac{card(\overline{C - \{c_j\}}D_i) - card(\overline{C}D_i)}{card(U) - card(\overline{C}D_i)}.$$

It is easy to see that these measures have the following properties:

1. $0 \leq m_i^+ \leq 1$;
   $m_i^+ = 0$ if the removal of condition attribute $c_j \in C$ does not change $Pos_C(D_i)$;
   $m_i^+ = 1$ if after the removal of condition attribute $c_j \in C$, $Pos_{C-\{c_j\}}(D_i) = \emptyset$ (i.e. D_i is internally $C - \{c_j\}$ undistinguishable).
2. $0 \leq m_i^- \leq 1$;
   $m_i^- = 0$ if the removal of condition attribute $c_j \in C$ does not change $\overline{C}D_i$;
   $m_i^- = 1$ if after the removal of condition attribute $c_j \in C$, $\overline{C - \{c_j\}}D_i = U$ (i.e. D_i is externally $C - \{c_j\}$ undistiguishable).

In the case of the second task the analysis of the corresponding decision table is reduced to:

- checking whether decision table is consistent,
- generation of the corresponding decision rules,
- finding the relative core and relative reducts for the set of condition attributes.

From the formal viewpoint it corresponds to the investigation of dependencies between the set of decision attributes D and the set of condition attributes C determined as $C \xrightarrow{k} D$, where $k = \gamma_c(D^*)$. Then the relative core $CORE_D(C)$ and relative reducts $RED_D(C)$, if they exist, are found for the set C of condition attributes.

5. Illustrative examples

The process of burning clinker is an example of industrial process for which the approach described above was successfully applied [2,3]. We present below the way how to pass from the description of stoker's behaviour when he controls a rotary kiln to the synthesis of decision rules for real-time computer control of the process.

5.1 Classification of situations

From the technological point of view, a rotary clinker kiln may be conventionally divided into some zones. The most significant of them is the burning zone. A stoker evaluates and classifies the state of burning zone and controls the kiln in order to obtain clinker with desired physical and chemical properties. Considered condition attributes are the following:
c_1 — burning zone temperature;
c_2 — burning zone colour;
c_3 — clinker granulation in burning zone;
c_4 — inside colour of the kiln.

In the language of stokers the characterictic states of a rotary clinker kiln are:
I — "very weak kiln",
II — "weak kiln",
III — "kiln weakening or becoming sharp",
IV — "sharp kiln",
V — "oversharpened kiln".

We shall assume that
$V_{c_1} = \{1,2,3,4\}$, where the numbers 1,2,3,4 denote temperature intervals, respectively:

1≙[1380°C–1420°C], 2≙[1420°C–1440°C], 3≙[1440°C–1480°C], 4≙[1480°C–1500°C],
$V_{c_2} = \{1,2,3,4,5\}$ where the numbers 1,2,3,4,5 denote burning zone colour, respectively: 1≙ scarlet, 2≙ dark pink, 3≙ bright pink, 4≙ decidedly bright pink, 5≙ rosy white;
$V_{c_3} = \{1,2,3,4\}$, where the numbers 1,2,3,4 denote clinker granulation in burning zone, respectively: 1≙ fines, 2≙ fines with small lumps, 3≙distinct granulation, 4≙ lumps;
$V_{c_4} = \{1,2,3\}$, where the numbers 1, 2, 3 denote the inside colour of the kiln, respectively: 1≙ distinct dark streaks, 2≙ indistinct dark streaks, 3≙ no dark streaks.

Table 2 presents an exemplary stoker classification protocol during one shift.

From the Table 2 we have $D^* = \{D_1, D_2, D_3, D_4, D_5\}$, where:
$D_1 = \{1,2,3,\}$;
$D_2 = \{4,5,6,7\}$;
$D_3 = \{8,9,10,11,12,123,14,15,16,17,18,19,44,45\}$
$D_4 = \{20,21,22,23,24,25,26,27,28,29,30,38,39,40,41,42,43\}$;
$D_5 = \{31,32,33,34,35,36,37\}$.
It is easy to compute that $\gamma_C(D^*) = 1$, where $C = \{c_1, c_2, c_3, c_4\}$.

Also the investigation of the influence of particular condition attributes from the set C on the accuracy of approximation $\mu_C(D_i)$ of sets D^* and on the quality of approximation $\gamma_C(D^*)$ and accuracy of approximation $\beta_C(D^*)$ of stoker's classification has its practical significance. It is illustrated in Table 3.

When we analyse the results in Table 3, we see that the set of condition attributes has three relative reducts if classification of D^* is concerned. They are

$$C_1 = \{c_1, c_2, c_3\}; \qquad C_2 = \{c_2, c_3, c_4\}; \qquad C_3 = \{c_2, c_3\}$$

With the aid of the relative reduct $C_3 = \{c_2, c_3\}$ one can present the stoker's inference model used during evaluation and classification of situations in rotary kiln as a tree of situations — Fig.1, [3,1].

5.2 Decisions made by the stoker

The control of rotary clinker kiln performed by a stoker consists in observing the values of condition attributes characterizing the state of the burning zone and in determination of such values of the decision attributes which will bring the rotary klinker kiln to a desired state, i.e."sharp kiln".

The following decision attributes vere distinguished in the discussed case: d_1 — kiln revolutions, d_2 — coal worm revolutions.

We shall assume that: $V_{d_1} = \{1,2\}$, where the numbers 1,2 denote the kiln revolutions, respectively 1≙0.9[rpm], 2≙1.22[rpm], $V_{d_2} = \{1,2,3,4\}$, where numbers 1,2,3,4 denote coal consumption measured in revolutions of the coal worm, respectively, 1≙0[rpm], 2≙15[rpm], 3≙20[rpm], 4≙40[rpm].

We display in Table 4 a control protocol registered during one shift and describing the stoker's behaviour in this period.

Analysing Table 4 it is easy to see that $D^* = \{D_1, D_2, D_3, D_4\}$, where:
$D_1 = \{58,59,60,61,62,63\}$;
$D_2 = \{1,3,11,12,13,15,16,17,29,30,31,32,33,44,45,46,47,56,57,64,66,67,68$
$69,70,\}$;

Situation number	Values of observable attributes				Number of characteristic state
	c_1	c_2	c_3	c_4	
1	1	1	1	1	I
2	1	1	1	1	I
3	2	1	1	1	I
4	2	2	1	1	II
5	2	2	1	1	II
6	1	2	1	1	II
7	2	2	1	1	II
8	2	2	2	1	III
9	1	2	2	1	III
10	2	2	2	1	III
11	2	2	2	1	III
12	3	2	2	1	III
13	3	2	2	2	III
14	2	2	2	2	III
15	2	2	2	2	III
16	1	2	2	1	III
17	2	2	2	1	III
18	2	2	2	1	III
19	3	2	2	1	III
20	3	3	2	1	IV
21	3	3	2	1	IV
22	4	3	3	1	IV
23	4	3	3	1	IV
24	3	3	3	1	IV
25	2	3	3	1	IV
26	2	3	3	1	IV
27	2	3	3	1	IV
28	3	4	3	1	IV
29	4	4	3	1	IV
30	4	4	3	1	IV
31	4	4	4	1	V
32	4	4	4	1	V
33	4	5	4	1	V
34	3	4	4	1	V
35	3	4	4	1	V
36	3	4	4	1	V
37	3	4	4	1	V
38	3	4	3	1	IV
39	3	3	3	1	IV
40	3	3	3	1	IV
41	2	3	3	1	IV
42	2	3	3	1	IV
43	2	3	3	1	IV
44	2	2	2	1	III
45	2	2	2	1	III

Table 2. An exemplary stoker's classification protocol.

Characteristic state number	Removed condition attribute					
	None	$\{c_1\}$	$\{c_2\}$	$\{c_3\}$	$\{c_4\}$	$\{c_1, c_4\}$
I	1.0	1.0	0.0	1.0	1.0	1.0
II	1.0	1.0	0.0	0.0	1.0	1.0
III	1.0	1.0	0.75	0.28	1.0	1.0
IV	1.0	1.0	0.79	0.56	1.0	1.0
V	1.0	1.0	1.0	0.1	1.0	1.0
$\gamma_C(D^*)$	1.0	1.0	0.75	0.49	1.0	1.0
$\beta_C(D^*)$	1.0	1.0	0.6	0.32	1.0	1.0

Table 3. $\mu_C(D_i), i = 1, 2, \cdots, 5, \gamma_C(D^*)$ and $\beta_c(D^*)$ when some condition attributes are removed.

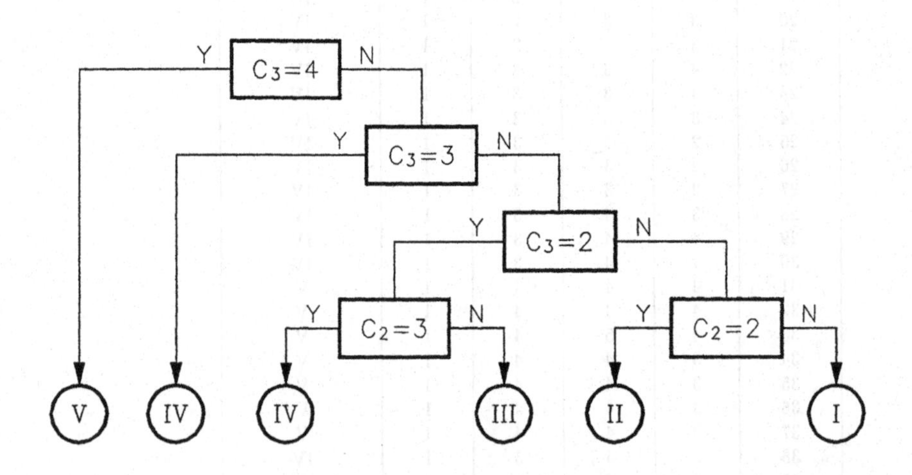

Figure 1. Tree of situations as classified by stoker.

Situation number	Condition attributes				Decision attributes		Situation number	Condition attributes				Decision attributes	
	c_1	c_2	c_3	c_4	d_1	d_2		c_1	c_2	c_3	c_4	d_1	d_2
1	3	1	3	2	2	4	36	4	2	3	3	2	2
2	3	2	3	2	2	3	37	4	2	4	3	2	2
3	3	1	3	2	2	4	38	4	1	4	3	2	2
4	4	2	3	2	2	2	39	4	1	3	3	2	2
5	4	2	4	2	2	2	40	4	1	3	2	2	2
6	4	2	4	3	2	2	41	4	1	3	2	2	2
7	4	1	4	3	2	2	42	4	3	3	2	2	2
8	4	1	3	3	2	2	43	4	3	3	2	2	2
9	4	1	3	2	2	2	44	3	3	2	2	2	4
10	4	3	3	2	2	2	45	3	3	2	2	2	4
11	3	1	3	2	2	4	46	3	1	2	2	2	4
12	3	1	3	2	2	4	47	3	2	2	2	2	4
13	3	1	3	2	2	4	48	3	2	3	2	2	3
14	3	3	3	2	2	3	49	3	2	3	2	2	3
15	3	3	2	2	2	4	50	3	2	3	3	2	3
16	3	1	2	2	2	4	51	3	2	3	3	2	3
17	3	2	2	2	2	4	52	3	2	3	3	2	3
18	3	2	3	2	2	3	53	4	1	3	3	2	2
19	3	2	3	2	2	3	54	4	3	3	3	2	2
20	4	2	3	2	2	2	55	3	3	2	3	2	3
21	4	2	4	2	2	2	56	3	3	2	2	2	4
22	4	2	4	3	2	2	57	3	3	2	2	2	4
23	4	1	4	3	2	2	58	2	3	2	2	1	4
24	4	1	3	3	2	2	59	2	3	2	1	1	4
25	4	1	3	2	2	2	60	2	3	1	1	1	4
26	4	3	3	2	2	2	61	2	1	1	1	1	4
27	3	3	3	2	2	3	62	2	2	1	1	1	4
28	3	3	3	2	2	3	63	2	2	2	1	1	4
29	3	3	2	2	2	4	64	3	2	3	1	2	4
30	3	3	2	2	2	4	65	3	2	3	2	2	3
31	3	1	2	2	2	4	66	3	1	3	2	2	4
32	3	2	2	2	2	4	67	3	1	3	2	2	4
33	3	2	2	2	2	4	68	3	1	3	2	2	4
34	3	2	3	2	2	3	69	3	1	3	2	2	4
35	3	2	3	3	2	3	70	3	1	3	2	2	4

Table 4. Control protocol of the klin stoker.

D^*	Number of elements	Number of elements of lower approximation	Number of elements of upper approximations	Accuracy of approximation
S_1	6	6	6	1.0
S_2	25	25	25	1.0
S_3	15	15	15	1.0
S_4	24	24	24	1.0

Table 5. Approximation D^* by condition attributes from the set C.

Attribute removed	none	c_1	c_2	c_3	c_4
Value $k = \gamma_C(D^*)$	1.0	0.38	0.71	0.70	0.71

Table 6. Influence of conditional attributes from set C on relation $C \xrightarrow{k} D$.

$D_3 = \{2, 14, 18, 19, 27, 28, 34, 35, 48, 49, 50, 51, 52, 55, 65\}$;
$D_4 = \{4, 5, 6, 7, 8, 9, 10, 20, 21, 22, 23, 24, 25, 26, 36, 37, 38, 39, 40, 41, 42, 43, 53, 54\}$.

We examine the approximation of family D^* by condition attributes from the set C. The results are displayed in Table 5.

In this case we have $\gamma_C(D^*) = 1$ and $\beta_C(D^*) = 1$, hence $C \longrightarrow D$, where $D = \{d_1, d_2\}$.

The influence of each condition attribute from set C on the dependency $C \stackrel{k=\gamma_C(D^*)}{\longrightarrow} D$ has been investigated and the results are shown in Table 6.

Data in Table 6 indicate that the condition attribute c_1 — burning zone temperature — has the greatest influence on dependency $C \xrightarrow{k} D$. The other attributes influence this dependency in a similar way.

Family D^* may be also used to interpret the way in which the stoker controls the rotary kiln. Situations in set D_1 correspond to a characteristic state of the rotary kiln to which the stoker refers as to a "very weak kiln". Situations in sets D_2, D_3, D_4 correspond to states "weak kiln", "kiln weakening or becoming sharp" , "sharp kiln".

We have investigated the influence of each condition attribute from the set C on approximation of particular characteristic states generated by partitionning of D^*; the results are in Table 7.

As it follows from Table 7, condition attribute c_1 — burning zone temperature — has substantional influence on the approximation of all characterictic states. The other attributes influence only "weak kiln" and "kiln weakening or becomming sharp" states.

5.3 Representation of the stoker influence model

The informations contained in Table 7 may be used to transform the resulting decision table written in Table 8 into its equivalent decision tree [2,6].

Name of a set from D^*	Attribute removed				
	none	c_1	c_2	c_3	c_4
D_1	1.0	0.38	1.0	1.0	1.0
D_2	1.0	0.27	0.43	0.4	0.46
D_3	1.0	0.04	0.20	0.2	0.13
D_4	1.0	0.27	1.0	1.0	1.0
$\gamma_C(^*)$	1.0	0.38	0.71	0.70	0.71

Table 7. Approximation and partioning when condition attributes are removed.

Decision number	Values of conditional attributes				Values of decision attributes	
	c_1	c_2	c_3	c_4	d_1	d_2
R_1	2	1	1	1	1	4
R_2	2	2	1	1	1	4
R_3	2	2	2	1	1	4
R_4	2	3	1	1	1	4
R_5	2	3	2	1	1	4
R_6	2	3	2	2	1	4
R_7	3	1	3	2	2	4
R_8	3	3	2	2	2	4
R_9	3	1	2	2	2	4
R_{10}	3	2	2	2	2	4
R_{11}	3	2	3	1	2	4
R_{12}	3	2	3	2	2	3
R_{13}	3	3	3	2	2	3
R_{14}	3	2	3	3	2	3
R_{15}	3	3	2	3	2	3
R_{16}	4	2	3	2	2	2
R_{17}	4	2	4	2	2	2
R_{18}	4	2	4	3	2	2
R_{19}	4	1	4	3	2	2
R_{20}	4	1	3	3	2	2
R_{21}	4	1	3	2	2	2
R_{22}	4	3	3	2	2	2
R_{23}	4	2	3	3	2	2
R_{24}	4	3	3	3	2	2

Table 8. Resulting decision table *DT*.

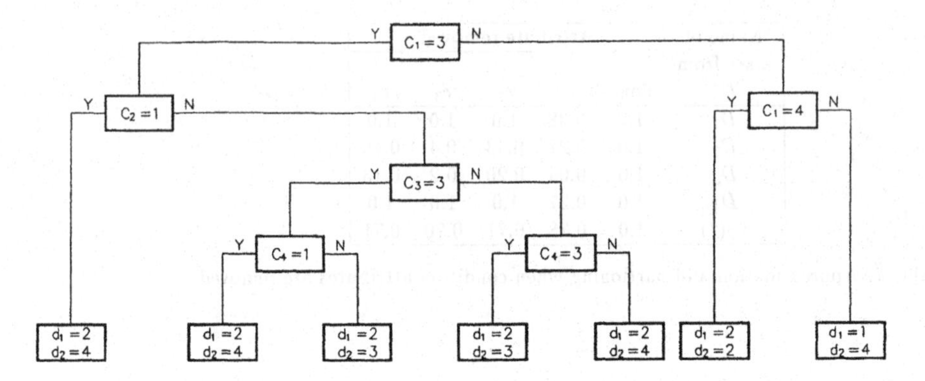

Figure 2. Decision tree equivalent to decision table DT.

This decision tree is presented in Fig.2.

The decision tree in Fig.2 presents the stoker's inference model during the control of rotary kiln control yielding from control protocol written in Table 4. When this control protocol was performed, the burning zone temperature was measured simultanously. Its values were memorized each one minute interval.

After a shift a corresponding program calculated the histogram of the distribution of burning zone temperature and determined the mean temperature $T_{bz} = 1434[^oC]$ and the standard deviation $\sigma_{T_{bz}} = 27.1[^oC]$. It is assumed that these parameters reflect the quality of the control.

5.4 Problems of computer implementation

The following measurable attributes were chosen:
c_1 — burning zone temperature, and additionally
c_5 — temperature of fumes.

Having measured these parameters, the values of other parameters were dermined:
c_7 — the derivate of temperature in burning zone,
c_8 — the derivate of temperature of gases leaving the zone.

The selected set of condition attributes (measurable and observable) was used to fill in the corresponding control protocol and then the decision rules were determined. It is briefly discussed in [5].

The results of real-time computer control based upon this set of decision rules and with experimentally chosen control cycle $\Delta T_c = 3[min]$ are also presented in [5].

The computer control results were evaluated using the same criteria as the control of human expert, i.e the temperature of burning zone as a function of time was registered during a shift, the histogram of its distribution was determined and then the mean and standard deviation were calculated. The values obtained were: $T_{bz} = 1435[^oC]$ and $\sigma_{bz} = 17.1[^oC]$.

Hence, the standard deviation of burning zone temperature was reduced by 10[°C] with the use of computer control. The fuel consumption was also reduced.

6. Conclusion

We propose a technique of real-time computer control of an industrial process. This approach is based upon computer implementation of decision rules characteristic for a human expert who controls the process.

The basic elements of this proposition are:

1. the notion of expert's inference model,
2. determination of the structure of control protocol which reflects the human expert behaviour,
3. The use of elements of rough set theory for selection of decision rules mode by the expert during the control and contained in the control protocol.

The example of rotary kiln control illustrates this proposition.

The accuracy of results, their relatively low cost and short time of the algorithm implementation indicate the usefulness of this approach, especially when it is difficult to obtain other control algorithms.

References

[1] Moret B.M.E., Decision trees and diagrams. *Computing Surveys*, **14**, (1982), 593.

[2] Mrózek A. Information systems and control algorithms. *Bulletin of the Polish Academy of Sciences T.S.*, **33**, (1985), 195.

[3] Mrózek A. Use of rough sets and decision tables for implementing rule-based control of industrial process. *Bulletin of the Polish Academy of Sciences T.S.*, **34**, (1986), 357.

[4] Mrózek A. Rough sets and some aspects of expert systems realization. *7-th International Workshop on Expert Systems and their Applications, Avignon*, (1987), 597.

[5] Mrózek A. Rough sets and dependency analysis among attributes in computer implementations of expert's inference models. *Int. J.Man-Machine Studies*, **30**, (1989), 457.

[6] Mrózek A. A new methods for discovering rules from examples in expert systems. *Int. J.Man-Machine Studies* (1991), (sent for publication).

Part I
Chapter 3

ANALYSIS OF DIAGNOSTIC SYMPTOMS IN VIBROACOUSTIC DIAGNOSTICS BY MEANS OF THE ROUGH SETS THEORY

Ryszard NOWICKI
Institute of Applied Mechanics,
Technical University of Poznań,
Piotrowo 3, 60-965 Poznań, Poland

Roman SŁOWIŃSKI Jerzy STEFANOWSKI
Institute of Computing Science,
Technical University of Poznań,
Piotrowo 3A, 60-965 Poznań, Poland

Abstract. The paper refers to problems of an application of the rough sets theory to diagnostic classification of mechanical objects. The use of the rough sets approach is shown on some practical examples. In particular, it concerns : evaluation of diagnostic capacity of symptoms, comparison of different methods of defining symptom limit values, reduction of the set of symptoms to a subset ensuring satisfactory evaluation of the object's technical state, creation of the classifier of the technical state. The analysed examples concern the evaluation of the technical state of rolling bearings.

1. Introduction

An efficient maintenance of machines in industrial processes requires reliable information about their technical state. This information is often extended by prediction of the change of this state [7,9]. It is particularly important for machines playing a critical role in the process [8]. For these machines, special stationary monitoring systems, often accompanied by diagnostic support systems, are introduced to minimize the outlay given for their maintenance. For machines having smaller influence on a final result of production, it is sufficient to perform a periodic control of their technical state using portable measurement instruments.

The technical state of working objects is non-measurable directly. Its control can be performed by evaluation of technical parameters of manufactured products or, more often, of residual (secondary) processes generated by machines during the production process. In industrial practice, measurements of physical quantities, which are changing in accordance with the technical state of a machine, are used for this evaluation [1,3]. These quantities are called symptoms of the technical state. Level of vibration for representative points of an object (measured in production conditions), level of noise near the working object, temperature, pressure or mutual position of particular parts in a machine are examples of typical symptoms of the technical state.

The choice of points and type of the measurements can follow from the specialists' knowledge a priori. It is realistic for critical machines, e.g. compressors or turbogenerator sets. However, in the case of many machines, in particular, complex ones (e.g. compact multistage transmission gears) this knowledge is very poor.

So, when setting up a diagnostic procedure, one often starts from greater number of symptoms, estimations of diagnostic signals, or measurement points than it is necessary. However, using the standard types of symptoms is not always satisfactory, so it is necessary to create new non-standard symptoms ensuring a reliable evaluation of the technical state.

Another essential issue in technical diagnostics refers to problems of defining symptom limit values [10,12]. The values of symptoms usually change monotonically with deterioration of the technical state. The symptom limit values are boundary values of a symptom which divide its domain into intervals corresponding to considered classes of the technical state (called conventional classes of the technical state). For some machines, there are known standards establishing not only the type of measurements but also defining intervals of symptom values. However, practical experience do not confirm these recommendations quite often. So, the problem of estimation of the symptom limit values is becoming important in development of diagnostic procedures for a specified type of machines. These problems were taken into account in theoretical and practical works in the 80's, particularly in the field of vibroacoustic diagnostics [10,11,12]. In consequence, several methods for defining symptom limit values were suggested. It was noticed, however, that they can lead to different evaluations of the technical state. Moreover, there are no clear indications for suitable range of their applications. So, it is desirable to have a methodology of comparing these methods for specified types of machines.

Evaluation of the technical state using the minimal number of symptoms is the most desired in practice because the lower is the number of symptoms used, the lower is usually the cost and time of the diagnostic investigation. However, rarely only one symptom is sufficient for this aim. So, a subset of symptoms must be used for a reliable evaluation. In technical diagnostics, the subset of symptoms with their limit values is called classifier of the technical state. In the case of solving diagnostic tasks, where the set of symptoms is extended by information about an objective technical state of objects, by the classifier of the technical state we understand the set of decision rules enabling the evaluation of the technical state on a base of values of symptoms from the considered subset. The latter meaning of the classifier is used in the present paper. However, it is also possible that there exist several subsets having good and similar diagnostic capacity. In such situations, a tool for reduction of the set of symptoms and comparison of several possible classifiers would be very useful.

The general aim of a majority of technical diagnostic investigations is to solve the following problems:

(a) evaluation of different methods of defining symptom limit values,
(b) evaluation of diagnostic capacity of particular symptoms,
(c) reduction of a set of symptoms to a minimal subset of symptoms ensuring satisfactory evaluation of the technical state,
(d) creation of the classifier of the technical state.

The above problems have not been solved yet in a satisfactory way. For example, problem (a) was undertaken in [10] for vibroacoustic diagnostics but was limited to analysis of simulation data. The problem (c) was considered more extensively in [2]. However, it was proved afterwards that the proposed procedure did not lead to good results in some cases.

The aim of the following paper is to show, on some practical examples, how to attain the mentioned aims of technical diagnostics by means of the rough sets approach.

The case of vibroacoustic diagnostics of mechanical objects is considered. Vibration and noise symptoms are very often used to evaluate the technical state. Their usefulness results from relatively high reliability of diagnostic information and facility of collecting such information.

The paper is based on encouraging results of two applications of the rough sets theory to evaluation of diagnostic capacity of vibroacoustic symptoms reported in [14,15,16].

Two examples concerning evaluation of the technical state of rolling bearings are analysed. The bearings are in one of two technical states : good and bad. They are described by symptoms resulting from measurements of noise and vibration of the bearing housings. In the first example, bearings were installed in a laboratory stand. In the other example, bearings were installed in a band conveyor and all measurements were taken in a real industrial environment. So, besides presentation of the application of the rough sets approach to particular steps of diagnostic research, it will be interesting to compare results of the analysis of two similar examples coming from different conditions of measurements.

In the next section, both analysed sets of data are described. Results of applications of the rough sets theory are presented in section 3. Conclusions are drawn in the final section. Considered methods of defining symptom limit values are given in the Appendix.

2. The problem definition

In this section, two considered diagnostic problems (called A and B) corresponding to different data sets are considered. The data sets give base to create the original information systems for each diagnostic case. It is known that in the rough sets approach, values of quantitative attributes are translated into some qualitative terms. This translation involves a division of the original domain into some subintervals and an assignment of qualitative codes to these subintervals. In technical diagnostics, attributes are symptoms of the technical state; using symptom limit values one can divide an original domain of a symptom into subintervals corresponding to conventional classes of the technical state [12]. As a result of this translation, coded information systems are obtained.

Several methods defining symptom limit values were proposed for diagnostic problems [1,3,4,5]. In this paper, we consider four of them. They are called L–, W–, P– and C–methods (formulae enabling determination of the limit values by means of these methods are presented in the Appendix).

2.1. Problem A – Rolling bearings examined in a laboratory

The analysed data set is composed of observations collected during a laboratory experiment with a set of 38 rolling bearings. The set of examined bearings is divided into two subsets. The first one consists of 19 bearings which were recognized to be good ones. At the end of their production, they were checked by a product quality control which proved that they were made according to a technical documentation. Other 19 bearings are in the bad technical state because some of their elements were artificially damaged (rolling elements or one of the bearing races).

The investigated bearings were successively assembled on a laboratory stand. Measurements of vibration and noise symptoms were collected in conditions of simulated working loads.

Vibration and noise levels of a bearing housing were taken as supposed symptoms of the technical state. In each case, measuring quantities were obtained as a result of band filtering of a signal for 6 different frequency bands. The filters were 1/3–octave filters with standard middle frequencies from range: 800 – 2500 Hz.

The data collected from the measurements set up the information system which is presented in Table 2.1. It contains information about 38 bearings described by means of 12 symptoms $s_1 - s_{12}$. Symptoms $s_1 - s_6$ are measured as accelerations of vibration [m/s^2] while symptoms $s_7 - s_{12}$ correspond to levels of noise in decibels [dB]. The information about each object is additionally extended by the two–valued decision attribute D. This attribute characterizes the real technical state of a bearing (0- for a good technical state, 1 - for a bad state). The information system presented in Table 2.1 is denoted by S1.

Table 2.1. Information system $S1$ (rolling bearings installed in a laboratory stand)

No.	s_1	s_2	s_3	s_4	s_5	s_6	s_7	s_8	s_9	s_{10}	s_{11}	s_{12}	D
1	0.50	0.79	1.41	0.94	0.79	1.77	79.5	81.5	72.0	72.5	76.5	75.5	0
2	0.89	1.00	1.25	0.94	0.59	1.77	74.0	78.0	75.0	74.5	66.5	78.0	0
3	1.49	1.49	1.25	0.70	0.53	1.33	85.0	88.0	79.5	79.5	74.0	68.5	0
4	1.67	2.51	2.11	1.12	1.25	1.58	93.0	92.0	83.0	72.0	72.0	69.0	0
5	3.16	3.16	2.66	1.41	0.56	1.77	81.5	84.0	81.5	96.5	96.0	91.5	0
6	0.50	0.70	0.79	0.53	0.63	1.18	78.0	83.0	80.5	96.0	94.0	86.5	0
7	0.79	0.89	1.77	0.44	0.70	1.18	80.0	76.0	82.0	78.5	94.5	92.0	0
8	1.41	2.11	2.51	1.58	1.12	3.16	80.0	80.0	79.5	83.5	79.5	94.0	0
9	1.05	0.56	0.56	1.33	1.05	0.39	85.5	79.5	74.5	74.0	77.5	71.0	0
10	0.63	1.12	0.70	0.79	0.70	0.39	78.0	75.0	78.0	77.5	75.0	76.5	0
11	0.63	1.12	0.70	0.79	0.70	0.39	75.5	71.0	82.0	78.5	78.0	78.0	0
12	3.98	2.81	1.33	1.05	1.33	1.41	69.5	75.5	75.0	85.0	77.5	80.0	0
13	2.23	1.58	1.05	1.12	2.11	3.16	69.5	70.0	76.0	80.5	79.5	80.5	0
14	2.23	2.51	1.33	0.63	0.74	0.56	68.0	75.0	69.0	71.5	80.0	85.5	0
15	1.18	0.74	0.44	0.39	0.70	0.66	63.0	63.0	70.0	70.0	64.0	75.5	0
16	1.41	1.41	1.33	1.05	1.77	1.67	60.0	68.0	72.5	79.0	71.0	72.0	0
17	1.88	2.66	1.25	1.58	3.54	1.77	67.5	60.0	79.0	76.0	77.0	74.0	0
18	1.25	1.18	0.50	1.00	2.66	1.05	60.0	60.0	60.0	62.0	60.0	66.5	0
19	1.67	1.41	0.89	1.49	2.98	0.75	67.0	69.0	66.0	60.0	68.0	67.0	1
20	0.39	0.56	0.28	0.14	0.11	0.11	79.5	81.5	72.0	72.5	76.5	75.5	1
21	0.23	0.33	0.29	0.18	0.22	0.26	78.0	75.0	74.5	66.5	78.0	75.5	1
22	0.43	0.21	0.16	0.10	0.35	0.29	79.5	79.5	74.0	68.5	67.0	68.0	1
23	0.28	0.31	0.18	0.10	0.10	0.13	72.0	72.0	69.0	71.5	73.5	76.5	1
24	0.25	0.31	0.31	0.26	0.23	0.33	80.0	78.0	77.0	73.0	83.0	85.5	1
25	0.10	0.10	0.12	0.10	0.15	0.22	72.5	72.0	72.0	70.0	74.0	80.0	1
26	0.23	0.37	0.35	0.22	0.15	0.22	75.5	72.5	74.5	69.5	75.0	78.5	1
27	0.22	0.26	0.20	0.15	0.15	0.20	71.0	73.0	69.0	69.0	76.0	82.0	1
28	0.26	0.44	0.50	0.39	0.37	0.53	71.0	77.0	67.0	66.5	78.0	76.0	1
29	0.23	0.26	0.16	0.10	0.10	0.10	66.0	66.5	68.0	64.0	71.5	71.5	1
30	0.10	0.11	0.10	0.10	0.10	0.10	72.5	72.0	72.0	70.0	74.0	80.0	1
31	0.18	0.23	0.16	0.13	0.11	0.20	75.5	72.5	74.5	69.5	75.0	78.5	1
32	0.12	0.22	0.16	0.10	0.10	0.15	70.5	73.5	69.0	69.0	76.5	82.0	1
33	0.31	0.63	0.22	0.12	0.12	0.17	71.0	77.0	67.0	66.5	78.5	76.0	1
34	0.10	0.14	0.14	0.10	0.10	0.14	67.0	70.5	66.0	63.0	64.5	68.0	1
35	0.33	0.42	0.70	0.35	0.33	0.50	70.0	68.5	75.0	69.5	79.5	80.5	1
36	0.20	0.21	0.35	0.12	0.14	0.14	70.0	68.0	72.0	63.5	76.0	74.0	1
37	0.11	0.15	0.10	0.10	0.10	0.10	65.5	72.0	71.0	67.0	66.0	71.0	1
38	0.15	0.15	0.16	0.15	0.11	0.10	70.0	74.0	70.0	67.0	77.5	82.0	1

Table 2.2 Symptom limit values for system $S1$

Symptom	Method	Symptom limit values			
		b1	b2	b3	b
s_1	L	0.22	0.34	0.46	0.58
	C	1.57	2.27	2.97	3.36
	W	1.34	1.78	2.21	2.64
	P	0.60	0.98	1.35	1.73
s_2	L	0.20	0.30	0.40	0.50
	C	1.60	2.26	2.93	3.60
	W	1.21	1.54	1.88	2.22
	P	0.61	1.01	1.42	1.82
s_3	L	0.18	0.26	0.34	0.42
	C	1.28	1.81	2.34	2.87
	W	0.96	1.23	1.50	1.77
	P	0.50	0.83	1.15	1.47
s_4	L	0.15	0.20	0.25	0.30
	C	0.96	1.35	1.73	2.12
	W	0.70	0.89	1.09	1.29
	P	0.39	0.64	0.88	1.12
s_5	L	0.21	0.32	0.43	0.54
	C	1.39	2.05	2.70	3.36
	W	1.35	1.86	2.36	2.86
	P	0.53	0.85	1.16	1.48
s_6	L	0.20	0.30	0.40	0.50
	C	1.42	2.05	2.69	3.32
	W	1.21	1.59	1.98	2.37
	P	0.54	0.89	1.23	1.58
s_7	L	79.60	99.20	118.80	138.40
	C	78.94	84.43	89.91	95.40
	W	21.14	35.16	49.18	63.20
	P	81.40	83.20	85.00	86.81
s_8	L	71.00	82.00	93.00	104.00
	C	79.56	84.81	90.07	95.32
	W	20.86	34.89	48.93	62.96
	P	75.03	79.06	83.09	87.12
s_9	L	75.10	90.20	105.30	120.40
	C	77.45	81.52	85.59	89.66
	W	19.44	33.60	47.76	61.92
	P	77.17	79.25	81.32	83.40
s_{10}	L	71.40	82.80	94.20	105.60
	C	78.97	85.23	91.49	97.75
	W	22.13	36.14	50.15	64.15
	P	75.51	79.62	83.73	87.83
s_{11}	L	76.90	93.80	110.70	127.60
	C	81.67	87.53	93.40	99.26
	W	21.60	35.59	49.59	63.58
	P	80.19	83.47	86.76	90.05
s_{12}	L	76.00	85.50	95.00	104.50
	C	82.73	88.03	93.33	98.63
	W	22.50	38.12	53.74	69.36
	P	79.59	83.19	86.78	90.37

The symptom limit values presented in Table 2.2. were used to translate the original values of symptoms $s_1 - s_{12}$ into coded values 1,2,3,4,5 corresponding to five subintervals (i.e. value of a symptom belonging to subinterval (0,b1] is coded by 1 etc.). As a result of this translation, we obtain the coded information system which is analysed afterwards.

2.2. Problem B – Rolling bearings installed in a band conveyor

The analysed data set is composed of observations collected from 55 rolling bearings installed in a band conveyor. The set of examined bearings was divided into two classes on a base of an expert evaluation. The first class consists of 34 bearings recognized to be good. The second class consists of 21 bearings being in different states of degradation.

Vibration and noise levels of bearing housings were taken as supposed symptoms of the technical state, similarly to the first diagnostic problem.

Measurements of vibration and noise levels were performed for different measuring frequency bands. The relatively low and high measuring frequency bands were chosen: from 500 to 2000 Hz and from 4 to 16 kHz for noise, and from 100 to 1000 Hz and from 1 to 11 kHz for vibration, respectively.

In addition, taking into account the possibly different propagation of vibrations from the same source in radial and axial directions of bearings, the measurements in both directions were taken.

The collected measurements created the data set which is presented in Table 2.3. It contains information about 55 objects described by 10 symptoms $s_1 - s_{10}$.
Symptoms $s_1 - s_2$ have the noise nature:

s_1 – the level of noise for frequency range 500 – 2000 Hz in [dB],
s_2 – the level of noise for frequency range 4 – 16 kHz in [dB],

Symptoms $s_3 - s_{10}$ have the vibration nature:

s_3 – the measurement of vibrations in axial direction and for frequency range 0.1 – 1 kHz,
s_4 – the measurement of vibrations in axial direction and for frequency range 1 – 11 kHz,
s_5 – the measurement of vibrations in radial vertical direction and for frequency range 0.1 – 1 kHz,
s_6 – the measurement of vibrations in radial vertical direction and for frequency range 1 – 11 kHz,
s_7 – the measurement of vibrations in radial horizontal direction and for frequency range 0.1 – 1 kHz,
s_8 – the measurement of vibrations in radial horizontal direction and for frequency range 1 – 11 kHz,
s_9 – the level of resultant vibrations for frequency range 0.1 – 1 kHz, calculated as $\sqrt{s_3^2 + s_5^2 + s_7^2}$,
s_{10} – the level of resultant vibrations for frequency range 1 – 11 kHz, calculated as $\sqrt{s_4^2 + s_6^2 + s_8^2}$.

All measurements of vibrations refer to acceleration of vibrations and are expressed in [m/s^2].

The information about each object is extended by the two-valued decision attribute D. It characterizes the real technical state of a bearing (0– for a good technical state, 1– for a bad state). The information system presented in Table 2.3 is denoted by $S2$.

Table 2.3. Information system $S2$ (rolling bearings installed in the band conveyer)

No.	s_1	s_2	s_3	s_4	s_5	s_6	s_7	s_8	s_9	s_{10}	D
1	97.0	85.8	8.0	17.0	31.0	30.0	17.0	24.0	36.2	42.0	0
2	97.2	83.1	11.0	22.0	25.0	21.0	14.0	15.0	30.7	33.9	0
3	109.6	99.0	34.0	43.0	120.0	110.0	56.0	53.0	136.7	129.5	0
4	109.5	97.5	33.0	42.0	165.0	115.0	58.0	55.0	178.0	134.2	0
5	109.2	95.0	28.0	36.0	95.0	95.0	48.0	43.0	110.1	110.3	1
6	92.9	82.6	5.5	7.5	17.0	12.5	12.5	12.5	21.8	19.2	0
7	97.8	95.0	4.5	6.5	11.5	9.5	8.5	8.5	15.0	14.3	0
8	97.2	82.6	9.5	14.5	31.0	22.0	23.0	17.0	39.8	31.4	0
9	97.9	85.4	6.5	7.5	23.0	16.5	14.5	10.5	28.0	20.9	0
10	89.4	85.1	7.5	8.0	26.5	17.0	14.5	11.0	31.1	21.8	0
11	103.6	92.1	17.0	23.0	65.0	45.0	31.0	27.0	74.0	57.3	1
12	109.0	104.4	22.0	27.0	75.0	55.0	36.0	37.0	86.1	71.6	1
13	97.7	86.3	8.5	14.5	22.5	17.0	15.0	13.0	28.3	25.9	0
14	101.7	94.8	9.5	8.5	29.0	14.0	17.5	15.0	35.2	22.2	0
15	107.1	96.7	20.0	29.0	78.0	75.0	40.0	40.0	89.9	89.8	1
16	107.7	97.7	23.0	31.0	95.0	75.0	46.0	43.0	108.0	91.8	1
17	114.0	113.3	32.0	55.0	150.0	130.0	58.0	80.0	164.0	162.2	0
18	94.2	79.4	6.3	2.8	14.5	5.2	15.5	4.5	22.1	7.4	0
19	94.4	75.8	6.5	3.1	11.5	5.7	18.0	4.2	22.3	7.7	0
20	94.9	77.6	6.0	4.5	12.5	6.8	20.5	5.5	24.7	9.8	0
21	95.3	78.8	6.3	3.7	15.0	4.8	21.0	5.5	26.6	8.2	0
22	93.0	71.7	8.5	1.1	17.0	3.5	22.0	1.8	29.1	4.1	0
23	95.1	73.9	7.8	1.3	17.5	3.8	22.5	1.8	29.6	4.4	0
24	94.3	76.8	6.3	2.8	11.0	5.5	19.5	4.8	23.3	7.8	0
25	95.1	76.6	6.5	2.5	12.0	5.5	11.5	4.5	17.8	7.5	0
26	97.1	82.0	9.5	11.5	37.0	13.5	23.0	12.5	44.6	21.7	0
27	100.8	86.4	12.0	23.0	35.0	35.0	28.0	37.0	46.4	55.9	1
28	107.1	92.4	23.0	28.0	95.0	80.0	32.0	42.0	102.8	94.6	1
29	107.5	100.1	22.0	25.0	80.0	80.0	45.0	37.0	94.4	91.6	1
30	99.9	95.3	5.6	11.0	16.5	17.0	9.5	15.5	19.8	25.5	0
31	102.4	102.1	5.0	17.0	17.0	25.5	9.5	26.5	20.1	40.5	0
32	98.3	94.8	10.0	23.5	26.5	30.0	21.5	32.0	35.6	49.8	0
33	100.4	97.1	12.0	23.0	30.0	33.0	20.5	38.0	38.3	55.3	1
34	107.7	97.3	27.0	33.0	90.0	81.0	47.0	46.0	105.1	98.8	1
35	108.4	95.0	30.0	65.0	105.0	90.0	55.0	45.0	122.3	119.8	0
36	106.0	91.0	33.0	30.0	90.0	100.0	70.0	56.0	118.7	118.5	0
37	104.0	91.0	28.0	30.0	90.0	70.0	46.0	50.0	104.9	91.1	1
38	103.0	92.0	23.0	25.0	75.0	40.0	37.0	47.0	86.7	66.6	1
39	102.0	91.0	22.0	23.0	56.0	48.0	34.0	48.0	69.1	71.7	1
40	100.0	88.0	14.0	16.0	54.0	46.0	22.0	38.0	60.0	61.8	1
41	100.0	89.0	13.0	19.0	56.0	34.0	30.0	27.0	64.8	47.4	1
42	102.0	90.0	18.0	20.0	48.0	54.0	34.0	30.0	61.5	64.9	1
43	103.0	92.0	15.0	26.0	48.0	52.0	28.0	42.0	57.6	71.7	1
44	108.0	99.0	32.0	68.0	160.0	140.0	66.0	130.0	176.0	202.8	0
45	112.0	114.0	32.0	52.0	110.0	100.0	58.0	66.0	128.4	130.6	0
46	95.0	84.0	8.0	15.0	26.0	36.0	22.0	34.0	35.0	51.7	0
47	101.0	96.0	9.5	44.0	54.0	110.0	25.0	92.0	60.3	150.0	0
48	104.0	96.0	20.0	34.0	76.0	92.0	38.0	72.0	87.3	121.7	1
49	104.0	92.0	22.0	13.0	76.0	46.0	40.0	28.0	88.7	55.4	0
50	111.0	102.0	44.0	66.0	220.0	210.0	92.0	110.0	242.5	246.1	0
51	112.0	100.0	36.0	64.0	170.0	180.0	84.0	110.0	193.0	220.4	0
52	101.0	94.0	24.0	48.0	42.0	82.0	36.0	76.0	60.3	121.7	0
53	103.0	93.0	14.0	34.0	50.0	66.0	34.0	54.0	62.1	91.8	1
54	104.0	98.0	28.0	25.0	82.0	94.0	42.0	54.0	96.3	111.3	1
55	103.0	104.0	20.0	27.0	76.0	76.0	38.0	52.0	87.3	96.0	1

Methods described in Appendix were used to calculate the symptom limit values for the data set from Table 2.3. Calculated values are presented in Table 2.4.

Table 2.4 Symptom limit values for system $S2$

Symptom	Method	Symptom limit values			
		b1	b2	b3	b
s_1	L	104.0	118.6	133.2	147.8
	C	106.34	110.81	115.29	119.77
	W	125.37	161.34	197.31	233.28
	P	104.63	107.4	110.17	112.94
s_2	L	91.0	110.3	129.6	148.9
	C	98.46	105.54	112.63	119.71
	W	128.1	184.51	240.91	297.31
	P	95.67	99.97	104.28	108.58
s_3	L	22.0	39.5	57.0	74.5
	C	25.12	33.04	40.97	48.9
	W	17.61	30.72	43.83	56.93
	P	20.6	24.01	27.43	30.84
s_4	L	23.0	44.9	66.8	88.7
	C	38.22	51.84	65.46	79.08
	W	28.39	55.69	82.98	110.29
	P	29.94	35.28	40.63	45.97
s_5	L	17.00	23.00	29.00	35.00
	C	97.69	134.26	170.83	207.40
	W	52.15	93.31	134.46	175.61
	P	74.85	88.58	102.30	116.03
s_6	L	37.00	72.20	107.40	142.60
	C	60.21	82.52	104.83	127.14
	W	39.75	77.71	115.66	153.62
	P	46.36	54.81	63.27	71.72
s_7	L	22.00	35.50	49.00	62.50
	C	47.73	62.24	76.75	91.26
	W	36.65	64.80	92.94	121.09
	P	39.62	46.03	52.43	58.83
s_8	L	37.00	72.20	107.40	142.60
	C	60.21	82.52	104.83	127.14
	W	39.75	77.71	115.66	153.62
	P	46.36	54.81	63.27	71.72
s_9	L	87.30	159.60	231.90	304.20
	C	111.73	151.13	190.53	229.93
	W	66.38	117.76	169.14	220.52
	P	87.93	103.53	119.12	134.72
s_{10}	L	71.70	139.3	206.90	274.50
	C	115.92	159.45	202.97	246.5
	W	72.60	141.11	209.61	278.12
	P	88.69	104.99	121.28	137.57

3. Rough sets analysis of the diagnostic problems

In this section, diagnostic problem A (rolling bearings in a laboratory stand) and diagnostic problem B (rolling bearings examined in industrial conditions) are analysed in the same way using the rough sets methodology. The analysis is organized in the way to solve problems (a) to (d) listed in the introduction.

3.1. Evaluation of different methods of defining symptom limit values

In the analysis of both diagnostic problems, all methods of defining symptom limit values were considered. The results, i.e. accuracies of approximations of each particular class of the technical state and quality of classification of the rolling bearings for the whole set of symptoms are presented in Table 3.1 for problem A and in Table 3.2 for problem B.

Using the criterion of the quality of classification, it is possible to rank the considered methods.

For diagnostic problem A, the C–method is the worst and should not be used for the evaluation of the technical state. All other methods: L, W and P, ensure the maximal value of the quality of classification. However, from the viewpoint of the number of atoms, the L–method is slightly better. A higher number of atoms enables better differentiation of considered bearings using their available description.

Comparing results obtained for diagnostic problem B, it can be noticed that the ranking of methods is different. The best rank is given to the L–method, the W–method is the second and the P–method is the third. The worst is the C–method, similarly to problem A. However, the qualities of classification obtained by the W– and P–methods are significantly lower than in the case of problem A and are unsatisfactory for evaluation of the technical state.

3.2. Evaluation of the diagnostic capacity of symptoms

The rough sets theory can also be used as a tool for evaluation of diagnostic capacity of symptoms. Its application will be shown first on the example of information system $S2$ were symptoms s_1 and s_2 are given both in two different scales, logarithmic and linear ones (cf. [16]). The corresponding versions of the information systems are denoted by $S2$ and $S2a$. The symptom limit values were calculated in the same way as before.

It is interesting indeed to check whether the different scale of noise symptoms infulences the evaluation of the technical state. Results of the application of the rough sets approach to analysis of information system $S2a$ are presented in Table 3.3.

When comparing the results presented in Tables 3.2 and 3.3., it can be noticed that slightly higher values of the quality are obtained for system $S2a$ but the difference between the results is rather small.

Another example illustrating the application of the rough sets theory to evaluation of the diagnostic capacity of newly constructed symptoms was described in [13]. This example concerns diagnosing of reducers which were initially analysed using traditional symptoms. In [13], the proposal of new symptoms was given. The rough sets analysis of this case gave the following results:

Table 3.1. Accuracies of approximations and quality of classification for information system $S1$

	Methods			
	L	C	W	P
Number of atoms	35	22	22	26
Class 0				
Lower approximation	19	18	19	19
Upper approximation	19	35	19	19
Accuracy of approx.	1.0	0.51	1.0	1.0
Class 1				
Lower approximation	19	3	19	19
Upper approximation	19	20	19	19
Accuracy of approx.	1.0	0.15	1.0	1.0
Accuracy of classification	1.0	0.38	1.0	1.0
Quality of classification	1.0	0.55	1.0	1.0

Table 3.2. Accuracies of approximations and quality of classification for information system $S2$

	Methods			
	L	C	W	P
Number of atoms	38	23	29	29
Class 0				
Lower approximation	32	12	12	13
Upper approximation	36	43	36	38
Occuracy of approx.	0.89	0.28	0.33	0.34
Class 1				
Lower approximation	19	12	19	17
Upper approximation	23	43	43	42
Accuracy of approx.	0.83	0.28	0.44	0.41
Accuracy of classification	0.86	0.28	0.39	0.38
Quality of classification	0.93	0.44	0.56	0.55

Table 3.3. Accuracies of approximations and quality of classification for diagnostic problem B and symptoms s_1 and s_2 in a linear scale (information system $S2a$)

	Methods			
	L	C	W	P
Number of atoms	43	21	33	29
Class 0				
Lower approximation	34	11	13	13
Upper approximation	34	43	36	39
Accuracy of approx.	1.0	0.26	0.36	0.33
Class 1				
Lower approximation	21	12	19	16
Upper approximation	21	44	42	42
Accuracy of approx.	1.0	0.27	0.45	0.38
Accuracy of classification	1.0	0.26	0.46	0.36
Quality of classification	1.0	0.41	0.67	0.53

- traditional symptoms do not give satisfactory quality of classification (close to 0.6),
- the new proposed symptoms have increased the quality of classification to 1.0; moreover, single new symptoms ensured quality in the range from 0.59 to 0.83 – it means that nearly each of them is better than all traditional ones together.

3.3. Reduction of symptoms

The set of all considered symptoms allows to evaluate with a certain quality, the technical state of objects from the information system. Very often, it is possible to reduce the number of symptoms without decreasing the quality. This leads to minimal subsets of symptoms.

For problem A, only methods L, W and P were considered, the C-method was rejected because of inadmissibly low quality of classification. For problem B, only method L was included to the analysis, others were rejected for the same reason. The results are summarized below.

Problem B, the L–method:

- the core is $\{s_5, s_6, s_7\}$
- 5 minimal subsets:

$\{s_5, s_6, s_7, s_8, s_9\}$
$\{s_2, s_4, s_5, s_6, s_7\}$
$\{s_2, s_3, s_5, s_6, s_7, s_9\}$
$\{s_4, s_5, s_6, s_7, s_8\}$
$\{s_1, s_5, s_6, s_7, s_8\}$

Problem A, the L–method:

- the core is empty,
- 17 minimal subset: two are singletons (symptoms s_1 and s_5), three subsets consist of two elements, 2 subsets of three elements, 2 subsets of two elements, other subsets of five elements;

Problem A, the W–method:

- the core is empty,
- 31 minimal subsets: one is a singleton (symptom s_1); 5 subsets consist of two elements, 15 subsets of three elements, 7 subsets of four elements and other subsets of five elements.

Problem A, the P-method:

- the core is symptom s_1 ,
- 10 minimal subsets: four of them are composed of three elements and other subsets of four elements.

Each of obtained minimal subsets can be used to represent the corresponding information system. A more detailed interpretation of the great number and structure of minimal subsets (which is interesting form the viewpoint of practical vibroacoustic diagnostics) was done in [15,16].

3.4. Creation of classifiers of the technical state

Reduction of the number of symptoms has not led to univocal and clear results. Each of the obtained minimal subsets ensures a satisfactory quality of classification and can be used to create the classifier of the technical state. As it is difficult to analyse a great number of possible classifiers, other criteria than the quality of classification must be employed. The minimum cardinality of a subset of symptoms and the minimal number of rules belonging to the classifier can be used as the secondary criteria. In practice, it is interesting to take into account also other criteria, e.g. easiness, cost and time of a measurement, or subjective preferences for certain symptoms.

Proceeding in this way, one can choose the best compromise minimal subset to create a classifier.

Let us consider first diagnostic problem A. It can be noticed that singleton $\{s_1\}$ was obtained by two methods : L and W. As a singleton is very attractive for the diagnosis, it will be used in the first order to create the classifier. Then two element minimal subsets can be considered. They are given below:
$S1$–L : $\{s_1, s_3\}$, $\{s_2, s_4\}$, $\{s_2, s_6\}$,
$S1$–W : $\{s_2, s_7\}$, $\{s_1, s_6\}$, $\{s_1, s_2\}$, $\{s_2, s_4\}$, $\{s_1, s_4\}$.

Considering problem B, it can be noticed that the core ensures the quality of classification equal to 0.65. This value can be increased to the level 0.89 (it means the value slightly lower than maximum: 0.93) by adding symptoms s_4 or s_8 . So, the subset s_4, s_5, s_6, s_7 can be chosen for further analysis.

After the choice of minimal subsets of symptoms with their limit values, one can derive decision algorithms creating classifiers. The classifier consists of decision rules which determine the assignment of an object to the technical state class basing on the values of symptoms represented in the classifier.

Examples of the most attractive classifiers are given below. One of them can be chosen to the final diagnostic procedure according to the secondary criteria described in section 3.4. and specialist's preferences.

Example 1.

For problem A (information system $S1$), subset $\{s_1\}$ with limit values defined by the L–method, the classifier is composed of the following decision rules :
if $(s_1 = 5)$ then (class=bad);
if $(s_1 = 4)$ then (class=bad);
if $(s_1 = 3)$ then (class=good);
if $(s_1 = 2)$ then (class=good);
if $(s_1 = 1)$ then (class=good);

Example 2.

For problem B (information system $S2$) and subset $\{s_4, s_5, s_6, s_7\}$ with limit values defined by the L–method, the classifier is composed of decision rules presented below :
if $(s_7 = 4)$ then (class=good);
if $(s_7 = 5)$ then (class=good);
if $(s_5 = 1)$ and $(s_7 = 1)$ then (class=good);
if $(s_5 = 3)$ and $(s_7 = 1)$ then (class=good);
if $(s_5 = 2)$ and $(s_7 = 1)$ then (class=good);
if $(s_5 = 2)$ and $(s_7 = 2)$ then (class=good);

if ($s_4 = 1$) and ($s_5 = 5$) and ($s_7 = 3$) then (class=good);
if ($s_4 = 4$) and ($s_5 = 5$) and ($s_7 = 3$) then (class=good);
if ($s_4 = 2$) and ($s_5 = 5$) and ($s_6 = 4$) and ($s_7 = 3$) then (class=good);
if ($s_5 = 5$) and ($s_7 = 1$) then (class=bad);
if ($s_4 = 2$) and ($s_5 = 5$) and ($s_7 = 3$) then (class=bad);
if ($s_4 = 1$) and ($s_5 = 5$) and ($s_6 = 2$) and ($s_7 = 2$) then (class=bad);
if ($s_4 = 2$) and ($s_5 = 5$) and ($s_6 = 3$) and ($s_7 = 2$) then (class=bad);
if ($s_4 = 2$) and ($s_5 = 5$) and ($s_6 = 2$) and ($s_7 = 2$) then (class=bad);
if ($s_4 = 1$) and ($s_5 = 4$) and ($s_6 = 1$) and ($s_7 = 1$) then (class=good or bad);
if ($s_4 = 1$) and ($s_5 = 4$) and ($s_6 = 1$) and ($s_7 = 2$) then (class=good or bad);
if ($s_4 = 1$) and ($s_5 = 5$) and ($s_6 = 1$) and ($s_7 = 2$) then (class=good or bad);
Values of condition symptoms refer to codes assigned to subintervals in domains of symptoms. Let us observe that the last three rules are non-deterministic.

Example 3. For problem A (information system $S1$) and $\{s_2, s_4\}$ with limit values defined by the L–method, the classifier is presented graphically in Figure 1.

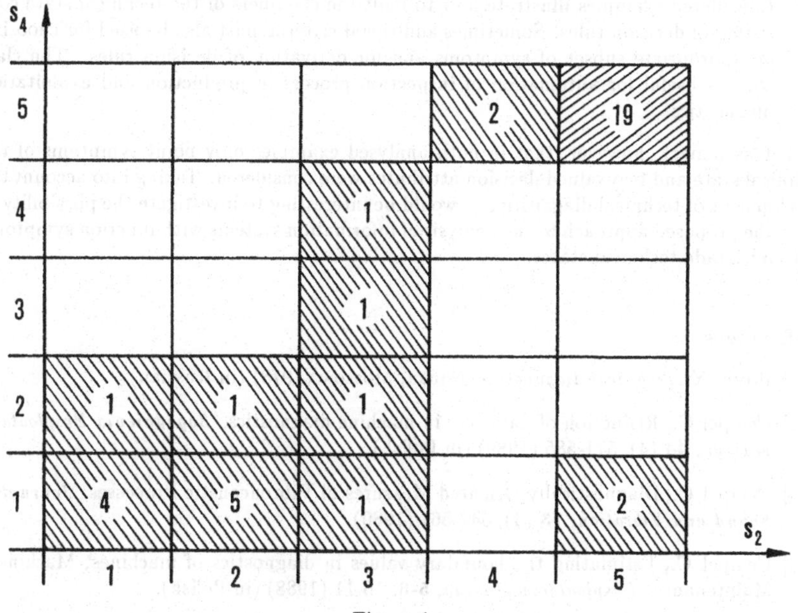

Figure 1.

Graphical representation of the decision algorithm for problem A, subset $\{s_2, s_4\}$ and the L–method:

////// – denotes bad technical state (class 1)
\\\\\\ – denotes good technical state (class 0)

(numbers in boxes give information about the number of objects which match the given combination of values of symptoms s_4 and s_2)

4. Conclusions

The analysis of the diagnostic problems considered in this paper demonstrates the usefulness of the rough sets approach for solution of several problems in technical diagnostics. In particular :

1. The rough sets approach seems to be a good tool for objective comparison of different methods of defining symptom limit values. In our study, the L-method was found the best in both diagnostics problems.
2. Using the criterion of the quality of classification it is possible to evaluate the diagnostic capacity of single symptoms or groups of symptoms of the technical state. This was shown on the example of symptoms of the same nature but expressed in different scales and on the example of newly introduced symptoms.
3. Reduction of symptoms, which is very desirable in practice, can be performed on the basis of minimal subsets of symptoms resulting from the rough sets analysis.
4. Considered examples illustrate how to built the classifiers of the technical state consisting of decision rules. Sometimes additional criteria must also be used for choosing an appropriate subset of symptoms and for derivation of decision rules. The classifier is useful for automation of inspection process in production and exploitation diagnostics.

Let us remark, however, that in both analysed examples only point symptoms of the technical state and two-valued decision attributes were considered. Taking into account the development of technical diagnostics, it would be interesting to investigate the possibility of using the proposed approach to the analysis of information systems with function symptoms and multigrade technical states.

References

[1] Birger Y., *Technical diagnostics.* Nauka, Moscow (1978) (in Russian).

[2] Cempel C., Reduction of data sets in machine diagnostics. *Zagadnienia Eksploatacji Maszyn*, **44** (4), 571-585 (1980) (in Polish).

[3] Cempel C., Diagnostically Oriented Measures of Vibroacoustic Processes. *Journal of Sound and Vibration*, **73** (4), 547-561 (1980).

[4] Cempel C., Estimating the boundary values in diagnostics of machines, Machinery Maintenance. *Eksploatacja Maszyn*, **5-6**, 10-11 (1988) (in Polish).

[5] Cempel C., Plant Determination Symptom Limit Value for Vibration Condition Monitoring, *Proc. of Conf. on Technical Diagnostics*, Prague, p 79-82 (1989).

[6] Cempel C., Limit values in the practice of machine vibration diagnostics. *Mechanical System and Signal Processing*, **5** no. 6, 483-493 (1990).

[7] Finley H., *Principles of optimum maintenance. Course materials.* The Howard Finley Corporation (1988).

[8] Kelly A., Harris M.J., *Management of industrial maintenance.* Newness-Butterworths, London (1978).

[9] Mitchell J.S., *An Introduction to Machinery Analysis and Monitoring*, PannWell Books Company, Tulusa, Oklahoma (1981).

[10] Nowicki R.,*Methods of evaluation of the technical state in vibroacoustic diagnostics of machines*, Ph.D Thesis, Technical University of Poznań (1985) (in Polish).

[11] Nowicki R., Statistical methods of evaluation of the technical state of machines. In *Proc of 7-th School on Diagnostics (DIAGNOSTYKA'85 - Poznań-Rydzyna)*, WPP, Poznań, 225-238 (1985) (in Polish).

[12] Nowicki R., Badly Conditioned Problems of Technical State of Mechanical Objects. *Zagadnienia Eksploatacji Maszyn*, **2** (1991) (to appear) (in Polish).

[13] Nowicki R., Słowiński R., Stefanowski J., Possibilities of an application of the rough sets theory to technical diagnostics. In *Proc. of IX National Symp. on Vibration Techn. and Vibroacoustics*, AGH Press, Kraków, 149-152 (1990).

[14] Nowicki R., Słowiński R., Stefanowski J., Rough Sets Based Diagnostic Classifier of Reducers. *International Journal of Man-Machine Studies* (1991), (submitted).

[15] Nowicki R., Słowiński R., Stefanowski J., Evaluation of Diagnostic Symptoms by Means of The Rough Sets Theory, *Computers in Industry* (1992) (to appear).

[16] Nowicki R., Słowiński R., Stefanowski J., Rough sets analysis of diagnostic capacity of vibroacoustic symptoms. *Journal of Mathematics and Computers with Applications*, (1992) (to appear).

Appendix

The symptom limit values can be defined in many different ways (cf. [1,4,5,10]). We shall use four methods described below :

The C–method :

$$b = \overline{s} + \sigma\sqrt{\frac{P_g}{2A}} \tag{1}$$

where:

$\overline{s}$ – mean value of a symptom, calculated as:

$$\overline{s} = \frac{\sum_{i=1}^{M} S_i}{M} \tag{2}$$

M – number of measurements of a symptom (number of observations);
S_i – result of measurement of a symptom,
σ – standard deviation of a symptom calculated as:

$$\sigma = \frac{\sqrt{\sum_{i=1}^{M}(S_i - \overline{s})^2}}{M} \tag{3}$$

P_g – the fiability index of an object (a ratio of the work time to the work time increased by the repair time), A – the permissible probability of superfluous repairs performed in order to avoid break–down;

The P–method :

$$b = (1 - \gamma^{-1})\overline{s}\,\sqrt[\gamma]{\frac{P_g}{A}} \tag{4}$$

where:

γ – Pareto's shape coefficient calculated as:

$$\gamma = 1 + \sqrt{1 + \left(\frac{\overline{s}}{\sigma}\right)^2} \tag{5}$$

The W–method :

$$b = s_{MIN} + (\overline{s} - s_{MIN})\Gamma^{-1}(1 + k^{-1})\sqrt[k]{\ln\left(\frac{P_g}{A}\right)} \tag{6}$$

where:

s_{MIN} – minimal observed value of a symptom ,

k - Weibul's shape coefficient calculated as:

$$k = \frac{\overline{s} - s_{MIN}}{\sigma} \tag{7}$$

$\Gamma(n)$ - the gamma function of the (n) argument,

The L–method :

$$b = 4s_M - 3s_{MIN} \tag{8}$$

where:

s_M – the mode value of an empirical distribution of results (of observations).

The value of b, calculated according to formulae (1),(4), (6) and (8) is treated as a threshold value (i.e. an 'alarm' value) which separates good and bad technical states. Three additional limit values b1, b2 and b3 are uniformly distributed over the range of symptom variability and are interpreted as 'alert' values of symptom [3]. They are defined as follows :

b1= $s^* + 0.25 * \mathrm{b} - s^*$ (9A)

b2= $s^* + 0.50 * \mathrm{b} - s^*$ (9B)

b3= $s^* + 0.75 * \mathrm{b} - s^*$ (9C)

where:

- in the case of the C–, P– and W–methods:

$$s^* = \overline{s}$$

- in the case of the L-method:

$$s^* = s_M \,.$$

Part I
Chapter 4

KNOWLEDGE-BASED PROCESS CONTROL USING ROUGH SETS

Adam J. SZLADOW
REDUCT Systems Inc.
P.O. Box 3570
Regina SK, S4P 3L7, Canada

Wojciech P. ZIARKO
Department of Computer Science
University of Regina
Regina SK, S4S 0A2, Canada

Abstract. A process model for heavy oil upgrading was developed using a machine learning system based on rough sets. The model has incorporated temporal patterns for control-loop responses and key relationships between the variables at low (feedback) and high (supervisory) control levels. The model predicted reactor temperature distribution with 90 to 95 percent accuracy. Accuracy depended on the number of training cycles and on the temperature resolution used. The key advantages of using the rough sets approach were: 1) it allowed the use of qualitative and quantitative process information in the model; 2) it provided a unified description of temporal events and patterns; and 3) it permitted the use of "raw" sensor data without preprocessing.

1. Introduction

The advances made in process control theory in the last 20 years have resulted in the development of robust, model-based adaptive control for a number of industrial applications. Practical methods have been developed for controller design and tuning, control optimization, etc., using a model-based approach.

The interest in process models came from an observation that operators with a correct mental model of a process are more successful in executing correct control actions. Also, with the introduction of computerized numerical techniques, the relationships between process variables were easier to model and the models were easier to implement in control design. The modelling techniques most often used have been differential equations and parametric correlations. Unfortunately, this has resulted in different degrees and forms of data representation at different control levels. For the control loop level, dynamic deterministic models are often used while for the process optimization level, static parametric models are easier to implement.

Adaptive process control based on deterministic models and statistical correlations has several disavantages, namely:

1. Data required for design of models are not always available; relationships between variables are not always known and may change with time.

2. Often there is no simple causal functional relationship between specific modelled variables and closed-loop trends, e.g., multivariable controllers or multiple faults.

3. Control based on deterministic and statistical models uses only a fraction of the information available on process dynamics and closed-loop responses.

Also, in building control algorithms from closed-loop data, two major difficulties must be addressed:

1. Noisy data must be filtered out in order to extract robust control information: however, excessive filtering may cause loss of information.

2. In multivariable controllers or for multiple faults diagnosis, one must assume certain functional dependencies in order to interpret information from closed-loop data.

Early research on implementation of AI knowledge-based methods in process control was aimed at support of algorithmic methods in order to overcome some of the above mentioned shortcomings. For example, Astrom (1986), Oyen (1986) and Buezli (1986) worked on incorporating knowledge-based information in controller monitoring and tuning. Birky et al. (1988) researched the use of knowledge-based systems in selection of controller configuration, and Garrison et al. (1986) studied the use of knowledge-based systems in control optimization.

These early knowledge-based process control applications required implementation teams which had in-depth knowledge of Artificial Intelligence methods, computer programming, process engineering and control theory. The problems studied were often either too complex and costly for the state-of-the-art technology or too simple to yield significant returns on research investment. Nevertheless, considerable research and development work continued at large universities and FORTUNE 500 companies in the mid and late 1980's. The introduction, in 1990/91, of knowledge-based expert tools such as Gensym G2, Honeywell TDC 3000 Expert, Talarian RTIME and Mitch RTAC created more interest in the application of AI methods in industrial process control. In addition, other AI methods such as machine learning emerged as viable tools for control applications in the space and defence areas - areas which traditionally have been at the forefront of AI developments. Information on recent AI research in process control can be found in monographs and articles by Stephanopoulos (1989), Mavrovouniotis (1989), Whitely and Davis (1990). Information on industrial applications appears in the work of Rowan (1989), Shum et al. (1989), and Stephanopoulos (1989).

This paper discusses the work performed at REDUCT Systems Inc. on the application of rough sets-based, machine learning for development of process models from process data. This research project is one of several underway at REDUCT on the development of knowledge-based computerized control methods for process control and optimization, fault detection and diagnosis, and for building of operator and robot controlled models. Previous work on the application of rough sets in process control is described in the work of Pawlak (1985), Mrozek (1986) and Ziarko et al. (1989).

2. Discovering Process Models From Process Data

Adaptive process control requires that robust and accurate process models representing the dynamics of the process's behavior be developed. These models are then used for performing real-time process control tasks such as:

- controller tuning, etc.
- assessment of load disturbance dynamics
- generation of hypotheses concerning the propagation of faults and testing of these hypotheses
- detection of performance deterioration, etc.

This section describes how process models can be derived from process data using the rough sets methods. The major advantages of using rough sets are:

1. They allow the use of "raw" process (sensor) data without the need for preprocessing such as filtering to eliminate noise, "finger printing" to capture events with different time scales, etc.
2. They permit the identification and capture of temporal trends of closed-loop responses (dead-time, time constants, sustained process disturbances, etc.)
3. They capture temporal patterns describing relationships between process variables for supervisory control.

The process modelled in this example was a heavy oil upgrading process for hydrocracking of vacuum bottoms from heavy oil to naphtha and gas oil. A brief description of the process follows.

2.1. HEAVY OIL HYDROCRACKER

Heavy oil hydrocracking is a crude oil upgrading process commercially applied at two crude refineries in western Canada. New higher efficiency hydrocracking processes are currently being developed in two demonstration plants, one of which, CANMET Hydrocracking, was modelled during this project. The advantages of applying hydrocracking processes are a high crude oil conversion and a high hydrocarbon liquid yield. The key to the performance of the process is optimal operation of the hydrocracking reactor.

The hydrocracking reactor in the plant modelled was a three-phase column reactor with fine particulates of catalyst dispersed in the liquid heavy oil phase (Figure 1). In normal operation, liquid feed containing fresh catalyst and recycle hydrogen enters the inlet nozzle in the reactor bottom and gradually rises to the top exit of the reactor. The overall heat of reaction of hydrogen with heavy crude is exothermic which means that either net heat must be removed or the reactor's contents must be quenched by a coolant. The reactor is, therefore, equipped with four quench nozzles located along the height of the reactor, where cold hydrogen gas is injected to control each zone temperature (Figure 2).

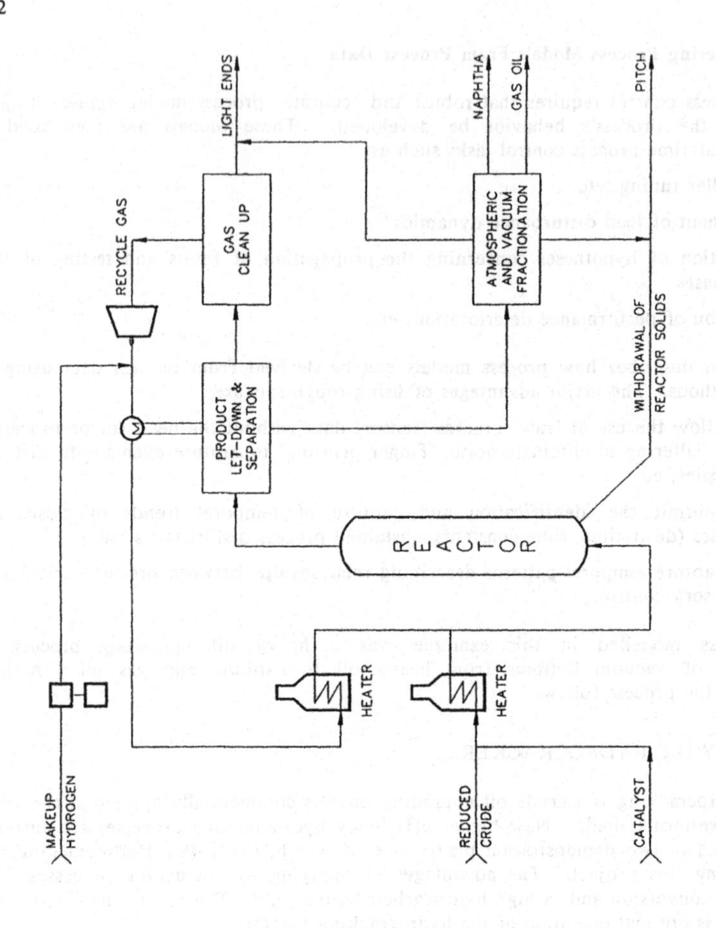

Figure 1. CANMET HYDROCRACKING PROCESS

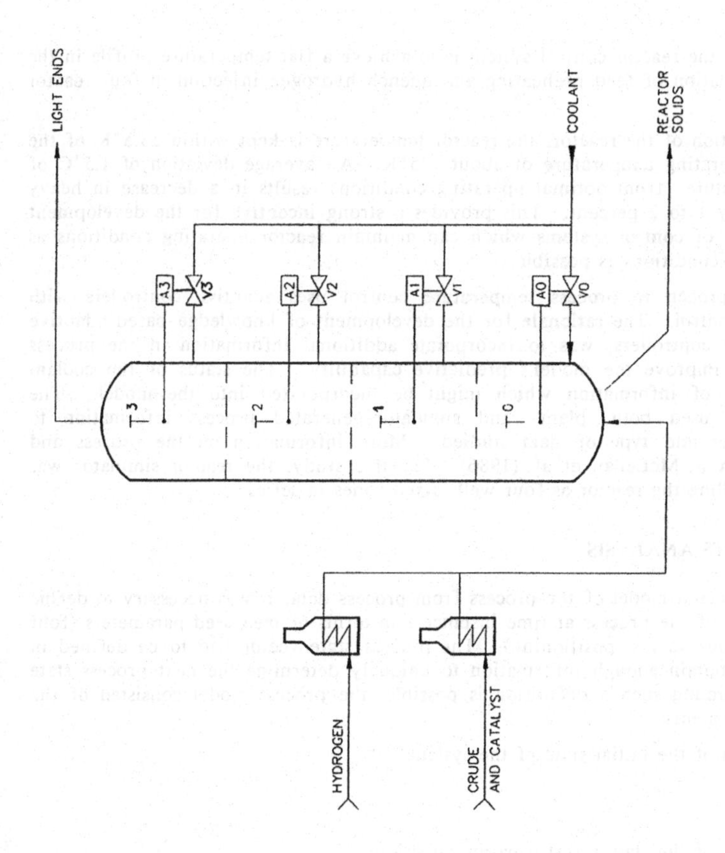

Figure 2. REACTOR ZONES AND QUENCH SYSTEM

The objective of the reactor control system is to achieve a flat temperature profile in the reactor by a combination of feed preheating and quench hydrogen injection in four reactor zones.

At normal operation of the reactor, the reactor temperature is kept within ±3.5°K of the nominal process operating temperature of about 725°K. An average deviation of 1.5°C of the reactor temperature from optimal operating conditions results in a decrease in heavy crude conversion by 1 to 2 percent. This provides a strong incentive for the development and implementation of control systems which can maintain reactor operating conditions as close to the optimal conditions as possible.

The current approach to process temperature control uses adaptive controllers with statistical process control. The rationale for the development of knowledge-based adaptive control for cascade controllers, was to incorporate additional information in the process model in order to improve the model's predictive capability. The status of the coolant actuators is typical of information which might be incorporated into the model. The development study used both, plant- and simulator-generated process information to increase the number and type of data studied. More information on the process and simulator was given in McLellan et al. (1986). For this study, the reactor simulator was simplified by modelling the reactor as four well mixed zones in series.

2.2. ROUGH SETS ANALYSIS

To discover a rule-based model of the process from process data, it was necessary to define the state vector S(t) of the process at time instance t in terms of measured parameters (four temperatures and four valves' positioning). The process state vector had to be defined in such a way as to contain enough information to uniquely determine the next process state vector S(t+1). Assuming such a definition is possible, the process model consisted of the following two components:

1. Specification of the initial state of the system:
   $S(t_0) = S_0$

and

2. Specification of the state transformation function:
   $F\colon\ S(t) \rightarrow S(t+1)$

The objective of the rough sets analysis of the above defined model was:

- to determine whether the adopted definition of the process state vector included sufficient and necessary information for derivation of a state transformation function
- to eliminate all redundant information (parameters) from process state vectors
- to generate a set of process model rules describing the state transformation function over the process domain.

The state transformation function, when fully specified in terms of rules, constituted a process model which describes the behavior of the system.

2.2.1. *Data Acquisition.* Four models were derived for control of four temperature zones. The objective of each model was to predict reaction temperature in subsequent time instances $(T_{(t+1)})^z$, where z is the reactor zone number, based on current and past process data for all temperatures $(T_t^0, T_t^1, T_t^2, T_t^3)$ and coolant valve actuator settings V_t^0, V_t^1, V_t^2, V_t^3. The process state vector $S(t)^z$ was defined as a time window of length n containing information about current temperatures, past temperatures, and actuators' settings ranging from the previous time instance t-1 through to the (-n)th past time instance t-n. For example, for zone=0, the state vector was defined:

$$
\begin{aligned}
S(t)^0 = \langle &(T_{(t-n)})^0, T_{(t-n)})^1, T_{(t-n)})^2, T_{(t-n)})^3), \\
&(V_{(t-n)})^0, V_{(t-n)})^1, V_{(t-n)})^2, V_{(t-n)})^3), \ldots \\
&\ldots (T_{(t-1)})^0, T_{(t-1)})^1, T_{(t-1)})^2, T_{(t-1)})^3) \\
&(V_{(t-1)})^0, V_{(t-1)})^1, V_{(t-1)})^2, V_{(t-1)})^3)\rangle
\end{aligned}
$$

To create the rule-based model of the system, the state vectors $S(t_W)$, $S(t_{w+1})$, ..., $S(t_m)$ were measured in the time interval from t_w to t_m, and stored in an attribute-value format. In a typical run, the time window length (w) was 20 resulting in 164 attributes in the tabular representation. As shown below, this window length was sufficient to provide discriminating and unambiguous process models.

The next objective was to discover predictive rules for each temperature zone based on the attribute-value format table for state vectors $S(t_w)$ to $S(t_m)$. Finding the rules characterizing each temperature zone $(T_0^0, T_0^1, T_0^2, T_0^3)$ was equivalent to finding the state transformation function F. To obtain the state transformation function, two analysis steps had to be implemented based on the theory of rough sets.

2.2.2. *Step 1. Data Quantization and Dependency Analysis.* One of the premises of the rough sets theory is that reducing the precision of data representation reveals data regularities (Pawlak, 1982). Consequently, the first step in the analysis of process data was data quantization, i.e., replacing actual temperatures with temperature ranges (two different temperature cuts were tested, 0.7°K and 1.4°K). However, replacing precise numeric values with ranges sometimes may cause information loss, which in this case would impede the model's ability to predict the reactor temperature based on information in the state vector. To determine whether or not the information contained in the state vectors after range conversion is sufficient to characterize the four temperatures T_0^0, T_0^1, T_0^2, T_0^3, a data dependency analysis was performed using REDUCT's rough-sets based data analysis and modelling software DataLogic/R+. The dependency was computed by treating each of the temperatures T_0^0, T_0^1, T_0^2, T_0^3 as a decision attribute and all past temperatures and control valve settings as conditions. The state vector S(t) was considered to contain a sufficient amount of information to construct the state model of the process when the dependency was fully functional for all temperatures. If the dependency was below 1, the size of the time window "w" was increased in steps until full dependency was reached. It was determined that window size w=20 ensured full functional dependency for all four temperatures.

2.2.3. *Step 2. Elimination of Redundant Parameters.* After finding sufficient definition of the state vector, the next problem was elimination of redundant state vector components (parameters). To eliminate redundant parameters from the state vector, an attribute reduct was computed with respect to every one of the four target temperatures, T_0^0, T_0^1, T_0^2, T_0^3. Because each of the computed reducts contains sufficient information to characterize exactly one temperature, their union is a minimal state vector characterizing all temperatures. The degree of reduction achieved over 1000 process records represented in the time window w=20 was approximately 80 percent.

2.3. RULE GENERATION AND TESTING

Following computation of the minimal state vector a set of rules was generated from the data table. These rules were restricted to the state parameters contained in the minimal stage vector. A subset of rules was produced for each temperature T_0^0, T_0^1, T_0^2, T_0^3 by using proprietary algorithms implemented in REDUCT's rough sets-based software. A typical format of the model-rules generated is depicted in Table 1. The rules are displayed in decision table format with ranges as rule outcomes, and inequalities and/or equalities as rule conditions.

Table 1. EXAMPLE OF MODEL RULES

Decision :: 72.1.60 < t03 <= 723.00 if:

0		v201=1 & t64>721.50 & t54<=723.00 & t13>721.50
	OR	
1		v93=1 & t64>721.50 & t33<=723.00 & t13>721.50
	OR	
2		v201=1 & t54<=723.00 & t33>723.00
	OR	
3		t74>721.50 & t33<=721.50
	OR	
4		t74<=721.50 & t54<=721.50 & t13>721.50

Decision :: 723.00 < t03 <= 724.40 if:

0		v141=1 & t74>723.00 & t54>723.00 & t54<=724.50
	OR	
1		v141=1 & t81>726.00 & t74>723.00 & t74<=724.50 & t54>723
	OR	
2		t81<=726.00 & t74<=724.50 & t64>723.00 & t54<=724.50
	OR	
3		v161=0 & v141=1
	OR	
4		t81<=727.50 & t64<=723.00 & t13>723.00
	OR	
5		t81<=727.50 & t74<=723.00 & t13>723.00

A number of test runs were performed to evaluate the accuracy of the models. The models were derived for training data bases varying from 100 to 1500 sampling points. This corresponded to an input of 1.7 to 27.1 cycles per training session (at 59 points per cycle). The models were then tested against the simulated reactor temperatures using identical input conditions (Figure 3).

As expected, the results varied depending on the temperature resolution or accuracy sought in the model. For a model with a temperature range of 1.4°K, the accuracy of the model exceeded 95 percent for 10 or more training cycles. For a temperature resolution of 0.7°K, the accuracy exceeded 90 percent for 25 or more training cycles. Since each cycle took about 6 minutes, an accurate process model could be generated within 1 to 2 hours of the reactor achieving quasi-steady-state conditions. This was sufficient to update the control model for changes in the feed composition and/or for major process disturbances.

3. Conclusions

The example of a heavy oil hydrocracker shows how rough sets can be applied for development of knowledge-based, adaptive process controllers. Incorporation of additional information on the status of the coolant valve actuators allowed for building more accurate process models directly from process data, without any need for complex data preprocessing. The key issue in rough sets application, the possibility of losing information due to data generalization (ranging), did not hinder the building of predictive models. The dependency analysis showed that for the two data precisions considered (0.7°K and 1.4°K), full functional dependencies were obtained for a time window size of 20 or more. This ability of rough sets-based machine learning system to identify temporal patterns in data, while at the same time ensuring full functional dependencies, is most useful in application of the method in real-time control.

The key characteristics of real-time control applications are:

1. *Temporal Reasoning*: Control actions and decisions depend on current and past data (trends).
2. *Changing Focus of Attention*: The invoked class of control rules often changes which requires unified representation of the control knowledge.

Time is the key variable in real-time systems. The system needs the ability to reason about past, present and future events as well as the sequences in which events happen. The ability of rough sets method to identify such temporal patterns and to ensure that their representation is based on the strongest, non-redundant variables, makes this method well suited for performing these tasks.

The rough sets-based approach also facilitates development of uniform knowledge representation for high and low control levels. The current approaches use deterministic dynamic representation at low level control and static parametric representation at high level control. A rough sets-based system uses rule-based representation of knowledge at both control levels. This results in:

1. unified description of temporal events at all levels of control (feedback, diagnosis, optimization, etc.)

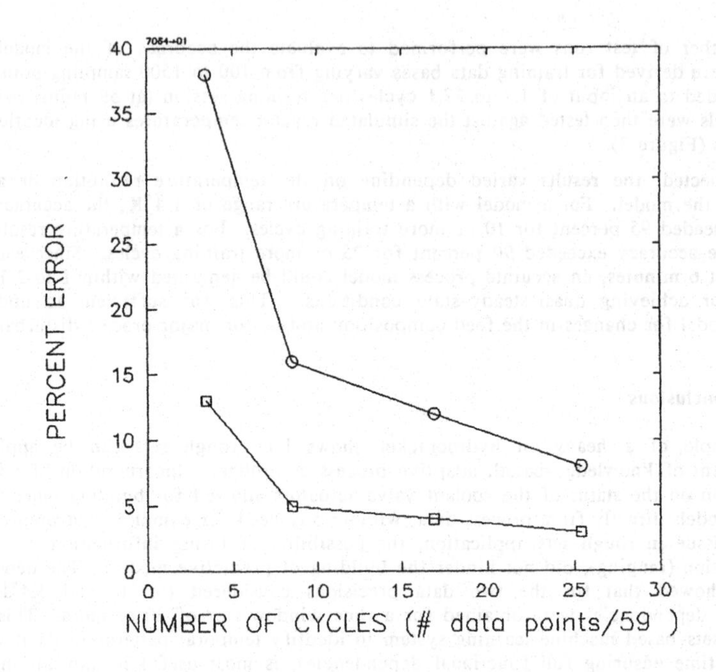

Figure 3. EFFECT OF TEMPERATURE RESOLUTION AND SIZE OF TRAINING SET ON MODEL ERROR
○ 0.7°K Resolution; □ 1.4°K Resolution

2. elimination of the assumptions about the relationships between low and high level control variables

3. use of qualitative and quantitative knowledge contained in the process trends.

REDUCT's future work in this area will focus, therefore, on development of a unified knowledge representation for feedback and supervisory control, using a rough sets-based approach, for real-time systems applications.

References

Astron, K.J. (1986) "Auto-tuning Adaptation and Smart Control", in Morari and McAroy (eds.) *Chemical Process Control - CPC III*, CACHE-Elsevier.

Birky, G.J., McAroy, T.J., and Modares, M. (1988) "An Expert System for Distillation Control Design", *Computers and Chemical Engineering*, *12*, 1045-63.

Buezli, C.W. Jr. (1986) "Summary of Research Activities", AAAI Workshop on AI for Process and Industrial Monitoring and Control, AAAI-86, Philadelphia, PA.

Cheung, J.T. and Stephanopoulos, G. (1989) "Representation of Process Trends, Part I and II", Technical Report, Laboratory of Intelligent Systems in Process Engineering, Department of Chemical Engineering, MIT.

Garrison, D.B., Pret, D.M., and Steachy, P.E. (1986) "Expert Systems in Process Control and Optimization - A Perspective", AAAI-86 Workshop, Philadelphia, PA.

Mavrovouniotis, M. (ed.) (1989) *Artificial Intelligence Applications in Process Engineering*. Academic Press.

McLellan, J., J. Tscheng, and Purvis, D.H. (1986) "A Real-Time Dynamic Simulator for the CANMET Hydrocracking Process", paper presented at 1986 Summer Computer Simulation Conference, Reno, Nevada.

Mrozek, A. (1986) "Use of Rough Sets and Decision Tables for Implementing Rule-Based Control of Industrial Processes", *Bull. Polish Acad. Sci. Tech. 34*, 357.

Oyen, R.A. (1986) "Summary of Research Activities", AAAI Workshop on AI for Process and Industrial Monitoring and Control, AAAI-86, Philadelphia, PA.

Pawlak, Z. (1985) "Decision Tables and Decision Algorithms", *Bull. Pol. Acad. Sci. (Tech. Sci.)*, *33*, No. 9-10, 487-494.

Pawlak, Z. (1982) "Rough Sets", *International Journal of Computer and Information Sciences*, *11*, 341-356.

Rowan, D.A. (1989) "On-Line Expert Systems in Process Industries", *AI Expert Magazine*, August.

Shum, S.K., Gandikota, M.S., Ramesh, T.S., McDowell, J.K., Myers, D.R., Whitely, J.R., and Davis, J.F. (1989) "A Knowledge-based System Framework for Diagnosis of Process Plants", Seventh Power Plant Dynamics, Control and Testing Symposium, Knoxville, TN.

Stephanopolous, G. (ed.) (1989) *CACHE Case Studies Series, Knowledge-Based Systems in Process Engineering*, Austin, Texas: CACHE Corporation.

Whitely, J.R. and Davis, J.F. (1990) "Application of Neural Networks to Qualitative Interpretation of Process Sensor Data." AICHE Spring National Meeting, Orlando.

Ziarko, W. and Katzberg, J. (1989) "Control Algorithm Acquisition" in *Knowledge-Based System Diagnosis, Supervision and Control*, Plenum Publishing Corp., pp. 167-178.

Part I
Chapter 5

ACQUISITION OF CONTROL ALGORITHMS FROM OPERATION DATA

Wojciech P. ZIARKO
Computer Science Department
University of Regina
Regina SK, S4S 0A2, Canada

Abstract. The article deals with automatic acquisition of control algorithms from process or device operation data. Our objective is to present a methodology for elimination of the mathematical modelling and programming stages in the control system development. In the presented approach these stages are replaced by a training stage followed by generation of decision rules equivalent to a Boolean network. The rules are produced from logged operation data obtained from experienced operators. The rules are subsequently converted into control program source code. The adapted training methodology has been developed within the framework of the theory of rough sets [1] and implemented as a commercial system for data analysis and rules extraction system by REDUCT Systems Inc., Regina, Canada. The presentation is illustrated with a comprehensive example demonstrating the generation of the control algorithms from simulated data representing movements of the robot arm.

1. Introduction

The traditional approach to the design of computerized control systems requires that the control strategy to achieve a specific goal be known in advance and clearly described in the form of a control algorithm. The control algorithm typically is a procedure associating observable sensor readings with some control actions. Control algorithms are easy to construct for simple, well defined problems such as, for example, controlling the operation of a photocopier. Most practical problems, however, are not that straightforward and the development of a control algorithm is a major chalenge involving complex and costly mathematical modelling. In addition to that, the mathematical models of many complex industrial or chemical processes are not known complicating the matter even further. On the other hand, despite tha lack of mathematical models, it is well known that experienced operators are capable of successfully controlling complex devices or processes to achieve desired device behavior or product quality. Attempts to convert such control knowledge of human operators into formalized control algorithms so that they could be used as a basis of automated control system were until recently largery unsuccessful. This situation has started to change in recent years with the introduction of mathematical theory of rough sets [1-5] and related data analysis and empirical modelling algorithms. It has become possible to provide a robust solution to the long-standing problem of automatic conversion of operator past experience, as represented in the log of process data, into computer processible control algorithms. What it means in practice is that control programs could be developed by training the computer rather than by programming it. The programming, or control algorithm development stage in the traditional sense, could be eliminated completely from the control system design process. The use of training instead of direct programming has a number of advantages, some of which are summarized below.

1. Control algorithms can be developed for previously untractable problems provided experienced operators are available.

2. The development cycle is short and the development cost is low due to elimination of the need for creation of a mathematical model and programming.

3. Control algorithms can be easily changed or updated, just by retraining them, e.g. to suit other applications.

4. Control systems can be easily customized by training them for a variety of different applications.

The primary application areas for this methodology are robotics and industrial process control. It should be emphasized at this point, however, that the presented approach is still at the early stages of development and further research and experimentation are needed to establish it as a practical method of control algorithm development for robots or industrial processes. In Section 3 a relatively simple example illustrating the application of the method to acquire control rules for robotic arm movement is presented. The sole purpose of this example is to familiarize the reader with the technique rather than trying to solve a robot arm control problem. It is the author's conviction, however, that the presented approach is applicable to complex real life control problems whose description and solution is beyond the scope of this article.

2. Major Stages in the Process of Control System Development From Experience

According to the proposed methodology in the process of control system development from past experience one can distinguish the following three major stages:

1. training
2. control code generation
3. operation by a computer with embedded control code

The control algorithm is generated automatically from training data accumulated during the training session. During such a training session, the system for which the controller is to be developed (the plant) is operated manually by one or several experienced operators who are capable of running the system in such a way that the predefined goals are met. While the system is being operated the data reflecting the control actions taken and the sensor readings sampled with a predefined frequency are being logged (Figure 1).

After the training stage the data are processed by the Control Algorithm Generator which will analyze them and produce a control algorithm. The control algorithm which is expressed in the form of decision rules equivalent to a Boolean network is then used as a basis of computer code generation. The code written either in assembler or the C programming language can then be incorporated into a control program (Figure 2).

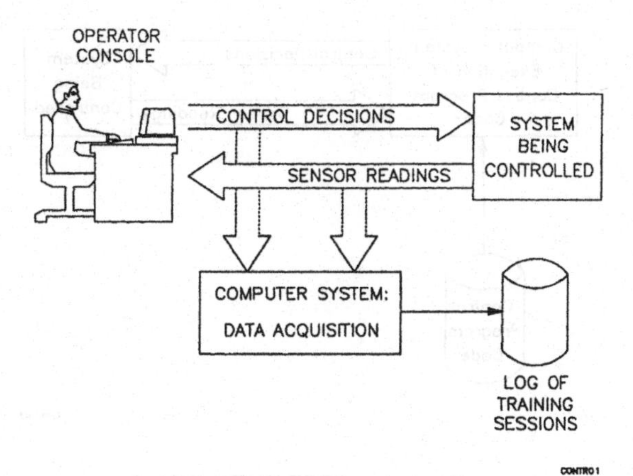

Figure 1. TRAINING STAGE

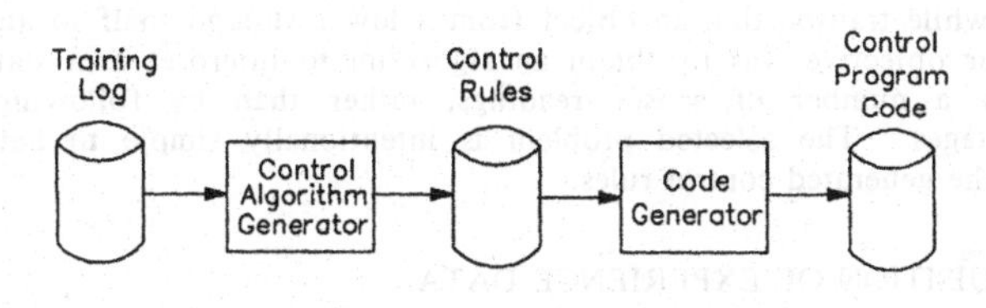

Figure 2. CONTROL CODE GENERATION

Finally during the operation phase, the control program is executed repeatedly to interpret sensor readings and to produce control signals activating system actuators (Figure 3). The control program has to be activated with the sensor sampling frequency to interpret the plant's state information within the brief time interval in between successive samples. This requirement is essential in order to ensure that generated control decisions correspond to the operator of the human operator.

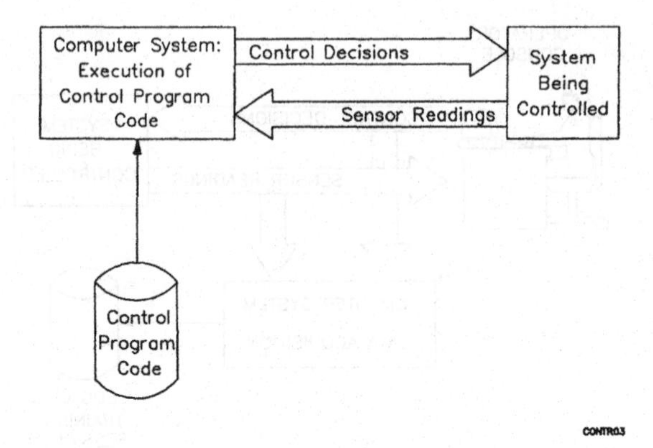

Figure 3. OPERATION PHASE

3. Generation of a Control Algorithm for the Movement of a Robotic Arm: An Example

To demonstrate the approach a simple problem of controlling simulated movements of a robotic arm while transporting an object from a lower storage shelf to an upper shelf was selected. Our objective was to obtain an algorithm to interpret arm state information as expressed by a number of sensor readings, rather than by following a sequence of predefined stages. The selected problem is intentionally simple to better illustrate the method and the generated control rules.

3.1. ACQUISITION OF EXPERIENCE DATA

The hypothetical arm used in our experiment is illustrated in Figure 4.

The arm is assumed to have three flexible joints, each equipped with a powerful engine to change the respective angles 1, 2 or 3 between arm sections. The operator can control these engines by turning them on for forward movement (FOR), reverse movement (REV) or turning them off. In addition to that the operator can activate an engine at "fingers" of the gripper to move them closer to each other or further apart, or to increase (decrease) the pressure on the object.

It is assumed that the speed of the joint engines is 10° per second and that the rate of pressure increase/decrease on the gripper is 2 psi per second.

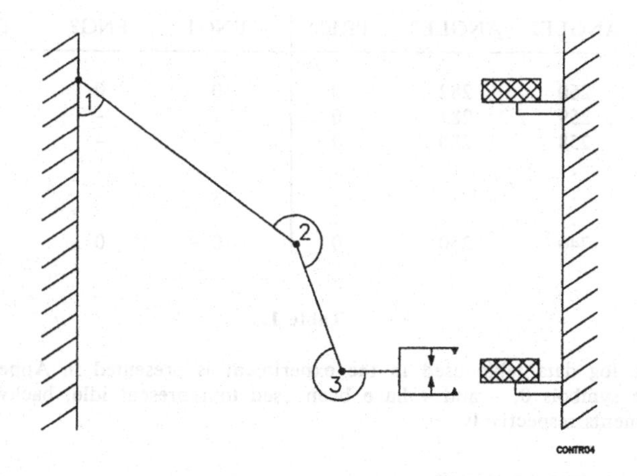

Figure 4. THE ROBOT ARM

In our simulated operation, the operator manipulates engines at the joints and the gripper fingers to grab the object, pulls it from the lower shelf, moves it up to the level of the upper shelf, places the object on the upper shelf and releases the grip. To capture the control knowledge of the operator five sensors have been attached to the arms: angle sensors measuring angles 1, 2 and 3, and 2 pressure sensors at the ends of the fingers. In addition to that, 4 sensors have been used to detect the movements of joint and gripper engines. While the arm is manipulated by the operator the readings from these sensors are sampled with the frequency of 10 samples per second, resulting in a log file of 199 training samples. That is, each training sample is a vector e_t recording sensor measurements taken at the time instance t:

e_t = (ANGLE1, ANGLE2, ANGLE3, PRESS, ENG1, ENG2, ENG3, GRIP)

where:

- ANGLE1-3 are readings reflecting angles 1-3 as illustrated in Figure 4.
- PRESS is a measured pressure at the ends of the arm fingers.
- ENG1-3 reflect the status of joint engines (forward movement, reverse movement or idle).
- GRIP represents the status of the gripper (opening, closing, idle).

The small excerpt from the training set is shown in Table 1.

ANGLE1	ANGLE2	ANGLE3	PRESS	ENG1	ENG2	ENG3	GRIP
68	260	282	0	0	0	0	0
67	259	281	0	-	-	-	0
66	258	280	0	-	-	-	0
.	.	.	.	.	.	.	.
.	.	.	.	.	.	.	.
.	.	.	.	.	.	.	.
86	246	250	0	0	0	0	-

Table 1.

The whole log data table used in the experiment is presented in Appendix 1. In the training table symbols 0, - and + have been used to represent idle, backward or forward engine movements respectively.

3.2. DATA PREPROCESSING

The main objective of this exercise was to use simulated operation log data to produce general control rules for each of the four manipulation engines controlled by the operator. The rules are supposed to link some specific conditions expressed in terms of sensor data ANGLE1-3 and PRESS with engine 1-3 or gripper activation commands such as move forward (FOR), reverse (REV), stop or no change (NC). Generally, each rule is a logical expression conforming to the following format:

<u>if</u> P(ANGLE1, ANGLE2, ANGLE3, PRESS) <u>then</u> COMMAND
where P(*) is a predicate of four respective sensors variables and COMMAND is any of FOR, REV, STOP or NC.

Before such rules can be generated it is necessary to identify control commands taken by the operator based on the engines activity data in the training table. This can be done by looking at changes in the engine activity at consecutive time instances. For example, if the engine was idle at time t, and running in reverse direction at time t+1 then the control command REV had to be issued at some point of time between t and t+1. Because our sampling rate of 10Hz is relatively high compared to the speed of the engines and expected activity of the operator one can assume, with a negligible error, that the command REV was issued at time t. Similarly, other commands can be associated with different time instances, in particular the null command NC (no change) will be assigned to a time instance t if the engine activity did not change within time interval <t, t+1>. By applying such command detection rules the original training table (Table 1) has been converted into a table in which engine status data have been substituted by control commands (Table 2).

The whole training table is shown in Appendix 2. The training table in such a converted format can be used to generate control rules. Separate set of control rules should be produced for each joint engine ENG1-3 and GRIP. For the purpose of this example the rules were extracted only for engine 1 because the rules for other engines are similar.

ANGLE1	ANGLE2	ANGLE3	PRESS	ENG1	ENG2	ENG3	GRIP
68	260	282	0	REV	REV	REV	NC
67	259	281	0	NC	NC	NC	NC
66	258	280	0	NC	NC	NC	NC
.	.	.	.	.	.	.	.
.	.	.	.	.	.	.	.
.	.	.	.	.	.	.	.
86	246	250	0	NC	NC	NC	NC

Table 2.

3.3. GENERATION OF CONTROL RULES

Following the initial data preprocessing stage the rule extraction software supplied by REDUCT Systems Inc. has been used to extract control rules for each of the three engines and the gripper. The rules have been generated according to a proprietary machine learning algorithm developed based on the idea of reduct [1]. The algorithm produces a non-redundant set of classification rules from both symbolic and numeric data. The main advantages of this algorithm are the completeness and the generality of the rules and the absence of the essential information loss in the process of rules generation. The generated rules are structured in the form of a decision tree in which internal nodes represent decision points, branches are associated with rule preconditions and leaves correspond to control engine commands. For instance, the rule structure for controlling engine 1 is given in Figure 5.

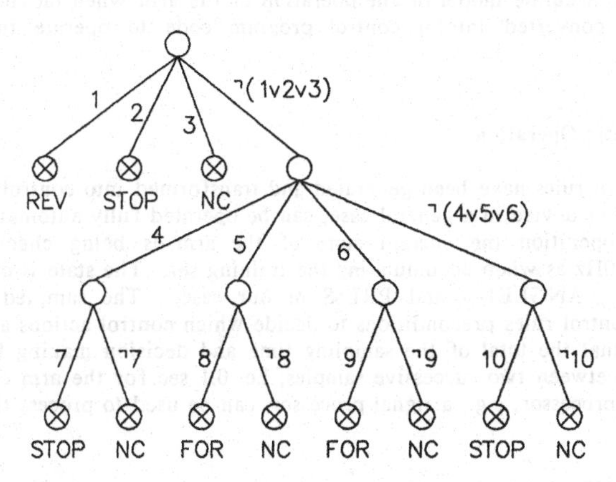

Figure 5. TREE STRUCTURED CONTROL RULES

where the following conditions 1-10 (or their negations denoted as $\neg$) are associated with branches of the tree:

1. 68≤ANGLE1≤69 AND PRESS<2
2. ANGLE3<244 AND 68≤ANGLE1≤ 69
3. (244≤ANGLE3≤274) OR
   ((ANGLE3<275 OR ANGLE3>284) AND PRESS<18) OR
   (218≤ANGLE2≤256 AND PRESS≥18) OR
   (256<ANGLE2<279 AND 69<ANGLE1<84) OR
   ((ANGLE2<249 OR ANGLE2>279) AND ANGLE1≥84) OR
   (275≤ANGLE3≤284 AND PRESS≥18 AND (ANGLE1<68 OR ANGLE1>69)) OR
   (66≤ANGLE1<68)
4. PRESS<17
5. ANGLE3<271
6. 282≤ANGLE3≤287
7. ANGLE3=277
8. ANGLE3<239 AND PRESS=20
9. ANGLE2=260
10. ANGLE2=278

Each path from the root of the tree to a leaf represents a control rule with a control command associated with the leaf of the tree. It can be verified that the example rules provide a 100% accurate model of the operation of the arm when moving the object. They can be easily converted into a control program code to operate the robot arm fully automatically.

4. Automatic Operation

After the control rules have been generated and transformed into control programs the robot arm, or any other device in a general case, can be operated fully automatically. During such an automatic operation the current state of the arm is being checked with the same frequency of 10Hz as when accumulating the training set. The state is checked by sampling sensor readings, ANGLE1-3 and PRESS in our case. The sampled readings are then compared to control rules preconditions to decide which control actions are to be taken next. It is essential that the total of the sampling time and decision making be significantly less than the time between two successive samples, i.e. 0.1 sec for the arm control. To make it possible a fast processor, e.g., a signal processor, can be used to process the decision making.

5. Conclusion

Creation of control algorithms for complex processes or devices is not a straight forward task. Nevertheless human operators, after some training, can control them quite effectively.

In our presentation, we have attempted to demonstrate a methodology for direct acquisition of control algorithms from logged device or process operation data, thus avoiding the mathematical modelling and programming steps. In the heart of the methodology is the utilization of a system for data analysis and rules extraction derived from the theory of rough sets. Although the approach looks promising, as demonstrated by the example of robot arm, more experimentation with actual mechanical devices or processes is needed before the methodology reaches its maturity. If successful, it might open up new application areas for automatic control, i.e., to deal with problems which are difficult to model using standard mathematical techniques.

6. Acknowledgements

The research reported in this paper has been partially supported by an operating grant from the Natural Sciences and Engineering Research Council of Canada. The support of REDUCT Systems Inc. in providing the software and the patience of Ms. Debbie Mills in preparing the manuscript is gratefully acknowledged. The preliminary version of this paper has been presented at the IASTED International Symposium on Modelling, Identification and Control, Innsbruck, Austria, 1992.

7. References

1. Pawlak, Z. (1991) *Rough Sets: Theoretical Aspects of Reasoning About Data*, Kluwer Academic Publishers, Dordrecht.

2. Grzymala-Busse, J. (1988) "Knowledge Acquisition Under Uncertainty - a Rough Set Approach", *Journal of Intelligent and Robotic Systems*, *1*, pp. 3-16.

3. Pawlak, Z. (1985), "Decision Tables and Decision Algorithms," *Bulletin of the Polish Academy of Sciences*, *33*, 9-10, pp. 487-494.

4. Ziarko, W., and Katzberg, J. (1989) "Control Algorithms Acquisition, Analysis and Reduction: Machine Learning Approach" in *Knowledge-Based System Diagnosis, Supervision and Control*, Plenum Press, pp. 167-178.

5. Mrozek, A. (1989) "Rough Set Dependency Analysis Among Attributes in Computer Implementation of Expert Inference Models," *International Journal of Man-Machine Studies*. *30*, pp. 457-473.

Appendix 1. Simulated Operation Log Data

68	260	282	0	0	0	0	0
67	259	281	0	–	–	–	0
66	258	280	0	–	–	–	0
65	257	279	0	–	–	–	0
64	256	278	0	–	–	–	0
63	255	277	0	–	–	–	0
63	254	276	0	0	–	–	0
63	253	275	0	0	–	–	0
63	252	274	0	0	–	–	0
63	251	273	0	0	–	–	0
63	250	272	0	0	–	–	0
63	249	271	0	0	–	–	0
63	248	270	0	0	–	–	0
63	247	269	0	0	–	–	0
63	246	268	0	0	–	–	0
63	245	267	0	0	–	–	0
63	244	266	0	0	–	–	0
63	243	265	0	0	–	–	0
63	242	264	0	0	–	–	0
63	241	263	0	0	–	–	0
63	240	262	0	0	–	–	0
63	239	261	0	0	–	–	0
63	238	260	0	0	–	–	0
63	237	259	0	0	–	–	0
63	236	258	0	0	–	–	0
63	235	257	0	0	–	–	0
63	234	256	0	0	–	–	0
63	233	255	0	0	–	–	0
63	232	254	0	0	–	–	0
63	231	253	0	0	–	–	0
63	230	252	0	0	–	–	0
63	229	251	0	0	–	–	0
63	228	250	0	0	–	–	0
63	227	249	0	0	–	–	0
63	226	248	0	0	–	–	0
63	225	247	0	0	–	–	0
63	224	246	0	0	–	–	0
63	223	245	0	0	–	–	0
63	222	244	0	0	–	–	0
63	221	243	0	0	–	–	0
63	220	242	0	0	–	–	0
63	219	241	0	0	–	–	0
63	218	240	0	0	–	–	0
63	217	239	0	0	–	–	0
63	216	238	0	0	–	0	0
63	215	238	0	0	–	0	0
63	214	238	0	0	–	0	0
63	213	238	0	0	–	0	0
63	212	238	0	0	–	0	0
63	211	238	0	0	–	0	0
63	210	238	0	0	–	0	0
63	210	238	1	0	0	0	+
63	210	238	2	0	0	0	+
63	210	238	3	0	0	0	+
63	210	238	4	0	0	0	+
63	210	238	5	0	0	0	+
63	210	238	6	0	0	0	+
63	210	238	7	0	0	0	+
63	210	238	8	0	0	0	+
63	210	238	9	0	0	0	+
63	210	238	10	0	0	0	+
63	210	238	11	0	0	0	+
63	210	238	12	0	0	0	+
63	210	238	13	0	0	0	+

63	210	238	17	0	0	0	+
63	210	238	18	0	0	0	+
63	210	238	19	0	0	0	+
63	210	238	20	0	0	0	+
64	211	239	20	+	+	+	0
65	212	240	20	+	+	+	0
66	213	241	20	+	+	+	0
67	214	242	20	+	+	+	0
68	215	243	20	+	+	+	0
68	216	244	20	0	+	+	0
68	217	245	20	0	+	+	0
68	218	246	20	0	+	+	0
68	219	247	20	0	+	+	0
68	220	248	20	0	+	+	0
68	221	249	20	0	+	+	0
68	222	250	20	0	+	+	0
68	223	251	20	0	+	+	0
68	224	252	20	0	+	+	0
68	225	253	20	0	+	+	0
68	226	254	20	0	+	+	0
68	227	255	20	0	+	+	0
68	228	256	20	0	+	+	0
68	229	257	20	0	+	+	0
68	230	258	20	0	+	+	0
68	231	259	20	0	+	+	0
68	232	260	20	0	+	+	0
68	233	261	20	0	+	+	0
68	234	262	20	0	+	+	0
68	235	263	20	0	+	+	0
68	236	264	20	0	+	+	0
68	237	265	20	0	+	+	0
68	238	266	20	0	+	+	0
68	239	267	20	0	+	+	0
68	240	268	20	0	+	+	0
68	241	269	20	0	+	+	0
68	242	270	20	0	+	+	0
68	243	271	20	0	+	+	0
68	244	272	20	0	+	+	0
68	245	273	20	0	+	+	0
68	246	274	20	0	+	+	0
68	247	275	20	0	+	+	0
68	248	276	20	0	+	+	0
68	249	277	20	0	+	+	0
68	250	278	20	0	+	+	0
68	251	279	20	0	+	+	0
68	252	280	20	0	+	+	0
68	253	281	20	0	+	+	0
68	254	282	20	0	+	+	0
68	255	282	20	0	+	0	0
68	256	282	20	0	+	0	0
68	257	282	20	0	+	0	0
68	258	282	20	0	+	0	0
68	259	282	20	0	+	0	0
68	260	282	20	0	+	0	0
69	261	283	20	+	+	+	0
70	262	284	20	+	+	+	0
71	263	285	20	+	+	+	0
72	264	286	20	+	+	+	0
73	265	287	20	+	+	+	0
74	266	288	20	+	+	+	0
75	267	289	20	+	+	+	0
76	268	290	20	+	+	0	0
77	269	290	20	+	+	0	0
78	270	290	20	+	+	0	0
79	271	290	20	+	+	0	0
80	272	290	20	+	+	0	0
81	273	290	20	+	+	0	0
82	274	290	20	+	+	0	0
83	275	290	20	+	+	0	0

86	280	290	20	0	+	0	0
86	281	290	20	0	+	0	0
86	282	290	20	0	+	0	0
86	283	290	20	0	+	0	0
86	284	290	20	0	+	0	0
86	285	290	20	0	+	0	0
86	286	290	20	0	+	0	0
86	287	290	20	0	+	0	0
86	286	289	20	0	-	-	0
86	285	288	20	0	-	-	0
86	284	287	20	0	-	-	0
86	283	286	20	0	-	-	0
86	282	285	20	0	-	-	0
86	281	284	20	0	-	-	0
86	280	283	20	0	-	-	0
86	279	282	20	0	-	-	0
86	278	281	20	0	-	-	0
86	277	280	20	0	-	-	0
86	276	279	20	0	-	-	0
86	275	278	20	0	-	-	0
86	274	277	20	0	-	-	0
86	273	276	20	0	-	-	0
86	272	275	20	0	-	-	0
86	271	274	20	0	-	-	0
86	270	273	20	0	-	-	0
86	269	272	20	0	-	-	0
86	268	271	20	0	-	-	0
86	267	270	20	0	-	-	0
86	266	269	20	0	-	-	0
86	265	268	20	0	-	-	0
86	264	267	20	0	-	-	0
86	263	266	20	0	-	-	0
86	262	265	20	0	-	-	0
86	261	264	20	0	-	-	0
86	260	263	20	0	-	-	0
86	259	262	20	0	-	-	0
86	258	261	20	0	-	-	0
86	257	260	20	0	-	-	0
86	256	259	20	0	-	-	0
86	255	258	20	0	-	-	0
86	254	257	20	0	-	-	0
86	253	256	20	0	-	-	0
86	252	255	20	0	-	-	0
86	251	254	20	0	-	-	0
86	250	253	20	0	-	-	0
86	249	252	20	0	-	-	0
86	248	251	20	0	-	-	0
86	247	250	20	0	-	-	0
86	246	250	20	0	-	0	0
86	246	250	18	0	0	0	-
86	246	250	16	0	0	0	-
86	246	250	14	0	0	0	-
86	246	250	12	0	0	0	-
86	246	250	10	0	0	0	-
86	246	250	8	0	0	0	-
86	246	250	6	0	0	0	-
86	246	250	4	0	0	0	-
86	246	250	2	0	0	0	-
86	246	250	0	0	0	0	-

Appendix 2. Training Data

68	260	282	0	REV	REV	REV	NC
67	259	281	0	NC	NC	NC	NC
66	258	280	0	NC	NC	NC	NC
65	257	279	0	NC	NC	NC	NC
64	256	278	0	NC	NC	NC	NC
63	255	277	0	STOP	NC	NC	NC
63	254	276	0	NC	NC	NC	NC
63	253	275	0	NC	NC	NC	NC
63	252	274	0	NC	NC	NC	NC
63	251	273	0	NC	NC	NC	NC
63	250	272	0	NC	NC	NC	NC
63	249	271	0	NC	NC	NC	NC
63	248	270	0	NC	NC	NC	NC
63	247	269	0	NC	NC	NC	NC
63	246	268	0	NC	NC	NC	NC
63	245	267	0	NC	NC	NC	NC
63	244	266	0	NC	NC	NC	NC
63	243	265	0	NC	NC	NC	NC
63	242	264	0	NC	NC	NC	NC
63	241	263	0	NC	NC	NC	NC
63	240	262	0	NC	NC	NC	NC
63	239	261	0	NC	NC	NC	NC
63	238	260	0	NC	NC	NC	NC
63	237	259	0	NC	NC	NC	NC
63	236	258	0	NC	NC	NC	NC
63	235	257	0	NC	NC	NC	NC
63	234	256	0	NC	NC	NC	NC
63	233	255	0	NC	NC	NC	NC
63	232	254	0	NC	NC	NC	NC
63	231	253	0	NC	NC	NC	NC
63	230	252	0	NC	NC	NC	NC
63	229	251	0	NC	NC	NC	NC
63	228	250	0	NC	NC	NC	NC
63	227	249	0	NC	NC	NC	NC
63	226	248	0	NC	NC	NC	NC
63	225	247	0	NC	NC	NC	NC
63	224	246	0	NC	NC	NC	NC
63	223	245	0	NC	NC	NC	NC
63	222	244	0	NC	NC	NC	NC
63	221	243	0	NC	NC	NC	NC
63	220	242	0	NC	NC	NC	NC
63	219	241	0	NC	NC	NC	NC
63	218	240	0	NC	NC	NC	NC
63	217	239	0	NC	NC	STOP	NC
63	216	238	0	NC	NC	NC	NC
63	215	238	0	NC	NC	NC	NC
63	214	238	0	NC	NC	NC	NC
63	213	238	0	NC	NC	NC	NC
63	212	238	0	NC	NC	NC	NC
63	211	238	0	NC	NC	NC	NC
63	210	238	0	NC	STOP	NC	FOR
63	210	238	1	NC	NC	NC	NC
63	210	238	2	NC	NC	NC	NC
63	210	238	3	NC	NC	NC	NC
63	210	238	4	NC	NC	NC	NC
63	210	238	5	NC	NC	NC	NC
63	210	238	6	NC	NC	NC	NC
63	210	238	7	NC	NC	NC	NC
63	210	238	8	NC	NC	NC	NC
63	210	238	9	NC	NC	NC	NC
63	210	238	10	NC	NC	NC	NC
63	210	238	11	NC	NC	NC	NC
63	210	238	12	NC	NC	NC	NC
63	210	238	13	NC	NC	NC	NC

63	210	238	17	NC	NC	NC	NC
63	210	238	18	NC	NC	NC	NC
63	210	238	19	NC	NC	NC	NC
63	210	238	20	FOR	FOR	FOR	STOP
64	211	239	20	NC	NC	NC	NC
65	212	240	20	NC	NC	NC	NC
66	213	241	20	NC	NC	NC	NC
67	214	242	20	NC	NC	NC	NC
68	215	243	20	STOP	NC	NC	NC
68	216	244	20	NC	NC	NC	NC
68	217	245	20	NC	NC	NC	NC
68	218	246	20	NC	NC	NC	NC
68	219	247	20	NC	NC	NC	NC
68	220	248	20	NC	NC	NC	NC
68	221	249	20	NC	NC	NC	NC
68	222	250	20	NC	NC	NC	NC
68	223	251	20	NC	NC	NC	NC
68	224	252	20	NC	NC	NC	NC
68	225	253	20	NC	NC	NC	NC
68	226	254	20	NC	NC	NC	NC
68	227	255	20	NC	NC	NC	NC
68	228	256	20	NC	NC	NC	NC
68	229	257	20	NC	NC	NC	NC
68	230	258	20	NC	NC	NC	NC
68	231	259	20	NC	NC	NC	NC
68	232	260	20	NC	NC	NC	NC
68	233	261	20	NC	NC	NC	NC
68	234	262	20	NC	NC	NC	NC
68	235	263	20	NC	NC	NC	NC
68	236	264	20	NC	NC	NC	NC
68	237	265	20	NC	NC	NC	NC
68	238	266	20	NC	NC	NC	NC
68	239	267	20	NC	NC	NC	NC
68	240	268	20	NC	NC	NC	NC
68	241	269	20	NC	NC	NC	NC
68	242	270	20	NC	NC	NC	NC
68	243	271	20	NC	NC	NC	NC
68	244	272	20	NC	NC	NC	NC
68	245	273	20	NC	NC	NC	NC
68	246	274	20	NC	NC	NC	NC
68	247	275	20	NC	NC	NC	NC
68	248	276	20	NC	NC	NC	NC
68	249	277	20	NC	NC	NC	NC
68	250	278	20	NC	NC	NC	NC
68	251	279	20	NC	NC	NC	NC
68	252	280	20	NC	NC	NC	NC
68	253	281	20	NC	NC	NC	NC
68	254	282	20	NC	NC	STOP	NC
68	255	282	20	NC	NC	NC	NC
68	256	282	20	NC	NC	NC	NC
68	257	282	20	NC	NC	NC	NC
68	258	282	20	NC	NC	NC	NC
68	259	282	20	NC	NC	NC	NC
68	260	282	20	FOR	NC	FOR	NC
69	261	283	20	NC	NC	NC	NC
70	262	284	20	NC	NC	NC	NC
71	263	285	20	NC	NC	NC	NC
72	264	286	20	NC	NC	NC	NC
73	265	287	20	NC	NC	NC	NC
74	266	288	20	NC	NC	NC	NC
75	267	289	20	NC	NC	NC	NC
76	268	290	20	NC	NC	STOP	NC
77	269	290	20	NC	NC	NC	NC
78	270	290	20	NC	NC	NC	NC
79	271	290	20	NC	NC	NC	NC
80	272	290	20	NC	NC	NC	NC
81	273	290	20	NC	NC	NC	NC
82	274	290	20	NC	NC	NC	NC
83	275	290	20	NC	NC	NC	NC

86	280	290	20	NC	NC	NC	NC
86	281	290	20	NC	NC	NC	NC
86	282	290	20	NC	NC	NC	NC
86	283	290	20	NC	NC	NC	NC
86	284	290	20	NC	NC	NC	NC
86	285	290	20	NC	NC	NC	NC
86	286	290	20	NC	NC	NC	NC
86	287	290	20	NC	REV	REV	NC
86	286	289	20	NC	NC	NC	NC
86	285	288	20	NC	NC	NC	NC
86	284	287	20	NC	NC	NC	NC
86	283	286	20	NC	NC	NC	NC
86	282	285	20	NC	NC	NC	NC
86	281	284	20	NC	NC	NC	NC
86	280	283	20	NC	NC	NC	NC
86	279	282	20	NC	NC	NC	NC
86	278	281	20	NC	NC	NC	NC
86	277	280	20	NC	NC	NC	NC
86	276	279	20	NC	NC	NC	NC
86	275	278	20	NC	NC	NC	NC
86	274	277	20	NC	NC	NC	NC
86	273	276	20	NC	NC	NC	NC
86	272	275	20	NC	NC	NC	NC
86	271	274	20	NC	NC	NC	NC
86	270	273	20	NC	NC	NC	NC
86	269	272	20	NC	NC	NC	NC
86	268	271	20	NC	NC	NC	NC
86	267	270	20	NC	NC	NC	NC
86	266	269	20	NC	NC	NC	NC
86	265	268	20	NC	NC	NC	NC
86	264	267	20	NC	NC	NC	NC
86	263	266	20	NC	NC	NC	NC
86	262	265	20	NC	NC	NC	NC
86	261	264	20	NC	NC	NC	NC
86	260	263	20	NC	NC	NC	NC
86	259	262	20	NC	NC	NC	NC
86	258	261	20	NC	NC	NC	NC
86	257	260	20	NC	NC	NC	NC
86	256	259	20	NC	NC	NC	NC
86	255	258	20	NC	NC	NC	NC
86	254	257	20	NC	NC	NC	NC
86	253	256	20	NC	NC	NC	NC
86	252	255	20	NC	NC	NC	NC
86	251	254	20	NC	NC	NC	NC
86	250	253	20	NC	NC	NC	NC
86	249	252	20	NC	NC	NC	NC
86	248	251	20	NC	NC	NC	NC
86	247	250	20	NC	NC	STOP	NC
86	246	250	20	NC	STOP	NC	REV
86	246	250	18	NC	NC	NC	NC
86	246	250	16	NC	NC	NC	NC
86	246	250	14	NC	NC	NC	NC
86	246	250	12	NC	NC	NC	NC
86	246	250	10	NC	NC	NC	NC
86	246	250	8	NC	NC	NC	NC
86	246	250	6	NC	NC	NC	NC
86	246	250	4	NC	NC	NC	NC
86	246	250	2	NC	NC	NC	NC
86	246	250	0	NC	NC	NC	NC

Part I
Chapter 6

ROUGH CLASSIFICATION OF HSV PATIENTS

Krzysztof SLOWIŃSKI
Department of Surgery
F.Raszeja Mem. Hospital
60-833 Poznań, Poland

Abstract. An information system containing 122 patients with duodenal ulcer treated by highly selective vagotomy (HSV) is analyzed with the concept of rough sets. Twelve attributes are used to describe the patiens: 11 attributes concern anamnesis and preoperative gastric secretion and 12th attribute defines classification of patients according to long term results of the operation in the Visick grading. Using the methodology based on the rough sets theory, the information system is reduced so as to get a minimum subset of attributes ensuring an acceptable quality of the classification. A "model" of patients in each class is constructed upon the analysis of values adopted by attributes from this subset. Then, the reduced information system is identified with a decision table, assuming that the attributes in the minimum subset are condition attributes and that the result of treatment is a decision attribute. From this table, a decision algorithm is derived, composed of 44 decision rules. The algorithm and the models are helpful in decision making concerning the treatment of new duodenal ulcer patients by HSV.

1. Introductory remarks

Highly selective vagotomy (HSV), also called proximal gastric vagotomy, is a newest and effective method of treatment of duodenal ulcer which consists in vagal denervation of the stomach area secreting hydrochloric acid [1,2]. In the Department of Surgery at the F. Raszeja Mem. Hospital in Poznań, 122 HSV patients took part in the follow-up program. They were described by 11 pre-operating attributes and classified from the viewpoint of long term results of the operation into 4 classes of the well known Visick grading. These data were collected in view of investigating dependencies existing among the pre-operating attributes and the result of operation with respect to an expert's opinion. To carry out this investigation, we have used the rough sets methodology (cf. [3, 4, 10]). First, we have found the minimum subset of attributes significant for high quality classification. Basing upon analysis of the distribution of values adopted by attributes from the subset in particular classes, we have constructed models of patients for each class. The models corresponding to good results of the operation have determined indications for treatment of duodenal ulcer by HSV. Moreover, a decision algorithm has been derived which, together with the models of patients, is helpful in decision making concerning the treatment of duodenal ulcer by HSV.

2. Material

2.1. Information system

The information system is composed of 122 patients with duodenal ulcer treated by HSV, described by 11 pre–operating attributes. Attributes 1–4 concern anamnesis, and the remaining attributes are related to pre–operating gastric secretion examined with the histaminic test of Kay [8]. The table representing the information system is given in the **Appendix**. The patients are classified according to a long term result of HSV, evaluated by a surgeon in the modified Visick grading. The grading was derived from the following definition [5]:

1. Excellent: absolutely no symptoms, perfect result.
2. Very good: patient considers result perfect, but interrogation elicits mild occasional symptoms easily controlled by a minor adjustment of diet.
3. Satisfactory: mild or moderate symptoms easily controlled by care, which cause some discomfort, but patient and surgeon are satisfied with result which does not interfere seriously with life or work.
4. Unsatisfactory: moderate or severe symptoms of complications which interfere with work or normal life; patient or surgeon dissatisfied with result; includes all cases with recurrent ulcer and those submitted to further operation, even though the latter may have been followed by considerable symptomatic improvement.

The Visick grading defines classification $\mathcal{Y}$ of set U composed of 122 patients, i.e. $\mathcal{Y} = \{Y_1, Y_2, Y_3, Y_4\}$ is a partition of U into four classes. In other words, considering the result of the operation as an extra attribute T, classification $\mathcal{Y}$ is equal to T^* – the family of equivalence classes of T.

All attributes, except 1 and 4, take arbitrary real values from intervals defined by extreme cases.

2.2. Values of attributes and their clinical norms

In clinical experience, exact values of the considered quantitative attributes are usually translated into qualitative terms, e.g. "low", "medium", "high" and "very high". This translation is due to some empirical norms defining intervals of attribute values corresponding to the qualitative terms. The terms are then coded by numbers 0, 1, 2, 3 which create the domain of coded attributes. The norms adopted are shown in Table 1.

3. Method and results

3.1. Reduction of attributes

Let us consider classification $\mathcal{Y}$ defined by the result of operation. Table 2 shows the accuracy of approximation of each particular class Y_i by the set of all the eleven attributes denoted by Q. It can be seen that class 3 is Q–definable and classes 1, 2 and 4 are roughly Q–definable in the information system, although the accuracy of their approximation is

Table 1. Norms for attributes

No.	Attribute (units)	Domain (code) 0	1	2	3	4	Remarks
1.	Sex	♂	♀	–	–	–	
2.	Age [years]	≤ 35	> 35	–	–	–	
3.	Duration of disease [years]	≤ 0.5	$(0.5, 3]$	> 3	–	–	
4.	Complication of ulcer	none	acute haemorrhage	multiple haemorrhages	perforation in the past	pyloric stenosis	
5.	HCL concentration [mmol HCL/100ml]	≤ 2	$(2, 4]$	> 4	–	–	basic secretion
6.	Volume of gastric juice per 1h [ml]	≤ 70	$(70, 150]$	> 150	–	–	
7.	Volume of residual gastric juice [ml]	≤ 50	$(50, 100]$	> 100	–	–	
8.	Basic acid output (BAO) [mmol HCL/h]	≤ 2	$(2, 3]$	> 3	–	–	
9.	HCL concentration [mmol HCL/100ml]	≤ 10	$(10, 15]$	> 15	–	–	secretion stimulated by histamine
10.	Volume of gastric juice per 1h [ml]	≤ 100	$(100, 250]$	> 250	–	–	
11.	Maximal acid output [mmol HCL/h]	≤ 15	$(15, 25]$	$(25, 40)$	> 40	–	

very high. The quality of classification by set Q equals 0.97. The information system is almost selective. The number of atoms is 116; 5 atoms are 2–element sets and all the remaining are single element sets. Moreover, only two 2–element atoms are composed of patients belonging to different classes. This proves that the norms for attributes are well–defined.

Table 2. Accuracy of approximation of aech class by Q

Class	Number of patiens card (Y_i)	Lower approx. card$(\underline{Q}Y_i)$	Upper approx. card$(\overline{Q}Y_i)$	Accuracy $\mu_Q(Y_i)$
1	81	79	83	0.95
2	19	18	20	0.9
3	8	8	8	1.0
4	14	13	15	0.87

In order to check whether a set of attributes is dependent or not, we have to remove one attribute at a time and compute the number of elementary sets for each case. If the set, say P, is independent, then the reduction of one attribute, say r, results in the equality of at least two rows of the reduced information system. These equal rows (P–elementary sets) are clustered together forming a $(P-\{r\})$–elementary set, and thus we get a smaller number of elementary sets. Let us notice, however, that if the clustered rows belong to the same P-lower approximation of set $Y_i (i = 1, ..., n)$, then $\alpha_P(Y_i) = \alpha_{P-\{r\}}(Y_i)$ for $i = 1, ..., n$, and $\beta_P(\mathcal{Y}) = \beta_{P-\{r\}}(\mathcal{Y})$, $\gamma_P(\mathcal{Y}) = \gamma_{P-\{r\}}(\mathcal{Y})$; otherwise, the signs of equality are replaced by $\geq$. Proceeding in this way, we have found out that set Q of all eleven attributes is dependent and has one following reduct (minimal set) including nine attributes:

$$\{2,3,4,5,6,7,9,10,11\}$$

Then, we removed the particular attributes from the reduct and observed the decrease in the accuracy of classes $Y_i (i = 1, ..., 4)$ and the quality of classification $\mathcal{Y}$. We did this to find the smallest set of relevant attributes which would give a satisfactory quality of classification.

Let an ordered subset of attributes be called a sequence of attributes. The elimination of attributes from set Q, according to a given sequence, consists in removing the first attribute, then the first and the second, and then the first, the second and the third, and so on, until all the attributes in the sequence have been removed. Using a trial–and–error procedure, we found three sequences of attributes, denoted by E, G and H, which are characterized by the least steep descent of the accuracy of classes and the quality of classification in the course of elimination. These sequences contain 6 attributes each. The accuracy of classes and the quality of classification approximated by the sets of 5 remaining attributes are given in Table 3. The results of the elimination of attributes according to sequences E, G and H are presented graphically in Figures 1, 2 and 3, respectively, in the system of coordinates where the abscissa corresponds to the removed attributes and the ordinate to the accuracy of particular classes and the quality of classification. The legend given in Figure 1 applies to all the three figures. Let us remark that the elimination of any attribute from outside the considered sequence causes a steep descent in the quality of classification (by at least 0.17 and by at most 0.43).

Table 3. Accuracy of classes and quality of classification approximated by a set of five attributes

Sequence	Removed attributes	Accuracy of classes				Quality of classification	Doubtful region of classification (no. of patients)
		1	2	3	4		
E	1 8 11 2 7 3	0.52	0.37	0.17	0.33	0.59	50
G	8 11 2 1 7 10	0.60	0.39	0.31	0.16	0.65	43
H	8 11 7 1 2 5	0.62	0.24	0.54	0.38	0.68	39

The analysis of Table 3 and Figures 1, 2 and 3 leads to the conclusion that set R of the most significant attributes ensuring a satisfactory quality of classification is composed of attributes from outside of the sequence H, i.e.

$$R = Q - H = \{3,4,6,9,10\}$$

The reasons for it could be summarized in the following points:

(*i*) The quality of classification for $Q - H$ (0.68) is higher then for $Q - E$ (0.59) or $Q - G$ (0.65).

(*ii*) The $(Q-H)$-doubtful region of classification $\mathcal{Y}$ is the smallest (39) in comparison to the $(Q-E)$-and $(Q-G)$-doubtful regions of $\mathcal{Y}$ (50 and 43, respectively).

(*iii*) The low accuracy of class 2 (0.24) for $Q-H$ is not disturbing in this case because in the $(Q-H)$-boundary of class 1 there are almost the same patients as in the $(Q-H)$-boundary of class 2, and both these classes correspond to positive results of the treatment by HSV. Indeed, in $Bn_R(Y_1)$ there are 37 patients - 23 of them also belong to$Bn_R(Y_2)$.

The list of attributes from set R, in the descending order of the influence on the quality of classification, is as follows:

- duration of the disease (3);
- complications of ulcer (4);
- basic volume of gastric juice per 1 hour (6);
- volume of gastric juice per 1 hour under histamine (10);
- HCl concentration under histamine (9).

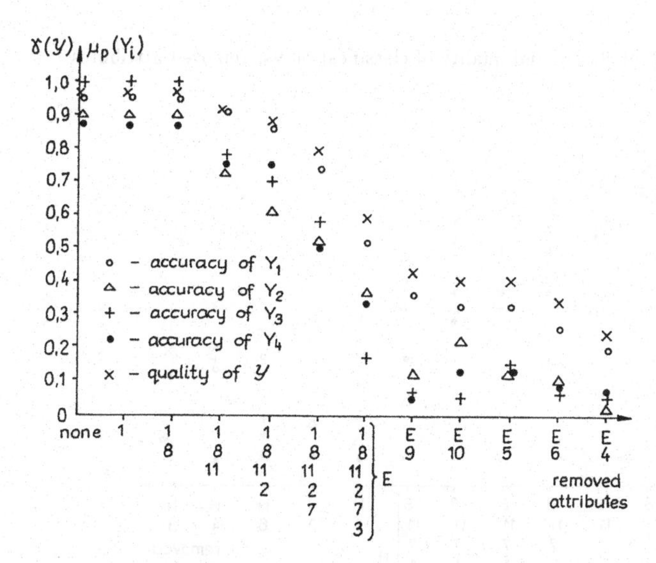

Figure 1. Accuracy of classes and quality of classification vs. removed attributes

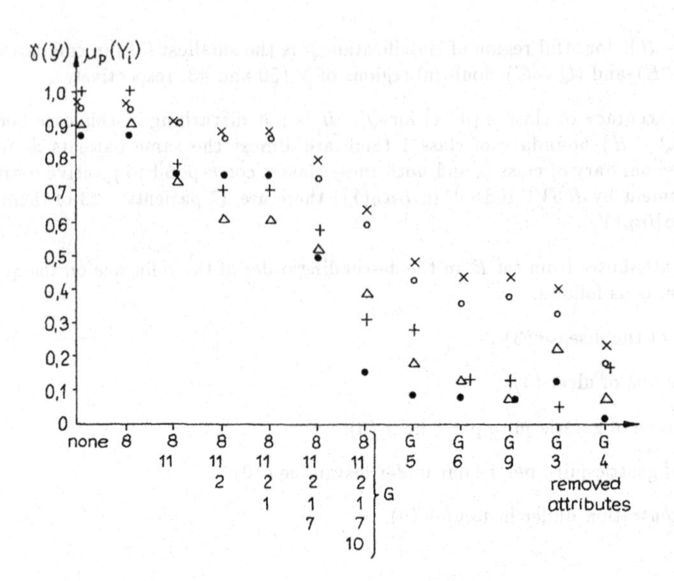

Figure 2. Accuracy of classes and quality of classification vs. removed attributes

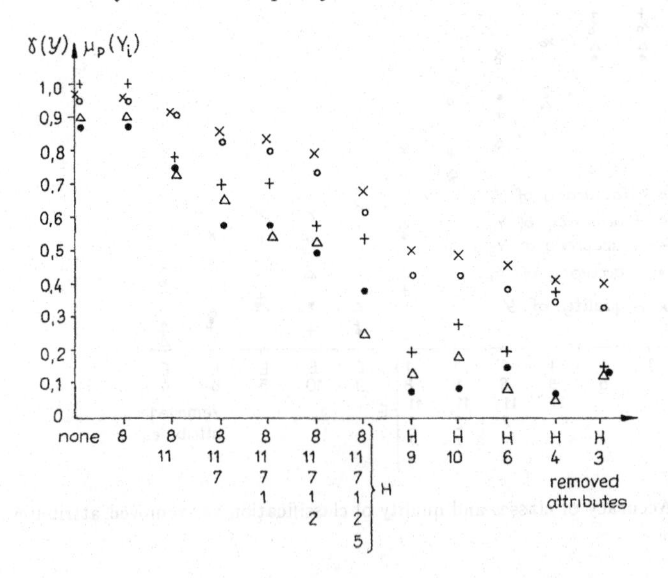

Figure 3. Accuracy of classes and quality of classification vs. removed attributes

3.2. Models of patients in each class

In this paragraph, we carry out an analysis of the distribution of values adopted by attributes from set R in particular classes in order to define the most characteristic values of these attributes for each class. The set of characteristic values for a given class defines a model of patients belonging to this class in terms of attributes from set R.

When analysing the distribution, we took into account the patients belonging to R–lower approximation of classes only. This distribution is described in Tables 4 to 8, respectively. The internal part of each table is composed of boxes including three numbers:

a_{ij}	
$b_{ij}[\%]$	$c_{ij}[\%]$

where

i – number of class,

j – value of an attribute analysed in a given table,

a_{ij} – number of patients belonging to $\underline{R}Y_i$ who take the same value j of the attribute under consideration,

b_{ij} – $a_{ij}/\sum_{i=1}^{4} card(\underline{R}Y_i)$,

c_{ij} – $a_{ij}/\sum_{i=1}^{4} a_{ij}$.

The last column in the tables shows characteristic values of the particular attributes following from the distribution. We avoid to overestimate high values of c_{ij} corresponding to $a_{ij} = 1$.

Table 4. Distribution of values taken by attribute 3

No. of class	Values of attribute 3						Card ($\underline{R}Y_i$)	Characteristic values of attribute 3
	0		1		2			
1	2		24		33		59	–long
	2.4%	40%	29%	69%	40%	77%		–medium
2	1		3		4		8	–long
	1.2%	20%	3.6%	8.6%	4.8%	9.3%		–medium
3	2		5		0		7	–medium
	2.4%	40%	6%	14%	0%	0%		–short
4	0		3		6		9	–long
	0%	0%	3.6%	8.6%	7.2%	14%		
Σ	5		35		43		83	

Table 5. Distribution of values taken by attribute 4

No. of class	Values of attribute 4										Card ($\underline{R}Y_i$)	Characteristic values of attribute 4
	0		1		2		3		4			
1	36		16		3		3		1		59	–lack of complication –acute haemorrhages
	43%	78%	19%	84%	3.6%	43%	3.6%	37%	1.2%	33%		
2	1		1		3		2		1		8	–multiple haemorrhages
	1.2%	2.2%	1.2%	5.3%	3.6%	43%	2.4%	25%	1.2%	33%		
3	4		1		0		2		0		7	–perforation of ulcer
	4.8%	8.7%	1.2%	5.3%	0%	0%	2.4%	25%	0%	0%		
4	5		1		1		1		1		9	–lack of complications
	6%	11%	1.2%	5.3%	1.2%	14%	1.2%	12.5%	1.2%	33%		
$\sum$	46		19		7		8		3		83	

Table 6. Distribution of values taken by attribute 6

No. of class	Values of attribute 6						Card ($\underline{R}Y_i$)	Characteristic values of attribute 6
	0		1		2			
1	19		29		11		59	–medium –small
	23%	76%	35%	76%	13%	55%		
2	1		6		1		8	–medium
	1.3%	4%	7.2%	16%	1.2%	5%		
3	3		1		3		7	–high –small
	3.6%	12%	1.2%	2.6%	3.6%	15%		
4	2		2		5		9	–high
	2.4%	8%	2.4%	5.3%	6%	25%		
$\sum$	25		38		20		83	

Table 7. Distribution of values taken by attribute 10

No. of class	Values of attribute 10						Card ($\underline{R}Y_i$)	Characteristic values of attribute 10
	0		1		2			
1	10		35		14		59	–medium
	12%	67%	42%	74%	17%	67%		
2	1		5		2		8	–medium
	1.2%	6.7%	6%	11%	2.4%	9.5%		
3	1		4		2		7	–high
	1.2%	6.7%	4.8%	8.5%	2.4%	9.5%		–medium
4	3		3		3		9	–small
	3.6%	20%	3.6%	6.4%	3.6%	14.3%		–high
$\sum$	15		47		21		83	

Table 8. Distribution of values taken by attribute 9

No. of class	Values of attribute 9						Card ($\underline{R}Y_i$)	Characteristic values of attribute 9
	0		1		2			
1	16		15		28		59	–high
	19%	62%	18%	68%	34%	80%		
2	4		2		2		8	–low
	4.8%	15%	2.4%	9%	2.4%	5.7%		–medium
3	3		2		2		7	–low
	3.6%	12%	2.4%	9.1%	2.4%	5.7%		–medium
4	3		3		3		9	–low
	3.6%	12%	3.6%	14%	3.6%	8.6%		–medium
$\sum$	26		22		35		83	

The set of characteristic values for a given class define a model of patients belonging to this class in terms of attributes from set R. These models are the following:

- Class 1
  - long or medium duration of the disease,
  - without complications of ulcer or acute haemorrhage from ulcer,
  - medium or small volume of gastric juice per 1 hour (basic secretion),
  - medium volume of gastric juice per 1 hour under histamine,
  - high HCl concentration under histamine.
- Class 2
  - long or medium duration of disease,
  - multiple haemorrhages,
  - medium volume of gastric juice per 1 hour (basic secretion),
  - medium volume of gastric juice per 1 hour under histamine,
  - low or medium HCl concentration under histamine.
- Class 3
  - medium or short duration of the disease,
  - perforation of ulcer,
  - high or small volume of gastric juice per 1 hour (basic secretion),
  - high or medium volume of gastric juice per 1 hour under histamine,
  - low or medium HCl concentration under histamine.
- Class 4
  - long duration of the disease,
  - without complications of ulcer,
  - high volume of gastric juice per 1 hour (basic secretion),
  - small or high volume of gastric juice per 1 hour under histamine,
  - low or medium HCl concentration under histamine.

3.3. Decision algorithm

The information system describing HSV patients can be identified with a decision table $DT = < U, C \cup D, V, f >$, where C is the set of condition attributes composed of all 11 pre–operating attributes and D is the set of decision attributes including only the result of the operation. As we have stated in paragraph 3.1, the information system is almost selective, thus, decision table DT is almost deterministic.

Using the procedure implemented in the microcomputer program RoughDAS [6], we shall derive the decision algorithm from the reduced decision table

$DT_R =< W, R \cup D, V, f >$, where W is the R–positive region of classification $\mathcal{Y}$ in R–representation of the information system, i.e. $W = POS_R(\mathcal{Y}) = \bigcup_{i=1}^{4} \underline{R}Y_i$, and $R = \{3,4,6,9,10\}$; $card(W) = 83$. Let us observe that decision table DT_R is deterministic.

It is known that, depending on the order of analysed attributes from set R, we can get slightly different decision algorithms. It is then natural to keep one with the smallest number of descriptors appearing in all decision rules. Using RoughDAS to decision table DT_R and different orders of analysed attributes, we obtained decision algorithms having from 44 to 48 decision rules. In Table 9, we present the best decision algorithm for the following order of attributes: 3 – 4 – 9 – 6 – 10

Table 9. Decision algorithm derived from R–representation of the HSV information system

Condition attributes					Class	Condition attributes					Class
3	4	6	9	10		3	4	6	9	10	
0	3				⇒1	2	3		2		⇒2
1	1		2		⇒1	1	2		2		⇒2
1	0		2		⇒1	2	2		1		⇒2
2	1		1		⇒1	1	2		0		⇒2
2	2		2		⇒1	0	1	1	0		⇒2
2	1		2		⇒1	1	3	1	0		⇒2
2	3		0		⇒1	2	4	0	0		⇒2
2	0	1	2		⇒1	2	0	2	1	2	⇒2
1	1	0	1		⇒1	0	0				⇒3
1	1	2	1		⇒1	1	3	0	1		⇒3
2	4	2	0		⇒1	1	3	2	0		⇒3
0	1	2	0		⇒1	1	0	0	1		⇒3
1	0	1	1		⇒1	1	1	2	0	2	⇒3
2	1	0	0		⇒1	1	0	2	0	2	⇒3
2	0	2	2	0	⇒1	2	4		2		⇒4
1	0	0	0	0	⇒1	1	2		1		⇒4
2	0	0	0	1	⇒1	1	3	2	1		⇒4
2	0	1	0	2	⇒1	2	1	1	0		⇒4
1	1	2	0	1	⇒1	2	0	0	0	0	⇒4
2	0	2	0	2	⇒1	2	0	2	2	2	⇒4
2	0	1	1	2	⇒1	1	0	2	0	1	⇒4
2	0	0	1	1	⇒1	2	0	0	1	0	⇒4

4. Discussion

The models of patients in each class as well as the decision algorithm need, however, a comment. The information system applies to patients who **have been** operated with HSV, thus, the surgeons expected good result of the operation on these patients, taking into account clinical experience. Hence, the distribution of patients in classes is not uniform. For this reason, the results of our analysis may be useful in establishing indications rather then contraindications for HSV. In other words, the most adequate models and decision rules are those concerning class 1 and class 2.

In decision making concerning the treatment of duodenal ulcer by HSV, the decision algorithm should be considered along with models of patients. It is connected with the fact that decision rules may be supported by different numbers of cases from the reduced information system; the distribution of this number is just amalgamated in the models.

In the literature concerning indications for treatment of duodenal ulcer by HSV, the anamnesis and secretion attributes are commonly taken into account. Some authors have considered more extensive anamnesis data [11] or other secretion tests [7]. The prevailing opinion is that the attributes have unequal influence on indication for treatment by HSV. It is generally agreed that the main indications is ineffectiveness of the conservative treatment of uncomplicated duodenal ulcer, while secretion criteria have not been generally agreed upon [7].

To date, retrospective studies based on traditional statistical methods, or only on clinical experience and intuition, have not revealed, however, the attributes to be the most important in predicting the outcome of the operation. The rough sets methodology give us, for the first time, a well founded answer on this pertinent question.

Taking into account indications following from models and decision rules concerning class 1 and 2, in years 1987 – 1991 we applied HSV to 70 new patients with duodenal ulcer. Fifty six of them are presently at least one year after operation.

The distribution of patients in particular classes of operational results is the following:

class 1 – 85.7% (48 patients)
class 2 – 7.1% (4 patients)
class 3 – 0.0%
class 4 – 7.1% (4 patients).

The above distribution is more advantageous than the one in the previous HSV information system which was the following:

class 1 – 66.4% (81 patients)
class 2 – 15.5% (19 patients)
class 3 – 6.6% (8 patients)
class 4 – 11.5% (14 patients).

This proves that indications for treatment by HSV resulting from the rough sets analysis have been satisfactorily verified in practice.

5. Final remarks

The results of the rough sets analysis of the information system including 122 HSV patients can be summarized in the following points:

1. The proposed norms for attributes define the domain ensuring a good classification of patients.

2. The information system was reduced from eleven to five significant attributes $\{3,4,6,9,10\}$ which ensure a satisfactory quality of classification.

3. Models of patients in each class were constructed in terms of the most significant attributes.

4. A decision algorithm was derived from the reduced representation of the information system.

5. The models of patients considered together with the decision algorithm provide indications for treatment of duodenal ulcer by HSV.

After this successful application, we have used the rough sets methodology to analysis of an information system concerning acute pancreatitis treated by peritoneal lavage [9,12]. The approach to the analysis is somewhat different from the HSV case [13].The difference follows from specificity of each treatment but also from different proportions between the numbers of patients and attributes. The information system describing HSV patients have a high ratio of the number of patients to the number of attributes, contrary to the information system concerning peritoneal lavage in acute pancreatitis; moreover, this kind of treatment is performed in many stages which yet multiplies the number of analytical attributes.

In both cases, using the rough sets analysis, we reduced the information system so as to get a minimal subset of attributes ensuring an acceptable quality of classification. Then, the reduced information system has been identified with a decision table.

In the case of the peritoneal lavage, the chosen subset of attributes gives the best description of the patient's state in the course of treatment. The decision algorithm represents in the most economical way all sorts of dependencies existing between the state of patients and the necessary number of peritoneal lavage stages.

As was mentioned above, the ratio of the number of patients to the number of attributes influences the way of the analysis. When this ratio is high (HSV), the number of minimal sets is small and the core is almost as large as the minimal sets; then, the minimal subset of attributes ensuring an acceptable quality of classification is obtained in result of removing the less significant attributes from the minimal sets. However, when the ratio is low (peritoneal lavage), the number of minimal sets is great and the core is very small; then the minimal subset of attributes ensuring an acceptable quality of classification is obtained in result of adding the attributes with the greatest discriminatory power to the core.

The above applications of the rough sets methodology show the suitability of the rough sets theory to the analysis of both typical and contrasted medical information systems.

References

[1] Alichniewicz A., Sołtysiak A., (1980). Highly Selective Vagotomy *(in Polish)*. Wydawnictwo Łódzkie, Łódź.

[2] Dunn D.C., Thomas W.E.G., Hunter, J.O. (1980). An evaluation of highly selective vagotomy in the treatment of chronic duodenal ulcer. *Surg. Gynecol. Obstet.* **150**, 845-849.

[3] Fibak J., Pawlak Z., Słowiński K., Słowiński R. (1986). Rough sets based decision algorithm for treatment of duodenal ulcer by *HSV*. *Bull. Polish Acad. Sci., Bio. Sci.,* **34** (10-12), 227-246.

[4] Fibak J., Słowiński K., Słowiński R. (1986). The application of rough sets theory to the verification of indications for treatment of duodenal ulcer by *HSV*. *Proc. 6th International Workshop on Expert Systems and their Applications,* Avignon, vol.1, 587-599.

[5] Goligher J.C., Hill G.L., Kenny T.E., Nutter E. (1978). Proximal gastric vagotomy without drainage for duodenal ulcer: results after 5-8 years. *British J. Surgery* **65**, 145-151.

[6] Gruszecki G., Słowiński R., Stefanowski J. (1990). *RoughDAS - Rough sets based data analysis software - User's Manual,* APRO S.A.,Warsaw.

[7] Hautefeuille P., Picaud R. (1983). *Les récidives aprés vagotomies pour ulcére duodénal,* Masson, Paris.

[8] Kay A. (1967). Memorial lecture. An evaluation of gastric acid secretion tests. *Gastroenterology* **53**, 834-852.

[9] McMahon M.J., Pickford J., Playforth M.J. (1980). Early prediction of severity of acute pancreatitis using peritoneal lavage. *Acta Chirurgica Scandinavica,* **146**, 171-175.

[10] Pawlak Z., Słowiński K., Słowiński R. (1986). Rough classification of patients after highly selective vagotomy for duodenal ulcer. *Int. J. Man-Machine Studies,* **24**, 413-433.

[11] Popiela T., Szafran Z. (1978). Surgical treatment of duodenal ulcer: pre-operating diagnosis and qualification criteria for the method of treatment *(in Polish). Proceedings of the Conferences and Courses at the Academy of Medicine in Cracow.*

[12] Ranson J.H, Spencer F.C. (1978). The role of peritoneal lavage in severe acute pancreatitis. *Annals of Surgery,* **187**, 565-575.

[13] Słowiński K., Słowiński R., Stefanowski J. (1988) Rough sets approach to analysis of data from peritoneal lavage in acute pancreatitis. *Med. Inform.,* **13**, 3 143-159.

Appendix

The HSV information system with original values of attributes

Patient	Pre–operating attributes											Class
no.	1	2	3	4	5	6	7	8	9	10	11	
1	1	46	12	0	5.6	79	50	4.4	19	119	22.6	1
2	0	27	3	1	12.5	58	15	7.3	26	120	31.2	1
3	0	25	6	0	11.5	77	15	8.9	16.1	93	15	1
4	0	48	3	0	15.6	29	2	4.5	28.7	186	53.4	1
5	1	26	0.5	0	7.6	80	45	6.1	17.1	101	17.2	3
6	0	32	5	1	11.9	56	100	6.7	13.6	150	20.4	1
7	0	26	2	1	6.1	19	8	1.2	14.8	58	8.6	1
8	1	28	2	1	6	36	40	2.2	20.4	65	13.3	1
9	0	55	30	0	16.8	118	12	19.8	40.4	172	69.6	1
10	0	21	5	3	20.9	111	32	23.2	34.5	270	93.1	2
11	0	37	2	0	12.6	152	30	19.2	38.7	202	78.2	1
12	0	48	5	2	2.3	73	6	1.7	5.5	199	10.9	2
13	0	43	20	3	8.1	97	32	7.8	11	120	13.2	2
14	0	30	2	0	10	15	15	1.5	18.8	121	22.7	1
15	0	49	14	2	11.7	118	38	13.8	23.2	266	52.5	1
16	0	27	3	1	9.5	154	25	14.6	13.5	141	19.1	1
17	0	28	10	0	20.9	178	26	36.1	23.3	214	49.8	1
18	1	40	4	0	8.1	62	17	5	5.6	41	2.3	4
19	0	60	20	0	13.4	107	27	14.3	19	335	63.5	1
20	0	22	4	0	3.5	176	40	6.1	5.6	190	10.6	2
21	0	21	4	0	1	155	66	1.6	2.6	160	4.2	1
22	0	21	6	4	4	360	210	14.4	3.4	211	7.1	1
23	0	28	0	1	6	152	15	9.2	9.8	227	22.3	1
24	0	31	2	3	1.8	60	10	1.1	12.3	117	14.4	3
25	0	37	3	0	8.5	94	20	8	17.3	188	32.6	1
26	0	22	2	0	8.3	111	28	9.2	20.8	192	39.8	1
27	0	43	5	0	1.9	401	53	7.5	16.3	94	15.2	1
28	1	59	1	0	4.8	30	12	1.4	9.3	27	5.2	1
29	0	32	3	0	2.8	164	35	4.5	10.3	178	18.3	1
30	0	34	8	0	6.3	82	13	5.2	7.4	130	9.6	1
31	0	51	1	0	8.6	87	25	7.5	13.7	230	31.4	1
32	0	41	20	0	2.6	29	15	0.8	6.1	108	6.6	1
33	1	50	5	1	2.5	44	120	1.1	4.2	49	2.1	1
34	0	24	2	0	14.1	160	22	22.5	21.2	209	44.4	1
35	0	32	3	0	9	122	45	10.9	15.7	223	35	1
36	0	30	8	0	8.5	121	26	10.3	5.7	261	11.4	1
37	0	63	2	0	5.8	60	34	3.5	8.7	133	11.5	1
38	0	30	2	1	1.7	171	60	2.8	4.7	139	6.6	1
39	0	21	4	0	14.7	182	31	26.8	27.5	379	104.2	4

The HSV information system with original values of attributes

Patient	Pre–operating attributes											Class
no.	1	2	3	4	5	6	7	8	9	10	11	
40	0	42	6	0	6.8	319	254	21.8	9.7	266	25.7	1
41	0	71	4	2	2	34	27	1.1	4.2	185	7.8	4
42	0	34	2	0	4.1	212	32	8.7	5.3	154	8.1	4
43	0	54	2	3	5.3	166	124	8.7	6.8	236	16	3
44	0	60	0	1	11.4	127	30	14.5	9.3	148	13.8	2
45	0	33	2	2	8.7	135	54	11.8	29	186	53.8	2
46	0	40	20	1	11.6	123	88	14.2	22	152	33.3	1
47	1	32	10	1	10.3	120	20	12.3	11.9	135	16.1	1
48	0	37	3	0	7.5	86	21	6.4	15	189	28.3	1
49	1	31	5	3	4	56	43	2.2	7.4	137	10.2	1
50	0	25	7	3	2.2	184	10	4.1	5.4	459	24.7	1
51	1	27	1	3	3.1	140	60	4.4	6.6	167	11	2
52	0	56	15	1	8.3	60	17	5	11.4	72	8.2	1
53	0	23	2	0	6	133	26	8	11.5	113	13	1
54	0	33	14	0	2.9	191	23	5.6	15.5	136	21.1	2
55	0	56	6	3	5.6	140	35	7.9	12.5	129	16.1	1
56	1	27	7	1	7.1	270	180	19.1	11.2	345	38.7	1
57	0	51	3	1	3.5	111	50	3.8	15.1	212	32	1
58	0	31	0.5	3	4.7	525	105	24.7	10.8	627	67.7	1
59	0	50	8	4	10.6	185	21	19.6	25.3	224	56.6	4
60	1	31	12	0	2	45	63	0.9	7.1	165	11.7	1
61	1	47	2	0	26.1	68	46	17.7	28	307	86	1
62	0	34	4	2	8.8	95	32	8.3	11.8	183	21.6	2
63	0	42	1	3	3.7	514	75	19.2	12.5	312	39.1	4
64	0	27	2	2	4	96	14	3.8	14.9	69	10.3	4
65	1	32	0.5	0	7.8	69	78	5.4	16.7	51	8.5	3
66	1	35	3	0	2.3	43	28	1	8.3	90	7.5	1
67	0	36	10	0	3.2	79	38	2.6	9.2	165	15.2	1
68	0	34	2	0	5.5	108	80	6	11.1	121	13.4	1
69	0	27	4	0	3.3	159	72	5.2	5	127	6.3	1
70	1	32	7	0	6.1	43	74	2.6	10.8	326	35.1	1
71	1	47	15	2	2.2	112	35	2.4	16.7	53	8.7	1
72	0	35	7	0	4.4	118	38	5.2	5.7	129	7.4	1
73	0	28	15	4	7.3	23	110	1.7	9.8	21	20.6	2
74	0	45	24	0	1.4	60	28	0.9	7.1	146	10.3	1
75	1	27	10	0	21	187	225	39.1	39.1	387	151.4	4
76	0	27	4	0	10.6	127	30	14	11	430	45.6	1
77	0	26	3	0	3.8	283	43	11	11.7	260	30.3	1
78	0	27	4	0	4.6	79	20	3.6	8.7	184	16.1	1
79	0	28	1	1	1	214	40	2.1	8.6	442	37.9	3
80	0	50	32	0	5.1	171	30	8.8	5.1	135	7	2

The HSV information system with original values of attributes

Patient no.	Pre-operating attributes 1	2	3	4	5	6	7	8	9	10	11	Class
81	0	28	11	0	4.3	145	65	6.3	6	196	11.8	4
82	0	27	4	0	6	225	50	13.6	18.8	129	24.2	3
83	0	48	10	0	11	102	20	11.2	16.3	142	23.2	1
84	0	30	10	0	9.4	249	70	23.5	18.6	194	36.1	1
85	1	34	15	0	15.9	136	60	21.6	17.8	184	32.8	1
86	0	22	3	0	10.6	198	30	20.9	11.9	188	22.4	2
87	0	30	5	0	8.6	155	37	13.3	13.9	232	32.1	2
88	0	51	1	1	14.9	80	20	11.9	20.7	128	26.5	1
89	1	30	10	0	6.8	136	100	9.3	20.7	128	26.5	1
90	0	30	5	0	7.4	213	90	15.7	10.5	266	28	2
91	0	35	4	0	3.8	57	116	2.2	10.4	191	19.8	1
92	0	30	10	0	7.6	158	22	12	12.1	169	20.4	4
93	0	43	6	0	3.1	122	15	3.8	1.6	208	3.4	1
94	0	42	10	0	11.7	159	132	18.6	19.6	127	24.9	1
95	1	45	12	0	5.2	53	32	2.7	13.8	286	39.5	4
96	0	34	1.5	1	4.5	104	70	4.6	12.4	263	32.6	2
97	0	36	5	0	7.1	110	26	7.9	13.5	277	37.4	1
98	0	30	9	0	4.3	134	55	5.8	8.8	336	29.6	1
99	0	31	5	2	2.5	19	134	0.48	9.1	149	13.5	1
100	0	25	9	0	8.2	60	78	4.9	14.2	151	21.4	1
101	0	30	10	1	1.5	122	80	1.9	5.3	220	11.6	4
102	0	33	5	2	5.7	68	10	3.9	6.4	245	15.6	1
103	0	32	2	0	6	187	60	11.2	11	285	31.4	2
104	0	45	22	2	8.7	80	90	7	42.3	270	114.3	1
105	0	38	2	0	5.8	58	8	3.4	7.1	148	10.6	4
106	1	56	0.83	0	8.8	73	30	6.4	20	68	13.7	1
107	1	45	11	0	6.3	50	105	3.1	13.2	91	12	4
108	0	32	2	1	8.9	143	75	12.8	10.9	280	30.4	1
109	0	60	2	0	4.2	195	50	8.1	6.5	265	17.3	3
110	1	44	3	2	3.7	86	5	3.2	7.7	170	13.1	2
111	0	49	4	0	6.3	180	15	11.4	21	115	84.2	1
112	1	28	10	0	9.5	98	60	9.3	14.7	134	19.7	1
113	0	26	2	0	8.3	82	60	6.8	26.3	330	86.9	1
114	0	39	5	0	7.5	137	14	10.3	10.7	160	17.1	1
115	0	49	9	0	3.1	150	40	4.6	7	261	18.4	1
116	0	30	1	1	17.4	76	29	13.2	24.8	229	56.7	1
117	0	52	4	1	5.7	45	27	2.6	15.4	242	37.2	1
118	0	45	3	0	5.2	67	128	3.5	11.8	230	27.1	3
119	0	53	7	0	7.4	68	30	5	8.7	140	12.2	1
120	0	29	6	0	15.7	120	40	18.8	12.3	220	27	2
121	0	28	4	0	8.9	88	28	7.8	12.3	163	20	2
122	0	38	5	2	1	128	6	1.3	5.8	145	8.4	1

Part I
Chapter 7

SURGICAL WOUND INFECTION – CONDUCIVE FACTORS AND THEIR MUTUAL DEPENDENCIES

Maciej KANDULSKI
Institute of Mathematics
A. Mickiewicz University
60-769 Poznań, Poland

Jacek MARCINIEC
Institute of Linguistics
A. Mickiewicz University
61-874 Poznań, Poland

Konstanty TUKALLO
Clinic of Surgery, Departament of Nursing
Medical Academy
60-479 Poznań, Poland

Abstract. The paper presents rough set analysis of commonly accepted factors of surgical wound infection. Two separate sets of data are compared and some possible transformations of the information system are proposed.

1. Introduction and preliminaries

In the years 1987-1988 we collected data concerning possible factors of surgical wound infection. Our aim was to establish a hierarchy of commonly accepted factors. The results obtained, published in [5], encouraged us to continue our investigation. In this paper we present the outcome of our research carried on between years 1988 and 1990 and compare the findings of both sessions. Taking into consideration the character of this volume we focus on possibilities of rough set analysis of medical data rather than on their clinical aspect.

The paper is organized as follows: below in this section we describe the motives of our work and sketch the formal apparatus employed. Section 2 concerns the methodological side of the experiment. The description of the information system itself is given in section 3. In section 4 a comparison of the two sets of data is presented. Hierarchies obtained during the second series of investigation are presented in section 5 with full details. Section 6 is devoted to various modifications of the information system. Final remarks close the paper.

One of the most common complication in the process of surgical wound healing is infection. Factors influencing infections are generally known but their significance is differently estimated by various authors. It is rarely considered however that those differences might be caused by distinct prophylactic routines in the centers where the investigations are conducted. Thus it seems reasonable to examine the problem of infection with reference to a particular surgical clinic or department. So far, the most often applied method for indicating connections between various factors has been a statistical analysis. Such a statistical analysis has also been employed to process data concerning wound infections cf.[2,11]. Rough set methods enable us to eliminate presuppositions which usually accompany statistical tests. The evaluation of the outcome can be done by comparing it with information presented in monographs like [10], in manuals like [3] or in other publications concerning only some of the factors cf. [4,7,8], . In this paper we can confront the results we just obtained with their counterparts obtained three years earlier. It should be mentioned that we did not cumulate the two sets of data but we conducted all calculations for years 1987-1988 and 1988-1990 separately. This is of course a much more challenging test for the stability of the results than if we only added new data to the old ones.

In our investigations we employ only a small part of rough set formalism which gives a possibility to determine the significance of attributes in the process of classification. We proceed as follows. For a given information system $S = \langle U, Q, V, \rho \rangle$, classification $\mathcal{X}$ of set U and an attribute $q \in Q$ we calculate the numbers $\beta_Q(\mathcal{X})$ and $\gamma_Q(\mathcal{X})$ and then $\beta_{Q-\{q\}}(\mathcal{X})$ and $\gamma_{Q-\{q\}}(\mathcal{X})$. The smaller the difference between $\beta_Q(\mathcal{X})$ and $\beta_{Q-\{q\}}(\mathcal{X})$ (resp. $\beta_Q(\mathcal{X})$ and $\beta_{Q-\{q\}}(\mathcal{X})$), the less important the attribute q is as far as the process of estimating $\mathcal{X}$ by elements of Q is concerned. We must stress, however, that in general the above mechanism reveals only the degree of co-occurrence of attribute q and the feature generating the classification $\mathcal{X}$. Whether it is a cause-result connection must be decided by material conditions of an experiment (cf. remarks concerning attribute 7 in [10]).

2. Description of the experiment

All data were collected in the same general surgery department composed of 70 beds. They concerned the healing of surgical wounds after procedures performed on neck, chest, abdomen and limbs. Operations of anus and of wounds caused by accidents were not taken into consideration.

Generally, patients admitted to the department were similarly prepared for an intervention. Only in some emergency cases (in which, for example, hemorrhage was involved) patients did not take a bath before the operation or had not the whole body washed. The skin in the area of incision was shaved and washed in the department on the day of the operation. Just before the intervention the operative site was prepared two more times: for the first time with an antiseptic agent Abacil, then with a 0.5% alcoholic solution of Hibitan. In contaminated and infected operations surgeons routinely changed gloves prior to sewing the wound up. At this stage of operation towels were replaced and skin margins were disinfected with Hibitan. After putting stitches in muscles but before closing skin, the wound was washed out with a 3% hydrogen peroxide solution. Surgeons were not strictly attributed only to a limited kind of interventions but still, some selection depending on the difficulty of an operation and on an operator's skills was made.The vast majority of operators were full-time surgeons of the Clinic. Sometimes however operations were executed by trainees or surgeons staying at training courses. Interventions were classified as clean, contaminated and infected. Clean operations were always performed in a strict aseptic operating room. Efforts were also made to execute contaminated and infected operations in, respectively, aseptic and septic operating rooms. The degree of a surgical wound infection is defined basing upon the course of healing. A wound is considered healed by first intention if the margins of the wound show no symptoms of inflammation, one does not observe an infiltration or a seropurulent discharge. A moderately infected wound is painful, reddened, there appears an inconsiderable, mostly serous, exudation from it. A wound is extensively infected if there is a discharge of pus from it and there are perceivable general symptoms of infection (fever, tachycardia, shiver etc.). If it was certain that an intervention would be performed in an infected area, antibiotics or other antiseptic drugs were prophylactically administered. This policy was continued after an intervention. As a result of our research carried out in the years 1988-1989 a decision was taken to divide the department into an aseptic and septic sections. Patients developing an infection of a wound were transferred to a septic section. Data for computation were recorded on charts, successively filled during a patient's stay in the department. Charts used in the second series (1988-1990) of research contained more entries than those from the first series (1987-1988). For numerical calculations a microcomputer was used.

3. Description of the information system

The information system described objects (patients) in terms of the following set of attributes:

1. type of intervention,
2. sex,
3. age,
4. weight,
5. personal cleanliness,
6. coexisting infectious focus,
7. time of stay in the hospital before the surgical intervention,
8. time of stay in the hospital after the surgical intervention,
9. presence of infectious cases in the hospital room,
10. preparation for the surgical intervention in the department,
11. diagnosis,
12. mode of intervention,
13. kind of intervention,
14. type of anesthesia,
15. duration of intervention,
16. surgeon,
17. type of operating room,
18. order of intervention,
19. wound suture,
20. antibiotics before the surgical intervention,
21. metronidazole before the surgical intervention,
22. antibiotics after the surgical intervention,
23. metronidazole after the surgical intervention,
24. type of antibiotics received,
25. coexisting diseases conducive to the development of infection.

The domains of attributes were as follows:

D_1 **clean, contaminated, dirty**
D_2 **male, female**
D_3 76 values (in years)
D_4 **deficiency, normal, mean adiposity, considerable adiposity**
D_5 **yes, no**

D_6 **yes, no**
D_7 **<12, 12–24, 24–48** (hours), **2–3, 3–6, >6** (days)
D_8 **3–6, 7–9, >9** (days)
D_9 **yes, no**
D_{10} **yes, no**
D_{11} 89 values
D_{12} **emergency, elective**
D_{13} 24 values
D_{14} **general, local, conduction**
D_{15} **<0.5, 0.5–1, 1–1.5, 1.5–2, >2** (hours)
D_{16} 22 values
D_{17} **strict aseptic, aseptic, septic**
D_{18} **first, second, next**
D_{19} **tight with a drain, tight without a drain, loose with a drain, loose without a drain**
D_{20} **yes, no**
D_{21} **yes, no**
D_{22} **yes, no**
D_{23} **yes, no**
D_{24} **aminoglycosides, cephalosporins, penicillins, tetracyclines, polymyxins, macrolides, sulfonamides** or their combinations
D_{25} **no, diabetes, other disease**

The classification $\mathcal{X}$ of the set of objects was given by sets X_1, X_2 and X_3 such that:

- X_1 consisted of patients whose surgical wound was healed by first intention,
- X_2 consisted of patients whose surgical wound was moderately infected,
- X_3 consisted of patients whose surgical wounds was extensively infected.

Attributes 1–19 were grouped in the following way: the first group included attributes which characterize a patient (attributes 2–8), the second group was composed of those attributes which characterize the hospital (attributes 9–19). Attribute 1 (type of intervention) can be added to both those groups: it obviously characterizes a patient but, being connected with a diagnosed disease as well as with the kind of intervention, it also characterizes the hospital. We denote the set of attributes 1–8 by P, and the set of attributes 9–19 together with attribute 1 by H. In this paper, especially in its next section, we will be discussing results obtained in the years 1987–1988 (cf.[5]) as well as new results from the years 1988–1990, which are presented for the first time. We denote the first and the second series of research by $\mathcal{A}$ and $\mathcal{B}$ respectively. The sets of attributes in $\mathcal{A}$ and $\mathcal{B}$ are different. In comparison with $\mathcal{A}$, the set of attributes in $\mathcal{B}$ was enriched with attributes 11, 13 and 20, 21, 22, 23, 24. Attributes 20–24 describe in short the course of antibiotic prophylaxis and treatment. In some cases we also changed the domains of attributes. Table 1 presents the alterations. We must explain the intentions of those changes. We decided to give values of attribute 3 (age) in distinct years in order to have the possibility of gluing them in age intervals. The value 'considerable deficiency' (attribute 3) never occurred in $\mathcal{A}$. As it was also the case in $\mathcal{B}$, we skipped this value in the domain of attribute 'weight'. The values of attribute 'time of stay in the hospital before the surgical intervention' in series $\mathcal{A}$ were distributed irregularly: we

Table 1.

attribute name	attribute number		domain	
	in $\mathcal{A}$	in $\mathcal{B}$	in $\mathcal{A}$	in $\mathcal{B}$
age	2	3	**≤20, 21–30, 31–40, 41–50, 51–60, 61–70, >70** (years)	76 values (in years)
weight	3	4	**considerable deficiency, deficiency, normal, mean adiposity, considerable adiposity**	**deficiency, normal, mean adiposity, considerable adiposity**
time of stay in the hospital before the surgical intervention	6	7	**<12, 12–24, 24–48, >48** (hours)	**<12, 12–24, 24–48** (hours), **2–3, 3–6, >6** (days)
time of stay in the hospital after the surgical intervention	7	8	**≤2, 3–6, 7–9, 10–14, 15–21, >21** (days)	**3–6, 7–9, >9** (days)
duration of intervention	12	15	**<0.5, 0.5–1.5, >1.5** (hours)	**<0.5, 0.5–1, 1–1.5, 1.5–2, >2** (hours)

had small intervals up to 48 hours and only one value for a stay longer than 48 hours. In this situation we added three new values describing longer stay. The attribute 'time of stay in the hospital after the surgical intervention' can not be treated as a factor inducing wound infections though undoubtedly this feature co-occurs with them very often (cf. our earlier remark concerning cause-result dependence). To eliminate the influence of this attribute we restricted its domain to only three values. Finally, the finer partition of the domain of the attribute 'duration of intervention' was introduced as requested by surgeons. The values of attributes 3 (age), 7 (time of stay in the hospital before the surgical intervention) and 15 (duration of intervention) in series $\mathcal{B}$ can be glued in such a way that the obtained intervals coincide with values of those attributes in series $\mathcal{A}$. The sets of attributes P and H in $\mathcal{B}$ with glued values for attributes 3, 7 and 15 will be denoted P_0 and H_0. Of course, from the theoretical point of view, the employed in $\mathcal{B}$ information systems with original and glued attributes are different. The matter however seems to be so obvious that we omit formalisms here. Having introduced the sets P_0 and H_0 we are able to compare the results of series $\mathcal{A}$ and $\mathcal{B}$.

Table 2.

$\mathcal{A}$		$\mathcal{B}$
303	X_1	836
34	X_2	107
7	X_3	39
343	overall	982

4. Comparison of hierarchies in $\mathcal{A}$ and $\mathcal{B}$.

The distribution of objects (patients) in sets X_1, X_2 and X_3 of the series $\mathcal{A}$ and $\mathcal{B}$ is presented in Table 2. Tables 3, 4 and 5 show the accuracy of approximation of both classifications by sets of attributes characterizing patients and the hospital.

Table 3.

$\mathcal{A}$	$\mathcal{B}$
$\beta_P(\mathcal{X}) = 0.82$	$\beta_{P_0}(\mathcal{X}) = 0.55$
$\beta_H(\mathcal{X}) = 0.97$	$\beta_{H_0}(\mathcal{X}) = 0.73$

Table 4.

$\mathcal{A}$ $(\mu_P(X_i))$		$\mathcal{B}$ $(\mu_{P_0}(X_i))$
0.89	X_1	0.71
0.43	X_2	0.17
0.60	X_3	0.14

Table 5.

$\mathcal{A}$ ($\mu_H(X_i)$)		$\mathcal{B}$ ($\mu_{H_0}(X_i)$)
0.98	X_1	0.83
0.84	X_2	0.40
1.0	X_3	0.40

As early as [5] we admitted that the values of accuracy of approximations of X_2 and X_3 by P in series $\mathcal{A}$ are too low. In series $\mathcal{B}$ we obtained even lower values of accuracy for sets P_0 and H_0 both in case of the classification $\mathcal{X}$ and in case of sets X_1, X_2, X_3 taken separately. So low values can bring the usefulness of further investigation in question. We decided to continue research for the following reasons:

- The aim of our work is to establish the hierarchy of *known* risk factors. The fact that the attributes insufficiently approximate our classification is a matter of quite different nature. One can suspect that the adopted norms for the values of domains are not properly set or one can look for additional features which could better characterize circumstances of infection origin. In the latter case, however, one should be very careful: although, for instance, in some research the positive result of a bacteriological examination in a wound prior to its closing turned out to be three times more significant than other factors of infection (cf.[2]) but this presence could hardly be admitted as a cause or a risk factor of infection. It is rather a prodrome of a coming infection and as such should influence the way of drug administration.
- As we will see in two next parts of the paper changes in domains of attributes in $\mathcal{B}$ (transition from P_0 and H_0 to P and H), as well as taking into consideration new factors improve values of the accuracy of approximation.
- The hierarchies of attributes in $\mathcal{A}$ and $\mathcal{B}$ turned out to be similar.

We pass to the comparison of hierarchies obtained in $\mathcal{A}$ and $\mathcal{B}$. As we said previously, in order to do it correctly we must employ the sets P_0 and H_0 in $\mathcal{B}$. For an attribute $q_i \in P$ we define a number $\Delta^i_P(\mathcal{X})$ in the following way:

$$\Delta^i_P(\mathcal{X}) = \beta_P(\mathcal{X}) - \beta_{P-\{q_i\}}(\mathcal{X})$$

In case of a selected class X_i of a classification $\mathcal{X}$ the above definition has the form

$$\Delta^i_P(X_k) = \mu_P(X_k) - \mu_{P-\{q_i\}}(X_k)$$

Naturally, both definitions can be applied to the remaining sets of attributes in $\mathcal{A}$ and $\mathcal{B}$. The number Δ^i_P informs us about the decrease of the accuracy of approximation (of a set or the whole classification) after skipping the attribute $q_i \in P$. Comparing hierarchies we set attributes in order according to values Δ^i_Q ($Q = P$, H in $\mathcal{A}$, $Q = P_0$, H_0 in $\mathcal{B}$) calculated for the classification $\mathcal{X}$. Table 6 shows hierarchies obtained in series $\mathcal{A}$ and $\mathcal{B}$.

We must explain the origin of the partition of attributes into groups of significance presented in this table. Two hierarchies can be compared in different ways: one can examine the presence of an attribute on exactly the same positions in both of them or one can admit a maximal difference between positions occupied by this attribute. One can also

Table 6.

$\mathcal{A}$			$\mathcal{B}$		
i attribute number	$\Delta_P^i(\mathcal{X})$	group of significance	group of significance	$\Delta_{P_0}^i(\mathcal{X})$	i attribute number
8	0.35	I	I	0.30	3
3	0.25	I	I	0.20	8
4	0.17	II	II	0.19	1
2	0.14	II	II	0.15	4
1	0.11	III	II	0.141	7
7	0.08	III	II	0.137	2
5	0	III	III	0.05	6
6	0	III	III	0.004	5

compare hierarchies by checking the number of common attributes in sets consisting of first two, three etc. attributes of each hierarchy. If the order of attributes is determined by a certain numerical coefficient then attributes can be grouped according to the value of this coefficient. We chose this last method. As the values of Δ_P^i and $\Delta_{P_0}^i$ are distributed rather regularly we decided to divide the intervals $\langle 0, 0.35\rangle$ and $\langle 0, 0.30\rangle$ into three equal parts (0.35 and 0.30 are maximal values of Δ_P^i and $\Delta_{P_0}^i$). Thus we created two partitions of P and P_0 into attributes of small, medium and great significance. This partition is shown in column 'group of significance' in Table 6. Having defined the method of dividing attributes, let us observe that six out of eight attributes characterizing a patient in $\mathcal{A}$ and $\mathcal{B}$ remain in their groups of significance. Moreover, the two attributes which left their

Table 7.

$\mathcal{A}$			$\mathcal{B}$		
i attribute number	$\Delta_H^i(\mathcal{X})$	group of significance	group of significance	$\Delta_{H_0}^i(\mathcal{X})$	i attribute number
15	0.1	I	I	0.45	16
16	0.08	I	II	0.20	18
9	0.04	II	III	0.12	1
18	0.03	III	III	0.09	19
14	0.02	III	III	0.06	15
17	0.2	III	III	0.05	17
10	0.0	III	III	0.044	10
12	0.0	III	III	0.043	14
19	0.0	III	III	0.027	9
1	0.0	III	III	0.012	12

places moved only to the adjoined group, thus there were no jumps of attributes from the group of small to the group of great significance and vice versa. In particular, attribute 1 (type of intervention) turned out to be of medium significance in $\mathcal{B}$. This result is more consistent with surgeons' experience than the small significance of the attribute revealed in $\mathcal{A}$. The influence of the second attribute which did not remain in its group - 7 (time of stay in the hospital before the surgical intervention) - is not categorically estimated so far as wound infections are concerned [3,1]. It is worth mentioning however that if we specify the values of this attribute more precisely, its significance will be greater (see the discussion in the next section). Let us compare now the hierarchies obtained in $\mathcal{A}$ and $\mathcal{B}$ for the set of attributes characterizing the hospital. The results are presented in table 7. One sees that when the same method is applied to determine the groups of significance, seven out of ten attributes remain in their previous groups. Unfortunately these seven attributes are of small influence. Concerning the three attributes which changed their significance, we can provide a reason only for the decrease in significance of attribute 9 (presence of infectious cases in the hospital room) from the second to the third group of significance. In our opinion this fact can be associated with the division of the department into a septic and aseptic part, as described in section 2. Let us note that although the places occupied by attribute 9 in both hierarchies are separated by six positions, the difference between values $\Delta_H^9(\mathcal{X})$ and $\Delta_{H_0}^9(\mathcal{X})$ which equal 0.04 and 0.027 respectively is not so dramatic. This is the result of the adopted method of dividing attributes into groups of significance. Attribute 15 (duration of intervention) which in $\mathcal{A}$ was the most important one belongs in $\mathcal{B}$ to the set of factors of small significance. A certain improvement of its position can be obtained by adding new values to its domain (we describe it in the next section). But still, the small importance of this factor, as revealed in series $\mathcal{B}$, is inconsistent with surgeons' experience. We are also lacking any foundation for a sudden increase in the role of attribute 18 (order of intervention).

5. Hierarchies for sets of attributes P and H in the second series of research

In this section we will deal with sets of attributes P and H containing full information about objects. The most disturbing feature of the information system employed in series $\mathcal{B}$ is the considerably great numbers of values of some attributes, like attributes 3 (age), 11 (diagnosis), 13 (kind of intervention). Such a proliferation can result in high values of the accuracy of approximation although in case of inadequate choice of attributes even the domains of large cardinality can not automatically ensure good results (we do not take into consideration extreme situations in which for example every object of a system stays in one-to-one relation with a certain element of a domain). On the other hand, if norms for values of attributes are not known, or if they are a matter of discussion, an information system with more values enables, by identifying those values, a verification of the correctness of adopted norms. In our case, as the aim of the work is to establish the hierarchy of risk factors, we will be interested in the problem of stability of obtained results with respect to variously defined domains of attributes. Let P_5 (resp. P_{10}) denote the set of attributes obtained from P by gluing values of attribute 3 (age) every 5 (resp. 10) years. Table 8 shows the increase in accuracy of approximation of classes X_1, X_2 , X_3 and $\mathcal{X}$ depending on the type of the set of patient attributes, whereas hierarchies obtained for the sets mentioned are represented in table 9. If we disregard attribute 8 (time of stay in the hospital after the surgical intervention), then those hierarchies turn out to be similar to one another. In particular, attribute 3 keeps its first position and the discriminating role it plays increases with the number of values it can take. The least important attributes, namely 2, 5 and 6, did not change their significance either. There is some change, however, as far as the middle

Table 8.

	$\mu_{P_0}(X_i)$	$\mu_{P_{10}}(X_i)$	$\mu_{P_5}(X_i)$	$\mu_P(X_i)$
X_1	0.71	0.77	0.85	0.96
X_2	0.17	0.22	0.34	0.71
X_3	0.14	0.32	0.51	0.95
	$\beta_{P_0}(\mathcal{X}) = 0.55$	$\beta_{P_{10}}(\mathcal{X}) = 0.63$	$\beta_{P_5}(\mathcal{X}) = 0.75$	$\beta_P(\mathcal{X}) = 0.93$

Table 9.

	P_0		P_{10}		P_5		P	
	i	$\Delta^i_{Pat_0}(\mathcal{X})$	i	$\Delta^i_{P_{10}}(\mathcal{X})$	i	$\Delta^i_{P_5}(\mathcal{X})$	i	$\Delta^i_P(\mathcal{X})$
1	3	0.30	3	0.32	3	0.42	3	0.60
2	8	0.20	7	0.23	7	0.19	7	0.10
3	1	0.19	8	0.17	8	0.154	4	0.071
4	4	0.15	1	0.16	1	0.147	1	0.069
5	7	0.141	4	0.15	4	0.146	8	0.066
6	2	0.137	2	0.11	2	0.10	2	0.04
7	6	0.05	6	0.04	6	0.03	5	0.008
8	5	0.004	5	0.007	5	0.009	6	0.006

group of attributes is concerned. The growth of importance of attribute 7 (time of stay in the hospital before the surgical intervention) is caused by the expansion of its domain during $\mathcal{B}$. Observe also the relatively stable position of attribute 1 (type of intervention) in all four hierarchies. Another regularity is indicated in table 10. It shows the hierarchy of

Table 10.

	X_1		X_2		X_3	
	i *attribute number*	$\Delta^i_P(X_1)$	i	$\Delta^i_P(X_2)$	i	$\Delta^i_P(X_3)$
1	3	0.43	3	0.66	3	0.92
2	7	0.06	7	0.18	7	0.48
3	4	0.044	4	0.17	1	0.25
4	8	0.041	1	0.164	4	0.19
5	1	0.039	8	0.158	8	0.17
6	2	0.024	2	0.10	2	0.09
7	5	0.005	5	0.03	5	0.0
8	6	0.004	6	0.02	6	0.0

attributes from set P calculated for each class X_1, X_2, X_3 exclusively . If we ignore attribute 8 again, then we see that the significance value $\Delta^i_P(X_j)$, determined for the first four most important attributes (3,7,4,1), increases with the growth of j, that is with the extent of wound infection. A similar dependence, though not as strong, can be observed for the most important hospital attributes. Table 11 shows accuracy of approximation of the sets X_1, X_2, X_3 and classification $\mathcal{X}$ by sets of attributes H_0 and H. The increase of accuracy for H

Table 11.

	$\mu_{H_0}(X_i)$	$\mu_H(X_i)$
X_1	0.83	0.92
X_2	0.40	0.65
X_3	0.40	0.60
	$\beta_{H_0}(\mathcal{X}) = 0.73$	$\beta_H(\mathcal{X}) = 0.87$

is obtained not only by expanding the domain of attribute 15 (duration of intervention) but also by adding two new attributes: 11 (diagnosis) and 13 (kind of intervention). However those additional attributes rank very low in the hierarchy for H as can be seen in table 12. The change of the domain of attribute 15 resulted in the shift from 5th position in set

Table 12.

	i	$\Delta_H^i(\mathcal{X})$
1	16	0.27
2	18	0.08
3	15	0.07
4	17	0.03
5	19	0.016
6	10	0.014
7	11	0.012
8	9	0.009
9	1	0.005
10	12	0.0
11	13	0.0
12	14	0.0

H_0 to 3rd in H. The most essential changes are reported in case of attribute 1 (type of intervention), which dropped 5 steps, though it is a factor of little significance both in H_0 and H. Attribute 1 and its connection with attributes 11 and 13 are discussed in section 6.1. Observe that the most significant attributes in H_0 — 16 (surgeon) and 18 (order of intervention) — also preserved their high positions in H and the least important ones — 14 (type of anesthesia), 9 (presence of infectious cases in the hospital room) and 12 (mode of intervention) — stay in the lower part in the hierarchy for H. The hierarchy of attributes from H calculated for all individual classes of classification $\mathcal{X}$ is presented in table 13. For attributes 16 (surgeon), 17 (type of operating room) and 18 (order of intervention), being among the most important ones, we observe a similar regularity as in the case of set P, namely the increase of the coefficient Δ_H^i together with the extent of wound infection.

Table 13.

	X_1		X_2		X_3	
	i	$\Delta^i_H(X_1)$	i	$\Delta^i_H(X_2)$	i	$\Delta^i_H(X_3)$
1	16	0.17	16	0.35	16	0.39
2	18	0.05	15	0.15	18	0.20
3	15	0.04	18	0.14	17	0.10
4	17	0.016	19	0.048	15	0.087
5	19	0.008	17	0.045	19	0.030
6	11	0.0078	10	0.044	10	0.030
7	10	0.0068	11	0.042	9	0.011
8	9	0.0058	9	0.021	1	0.0
9	1	0.0033	1	0.019	11	0.0
10	12	0.0	12	0.0	12	0.0
11	13	0.0	13	0.0	13	0.0
12	14	0.0	14	0.0	14	0.0

6. Transformations of the information system

6.1. Type of intervention *vs.* diagnosis and kind of intervention

In $\mathcal{A}$ attribute 1 (type of intervention), belonging both to set P and H, turned out to be a factor of small influence on wound infection (though in P it was the most important attribute among those of small significance, decreasing the accuracy of approximation by 0.11). This result does not agree with the experience of clinicians and with publications on the topic, and in [5] we could not find any reasonable explanation for that situation. In $\mathcal{B}$ attribute 1 occupied the third position in the hierarchy for P_0, belonging thus to the set of attributes of medium influence, whereas in the sets H_0 and H it was an attribute of small influence.

As we have mentioned already, attribute 1 is a factor characterizing a patient and also, due to its connection with diagnosis and kind of intervention, refers to a hospital. The following two arguments incline us to regarding attribute 1 as a factor characterizing a patient rather than a hospital:

- not only attribute 1 but also attributes 11 and 13 associated with it have small significance in H.
- if we replace in P attribute 1 by attribute 11 or 13 or both, then the accuracy of approximation by such a set of attributes will be close to that calculated for set P.

Table 14 illustrates the described process of replacement (by P^-, P^-_{11}, P^-_{13}, $P^-_{11,13}$ we denote $P-\{1\}$, $P-\{11\}$, $P-\{13\}$, $P-\{11,13\}$ respectively). If a further study confirm our conjecture then it will be possible to exclude from the information system attributes 11 and 13 and take into account only attribute 1 with has much smaller domain.

6.2 The expansion of sets P and H.

By P_{25} we denote set P with added attribute 25 and we calculate the accuracy of approximation by P_{25}. It is clear from table 15 that the addition of attribute 25 does not change

Table 14.

	$\mu_P(X_i)$	$\mu_{P^-}(X_i)$	$\mu_{P^-_{11}}(X_i)$	$\mu_{P^-_{13}}(X_i)$	$\mu_{P^-_{11,13}}(X_i)$
X_1	0.96	0.92	0.97	0.96	0.97
X_2	0.71	0.54	0.82	0.74	0.82
X_3	0.95	0.70	1.0	0.86	1.0
	$\beta_P(\mathcal{X})$	$\beta_{P^-}(\mathcal{X})$	$\beta_{P^-_{11}}(\mathcal{X})$	$\beta_{P^-_{13}}(\mathcal{X})$	$\beta_{P^-_{11,13}}(\mathcal{X})$
	0.93	0.86	0.96	0.93	0.96

Table 15.

	$\mu_P(X_i)$	$\mu_{P_{25}}(X_i)$
X_1	0.96	0.962
X_2	0.71	0.73
X_3	0.95	0.95
	$\beta_P(\mathcal{X}) = 0.93$	$\beta_{P_{25}}(\mathcal{X}) = 0.933$

the accuracy much. The small significance of this attribute, confirmed also by table 16, might be caused by the rather small number of patients who had coexisting diseases (52 objects) or by a neutralizing influence of antibiotic treatment.

Table 16.

	i	$\Delta^i_{P_{25}}(\mathcal{X})$
1	3	0.56
2	7	0.076
3	8	0.066
4	4	0.057
5	1	0.051
6	2	0.033
7	5	0.008
8	25	0.008
9	6	0.006

An essential improvement of approximation is obtained after adding attributes 20–24 (describing antibiotic therapy) to set H. The new set of attributes will be denoted H_A. Tables 17 and 18 show the accuracy of set H_A and hierarchies of its attributes respectively. As for additional attributes, it is attribute 22 (antibiotics after the surgical intervention) which turned out to play the strongest discriminant role in the approximation. We cannot however interpret this result unequivocally, because the reasons for which antibiotics are administered after operation may be totally different. It might be done when the symptoms of infection are already present, or when the probability of appearance of infection is exceptionally high (e.g. in case of elderly people). In neither of those cases can we say that administering antibiotics causes wound infection. We can at most speak about the

Table 17.

	$\mu_H(X_i)$	$\mu_{H_A}(X_i)$
X_1	0.92	0.986
X_2	0.65	0.90
X_3	0.60	0.93
	$\beta_H(\mathcal{X}) = 0.87$	$\beta_{H_A}(\mathcal{X}) = 0.97$

Table 18.

	i	$\Delta^i_{H_A}(\mathcal{X})$
1	16	0.13
2	22	0.05
3	15	0.039
4	18	0.034
5	19	0.01
6	11	0.008
7	24	0.008
8	17	0.006
9	1	0.004
10	9	0.002
11	10	0.002
12	12	0.002
13	13	0.002
14	14	0.002
15	20	0.002
16	21	0.002
17	23	0.002

coexistence of the two attributes. This however may only confirm an obvious fact that if a wound is infected then antibiotics are administered or it may ascertain the ineffectiveness of preventive usage of antibiotics.

6.3 Reduction of the number of objects in the information system

In order to eliminate the influence of accidental factors, we excluded from our system those objects for which the value of attribute 15 (duration of intervention) appeared in the system less than six times (rare interventions). For the information system obtained this way, consisting of 964 objects, we carried out all usual calculations. As follows from table 19, the omission of rare surgical interventions does not influence much the value of approximation both for sets P and H.

We employed the procedure described above with reference to attribute 16 (surgeon). In this case we left in the system only these objects for which the value of attribute 16 appears in the system more than 10 times (971 objects). Table 20 shows the accuracy of approximation calculated for such a system. It is clear that the differences in approximation

Table 19.

P			H	
full system	reduced system		full system	reduced system
$\mu_P(X_i)$	$\mu_P(X_i)$		$\mu_H(X_i)$	$\mu_H(X_i)$
0.96	0.95	$\lhd X_1 \rhd$	0.92	0.92
0.71	0.698	$\lhd X_2 \rhd$	0.65	0.644
0.95	0.949	$\lhd X_3 \rhd$	0.60	0.596
$\beta_P(\mathcal{X})$	$\beta_P(\mathcal{X})$		$\beta_H(\mathcal{X})$	$\beta_H(\mathcal{X})$
0.93	0.924		0.87	0.869

Table 20.

P			H	
full system	reduced system		full system	reduced system
$\mu_P(X_i)$	$\mu_P(X_i)$		$\mu_H(X_i)$	$\mu_H(X_i)$
0.96	0.96	$\lhd X_1 \rhd$	0.92	0.92
0.71	0.716	$\lhd X_2 \rhd$	0.65	0.649
0.95	0.947	$\lhd X_3 \rhd$	0.60	0.588
$\beta_P(\mathcal{X})$	$\beta_P(\mathcal{X})$		$\beta_H(\mathcal{X})$	$\beta_H(\mathcal{X})$
0.93	0.93		0.87	0.87

between the full system and the reduced one are minimal.

7. Final remarks and conclusions

We tried to show in this paper how using even a small part of rough set theory can bring clinically interesting results. The information system we presented is not a perfect one. We still cannot find a factor characterizing the elements from classes X_2 and X_3; furthermore high values of accuracy of approximation are obtained through the use of attributes with a large number of values. However, such a situation is determined by the fact that the risk factors of postoperative wound infection have not been established definitively and their influence may vary depending upon specific conditions of particular hospitals. In addition, there is no general agreement as to the norms for attribute values. Thus our approach seems to be justified. Because our system is comprised of all data available at the moment, we are able to eliminate redundant attributes or glue values. The elimination has proved to be successful in the case of attributes 11 and 13 which could be replaced by attribute 1.

Finally, let us point out once more that our general aim is to establish the hierarchy of *known* risk factors. The fact that the results obtained previously (described in [5]) and presented here appear to be stable encourages us to continue the research.

References

[1] S. Danielsen, P.F. Amland, K. Ardenaes, *Infections problems in elective surgery of the alimentary tract: The influence of pre-operative factors*, Curr. Med. Res. Opin., **11**(1988), 171–178.

[2] A. J. G. Davidson, C. Clark,G. Smith, *Postoperative wound infection: a computer analysis*, British J. Surgery, **58** (1971), 333–337.

[3] S. L. Gorbach, J. G. Bartlett, R. L. Nichols, *Manual of Surgical Infections*, Little, Brown and Company, Boston 1984.

[4] M. Jurkiewicz, J. Garret, *Studies on the influence of anemia on wound healing*, American Surgeon, **32** (1966), 442–445.

[5] M. Kandulski, B. Litewka, A. Mrózek, K. Tukałło, *An attempt to establish the hierarchy of factors inducing surgical wound infections by means of the rough set theory*, Bull. Pol. Ac.:Biol., **39** (1991), 5–18.

[6] Z. H. Krukowski, N. A. Matheson, *Ten-year computerized audit of infection after abdominal surgery*, British J. Surgery, **75** (1988), 857–861.

[7] W. Lau, S. Fan, K. Chu, W. Yip, W. Yuen, K. Wong, Influence of surgeons' experience on postoperative sepsis, Am. J. Surgery, bf 155 (1988), 322–326.

[8] D. E. Lilienfeld, D. Vlahov, J. H. Tenney, J. S. McLaughlin, *Obesity and diabetes as risk factors for postoperative wound infection after cardiac surgery*, Am. J. Infect. Control, **16** (1983), 3–6

[9] S. F. Mishriki, D. J. Law, P. J. Jeffery, *Factors affecting the incidence of postoperative wound infection*, J. Hosp. Infect. **16(3)** (1990), 223–230.

[10] H. C. Polk Jr (ed.), *Infections and the Surgical Patient*, Churchill Livingstone, Edinburgh, 1982.

[11] A. P. Wilson, C. Weavill, J. Burridge, M. C. Kelsey, *The use of the wound scoring method 'ASEPSIS' in postoperative wound surveillance*, J. Hosp. Infect. **16(4)** (1990), 297–309.

Part I
Chapter 8

FUZZY INFERENCE SYSTEM BASED ON ROUGH SETS AND ITS APPLICATION TO MEDICAL DIAGNOSIS

Hideo TANAKA Hisao ISHIBUCHI Takeo SHIGENAGA
Department of Industrial Engineering
University of Osaka Prefecture
Sakai, Osaka 591, Japan

Abstract. In this paper, a method of constructing fuzzy if–then rules in an expert system is described by the proposed approach, where the lower approximations in rough sets are used to extract if–then rules from the given information system and fuzzy if–then rules are constructed from the extracted if–then rules.

1. Introduction

Z.Pawlak(1982) has proposed a new concept of approximate data analysis based on rough sets. An application of rough sets to medical analysis for heart disease has been done also by Z.Pawlak(1984). In this analysis, inconsistency of medical test data and expert's diagnoses can be clarified. A.Mrozek (1989) has constructed if–then rules by rough sets to control a rotary clinker kiln in a cement plant. This research was concerned with constructing an expert's inference model.

In this paper, we propose a new method of reducing information systems by considering the classification given by experts, which is based on Z.Pawlak's suggestion in Z.Pawlak(1984). Our proposed method can reduce more attributes than Z.Pawlak's method described in Z.Pawlak(1984). Also, a method of constructing a fuzzy inference system is described by introducing fuzzy intervals represented as fuzzification of attribute values. The reason for using fuzzy intervals is that fuzzy if–then rules can cover the whole space of attribute values. In other words, coventional if–then rules derived from only lower approximations can not cover the whole space because the number of the derived rules is rather small comparing with the number of possible rules constructed by the combination of all intervals of each attribute.

H.Tanaka *et al.* (1990) have already constructed a fuzzy expert system based on the reduced information system and applied it to a diagnosis problem. This paper aims to improve the performance of the fuzzy expert system in H.Tanaka *et al.*(1990) by proposing a new fuzzy inference model. This fuzzy inference model is constructed by considering all fuzzy if–then rules whose conclusions are the same, although in usual fuzzy inference models, the inference result depends on the most fitting fuzzy if–then rule. In order to consider all fuzzy if–then rules inferring the same conclusion, we use the average–product combination in the classification process. The classification power of the proposed inference model is demonstrated by an application to the same diagnosis problem as in H.Tanaka *et al.*(1991).

2. Reduction of Information Systems by Classification

An approximation space A is defined as $A = (U, R)$, where U is a set called the universe and R is an equivalent relation. Equivalence classes of the relation R are called elementary sets in A. The empty set is assumed to be also elementary. Any finite union of elementary sets in A is called a definable set in A. Let $X \subset U$. An upper approximation of X in A denoted as $A^*(X)$ is defined by the least definable set in A, containing X. A lower approximation denoted as $A_*(X)$ is defined by the greatest definable set in A, contained in X.

An accuracy measure of X in $A = (U, R)$ is defined as

$$\alpha_Q(X) = Card(A_*(X))/Card(A^*(X)) \tag{1}$$

where $Card(B)$ is the cardinal number of the set B. Let $F = \{X_1, \ldots, X_n\}$ be a classification of U, i.e. $X_i \cap X_j = \phi$ for every $i \neq j$ and $\cup X_i = U$. Then, F is called a partition of U and X_i is called a class. An accuracy measure of F in A is defined as

$$\beta_Q(F) = Card(\cup A_*(X_i))/Card(U) \tag{2}$$

Z.Pawlak's method does not consider the given classification. A new method for reducing attributes in an information system is proposed here, considering the given classification. Let $S = (U, Q, V, \psi, F)$ be an information system, where U is the universe of S, an element of U is called an object, Q is a set of attributes, $V = \cup V_q$ is a set of values of attributes, $\psi : U \times Q \rightarrow V$ is a description function and F is a classification of U.

[Theorem 1] Given a classification F, it follows for the subset $P' \subset P$ that

$$\beta_{P'}(F) \leq \beta_P(F) \tag{3}$$

where $\beta_P(F)$ is the accuracy measure of F, defined by (2) with regard to the subset P of the attributes Q.

The followings can be defined from Theorem 1.

[Definition 1]

i) Let P be a subset of Q. A subset P is said to be independent in an information system S if and only if

$$\beta_{P'}(F) < \beta_P(F) \text{ for all } P' \subset P \tag{4}$$

Also, P is said to be dependent in S if and only if there is $P' \subset P$ such that

$$\beta_{P'}(F) = \beta_P(F) \tag{5}$$

ii)Let $P' \subset P$ and $P" = P - P'$. A subset P' is said to be superfluous in P if and only if

$$\beta_{P"}(F) = \beta_P(F) \tag{6}$$

[Theorem 2] If $P' \subset P$ is superfluous in P and $\{p_i\}$ is superfluous in $P - P'$, then $P' \cup \{p_i\}$ is superfluous in P.

[Definition 2] A subset $P \subset Q$ is called reduct of Q in S if and only if $Q - P$ is superfluous and P is independent in S.

An algorithem for obtaining reduct of Q can be described from Theorem 2 and Definition 2 as follows.

[Algorithm]

Step 1: Find a superfluous attribute, say p_i, in Q. If there is not such a p_i, go to Step 3.

Step 2: Set $Q = Q - \{p_i\}$ and go to step 1.
Step 3: End. Q is a reduct of the given attributes.

Using the above algorithm, superfluous attributes can be taken away one by one. Thus, reduct of Q can be obtained by the algorithm.

[Example 1]

Let us consider the example shown in Table 1. The reducts of the given attributes obtained by Z.Pawlak's method are as follows.

$$\{q_1, q_2, q_3\},\ \{q_1, q_3, q_4\},\ \{q_2, q_3, q_4\} \tag{7}$$

On the other hand, the proposed method based on the accuracy measure can generate the following reduct.

$$\{q_3, q_4\} \tag{8}$$

In what follows, let us show that $\{q_3, q_4\}$ is a reduct of the given set of attributes. The lower approximations of X_1 and X_2 by $Q = \{q_1, \ldots, q_4\}$ are as follows.

$$A_*(X_1) = \{x_1, x_2\},\ A_*(X_2) = \{x_5, x_6\} \tag{9}$$

Thus,

$$\beta_Q(F) = 4/6 = 0.67 \tag{10}$$

Considering $P = \{q_3, q_4\}$, we have the same lower approximations of X_1 and X_2 by P as in (9). Therefore, $\beta_P(F) = 0.67$, which means that the accuracy measure is not reduced by removing two attributes. Then, let $P' = \{q_3\}$ and $P" = \{q_4\}$. Thus we have

$$\beta_{P'}(F) = 2/6 = 0.33 < \beta_P(F) \tag{11}$$

$$\beta_{P"}(F) = 1/6 = 0.17 < \beta_P(F) \tag{12}$$

which lead to the fact that $P = \{q_3, q_4\}$ is a reduct.

Comparing (8) with (7), we can conclude that our proposed method for reducing attributes is more effective than Z.Pawlak's method.

TABLE 1 . An example of information system

F	U	q_1	q_2	q_3	q_4
	x_1	1	1	1	1
X_1	x_2	1	2	1	3
	x_3	2	2	2	2
	x_4	2	2	2	2
X_2	x_5	1	2	2	3
	x_6	2	3	2	3

3. Fuzzy Inference Model

Integers are used as values of attributes in Section 2. In general, a value of attribute is obtained as a real number. Thus, a real number is assigned to some integer as shown in Tabel 1. Given $F = \{X_1, \ldots, X_n\}$ derived from expert knowledge, the lower approximation of each X_i can be obtained in the reduced system. From the example of Table 1, we have three rules constructed by the lower approximations of X_1 and X_2 as follows.

$$\begin{aligned} &\text{If } \{y_3 = 1\} \wedge \{y_4 = 1\} \text{ then } \mathbf{y} \in X_1 \\ &\text{If } \{y_3 = 1\} \wedge \{y_4 = 3\} \text{ then } \mathbf{y} \in X_1 \\ &\text{If } \{y_3 = 2\} \wedge \{y_4 = 3\} \text{ then } \mathbf{y} \in X_2 \end{aligned} \tag{13}$$

where reduct of Q is $\{q_3, q_4\}$ and $\mathbf{y} = (y_3, y_4)^t$ is a pair of integers representing attribute values of q_3 and q_4 of an inferred object. It should be noted that only the lower approximations of F are used to construct if–then rules. Therefore only the data being consistent with expert's diagnoses are used. In the example of Table 1, the following if–then rule is taken away.

$$\text{If } \{y_3 = 2\} \wedge \{y_4 = 2\} \text{ then } \mathbf{y} \in X_1 \text{ or } \mathbf{y} \in X_2 \tag{14}$$

Since an input vector $\mathbf{y}$ into our inference model is a real number vector, if–then rules (13) are fuzzified by introducing fuzzy intervals. For example, the first rule in (13) can be written as

$$\text{If } y_3 = \tilde{1} \wedge y_4 = \tilde{1} \text{ then } \mathbf{y} \in X_1 \tag{15}$$

where $\tilde{1}$ is a fuzzy interval as shown in Figure 1. In general, the j–th fuzzy if–then rule for inferring X_i from an input vector $\mathbf{y} = (y_1, \ldots, y_m)^t$ is represented as

$$\text{If } y_1 = A^i_{1j} \text{ and } \cdots \text{ and } y_m = A^i_{mj} \text{ then } \mathbf{y} \text{ is } X_i \tag{16}$$

where A^i_{kj} is a fuzzy interval specified by a membership function $\mu^i_{kj}(y_k)$.

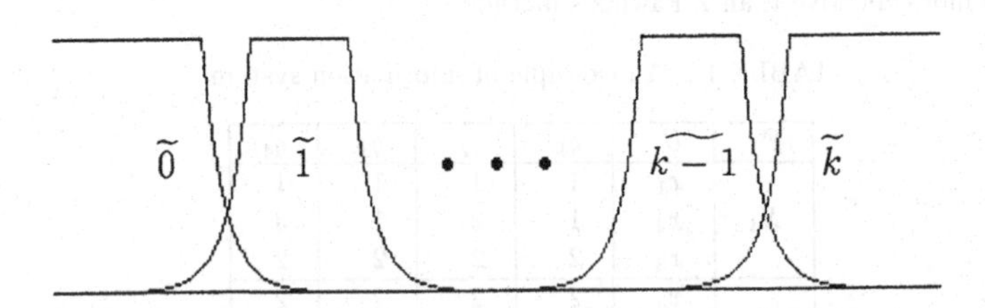

Figure 1 . Fuzzy division of attribute value

Let $\pi_{X_i}(\mathbf{y})$ be a grade to which $\mathbf{y}$ is belonging to X_i. Using the concept of fuzzy inference, three types of $\pi_{X_i}(\mathbf{y})$ can be considered as follows.

i) Max · Min composition:

$$\pi_{Xi}(\mathbf{y}) = \bigvee_j (\mu^i_{1j}(y_1) \wedge \cdots \wedge \mu^i_{mj}(y_m)) \tag{17}$$

which is widely used in fuzzy control and also was used in H.Tanaka *et al.*(1991).

ii) Max · Product composition:

$$\pi_{Xi}(\mathbf{y}) = \bigvee_j (\pi^i_{1j}(y_1) \times \cdots \times \pi^i_{mj}(y_m)) \tag{18}$$

iii) Average · Product composition:

$$\pi_{Xi}(\mathbf{y}) = \frac{\sum_j (\pi^i_{1j}(y_1) \times \cdots \times \pi^i_{mj}(y_m))}{\text{number of rules infering } X_i} \tag{19}$$

which is proposed in this paper to use for medical diagnosis.

It should be noted that the grade of $\mathbf{y}$ fitting to all the fuzzy if–then rules infereing X_i is defined as iii). In i) and ii), the value of $\pi_{Xi}(\mathbf{y})$ is determined by only one fuzzy if–then rule so that it can be said that all the fuzzy if–then rules are independent each other.

The decision rule can be written as

$$\text{If } \pi_{Xk}(\mathbf{y}) = \pi_{X1}(\mathbf{y}) \vee \cdots \vee \pi_{Xn}(\mathbf{y}), \text{ then } \mathbf{y} \in X_k \tag{20}$$

where $F = \{X_1, \ldots, X_n\}$ is given by expert's diagnoses. These three types of $\pi_{Xi}(\mathbf{y})$ are compared with each other in the following application.

4. Fuzzy Inference System for Medical Diagnosis

Our proposed method described in Section 2 and 3 is applied to medical diagnosis of hepatic disease. All test data are obtained in Kawasaki Medical College as follows:

i) A classification $F = \{X_1, \ldots, X_5\}$ obtained by experts:

X_1 = Healthy person, X_2 = Hepatoma, X_3 = Acute hepatitis
X_4 = Chronic hepatitis, X_5 = Liver cirrhosis

ii) The number of data is 568 and the data are divided into two parts: data for modelling(468 persons) and data for checking (100 persons). Data for modelling are used for constructing fuzzy if–then rules and each person in the data for checking is assigned to one of the five classes using the constructed fuzzy if–then rules. Fuzzy if–then rules are evaluated by the data for checking.

iii) The number of attributes (medical inspections) is 20. Values of each attribute are divided into 3 to 6 intervals as shown in Table 2, where the integers corresponding to **** are not used.

A missing value is assigned to 0 whose value of membership function is 1 over the domain. This means that an attribute q_i is disregarded in fuzzy if–then rules if the value of q_i is missing.

TABLE 2 . Division of medical test data

	Medical	Integers of attribute values					
	inspection	1	2	3	4	5	6
q_1	SP	~ 5.5	5.6 ~ 6.5	6.6 ~ 7.5	7.6 ~ 8.5	8.6 ~	****
q_2	II	~ 4	5 ~ 6	7 ~ 9	10 ~	****	****
q_3	TBil	~ 1.0	1.1 ~ 5.0	5.1 ~ 10.0	10.1 ~ 20.0	20.1 ~	****
q_4	DBil	~ 40	41 ~ 60	61 ~ 80	81 ~ 100	****	****
q_5	Alp	~ 80	81 ~ 200	201 ~ 300	301 ~ 400	401 ~	****
q_6	G.GTP	~ 30	31 ~ 100	101 ~ 200	201 ~ 300	301 ~	****
q_7	LDH	~ 100	101 ~ 250	251 ~ 500	501 ~ 1000	1001 ~	****
q_8	Alb-G	~ 2.0	2.1 ~ 3.0	3.1 ~ 4.0	4.1 ~ 5.0	5.1 ~	****
q_9	ChE	~ 100	101 ~ 150	151 ~ 200	201 ~ 250	251 ~ 500	501 ~
q_{10}	GPR	~ 25	26 ~ 100	101 ~ 200	201 ~ 500	501 ~ 1000	1001 ~
q_{11}	GOT	~ 20	21 ~ 100	101 ~ 200	201 ~ 500	501 ~ 1000	1001 ~
q_{12}	BUN	~ 9	10 ~ 20	21 ~ 30	31 ~ 40	41 ~	****
q_{13}	UrA	~ 2.7	2.8 ~ 8.5	8.6 ~	****	****	****
q_{14}	Retic	~ 1.5	1.6 ~ 3.0	3.1 ~ 6.0	6.1 ~	****	****
q_{15}	Plt	~ 1.0	1.1 ~ 5.0	5.1 ~ 10.0	10.1 ~ 15.0	15.1 ~ 35.0	35.1 ~
q_{16}	Lympho	~ 20.0	20.1 ~ 40.0	40.1 ~ 60.0	60.1 ~	****	****
q_{17}	Fibrino	~ 200	201 ~ 400	401 ~	****	****	****
q_{18}	Alb-%	~ 45.0	45.1 ~ 65.0	65.1 ~	****	****	****
q_{19}	Al-%	~ 2.5	2.6 ~ 3.7	3.8 ~ 5.0	5.1 ~	****	****
q_{20}	AFP	~ 20	21 ~ 100	101 ~ 200	201 ~ 1000	1001 ~	****

First, let us define the ratio of correct inference. If **y** is inferred correctly, **y** is given one point and if **y** is inferred erroneously, **y** is given zero point. When **y** is inferred more than or equal to two classes, **y** is given 1/(the number of classes) point if the correct class is included in the inferred classes and otherwise, given zero point. Then, the ratio of correct inference is defined by

$$\frac{\text{Sum of points to checking data}}{\text{The number of checking data}} \times 100(\%) \qquad (21)$$

The results are as follows:

i) The number of attributes in the reduct of Q is 7. The seven attributes are used for constructing fuzzy if–then rules and comparing three fuzzy inference methods.

ii) There are 367 if–then rules constructed by the low approximations of X_i, $i = 1, \ldots, 5$.

iii) The ratios of correct inference to checking data by different three types of compositions are shown in Table 3. It can be concluded from Table 3 that Average · Product composition is the best in the sense of the highest correct ratio and no multiple inference.

TABLE 3 . The ratios of correct inference to checking data

Type of composition	Correct ratio	Number of correct inferences	Number of error inferences	Number of multiple inferences
Max · Min	67.5	62	25	13
Max · Product	68.2	66	29	5
Average · Product	74.0	74	26	0

References

[1] Mrozek,A.(1989) 'Rough sets and dependency analysis among attributes in computer implementations of expert's inference models', *Int. J. of Man-Machine Studies*, **30**, 457-473.

[2] Tanaka,H. , Ishibuchi,H. and Matsuda,N. (1990) 'Reduction of information systems based on rough sets and its application to fuzzy expert systems', *Proceedings of Sino-Japan Joint Meeting on Fuzzy Sets and Systems*, C2-5, October 15-18, Beijing, China.

[3] Tanaka,H. , Ishibuchi,H. and Matsuda,N. (1991) 'Reduction of information systems based on rough sets and its application to fuzzy expert systems', *SICE*, **27**, **3**, 357-364 (in Japanese).

[4] Pawlak,Z.(1982) 'Rough sets', *Int. J. of Information and Computer Sciences*, **11-5**, 341-356.

[5] Pawlak,Z.(1984) 'Rough classification', *Int. J. of Man-Machine Studies*, **20**, 469-485.

Part I
Chapter 9

ANALYSIS OF STRUCTURE – ACTIVITY RELATIONSHIPS OF QUATERNARY AMMONIUM COMPOUNDS

Jerzy KRYSIŃSKI
Medical Academy
Chair of Pharmaceutical Technology
60-780 Poznań, Poland

Abstract. Relationships between chemical structure and antimicrobial activity of quaternary imidazolium, quinolinium and isoquinolinium compounds is analysed. The compounds are described by a set of condition attributes concerning structure and by a set of decision attributes concerning activity. The description builds up an information system. Using the rough sets approach, a smallest set of condition attributes significant for a high quality of classifications has been found. The analysis of distributions of values of significant attributes in the best and worst class led to the definition of typical representatives of the best and worst imidazolium compounds in the terms of the significant condition attributes. A decision algorithm has been derived from information system, showing up important relations between structure and activity. This may be helpful in supporting decisions concerning synthesis of new antimicrobial compounds.

1. Introduction

One of the most important issues in search for new chemical compounds with an expected biological activity is to settle the dependence between the chemical structure and biological activity. The knowledge of dependence between the structure and activity enables one to foresee the biological activity of compounds and, in this sense to lead the research in a right direction. Moreover, the chance of synthesis of new biologically effective preparations increases. Until now, all analysis of dependence between structure and efficiency of antimicrobial agents has been based on statistic methods such as regression, correlation and discriminat analysis. These methods are usually summarized under the heading *"quantitative structure-activity relationships"* (*QSARs*). In the most general sense, *QSARs* are mathematical expressions which describe the dependence of biological activity of chemical compounds in terms of suitable molecule parameters. All these methods use either structural parameters or parameters concerning physical chemistry properties of a molecule (Franke et al. (1973, 1980), Gäbler et al.(1976), Weuffen et al. (1981)). The number and values of applied parameters differ from one method to another. For exemple in the *Hansch* approach, activity and structure parameters are use together with parameters derived from hydrophobic, electronic and steric substituent effects (Hansch et al. (1971, 1972, 1973),

Weuffen et al. (1981)). Similary, non-elementary discriminant functions used in the discriminant analysis, describe a distibution of compounds over classes of different biological activity in terms of the same hydrophobic, electronic and steric molecule parameters used in the *Hansch* approach (Franke (1980), Weuffen et al. (1981)). Another well known method by *Free-Wilson* is based on the analysis of activity and structure parameters (Free and Wilson (1964), Gäbler et al. (1976).

As was mentioned in other chapters, the statistic methods are usually appropriate for the analysis of big populations in which the attributes have quantitative character and their values are continuous. Otherwise, application of these methods would be impossible or would give great computational difficulties. Moreover, the statistic methods used by *QSARs* require a relatively high expenditure of experimental work connected with calculation of the parameter values.

The method based on the theory of rough sets is alternative to the statistic methods of analysing the relationship between the structure and activity of chemical compounds. This method does not require great expenditure of experimental work, it takes easily into account discrete attributes and it can be used for small populations as well (Słowiński et al. (1988), Krysiński (1990, 1991)).

Quaternary ammonium compounds, such as imidazolium, quinolinium and isoquinolinium chlorides, are being used for disinfection and antiseptics. They are characterized by a great antimicrobial activity. Their usage is limited however, because new resistant bacteria appear, mainly gram negative ones. Therefore it is necessary to look for new active antimicrobial compounds which would replace unactive, preparations. The result of the rough sets analysis gives a better chance for synthesis of new ammonium compounds with strong bacteriostatic activity.

In this chapter, we describe an application of the rough sets method to the analysis of the relationship between structure and antimicrobial activity of 247 quaternary imidazolium chlorides and of 72 quaternary quinolinium and isoquinolinium chlorides.

2. Quaternary Imidazolium Chlorides

2.1. Information system

The basis for analysis of structure activity relationship of quaternary imidazolium chlorides, is building of an information system. The information system is a set of objects (imidazolium chlorides), which is described by a set of attributes. The set of attributes includes eight condition attributes and one decision attributes. The condition attributes describe the object's structure (substituent in imidazole), whereas the decision attribute define the object's classification according to the value of minimal inhibitory concentration (*MIC*).

In table 1, the domain of condition attributes of imidazolium chlorides is presented. Conventional code numbers correspond with substituted imidazole.

Imidazolium compounds are divided into 5 classes of antimicrobial activity ($Y_1 \dots Y_5$). The classes correspond to the following ranges of the minimal inhibitory concentration (decision attribute):

Table 1. Domains of attributes of imidazolium chlorides

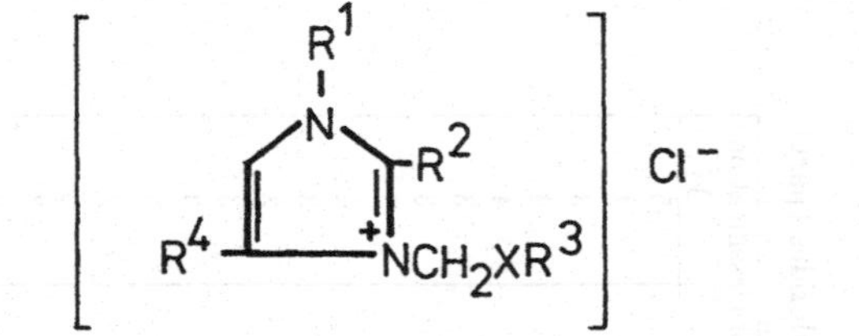

No.	Attribute	Values (code number)														
	Kind of	0	1	2	3	4	5	6	7	8	9	10	11	12	14	16
1	substituent R^1	without	alkyl	alkoxymethyl	–	–	–	–	–	–	–	–	–	–	–	–
2	alkyl in R^1	without	methyl	ethyl	–	butyl	–	hexyl	–	oktyl	–	decyl	–	dodecyl	–	–
3	alkoxy–methyl in R^1	without	–	ethoxy	propoxy	–	–	–	–	–	–	–	–	–	–	–
4	substituent R^2	without	phenyl	alkyl	–	–	–	–	–	–	–	–	–	–	–	–
5	alkyl in R^2	without	methyl	ethyl	propyl	isopropyl	–	–	–	–	–	–	–	–	–	–
6	substituent X	oxygen	sulphur	–	–	–	–	–	–	–	–	–	–	–	–	–
7	residue R^3	benzyl	–	ethyl	propyl	butyl	pentyl	hexyl	heptyl	oktyl	nonyl	decyl	undecyl	do decyl	–	–
8	substituent R^4	without	chloride	–	–	–	–	–	–	–	–	–	–	–	tetradecyl	hexadecyl

Class Y_1 – very good – $MIC \leq 10$ mg/dm^3
Class Y_2 – good – $10 < MIC \leq 50$ mg/dm^3
Class Y_3 – middle – $50 < MIC \leq 200$ mg/dm^3
Class Y_4 – weak – $200 < MIC < 500$ mg/dm^3
Class Y_5 – very weak – $MIC \geq 500$ mg/dm^3

The classification was made for two bacteria, *Pseudomonas aeruginosa NCTC 6749* which showed the greatest resistance against imidazolium compounds (classification $\mathcal{Y}_1$) and *Escherichia coli NCTC 8196* (calssification $\mathcal{Y}_2$).

In table 2, a part of information system is shown where lines corresponding to the objects and the columns to the attributes. 247 objects were hold in the information system.

Table 2. Information system of imidazolium chlorides (part)

Objects	Attribute								Class of classification	
	1	2	3	4	5	6	7	8	$\mathcal{Y}_1$	$\mathcal{Y}_2$
1	0	0	0	1	0	0	2	0	5	5
2	0	0	0	1	0	0	4	0	5	4
3	0	0	0	1	0	0	6	0	5	3
4	0	0	0	1	0	0	8	0	5	4
5	0	0	0	1	0	0	10	0	5	4
6	0	0	0	1	0	0	12	0	5	3
7	0	0	0	1	0	0	14	0	5	3
8	0	0	0	1	0	0	16	0	5	3
9	1	4	0	1	0	0	2	0	5	4
10	1	4	0	1	0	0	4	0	5	3
11	1	4	0	1	0	0	6	0	5	3
12	1	4	0	1	0	0	8	0	5	3
13	1	4	0	1	0	0	10	0	5	3
14	1	4	0	1	0	0	12	0	5	3
15	1	4	0	1	0	0	14	0	5	2
16	1	4	0	1	0	0	16	0	5	2
17	2	0	2	0	0	1	4	0	5	3
18	2	0	2	0	0	1	6	0	4	2
19	2	0	2	0	0	1	8	0	3	1
20	2	0	2	0	0	1	10	0	4	2
...	.	.	.	.	.	.	..	.	.	.
247	1	12	0	1	0	0	16	0	5	5

2.2. Reduction of attributes and the quality of classification

Initial quality of classification $\mathcal{Y}_1$ and $\mathcal{Y}_2$, i.e. for the whole set of condition attributes, was qual to 1. From the set of 8 condition attributes were received in both classifications per one minimal set diminished by attribute 1. In order to find the minimal set of condition attributes which are important for the quality of classification and accuracy of classes a computational test was done based on removal of respective attributes from the minimal set and observation of the quality of classification and accuracy of classes. Fig. 1 and 2 show the ways of attributes reduction in the classification $\mathcal{Y}_1$ and $\mathcal{Y}_2$, respectively. These ways are characteristic of the lowest starting downfall of the classification quality of all the examined ways. Removing attributes 1, 3, 5 and 8 in both classifications still gave a high quality in $\mathcal{Y}_1$ equal to 0.85 and in $\mathcal{Y}_2$ equal to 0.82. Reduction of attribut 4 diminished the classification quality in $\mathcal{Y}_1$ to 0.59 and in $\mathcal{Y}_2$ to 0.53. After removing attributes 2, 6 and 7 separately, the quality of classifications and accuracy of classes diminished still furether (fig. 1 and 2). Thus for the high quality of classifications $\mathcal{Y}_1$ and $\mathcal{Y}_2$ and for the accuracy of classes appeared essential:

attribute 2 – kind of alkyl in substituent in 1st imidazole position
attribute 4 – kind of subtituent in 2nd imidazole position
attribute 6 – kind of X in substituent in 3st imidazole position
attribute 7 – kind of alkyl in substituent in 3st imidazole position.

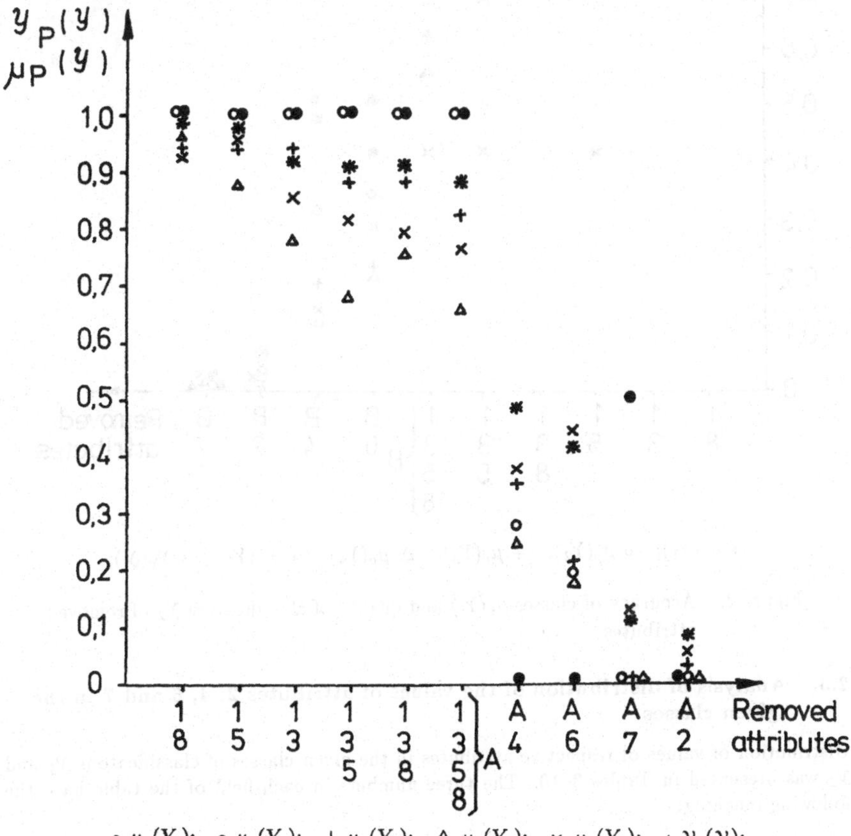

• $\mu_p(Y_1)$; ○ $\mu_p(Y_2)$; + $\mu_p(Y_3)$; △ $\mu_p(Y_4)$; × $\mu_p(Y_5)$; ∗ $\mathcal{Y}_p(\mathcal{Y})$;

Figure 1. Accuracy of classes $\mu_p(Y_i)$ and quality of classification $\mathcal{Y}_1$ of removed attributes.

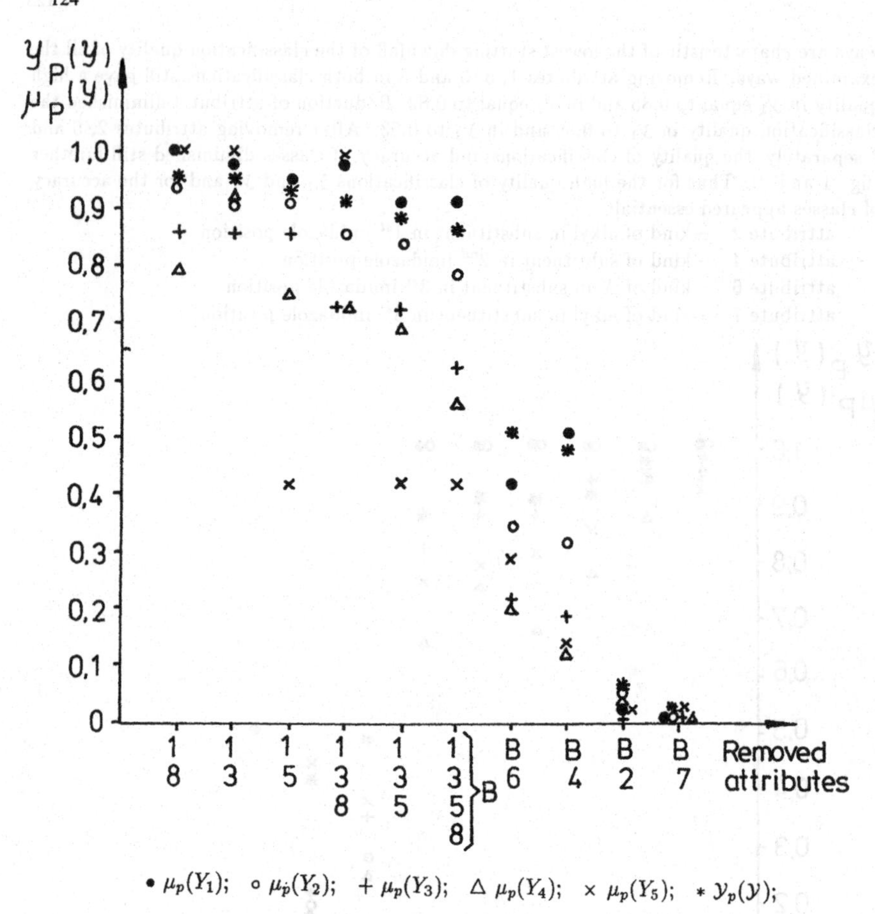

Figure 2. Accuracy of classes $\mu_p(Y_i)$ and quality of classification $\mathcal{Y}_2$ of removed attributes.

2.3. Analysis of distribution of the values of attributes 2, 4, 6 and 7 in the given classes

Distribution of values of respective attributes in the given classes of classificatons $\mathcal{Y}_1$ and $\mathcal{Y}_2$ was presented in Tables 3–10. The three numbers in each field of the table have the following meaning:

<table>
<tr><td colspan="2">a_{ij}</td></tr>
<tr><td>b_{ij}[%]</td><td>c_{ij}[%]</td></tr>
</table>

Table 3. Distribution of values taken by attribute 2 of y_1

No.of class	Values of attribute 2								Lawer approx.	Characteristic value of attribute 2
	0	1	2	4	6	8	10	12		
1	0	0	0	0	0	0	6	3	9	Decyl and dodecyl chain
							2.8 17	1.4 8		
2	0	1	1	2	2	12	5	6	29	–
		0.5 4	0.5 11	0.9 7	0.9 15	5.5 36	2.3 14	2.8 17		
3	8	9	3	5	5	12	11	5	58	–
	3.7 23	4.1 32	1.4 33	2.3 18	2.3 38	5.5 36	5.0 31	2.3 14		
4	7	5	2	4	3	9	8	8	46	–
	3.2 21	2.3 18	0.9 22	1.8 14	1.4 23	4.1 27	3.7 22	3.7 22		
5	19	13	3	17	3	0	6	14	75	without alkyl or methyl and butyl chain
	8.8 56	6.0 46	1.4 33	7.8 61	1.4 23		2.8 17	6.5 39		
Total	34	28	9	28	13	33	36	36	217	

Table 4. Distribution of values taken by attribute 7 of y_1

No.of class	Values of attribute 7														Lower approx	Characteristic value of attribute 7
	0	2	3	4	5	6	7	8	9	10	11	12	14	16		
1	0	0	0	2	0	2	0	2	0	1	0	1	1	0	9	–
				0.9 7		0.9 6		0.9 7		0.5 4		0.5 3	0.5 6			
2	0	0	0	1	2	5	1	7	1	7	0	4	1	0	29	–
				0.5 3	0 9 22	2.3 16	0.5 14	3.2 24	0.5 20	3 2 29		1.8 14	0.5 6			
3	2	1	0	6	3	6	0	11	0	8	1	16	3	1	58	–
	0.9 50	0.5 20		2.8 19	1.4 33	2.8 19		5.0 38		3.7 33	0.5 33	7.4 55	1.4 18	0.5 6		
4	2	1	2	3	1	8	4	6	1	5	0	3	5	5	46	–
	0.9 50	0.5 20	1.4 33	1.4 10	0.5 11	3.7 25	1.8 57	2.8 21	0.55 20	2 3 21		1.4 10	2.3 29	2.3 31		
5	0	3	4	19	3	11	2	3	3	3	2	5	7	10	75	ethyl, propyl, butyl,nonyl, undecyl,hexa-decyl chain
		1 4 60	1 8 67	8.8 61	1 4 33	5.0 34	0.9 29	1.4 10	1.4 60	1 4 13	0 9 67	2.3 17	3.2 37	4.6 63		
Total	4	5	6	31	9	32	7	29	5	24	3	29	17	16	217	

Table 5. Distribution of values taken by attribute 4 of $\mathcal{Y}_1$

No. of class	Values of attribute 4						Lower approx.	Characteristic value of attribute 4
	0		1		2			
1	9		0		0		9	without substituent
	4.1	9						
2	20		8		1		29	without substituent
	9.2	20	3.7	8	0.5	6		
3	26		27		5		58	–
	12	26	12.4	27	2.3	31		
4	24		21		1		46	–
	11	24	9.7	21	0.5	6		
5	22		44		9		75	alkyl or phenyl residue
	10	22	20.3	44	4.1	56		
Total	101		100		16		217	

Table 6. Distribution of values taken by attribute 6 of $\mathcal{Y}_1$

No. of class	Values of attribute 6				Lower approx.	Characteristic value of attribute 6
	0		1			
1	0		9		9	sulphur atom
			4.1	10		
2	13		16		29	–
	6	10	7.4	18		
3	27		31		58	–
	12.4	21	14.3	35		
4	27		19		46	–
	12.4	21	8.8	21		
5	61		14		75	oxygen atom
	28.1	48	6.5	16		
Total	128		89		217	

Table 7. Distribution of values taken by attribute 2 of $\mathcal{Y}_2$

No.of class	Values of attribute 2: 0	1	2	4	6	8	10	12	Lower approx.	Characteristic value of attribute 2
1	4	4	2	5	5	20	18	6	64	oktyl and decyl chain
	1.9 10	1.9 25	0.9 22	2.4 18	2.4 38	9.5 61	8.5 50	2.8 17		
2	14	8	5	7	6	12	16	13	81	–
	6.6 35	3.8 50	2.4 56	3.3 25	2.8 46	5.7 36	7.6 44	6.2 36		
3	12	3	0	11	2	1	2	9	40	–
	5.7 30	1.4 19		5.2 39	0.9 15	0.5 3	0.9 6	4.3 25		
4	8	1	0	5	0	0	0	6	20	–
	3.8 20	0.5 6		2.4 18				2.8 17		
5	2	0	2	0	0	0	0	2	6	ethyl chain
	0.9 5		0.9 22					0.9 6		
Total	40	16	9	28	13	33	36	36	211	

Table 8. Distribution of values taken by attribute 7 of $\mathcal{Y}_2$

No.of class	Valuesof attribute 7: 0	2	3	4	5	6	7	8	9	10	11	12	14	16	Lower approx.	Characteristic value of attribute 7
1	2	0	0	7	3	11	2	14	1	11	1	11	1	0	64	benzyl and oktyl residue
	0.9 50			3.3 26	1.4 27	5.2 39	0.9 40	6.6 52	0.5 20	5.2 39	0.5 33	5.2 38	0.5 6			
2	2	2	2	5	3	10	2	9	1	13	1	14	11	6	81	benzyl, decyl dodecyl and tetradecyl residue
	0.9 50	0.9 40	0.9 33	2.4 19	1 4 27	4.7 36	0.9 40	4.3 33	0.5 20	6.2 46	0.5 33	6.6 48	5.2 65	2.8 38		
3	0	0	1	10	1	6	1	3	3	2	0	2	4	7	40	nonyl residue
			0.5 17	4.3 37	0.5 9	2.8 21	0.5 20	1.4 11	1.4 60	0.9 7		0.9 7	1.9 24	3.3 44		
4	0	2	2	5	3	0	0	1	0	2	1	2	1	1	20	–
		0.9 40	0.9 33	2.4 19	1 4 27			0.5 4		0.9 7	0.5 33	0.9 7	0.5 6	0.5 6		
5	0	1	1	0	1	1	0	0	0	0	0	0	0	2	6	ethyl and propyl residue
		0.5 20	0.5 17		0 5 9	5.0 4								0.9 12		
Total	4	5	6	27	11	28	5	27	5	28	3	29	17	16	211	

Table 9. Distribution of values taken by attribute 4 of $\mathcal{Y}_2$

No. of class	Values of attribute 4						Lower approx.	Characteristic value of attribute 4
	0		1		2			
1	35		27		2		64	without substituent
	16.6	34	12.8	27	0.9	25		
2	41		35		5		81	alkyl residue
	19.4	40	16.6	35	2.4	62		
3	18		22		0		40	–
	8.5	17	10.4	22				
4	7		12		1		20	–
	3.3	7	5.2	12	0.5	12		
5	2		4		0		6	phenyl residue
	0.9	2	1.9	4				
Total	103		100		8		211	

Table 10. Distribution of values taken by attribute 6 of $\mathcal{Y}_2$

No. of class	Values of attribute 6				Lower approx.	Characteristic value of attribute 6
	0		1			
1	28		36		64	sulphur atom
	13.3	23	17.1	41		
2	43		38		81	–
	20.4	35	18	44		
3	31		9		40	oxygen atom
	14.7	25	4.3	10		
4	18		2		20	oxygen atom
	8.5	14	0.9	2		
5	4		2		6	–
	1.9	3	0.9	2		
Total	124		87		211	

a_{ij} – number of objects in class Y_i which take the attribute value equal to "j";
$b_{ij}[\%]$ – the ratio of a_{ij} to the total number of properly classified objects;
$c_{ij}[\%]$ – the ratio of a_{ij} to number of objects properly classified which take the attribute value equal to "j".

As it comes out from the presented tables, the characteristic value of the respective attributes can be obtained only in the case of some extreme classes (best of all Y_1 and worst of all Y_5). In the case of intermediate classes it is difficult to differentiate the characteristic values of attributes and therefore they were not presented in the tables. The distribution of values of those attributes in various classes of classification $\mathcal{Y}_1$ is shown in Tables 3-6. As a result of distribution of the attributes values we obtained the best (class Y_1) and the worst (class Y_5) imidazolium compounds for the classifications $\mathcal{Y}_1$ and $\mathcal{Y}_2$.

The typical compounds belonging to class Y_1 of classification $\mathcal{Y}_1$ *(Pseudomonas aeruginosa NCTC 6749)*:

- have a decyl or dodecyl substituent in the first imidazole position,
- have no substituent in the second imidazole position,
- have a substituent with a sulphur atom in the third imidazole position

The typical compounds belonging to class Y_5 of $\mathcal{Y}_1$:

- have no substituent or have a methyl or butyl residue in the first imidazole position,
- have a phenyl or alkyl residue in the second imidazole position,
- have a substituent composed of an oxygen atom and an ethyl, propyl, butyl, nonyl, undecyl and hexadecyl residue in the third imidazole position.

However in Tables 7-10 were presented the distribution of the values of respective attributes in given classes for classification $\mathcal{Y}_2$.

The typical compounds belonging to class Y_1 of classification $\mathcal{Y}_2$ *(Escherichia coli NCTC 8196)*:

- have an octyl or decyl residue in the first imidazole position,
- have no substituent in second imidazole position,
- have a substituent composed of a sulphur atom and an oktyl and benzyl residue in the third imidazole position.

The typical compounds belonging to class Y_5 of $\mathcal{Y}_2$:

- have an ethyl residue in the first imidazole position,
- have a phenyl residue in the second imidazole position,
- have an ethyl and propyl chain in the third imidazole position.

2.4. Decision algorithm

In the classification case $\mathcal{Y}_1$ – 168 decision rules were obtained among them 13 non-deterministic ones. In the $\mathcal{Y}_2$ classification, 162 decision rules were obtained among them 14 non–deterministic ones. From the practical point of view, the deterministic rules which lead to the best and the worst compounds are the most interesting ones. Therefore in tables 11 and 12 the decision algorithm for the class Y_1 and Y_5 of imidazolium chlorides are shown, respectively. The absence of values in given rules means that they can be arbitrary.

The obtained algorithms are an objective picture of an experiment connected with synthesis and antimicrobial activity of imidazolium compounds. To the most active compounds in the classification $\mathcal{Y}_1$ (class Y_1) belong chlorides with decyl in the 1-position, without substituent in the 2-position and with sulphur atom in the substituent in the 3-position and such compounds which in the 1-position have a dodecyl chain, in the 2-position have no substituent and in the 3-position have sulphur atom together with butyl, hexyl or oktyl chain.

To the worst compounds (class Y_1) of $\mathcal{Y}_1$ classification belong chlorides which in the 1-position of imidazole have no substituent or have a short (methyl, ethyl, butyl) or long (decyl, dodecyl) alkyl chain; and in the 3-position have the oxygen atom with a short (ethyl, propyl, butyl, pentyl, hexyl) or a long (undecyl, dodecyl, tetradecyl, hexadecyl) alkyl chain and in the 2-position have phenyl residue.

To the most active compounds (class Y_1) of $\mathcal{Y}_2$ classification belong prevailingly chlorides with oktyl, decyl and dodecyl chain in the 1-position of imidazole, without substituent in the 2-position, and in the 3-position with the sulphur atom and a oktyl, decyl and dodecyl chain.

To the worst compounds (class Y_5) of $\mathcal{Y}_2$ classification belong chlorides which in the 1-position have no substituent or have ethyl chain, in the 2-position have no substituent or have phenyl residue and in the 3-position have oxygen atom and ethyl, propyl, pentyl, hexyl or hexadecyl chain.

The obtained decision algorithms are an objective picture of the experiment connected with the synthesis and antimicrobial activity of imidazolium chlorides. It represents, at the same time, the possibilities of real relationships between the chemical structure and biological activity of imidazolium chlorides. From the relationships results that the greatest influence on the antimicrobial activity of imidazolium chlorides has the length of alkyl chain in the 1-position and a kind of substituent in the 3-position of imidazole.

Table 11. Decision algorithm for classification $\mathcal{Y}_1$ of imidazolium chlorides

Attribute				Class	Attribute				Class
2	4	6	7		2	4	6	7	
10	0	1		1	0	1		14	5
12	0	1	4	1	0	1		5	5
12	0	1	6	1	0	1		7	5
12	0	1	8	1	0	1		9	5
0			16	5	0	0		4	5
4			2	5	10	1		14	5
4			4	5	10	1		16	5
4			14	5	12	1		12	5
4			16	5	12	1		14	5
0			3	5	0	1	0	4	5
4			3	5	0	1	0	6	5
4			5	5	0	1	0	8	5
4			7	5	0	1	0	10	5
4			9	5	0	1	0	12	5
2			5	5	4	1	0	6	5
2			6	5	4	1	0	8	5
1			4	5	4	1	0	10	5
6			4	5	4	1	0	12	5
12			16	5	1	0	0	6	5
10			11	5	6	0	0	6	5
12			2	5	1	2	0	6	5
12			3	5	10	1	0	12	5
12			9	5	12	1	0	4	5
12			11	5	12	1	0	8	5
0	1		2	5	12	1	0	10	5

Table 12. Decision algorithm for classification $\mathcal{Y}_2$ of imidazolium chlorides

Attribute				Class	Attribute				Class
2	4	6	7		2	4	6	7	
8			6	1	8	1	1	4	1
8			8	1	8	1	1	12	1
8			10	1	1	2	1	8	1
8			7	1	4	0	1	8	1
8			9	1	10	0	1	10	1
8			11	1	12	0	1	6	1
2			12	1	4	0	0	10	1
6			10	1	8	0	0	12	1
6			12	1	8	0	0	14	1
10			4	1	10	0	0	8	1
10			6	1	10	0	0	10	1
10			5	1	12	0	0	4	1
10			0	1	12	0	0	6	1
10			7	1	12	0	0	8	1
0		1	8	1	12	0	0	10	1
4		1	10	1	12	1	1	4	1
4		1	12	1	2			5	5
8		1	5	1	0	1		2	5
1		1	12	1	0	1		3	5
6		1	8	1	12	1		16	5
10		1	8	1	2	0	0	6	5
0	1	1	6	1					

3. Quaternary Quinolinium and Isoquinolinium Chlorides

The analysis of relationships between structure and activity of quaternary quinolinium and isoquinolinium chlorides was led up in the same way as the analysis of quaternary imidazolium chlorides.

3.1. Information system

The information system was built on the basis of the set of 72 objects (quinolinium and isoquinolinium chlorides) described by 11 condition attributes (substituent in quinoline and isoquinoline) and 1 decision attributes.

In the table 13 was presented the domain of condition attributes. Likewise as in the analysis of imidazolium chlorides 2 classifications were driven for 2 bacteria for *Pseudomonas aeruginosa NCTC 6749* (classification $\mathcal{Y}_1$) and for *Escherichia coli NCTC 8196* (classification $\mathcal{Y}_2$).

The quinolinium and isoquinolinium chlorides were divided into 3 antimicrobial activity classes (Y_1 , Y_2 and Y_3). The following ranges of the minimal inhibitory concentraton (MIC) correspond to the classes:
for the classificaton $\mathcal{Y}_1$:

Class Y_1 – very good – $MIC \leq 200\text{mg/dm}^3$
Class Y_2 – middle – $200 < MIC < 500\text{mg/dm}^3$
Class Y_3 – weak – $MIC \geq 500\text{mg/dm}^3$.

For the classification $\mathcal{Y}_2$:

Class Y_1 – very good – $MIC \leq 50\text{mg/dm}^3$
Class Y_2 – middle – $50 < MIC < 300\text{mg/dm}^3$
Class Y_3 – weak – $MIC \geq 300\text{mg/dm}^3$.

3.2. Reduction of attributes and the quality of classification

From a set of 11 condition attributes were obtained in both classifications per one minimal set reduced by attributes 1 and 6. Then a reduction of succeeding attributes was led from the minimal set, observing the classification quality and the accuracy of classes. After the reduction of condition attributes it became clear that for a high quality of classification $\mathcal{Y}_1$ and accuracy of classes are important 5 following attributes:

attribute 2 – kind of X in substituent R^1
attribute 3 – length of alkyl in R^1
attribute 4 – kind of substituent R^2
attribute 7 – length of alkyl in R^2
attribute 10 – kind of substituent R^5.

However for a high quality of classification $\mathcal{Y}_2$ and accuracy of classes are important 6 following attributes:

Table 13. Domains of attributes of quinolinium and isoquinolinium chlorides

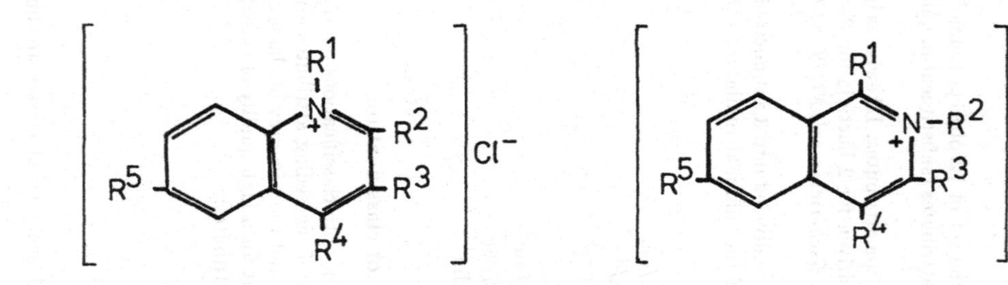

No.	Attribute	Values (code number)											
	Kind of	0	1	2	4	5	6	7	8	9	10	11	12
1	substituent R^1	without	group with X*	–	–	–	–	–	–	–	–	–	–
2	X in R^1	without	sulphur	oxygen	–	–	–	–	–	–	–	–	–
3	alkyl in R^1	without	–	–	butyl	pentyl	hexyl	heptyl	oktyl	nonyl	decyl	undecyl	dodecyl
4	substituent R^2	without	methyl	group with X*	–	–	–	–	–	–	–	–	–
5	X in R^2	without	sulphur	oxygen	–	–	–	–	–	–	–	–	–
6	residue in R^2	without	alkyl	benzyl	–	–	–	–	–	–	–	–	–
7	alkyl in R^2	without	–	–	butyl	pentyl	hexyl	heptyl	oktyl	nonyl	decyl	undecyl	dodecyl
8	substituent R^3	without	methyl	bromide	–	–	–	–	–	–	–	–	–
9	substituent R^4	without	methyl	hydroxy	–	–	–	–	–	–	–	–	–
10	substituent R^5	without	methyl	metoxy	–	–	–	–	–	–	–	–	–
11	substituent R^6	without	hydroxy	–	–	–	–	–	–	–	–	–	–

*Group with X: CH_2XAlkyl

attribute 2 – kind of X in substituent R^1
attribute 3 – length of alkyl in R^1
attribute 4 – kind of substituent R^2
attribute 7 – length of alkyl in R^2
attribute 8 – kind of substituent R^3
attribute 10 – kind of substituent R^5.

3.3. Decision algorithm

We presented in tables 14 and 15 the decision rules obtained after the removal of unimportant attributes for the classifications quality and accuracy of classes. In classification $\mathcal{Y}_1$ 32 decision rules were obtained, among them 8 non–deterministic ones, and in classification $\mathcal{Y}_2$ – 49 decision rules among them 10 non–deterministic. From the practical point of view, as it was mentioned before, we are interested in deterministic rules leading to the best and the worst chemical compounds.

In the classification $\mathcal{Y}_1$ the most active (class Y_1) are the isoquinolinium chlorides which in the 2-position have a nonyl, decyl and undecyl chain. From all quinolinium chlorides the most activ are those with methyl or methoxy group in the 6-position and with octyl, decyl and dodecyl chain in the 1-position and these with the hexyl-, octyl- and dodecylthiomethyl substituent in the 1-position. The worst activity (class Y_3) have the isoquinolinium chlorides with the substituent of butyl, pentyl, hexyl and benzyl in the 2-position. However, from all quinolinium chlorides which belong to this class we perceive all compounds with methyl chain in the 2-position and these which in the 1-position have butyl and pentyl chain or hexylthiomethyl and hexyloxymethyl substituent as well.

In the clssification $\mathcal{Y}_2$ to the best class Y_1 of isoquinolinium compounds belong chlorides with heptyl, nonyl, decyl and undecyl chain in the 2-postion and these which have in the 2-position heptyl, octyl and dodecyl chain together with the methyl group in the 3-position. From among the quinolinium chlorides belong to the Y_1 class these with the methyl and methyloxy group in the 6-position or those with bromide atom in the 3-position and the lengh of which chain in the 1-position is from hexyl to dodecyl. To the Y_3 class belong the isoquinolinium compounds with the butyl chain in the 2-position. However from the quinolinium compounds to this class belong all chlorides with methyl group in the 2-position and with oxygen atom in the substituent in the 1-position. In this class are also the quinolinium chlorides with sulphur atom which have the alkyl chain from hexyl to dodecyl in the 1-position.

The obtained decision algorithms represent important relationships between the chemical structure and antimicrobial activity of quaternary quinolinium and isoquinolinium chlorides. These relationships show that alkyl chain length in the substituent in the 1-position of quinoline and in the 2-position of the isoquinoline as well as the substituent in the 6-position of quinoline have the highest influence upon the antimicrobial activity of the quaternary quinolinium and isoquinolinium chlorides.

From the point of view of antimicrobial activity the proposed attributes make sure a good compounds classification.

We should pay attention to the fact that the descriptors number which appear in the decision algorithm of imidazolium chlorides is only 25% (classification $\mathcal{Y}_1$) or 27% (classification $\mathcal{Y}_2$) of the descriptors number which appeared in the initial information system. However in the decision algorithm of quinolinium and isoquinolinium chloides the descrip-

Table 14. Decision algorithm for classification $\mathcal{Y}_1$ of quinolinium and isoquinolinium chlorides

Attribute 2	Attribute 3	Attribute 4	Attribute 7	Attribute 10	Class
		2	9		1
		2	10		1
		2	11		1
	10	0	0	1	1
	8	0	0	2	1
	10	0	0	2	1
	12	0	0	2	1
1	6	0	0	1	1
1	8	0	0	1	1
1	12	0	0	1	1
0	0	2	12	0	1 or 3
1	8	0	0	0	1 or 3
1	12	0	0	0	1 or 3
2	10	0	0	0	1 or 3
		2	8		2
		2	7		2
2	8	0	0	1	2
2	12	0	0	1	2
1	6	0	0	2	2
2	6	0	0	0	2 or 3
2	8	0	0	0	2 or 3
2	12	0	0	0	2 or 3
1	10	0	0	0	2 or 3
		1			3
		2	4		3
		2	5		3
		2	6		3
		2	0		3
	4	0	0		3
	5	0	0		3
1	6	0	0	0	3
2	6	0	0	1	3

Table 15. Decision algorithm for classification $\mathcal{Y}_2$ of quinolinium and isoquinolinium chlorides

Attribute 2	Attribute 3	Attribute 4	Attribute 7	Attribute 8	Attribute 10	Class
		2	7			1
		2	9			1
		2	10			1
		2	11			1
	6	0	0		1	1
	8	0	0		1	1
	12	0	0		1	1
	10	0	0		1	1
	6	0	0		2	1
	8	0	0		2	1
	10	0	0		2	1
	12	0	0		2	1
	0	2	6	1	0	1
	0	2	8	1	0	1
	0	2	12	1	0	1
	6	0	0	2	0	1
	10	0	0	2	0	1
0	0	2	12	0	0	1 or 2
1	8	0	0	0	0	1 or 3
1	12	0	0	0	0	1 or 3
2	6	0	0	0	0	1 or 2
2	8	0	0	0	0	1 or 2
2	10	0	0	0	0	1 or 2
2	12	0	0	0	0	1 or 2
			8	0		2
			4	1		2
	5		0	0		2
	0		0	0		2
	4		0	2		2
	8		0	2		2
	4		0	0	1	2
	4		0	0	2	2
1	10		0	0	1	2
1	4	1	0	0	0	2
1	6	1	0	0	1	2
1	8	1	0	0	1	2
1	12	1	0	0	1	2
0	0	2	5	0	0	2 or 3
0	0	2	6	0	0	2 or 3
1	4	0	0	0	0	2 or 3
			4	0		3
2		1	0	0	0	3
2		1	0	0	1	3
1	6	0	0	0	0	3
1	6	1	0	0	0	3
1	8	1	0	0	0	3
1	10	1	0	0	0	3
1	12	1	0	0	0	3
1	10	0	0	0	0	3

tors number makes adequately 15% (classification $\mathcal{Y}_1$) or 29% (classification $\mathcal{Y}_2$) descriptors numbers which arise in the initial information system.

The method based on the rough sets theory is characteristic because it does not correct non-deterministic relationships but presents the real conditions of these relationships. Moreover in comparison with the initial information system the obtained decision algorithms are deprived of all redundand, unimportant informations which put shade on the picture of these relationships.

The obtained decision algorithms give distinct indications which may be useful in the synthesis of new activ antimicrobial quaternary imidazolium, quinolinium and isoquinolinium chlorides.

The method based on the theory of rough sets may be also used for the analysis of relationship between chemical structure and biological activity of other chemical compounds. Comparing this to the statistic methods used in $QSARs$, it seems to be a simpler and more accessible method.

References

[1] Franke, R. and Oehme, P. (1973) "Aktuelle Probleme der Wirkstoffforschung, 1.Mitt.: Ermittlung quantitativer Struktur–Wirkung–Beziehungen bei Biowirkstoffen; theoretische Grundlagen, Durchführung, Voraussetzungen und gegenwärtiger Stand", Pharmazie **28**, 489–508.

[2] Franke, R. (1980) Optimierungsmethoden in der Wirkstoffforschung, Akademie Verlag, Berlin.

[3] Franke, R., Barth, A., Dove, S. and Laass, W. (1980) "Qantitative structure–activity relationship in piperidinoacetanilides for different biological effects", Pharmazie **35**, 181–182.

[4] Free, S.M. and Wilson, J.W. (1964) "A mathematical contribution to structure–activity studies", J. Med. Chem. 7, 395–399.

[5] Gäbler, E., Franke, R. and Oehme P. (1976) "Aktuelle Probleme der Wirkstoffforschung, 9.Mitt.:Parameterfreie Verfahren für die Struktur–Wirkungs–Analyse", Pharmazie **31**, 1–14.

[6] Hansch, C. and Lien, E. (1971) "Structure–activity relationship in antifungal agents", J. Med. Chem. **14**, 653–670.

[7] Hansch, C. and Dunn, W.J. (1972) "Linear relationship between lipophilic charcter and biological activity of drugs", J. Pharm. Sci. **61**, 1–19.

[8] Hansch, C., Nakamoto, K., Gorin, M., Denisevich, P., Garrett, E.R., Heman–Ackah, S.M. and Won, C.H. (1973) "Structure–activity relationship of chloramphenicols", J. Med. Chem. **16**, 917–922.

[9] Krysiński, J. (1990) "Rough sets approach to the analysis of the structure–activity relationship of quaternary imidazolium compounds", Arzneim.–Forsch. **40**, 795–799.

[10] Krysiński, J. (1991) "Grob–Mengen–Theorie in der Analyse der Struktur–Wirkung Beziehungen von quartären Pyridiniumverbindungen", Pharmazie **46**, 878–881.

[11] Krysiński, J. (1991) "Grob–Mengen–Theorie in der Analyse der Struktur–Wirkung Beziehungen von quartarer Chinolinium– und Isochinoliniumverbindungen", Arch. Pharm. (Weinheim) **324**, 827–832.

[12] Słowiński, K., Słowiński, R. and Stefanowski, J. (1988) "Rough sets approach to analysis of data from peritoneal lavage in acute pancreatitis", Med.Inform. **13**, 143–159.

[13] Weuffen, W., Kramer, A., Groschel, D., Berensci, G. and Bulka, E. (1981) Handbuch der Antiseptik, B.I, T.2, VEB Velag Volk und Gesundheit, Berlin.

Part I
Chapter 10

ROUGH SETS-BASED STUDY OF VOTER PREFERENCE IN 1988 U.S.A. PRESIDENTIAL ELECTION

Michael HADJIMICHAEL
Department of Computer Science
University of Regina
Regina SK, S4S 0A2, Canada

Anita WASILEWSKA
Department of Computer Science
State University of New York
Stony Brook, NY, 11794, U.S.A.

Abstract. We present here an interactive probabilistic inductive learning system and its application to a set of real data. The data consists of a survey of voter preferences taken during the 1988 presidential election in the U.S.A. Results include an analysis of the predictive accuracy of the generated rules, and an analysis of the semantic content of the rules.

1. Introduction

We present here an interactive probabilistic inductive learning system CPLA/CSA that has been implemented by the first author and discuss its application to a set of real data.

To build our system we extend the rough set-based learning algorithms described by Wong & Wong in [9]; Wong & Ziarko in [10]; Pawlak, Wong, & Ziarko in [6]; and Wong, Ziarko & Ye in [11] to a conditional model as described by Wasilewska in [8] and Hadjimichael & Wasilewska in [2], and formulate our Conditional Probabilistic Learning Algorithm (CPLA), applying conditions to a probabilistic version of the work [9] by Wong & Wong. We propose the Condition Suggestion Algorithm (CSA) as a way to use the syntactic knowledge in the system to generalize the familly of decision rules.

Our system allows for semantic knowledge to be deduced from the database in ways not previously explored by other systems. It is distinctive in that it includes the conditions feature which allows user control over sets of attribute values, and thus allows a greater flexibility of analysis. More specifically, conditions specify equivalences on sets of attribute values, such that objects may become indistinguishable.

Moreover, the system is distinctive in that it defines a feedback relationship between the user and the learning program, unlike other inductive learning systems which simply input data and output a decision tree. This interactiveness allows for rule (tree) compaction and generalization.

Finally, our system is also distinctive in that it yields a family of production rules using easily comprehended descriptions formed of attribute-value pairs. Compared to the decision trees used by many other inductive learning systems, this format is more easily understood and manipulated by humans.

The *Conditional Probabilistic Learning Algorithm* (CPLA), which form a part of our system is founded on the model of Wong and Ziarko's INFER ([10]). It uses the probabilistic information inherent in a database and generates a family of probabilistic decision rules based on a minimized set of object attributes. We generalize further on Wong and Ziarko's model by adding *conditions* to the system. Conditions are a form of user input which make the system interactive and can reduce the size of the rule family. Also, the decision rules are input to the *Condition Suggestion Algorithm* (CSA), which generates suggested conditions. Conditions, applied to a decision rules, generalize them, resulting in a smaller and more concisely described family of rules. The entire system is graphically described below:

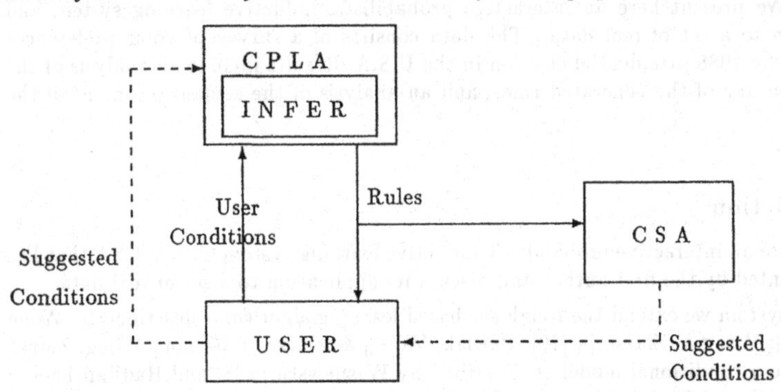

The diagram above shows how the three elements of the system, CPLA, CSA, and the user, form a cycle in which the user moderates the feedback from CSA. The cycle begins by taking any (possibly empty) set of user-supplied conditions and running CPLA on the examples database with those conditions. CPLA, by definition, will remove the statistical functional dependencies. It will then output a family of decision rules which will then be passed to CSA, where suggested conditions will be generated. The user can examine the suggested conditions and select a subset of them to feed back into CPLA for another pass, beginning the cycle again. The suggested conditions may affect the rule family in two ways (in addition to reducing the number of rules generated). They may introduce superfluous attributes, and they may change decision rule certainties. The cycle continues until the *user* is satisfied with the final generation of decision rules.

The model and system we use, is described in detail in Hadjimichael & Wasilewska in [3].

This paper presents the algorithms which form the basis of the system and describes its application as a learning system to real data. The results discussed are based on data from a survey of voters in the 1988 U.S. presidential election.

We observe that the conditions feature was able to yield a roughly 47% decrease in the number of rules, and up to 88% predictive accuracy.

2. Formal Basis

The goal of inductive learning is to automatically infer the decision rules from specific examples given by an expert. The automated inference of decision rules is the subject of the works of many authors (among which are Michalski [4], Quinlan [7]). Descendants of these systems improved their decision tree output, but still maintain the simple data-in/tree-out format.

Wong & Wong presented in [9] an inductive learning algorithm, ILS, and compared it favorably to those systems. In particular, they showed this to be an improvement over the earlier systems of Michalski (AQ11) (1978) and Quinlan (ID3) (1983), because it allowed for shorter descriptions (and thus shorter decision trees) based on a smaller set of attributes. Furthermore, the output was in the more comprehensible *production rule* format. ILS improved on the above-mentioned methods by using Rough Set theory (Pawlak, [5]), as did INFER (Wong & Ziarko, [10]). INFER also took advantage of the probabilistic information inherent in databases to generate probabilistic decision rules.

The probabilistic approach allows us to retrieve some of the information discarded by the deterministic approach by attaching a degree of certainty to probabilistic rules which would have not existed in the earlier, deterministic case (see also [10], [9], [6], and [12]).

The model we propose here expands on ILS and INFER of Wong & Wong [9] and Wong & Ziarko [10], respectively, by generalizing and adding *conditions* to the system. Conditions introduce into the model a new form of generalization, and a channel for user feedback, allowing for a more powerful analysis of the dataset.

Given a knowledge representation system $K = (OBJ, AT, E, VAL, f)$ as in the works above.

We formally define conditions as family of equivalence relations $\{cond_a\}_{a \in AT}$ defined in the set VAL_a. I.e., for each $a \in AT$, $cond_a \subseteq VAL_a \times VAL_a$, and $cond_a$ is an equivalence relation. By $cond_a(v)$ we mean $\{v' : (v, v') \in cond_a\}$.

Notational Remark: We will list only pairs which define the conditions explicitly. We will not list pairs which assure the reflexive, symmetric, and transitive properties.

We also generalize the notion of indiscernibility, by adding to it the notion of conditions, and we define, for any $A \subseteq AT$, and any family of conditions $\{cond_a\}_{a \in A}$, a family of binary relations $R(A)$ on OBJ as follows:

$$o_1 R(A) o_2 \text{ iff } (\forall a \in A)((f(o_1, a), f(o_2, a)) \in cond_a).$$

We will call the *identity conditions* the set of conditions $cond_a = \{(v, v) : v \in VAL_a\}$.

Notational Remark: Given $A \subseteq AT$, if we define only $cond_a$ for a certain attribute $a \in A$, then we mean that $cond_b$ for all $b \neq a, b \in A$, are identity conditions.

We incorporate probability into our system by extending the models of the previously mentioned works. Let A be any subset of AT. Let $R(A)^* = \{A_1, A_2, \ldots, A_n\}$ denote the partition induced by $R(A)$ on OBJ, where A_i is an equivalence class of $R(A)$. Let $R(E)^* = \{E_1, E_2, \ldots, E_m\}$ (where E is the set of expert attributes) denote the partition induced by $R(E)$ on OBJ, so that each element of $R(E)^*$ corresponds to one of the expert-defined concepts. Given a relation $R(A)$, and the partitions $R(A)^* = \{A_1, \ldots, A_n\}$ and $R(E)^* = \{E_1, \ldots, E_m\}$, we let P denote the *conditional probability*, $P(E_j|A_i) = \frac{P(E_j \cap A_i)}{P(A_i)}$, where $P(E_j|A_i)$ denotes the probability of occurrence of event E_j conditioned on event A_i.

Given a set of equivalent objects, $A_i \in R(A)^*, A \subseteq AT$, we may describe those objects in the following standard way:

$$\bigwedge\nolimits_{a \in A} (a, v),$$

where (a, v) are pairs such that $f(o, a) = v$, for $o \in A_i$.

If we also have a family of conditions $\{cond_a\}_{a \in A}$, we define the *conditioned description* (from now on referred to simply as the *description*) of any equivalence class $A_i \in R(A)^*$:

$$\bigwedge_{a \in A} (a, cond_a(f(o, a))).$$

We will use $des(A_i)$ as a shorthand notation for the description. When it is not obvious from the context, we will use $des_A(A_i)$ to indicate the attributes from which the description is formed.

Notational Remark: Note that in a description, if the only condition on the value of an attribute is that it is equivalent to itself, then in the description we write the pair (*attribute*, *value*) rather than (*attribute*, $\{$*value*$\}$).

We add to a knowledge representation system K conditions and conditional probabilities, the result is a conditional probabilistic knowledge representation system, *CPK*. I. e. we formally define CPK,

$$CPK = (K, \mathcal{A}_P),$$

where K is the deterministic knowledge representation system and $\mathcal{A}_P$ is the conditional probabilistic approximation space,

$$\mathcal{A}_P = (\{(VAL, \{cond_a\}_{a \in A}, R(A))\}_{A \subseteq AT}, P),$$

respectively.

2.1. Learning

Inductive learning consists of generalization from a set of examples. Therefore, when given an example object, we generalize its description by removing superfluous attributes from the description. This yields a minimal-length description, i.e. a description based on just enough attributes to distinguish the objects in the set from those outside it. The minimal length description will be the most general, as it involves the fewest attributes, and thus the fewest constraints.

By the nature of CPLA, all superfluous attributes are removed, so that the resulting set of rules contain the minimal number of attributes required to distinguish the objects in the concept from those outside it (Hadjimichael & Wasilewska, [3]). The result of the learning process is a set of rules which together describe the concepts learned, in terms of generalized object descriptions, as defined above. Each rule is a mapping from a description to a concept. The collection of rules which all map to the same concept can be considered to compose the description of the concept. The goal of CPLA is to generate such rules, as general as possible, and thus, as few as possible, with the shortest possible descriptions.

2.2. The Algorithms

The Conditional Probabilistic Learning Algorithm traces its roots to the papers of Wong & Ziarko ([10]), Pawlak, Wong, & Ziarko ([6]), and Wong, Ziarko, & Ye ([11]). From Wong, Ziarko, & Ye ([11]) it inherits the procedural structure of the algorithm. As an extension of Wong and Ziarko's INFER algorithm ([9]), it maintains the property that the output will have no superfluous attributes. It takes the most from Pawlak, Wong and Ziarko's 1988 paper [6], however, as it utilizes the entropy function suggested in their paper to calculate attribute dependencies, and it uses the probabilistic rules proposed in their paper, i.e. we define, after [6], the family of decision rules $\{r_{i,j}\}$ as:

1. $des(A_i) \stackrel{c}{\Longrightarrow} des(E_j)$ if $P(E_j|A_i) > 0.5$
2. $des(A_i) \stackrel{c}{\Longrightarrow} NOTdes(E_j)$ if $P(E_j|A_i) < 0.5$
3. $des(A_i) \stackrel{c}{\Longrightarrow} unknown(E_j)$ if $P(E_j|A_i) = 0.5$

where, the *certainty* of a rule is defined as:

$$c = \max(P(E_j|A_i), 1 - P(E_j|A_i))$$

The algorithm is: (as presented in detail in (Hadjimichael & Wasilewska, [3]))

- **Input a Conditional Probabilistic Knowledge Representation System, $(K, \mathcal{A}_\mathcal{P})$, where** $K = (OBJ, AT, E, VAL, f)$ **and** $\mathcal{A}_\mathcal{P}$ **is the conditional probabilistic approximation space.**

- **Let** $OBJ' = OBJ$, $A = \phi$, $B = AT$.

- **Repeat until** $OBJ' = \phi$ **or** $B = \phi$.

  Loop: **Find** $a \in B$ **such that** $H(R(E)^*|R(A \cup \{a\})^*)$ **is** ***minimum*** **for** OBJ'.

  If $A \cup \{a\}$ **is statistically dependent on** A, **then let** $B \leftarrow B - \{a\}$, **goto** ***Loop***.

  Let $B \leftarrow B - \{a\}$.

  Let $A \leftarrow A \cup \{a\}$.

  For each $E_j \in R(E)^*$ **from** OBJ'

  For each A_i **such that** $P(E_j|A_i) = 1.0$

  Output $des(A_i) \stackrel{1.0}{\Longrightarrow} des(E_j)$

  Let $OBJ' \leftarrow OBJ' - A_i$

- **If** $OBJ' \neq \phi$

  For each $E_j \in R(E)^*$

  For each A_i **such that** $A_i \cap E_j \neq \phi$

  Calculate $p_{i,j} = P(E_j|A_i)$.

  Calculate $c = max(p_{i,j}, 1 - p_{i,j})$.

  If $p_{i,j} > 1 - p_{i,j}$

  $\longrightarrow$ **Output** $des(A_i) \stackrel{c}{\Longrightarrow} des(E_j)$

  If $p_{i,j} < 1 - p_{i,j}$

  $\longrightarrow$ **Output** $des(A_i) \stackrel{c}{\Longrightarrow} NOTdes(E_j)$

  If $p_{i,j} = 1 - p_{i,j}$

  $\longrightarrow$ **Output** $des(A_i) \stackrel{0.5}{\Longrightarrow} unknown(E_j)$.

- **End.**

Often, a system may contain attributes whose values are irrelevant in determining the expert classification of the objects. These attributes are not conditionally statistically superfluous, but the information they supply has no effect on the expert's global decision. The problem is one of determining which attributes contribute nothing to the expert's classification. There are two possible indicators. (1) Two widely separated values (assuming ordered values) of an attribute can be unified through conditions without decreasing the accuracy of the system. (2) *All* values of an attribute may be unified without loss of accuracy. These indicators can lead us to conclude that the attribute to which these values belong must not be significant to the expert's classification, since differences in the attribute's values do not play a part in the final classification.

This kind of syntactic information is extracted by the Condition Suggestion Algorithm. The main idea behind Condition Suggestion is the collapsing of *similar* rules into one rule. Or, equivalently, the generalization from a set of specific rules to a more general rule.

To proceed with our definition of similarity, we introduce a further notational shorthand:

$$\begin{aligned} DES_1(E_j) &= des(E_j) \\ DES_2(E_j) &= NOTdes(E_j) \\ DES_3(E_j) &= unknown(E_j). \end{aligned}$$

So all rules may now be denoted by:

$$des(A_i) \stackrel{c}{\Longrightarrow} DES_k(E_j).$$

The principle upon which the Condition Suggestion Algorithm (CSA) is based is the idea of ***rule similarity*** as appears in (Hadjimichael, 1989). Specifically, given a conditional probabilistic knowledge representation system, and a set of of attributes, $A \subseteq AT$, let $r_{i,j}$ and $r_{p,q}$ be some two rules from the family of rules $\{r_{i,j}\}$:

$$r_{i,j} : des(A_i) \stackrel{c_1}{\Longrightarrow} DES_k(E_j)$$

$$r_{p,q} : des(A_p) \stackrel{c_2}{\Longrightarrow} DES_r(E_q)$$

(where $A_i, A_p \in R(A)^*$, $E_j, E_q \in R(E)^*$, $1 \leq k, r \leq 3$). We define a similarity relation, **sim**, as follows:

$$r_{i,j} \ \textbf{sim} \ r_{p,q} \ \text{ iff } \ c_1 = c_2, \ j = q, \ k = r.$$

However, to avoid over-generalizing, we choose as Suggested Conditions, conditions based on **sim**$'$. For: $r_{i,j}, r_{p,q} \in \{r_{i,j}\}$,

$$r_{i,j} : des_B(A_i) \stackrel{c_1}{\Longrightarrow} DES_k(E_j)$$

$$r_{p,q} : des_C(A_p) \stackrel{c_2}{\Longrightarrow} DES_r(E_q)$$

$$r_{i,j} \ \textbf{sim}' \ r_{p,q} \ \text{ iff } \ c_1 = c_2, j = q, k = r, B = C.$$

We use the **sim**$'$ relation to determine which rules we would like to merge together. Conditions are suggested which would aid in such a merging. The procedure is described in [3].

The alhorithm is:

- **Input** A family of rules, $\{r_{i,j}\}$.

- Generate the suggested new conditions, $COND'$

  let $COND' = \phi$.

  for each rule, $r_{i,j} : des_A(A_i) \stackrel{c}{\Longrightarrow} DES_k(E_j) \in \{r_{i,j}\}$

  Search $\{r_{i,j}\}$ for all similar rules $r'_{l,j} : des_{A'}(A_l) \stackrel{c'}{\Longrightarrow} DES_k(E_j)$, A = A'. Let $D = \{A_l\}$, the set of domains of those rules.

  Update $COND'$ with the conditions:

$$cond_a = \{(v, v') : (\forall o \in D)\ (\exists o' \in D)(v = f(o, a)\ \wedge\ v' = f(o', a))\}.$$

  output $COND'$.

- **End.**

3. Application

We have applied CPLA to a voter survey data from the 1988 U.S.A. presidential election[1]. Our analysis was concerned with:

1. the semantic analysis of rules and conditions – What information about the election itself could we extract from the rules and suggested conditions?
2. the effect of conditions on the number of rules – How did conditions affect the number of rules output by CPLA?
3. the predictive accuracy of the rules – How did the conditions affect the ability of the rules to classify previous unseen examples?

Item (1) consists of an analysis of CPLA results. This is a non-standard analysis, since we are discussing the semantic content of the rules themselves. Items (2) and (3) are more traditional analyses of performance issues.

3.1. The Data

In the database of election survey responses, example objects were the responses of an individual to a set of questions. Each question is considered an attribute, and each answer, a value. The attributes, AT, and their corresponding values, $VAL_a(a \in AT)$ are listed in Table 1. The expert attribute was the survey respondent's vote, in this case, George Bush or Michael Dukakis.

[1]Data acquired courtesy of Dr. Taber and Dr. Lodge, Department of Political Science, SUNY at Stony Brook.

	AT	VAL_a
a_1	Party Identification	$\{1,\ldots,7\}$
a_2	Ideological Distance from Dukakis	$\{0,\ldots,7\}$
a_3	Ideological Distance from Bush	$\{0,\ldots,7\}$
a_4	Bush Issue-distance: Government Services	$\{0,\ldots,7\}$
a_5	Bush Issue-distance: Defense Spending	$\{0,\ldots,7\}$
a_6	Bush Issue-distance: Health Insurance	$\{0,\ldots,7\}$
a_7	Bush Issue-distance: Standard of Living	$\{0,\ldots,7\}$
a_8	Dukakis Issue-distance: Government Services	$\{0,\ldots,7\}$
a_9	Dukakis Issue-distance: Defense Spending	$\{0,\ldots,7\}$
a_{10}	Dukakis Issue-distance: Health Insurance	$\{0,\ldots,7\}$
a_{11}	Dukakis Issue-distance: Standard of Living	$\{0,\ldots,7\}$
a_{12}	Race	$\{-1,0,1\}$
a_{13}	Sex	$\{0,1\}$
a_{14}	Age	$\{18,\ldots,100\}$
a_{15}	Education	$\{1,\ldots,7\}$
a_{16}	Type of Community raised in	$\{1,\ldots,7\}$
a_{17}	Rate Intelligence: Bush	$\{1,\ldots,7\}$
a_{18}	Rate Intelligence: Dukakis	$\{1,\ldots,7\}$
a_{19}	Rate Compassion: Bush	$\{1,\ldots,7\}$
a_{20}	Rate Compassion: Dukakis	$\{1,\ldots,7\}$
a_{21}	Rate Morals: Bush	$\{1,\ldots,7\}$
a_{22}	Rate Morals: Dukakis	$\{1,\ldots,7\}$
a_{23}	Rate as Inspiring: Bush	$\{1,\ldots,7\}$
a_{24}	Rate as Inspiring: Dukakis	$\{1,\ldots,7\}$
a_{25}	Rate as Leader: Bush	$\{1,\ldots,7\}$
a_{26}	Rate as Leader: Dukakis	$\{1,\ldots,7\}$
a_{27}	Rate as Decent: Bush	$\{1,\ldots,7\}$
a_{28}	Rate as Decent: Dukakis	$\{1,\ldots,7\}$
a_{29}	Rate as Caring: Bush	$\{1,\ldots,7\}$
a_{30}	Rate as Caring: Dukakis	$\{1,\ldots,7\}$

Table 1: Attributes

The database consists of 444 records. In order to discuss both rules generation, and predictive accuracy, we randomly split the database into two parts. Two thirds of the data were designated as "training data," and used by CPLA to inductively generate rules. The remaining third was designated as "test data," and was used to measure predictive accuracy, as described in Section 3.5. A small portion of the database is given below.

Objects	Attributes				Expert Attribute
Respondent	Party Id	Ideological Dist. Duk.	...	Caring: Dukakis	Vote
1	independent	2	...	7	Dukakis
2	liberal	1	...	5	Dukakis
⋮	⋮	⋮	⋮	⋮	⋮
444	conservative	6	...	3	Bush

3.2. Application Process

In the application of CPLA to the data there were three phases:

A: Application to the training set with all 30 attributes, and no initial conditions, i.e. $\{cond_a\}_{a \in AT} = \phi$.

B: Application to the training set with all 30 attributes, and with the condition that ages (a_{14}) are grouped into intervals of 10. This was done to compensate for the fact that the age attribute yields a very fine partition and thus very specific rules.

C: Application to the training set using only the four attributes selected as non-superfluous in phase A. Various sets of 4 attributes chosen are:
1. *party identification, age, health insurance: Dukakis, ideology: Dukakis* $(a_1, a_{14}, a_{10}, a_2)$.
2. *party identification, ideology: Bush, age, education* $(a_1, a_3, a_{14}, a_{15})$.
3. *party identification, age, government services: Bush, ideology: Dukakis* (a_1, a_{14}, a_4, a_2).
$\{cond_a\}_{a \in AT} =$ *ages grouped into groups of 10*
and *selected conditions suggested by CSA.*

3.3. Semantic Analysis

From our analysis, we conclude that CPLA/CSA has done a fair job of generating a set of useful rules describing the concepts "voter for Bush" and "voter for Dukakis." We are also able to draw other conclusions, both about voting patterns, and about the relevant survey questions. It is interesting to note that all these conclusions seem reasonable, and have been made by computer scientists. We must now compare our conclusions to an analysis via standard techniques by political scientists. Such a comparison will be the subject of a separate paper.

By examining features such as choice of attributes, order of attribute selection, effects of conditions on performance, we have been able to make the following observations:

- *Party Identification* (a_1), in every trial, is always the first attribute chosen by CPLA. This leads to our conclusion that party affiliation is the most important deciding attribute.
- Almost all attribute value ranges were between 0 and 7. But when conditions eliminating distinctions between 0, 1, 2 and 5, 6, 7 were imposed, the performance of the system did not decrease. A logical conclusion is that such distinctions are unnecessary.
- The attributes *Party Identification, Age,* and *Issues* ($a_1, a_{14}, a_4 - a_{11}$) were chosen to create the rules, while personal feelings ($a_{17} - a_{30}$) were discarded as superfluous, indicating perhaps that personal feelings were not reliable predictors of voting patterns.
- The most frequently chosen non-superfluous attributes related to Dukakis. We might conclude therefore that voters were reacting to Dukakis rather that voting for Bush.
- From the rules we note that personal feelings about issues ($a_4 - a_{11}$) were often at odds with party affiliation, and that voters voted often according to their party affiliation, despite their personal feelings about issues. This also shows the importance of party affiliation.
- The attribute *Ideological Distance from Dukakis* indicated that (1) respondents far from Dukakis in ideology (value of $a_2 \geq 5$) usually voted for Bush, and (2) respondents close to Dukakis in ideology (value of $a_2 \leq 3$) voted for both candidates. We might conclude from this that ideology was not strong enough a factor to make Dukakis sympathizers vote for Dukakis, implying that perhaps there were more more important factors involved for those voters.
- The *age* attribute (a_{14}) is important even when grouped by conditions into 10's. However, there was no obvious trend connecting age to voting patterns.

3.4. Performance Issues

Initial results show a significant decrease in the number of rules (see Table 2). The initial family of rules generated was entirely deterministic. After adding conditions, the resulting rules were still all entirely deterministic - indicating a good choice of conditions. Furthermore, in all cases, approximately 85% of attributes were discarded as superfluous. Although in most trials the exact attributes chosen were not exactly the same, they were similar enough to indicate a trend.

Phase	attributes in training set	non-superfluous attributes found	Conditions used	Rules	Improvement (rule reduction)
A	30	4	none	168	-
B	30	5	by user	89	-47%
C	4	4	by CSA&user	99	-41%

Table 2: Rule reduction results.

3.5. Predictive Accuracy

Because the ultimate goal of inductive learning is to use the knowledge acquired to recognize (classify) previously unseen objects, we are interested in measuring this ability. Predictive accuracy is the standard measurement tool. It is a measure of the "usefulness" of the rule family. Therefore, after CPLA generated a family of rules, we tested these rules on the test data to see how well they would predict the expert classification of each test object.

Because we have begun with a well-defined formal model, we can define formally the standard notions used above: *classification*, and *prediction.*

Object, o, *satisfies* a_i *in a rule* $r : des_A(A_j) \Longrightarrow des(E_k)$ if, for a pair $(a_i, cond_{a_i}(v_i))$ in $des_A(A_j)$, $f(o, a_i) \in cond_{a_i}(v_i)$).

Object, o, and rule, $r : des_A(A_j) \Longrightarrow des(E_k)$, are an *exact match* if o satisfies all $a_i \in A$.

A rule $r : des_A(A_j) \Longrightarrow des(E_k)$, *classifies* an object, o, into the expert class, $des(E_k)$, if r is an exact match.

Given a rule, $r : des_A(A_j) \Longrightarrow des(E_k)$, classifying o, $des(E_k)$ is called the rule's *prediction.*

A rule classifies an object, o, *correctly* if its prediction, for o, $des(E_k)$, is equal to $f(o, E)$, where E is the expert attribute. Otherwise, the classification is *incorrect.*

Predictive accuracy of a family of rules is defined, for a test database, as the percentage of test objects correctly classified by the rules.

If there is no rule which matches an object exactly, then there is no prediction for that object, and the object is *unclassified.*

We have dealt with the problem of unclassified objects by creating a *guessing* heuristic. When we use this heuristic, no test object is ever left unclassified.

For every unclassified object, o, we determine $\rho(o, r)$ for each rule, r:

$$\rho(o, r) = \frac{\text{number of attributes satisfied by } o \text{ in rule } r}{\text{number of attributes in description of rule } r}.$$

We "guess" that object o satisfies rule r if $m = \max_r \rho(o, r)$. This heuristic yields the

	Predictive Accuracy		Improvement	
Phase	exact match	with guessing	exact match	with guessing
A	55%	77%	-	-
B	77%	84%	41%	9%
C	84%	88%	52%	14%

Table 3: Predictive Accuracy

"closest" match of an object and description, when no exact match is available. Results show this to be a useful technique.

3.6. Performance Conclusions

Overall, for this dataset, *conditions* proved themselves to be useful for system optimization. First, they decreased the number of rules without introducing any probabilistic factors into the rule family. Second, they increased the predictive power of the rule family. Table 3 shows the predictive accuracy of the rules without guessing ("exact match"), and with guessing. The improvement described shows the increase in predictive accuracy of tests **B** and **C** as compared to **A**, both with and without the guessing heuristic.

A: This case obviously yields the poorest results since there has been no attempt at generalization. CSA suggests (among others) conditions which in general make "highly-valued" scales, such as the range 1-7, into "fewer-valued" scales, such as $\{1,2,3\},\{4\},\{5,6,7\}$.

B: In this case we have partitioned VAL_{age} into groups of 10, to make that attribute yield a coarser partition, and thus more general rules. Immediately, more general, and therefore fewer, rules are generated. By making the rule family more general, we have increased the number of possible objects which might satisfy a rules description, and thus the number of correctly classified objects (and incorrect) increases, as can be seen in Table 3.

C: In this case, with the CSA-suggested conditions, we've still decreased the number of rules from **A**, although not as much, but we've significantly increased the predictive accuracy of the rules. Thus, with the "guessing" heuristic in place, we achieve a respectable 88% predictive accuracy (for this database).

Thus we see that conditions have yielded a significant improvement. Conditions alone (no guessing) took the predictive accuracy from 55% to 84%. Furthermore, it is interesting to note that as conditions were added to the system, they seemed to "replace" the guessing algorithm. While in Phase **A** guessing improves accuracy by 20 percentage points, in Phase **C** guessing only yields 4 points improvement. This phenomenon can be explained by noting that both conditions and guessing are a form of generalization. In the first case, generalization is by making equivalence classes of attributes, while in the second case generalization is by dropping a clause (somewhat arbitrarily) from the description.

Together, these techniques have improved predictive accuracy from 55% (exact match, no guessing), to 88% (conditions + guessing), a 60% improvement.

In this study, we have also noted that CSA conditions may blur distinctions too much. In an attempt to merge many rules, CSA may suggest many conditions. Application of too many conditions leads to overgeneralization. Furthermore, rules need to be assigned some sort of *strength*, indicating how much evidence supports them. "Weakly" supported rules can confuse conclusions. Nevertheless, conditions are still necessary to make sense of such data. In our study conditions helped indicate the usefulness of various gradations in possible attribute values.

4. Conclusion

We have presented a probabilistic inductive learning system built on the ILS model of Wong & Wong ([9]), and the INFER model of Wong & Ziarko ([10]). We have demonstrated that the resulting system can be effective in the inductive learning task, as well as the task of semantically analyzing the training database. The Conditional Probabilistic Learning Algorithm (CPLA) incorporates the concept of *conditions* and allows for direct user interaction with the Data. The Condition Suggestion Algorithm (CSA) extracts syntactic knowledge from the knowledge representation system and allows the user to translate it to semantic knowledge. This syntactic knowledge is presented as suggested conditions which generalize attribute values, and thus generalize decision rules.

The application of CPLA/CSA to the 1988 U.S.A. election survey data demonstrates our learning system can not only result in a smaller, more efficient set of decision rules to describe a concept, but can also allow a non-domain expert to extract useful semantic meaning from the data.

References

[1] HADJIMICHAEL, M. (1989) Conditions Suggestion Algorithm for Knowledge Representation Systems. *Proceedings of the Fourth International Symposium on Methodologies for Intelligent Systems: Poster Session Program*, Charlotte, NC, ORNL/DSRD-24.

[2] HADJIMICHAEL, M., WASILEWSKA, A. (1990) Rules Reduction for Knowledge Representation Systems. *Bulletin of Polish Academy of Science* Technical Sciences, Vol. 38, No. 1-12. 1990, pp. 113 -120.

[3] HADJIMICHAEL, M., WASILEWSKA, A. (1992) Interactive Inductive Learning. To appear in *International Journal of Man-Machine Studies.*

[4] MICHALSKI, R. S., LARSON, J. B. (1978) Selection of most representative training examples and incremental generation of VL1 hypothesis: the underlying methodology

and the description of programs ESEL and AQ11. *Report No. 867, Department of Computer Science, University of Illinois*, Urbana, Illinois.

[5] PAWLAK, Z. (1982) Rough sets. *International Journal of Information and Computer Science*, 11, pp. 344-356.

[6] PAWLAK, Z., WONG, S. K. M., ZIARKO, W. (1988) Rough Sets: Probabilistic Versus Deterministic Approach. - *International Journal of Man-Machine Studies*, 29, pp. 81-95.

[7] QUINLAN, J. R. (1983) Learning Efficient Classification Procedures and their Application to Chess End Games, *Machine learning, an Artificial Intelligence Approach.* (eds. R. S. Michalski, J. G. Carbonell, and T. M. Mitchell, Morgan Kaufmann Publishers, Inc., San Mateo.

[8] WASILEWSKA, A. (1990) Conditional Knowledge Representation System - Model for an Implementation. *Bulletin of the Polish Academy of Science*, vol. 37, 1-6, pp. 63-69.

[9] WONG, S. K. M., WONG, J. H. (1987) An Inductive Learning System - ILS. *Proceedings of the 2nd ACM SIGART International Symposium on Methodologies for Intelligent Systems, in Charlotte, North Carolina*, North Holland, pp. 370-378.

[10] WONG, S. K. M., ZIARKO, W. (1986) INFER- an Adaptive Decision Support System Based on the Probabilistic Approximate Classification. *The 6th International Workshop on Expert Systems and Their Applications.* Avignon, France. vol.I, pp. 713-726.

[11] WONG, S. K. M., ZIARKO, W., YE, R. L. (1986a) Comparison of rough set and statistical methods in inductive learning. *International Journal of Man-Machine Studies* 24, pp. 53-72.

[12] WONG, S. K. M., ZIARKO, W., YE, R. L. (1986b) On Learning and Evaluation of Decision Rules in Context of Rough sets. *Proceedings of the first ACM SIGART International Symposium on Methodologies for Intelligent Systems, Knoxville, Tenn*, pp. 308-324.

Part I
Chapter 11

AN APPLICATION OF ROUGH SET THEORY IN THE CONTROL OF WATER CONDITIONS ON A POLDER

Andrzej REINHARD
Academy of Agriculture in Wrocław
Institute of Agriculture and Forest Improvement
50-356 Wrocław, Poland

Urszula WYBRANIEC-SKARDOWSKA
Bogusław STAWSKI Tomasz WEBER
College of Education in Opole
Institute of Mathematics
45-951 Opole, Poland

Abstract. In this work results of two experiments on application of rough set theory in land improvement sciences are included.

1. Introduction

It is generally known, that rational exploitation of water resources both on the micro as well as on the macroregional scale has a fundamental effect on the level of living and managing the natural resources by the inhabitants of a given region. Hydrogeological and land improvement investments disturb in a permanent manner the natural environment and are very capital-intensive. The effects of erroneous investment and exploitation decisions are usually difficult to remove and they create many ecological hazards.

To these belong also the hazards existing in Poland, in particular, such as:

* pollution of drinking water as the effect of industrial contamination and plant protection agents penetration into ground waters and underflows (for instance — Siechnice near Wrocław);

* filtration of brine from mines and floating tailings into surface underflow (for instance — Lubin Basin);

* degradation of large areas of cultivated land due faulty designing of drainage facilities (for instance the region of Greater Poland acquiring the characteristics of a steppe) or as a result of too close a location of open-cast mines (for instance - Bełchatów and Konin regions).

Therefore, in many scientific centers research has been carried out to find suitable methodology, which would make it possible, in a simple way, to formalize or describe the filtration and transport of ground and surface water phenomena.

These investigations have been carried out not only in the direction of constructing models which would allow to better describe the physical nature of these phenomena, but also in the one of obtaining models, which can be easily adapted for monitoring, controlling and computer assisting in taking an economic as well as exploitation decisions.

This work attempts to verify the question if by utilizing the rough set methodology [3], it is possible to find some basic relationships between selected input parameters of a river basin and a land improvement object (such as, rainfall, evaporation, air temperature, water level and others) and the starting (output) state described by one essential parameter (height of water rain off or the state of soil humidity, respectively) and thus if this methodology can be used to build expert systems assisting the control of the object.

Calculations are based, on the one hand, on the measurement results carried out during 100 days (from March 7, 1984 till June 14, 1984) at the experimental river basin of Ciesielska Woda (a tributary of the Polish river Widawa) by the Institute of Land Improvement and Grassland of the Academy of Agriculture in Wrocław [1] and, on the other hand, on the results of measurements carried out in the years 1976-1979 at Piwonia Górna object (region of Wieprz-Krzna canal in Poland) by the Institute of Land Improvement and Grassland at Lublin [8].

2. Ciesielska Woda

The experimental river basin Ciesielska Woda is provided with a weather station, which carries out measurement of rainfall, air temperature, wind velocity and direction, air humidity, insolation, atmospheric pressure and evaporation. In the terminology of rough set theory, the measurement set can be treated as the information system (see [3], [7], [6]).

Let us consider *the information system* constructed on the basis of measurement results, made in connection with the experiment (cf. [1], [2], [4]):

$$S = \langle U, Q, V, F \rangle$$

where

$U = \{1, \dots, 100\}$ is *a set of objects* composed of numbers assigned to succeeding measuring days;

$Q = \{q1, q2, q3, q4, q5, q6, q7, q8\}$ is *a set of attributes* understood as physical quantities taken into account in the experiment:

- q_1 – wind velocity;
- q_2 – evaporation;
- q_3 – rainfall measured during the day of water rain off measurement;
- q_4 – rainfall measured one day before water rain off measurement;
- q_5 – rainfall measured two days before water rain off measurement;
- q_6 – rainfall measured three days before water rain off measurement;
- q_7 – rainfall measured four days before water rain off measurement;
- q_8 – water rain off from the river basin;

$V = \bigcup\{V_{q_i} : q_i \in Q\}$, where V_{q_i} is *the domain of attribute* q_i, i.e. the set of identifiers of classes discriminated from *the measuring range* of possible values of q_i attribute (their

number is the consequence of the maximum, permissible from the system user's point of view, indiscernibility of system states);
$f: U \times Q \rightarrow V$ is the *function information* i.e. *measurement function* assigning to each measured quantity at given day a corresponding identifier of the indiscernibility class of the measured results.

The accepted, in the calculations, division of measuring ranges of particular attributes into classes and domain of attributes are presented in Table 1.

TABLE 1. Division into classes of measuring ranges of the attributes and determination of their domains.

Attribute	Measuring range	Division into	Identifier	Domain of attributes
q_1[m/s]	1-14	1-2.99	1	{1,2,3,4}
		3-5.99	2	
		6-9.99	3	
		10-14	4	
q_2[mm]	0-7	0-0.99	1	{1,2,...,8}
		1-1.49	2	
		1.5-1.99	3	
		2-2.99	4	
		3-3.99	5	
		4-4.99	6	
		5-5.99	7	
		6-7	8	
q_3, q_4	0-27	0-0.9	1	{1,2,...,11}
q_5, q_6		1-1.9	2	
q_7[mm]		2-2.9	3	
		3-3.9	4	
		4-4.9	5	
		5-5.9	6	
		6-6.9	7	
		7-9.9	8	
		10-14.9	9	
		15-19.9	10	
		20-27	11	
q_8 [l/s]	76-103	76-79.9	1	{1,2,3}
		80-94.9	2	
		95-103	3	

Distinguishing in set Q of all attributes of the information system S one or some attributes and treating them as the decision attributes, *a decision table* $DT = \langle U, C \cup D, V, f \rangle$ can be formed, where C is the *set of condition attributes* and D the distinguished *set of decision attributes* ($C \cup D = Q$). Analysis of the decision table makes it possible to find, in a simple manner, the decision rules and by this to build an expert system, which can be used for object control assistance.

Because in the considered case the set of experimental data did not promise a chance of building a reliable expert system (too small a number of data) for verification to what extent the decision attribute (in water rain off experiment) can be described by means of conditional attributes, there was used coefficient $\gamma_C(D^*)$ of *quality of approximation of D^* classification by C* and *determinant $\delta_B(D^*)$ of a measure of significance* for one element subsets B of set C considering this, induced by decision attribute, classification. Determinant $\delta_B(D^*)$ is set as a difference:

$$\delta_B(D^*) = \gamma_C(D^*) - \gamma_{C-B}(D^*), \qquad B \subseteq C.$$

Calculation results for two decision tables:

$$DT_1 = (U, C_1 \cup D, V, f) \qquad \textit{and} \qquad DT_2 = (U, C_2 \cup D, V, f)$$

where

$$C_1 = \{q_1, q_2, q_3\}, \qquad C_2 = \{q_1, q_2, q_3, q_4, q_5, q_6, q_7\} \quad \textit{and} \quad D = \{q_8\}$$

are presented in Table 2.

TABLE 2. Accuracy of approximation and measure of significance.

$\gamma_{C_1}(D^*)$	$\gamma_{C_2}(D^*)$	B	$\gamma_{C_2-B}(D^*)$	$\delta_B(D^*)$
0.47	0.86	$\{q_1\}$	0.79	0.07
		$\{q_2\}$	0.80	0.06
		$\{q_3\}$	0.85	0.01
		$\{q_4\}$	0.81	0.05
		$\{q_5\}$	0.86	0.00
		$\{q_6\}$	0.83	0.03
		$\{q_7\}$	0.84	0.02

3. Piwonia Górna

The Piwonia Górna object is a typical polder object on which water-air relations in the soil can be controlled by changing in a proper way the water surface level in the surrounding land improvement ditches. Objects of this type are characterized either by constant or long term excess of humidity in the zone of root layer of the meadow plants, or by long-term of humidity deficiency. This does not promote high crop of these plants. At Piwonia Górna organic soils Mt II ba, well friable, containing in the layer down to 30cm from 16.9% to 26.6% of macropores are predominating. Good capillary conduction is obtained due to the quantity of mesopores and micropores, which oscillates in the range from 38.9% to 48.8% of the volume for mesopores and from 20.9% to 26.9% of the volume for micropores.

Dynamics of fluctuation of ground water level and the water level in the drainage system was studied in the period from April to October during 5 years. These measurements were made in control wells and on water gauges. In 1976 measurements were made once a week and in the remaining years three times a month, on the average. Samples of soil formation were taken, counting from its surface to ground water level at the following depths: 5–10 cm, 15-20 cm, 25-30 cm, 35-40 cm, 55-60 cm, 75-80 cm and 95-100 cm. Soil humidity was determined by the drier-weight method (per cent of soil volume).

In the below calculations only average humidity was taken into account, which was in the meteorological conditions on the objects measurements carried out by the meteorological station, located at the nearby place Sosnowica were utilized. In the investigation the total rainfall in decade periods and average temperatures in particular decades were taken into account.

It is assumed that the optimal arrangement of water-air relations in the layer from 0 to 30 cm containing the essential part of grass roots is within the limits determined by two critical humidity states:

- state of maximal humidity at which air content is 6% of the volume, and
- state of minimal humidity corresponding to the suction pressure pP 2.7, at which the soil humidity is approximate to the limiting level of water easily accessible for plants.

Conducted parallel on the considered polder field and laboratory investigations showed, that the best conditions for correct grow of plants are when the coefficient of soil humidity was in the range from 42% to 58%. Therefore the *measuring range* of the decision attribute (soil humidity) was divided into three classes of value with identifiers:

1 – when the soil humidity is greater than 58% (too great an amount of water in soil);

2 – when humidity coefficient is within the range from 42% to 58% (optimal conditions of meadow plants growth);

3 – when humidity coefficient is smaller than 42% (too small quantities of water in soil).

Experimental investigations concerning the discussed polder (cf. [5]) led to construction of decision table:

$$DT_3 = (U', C_3 \cup D_3, V', f')$$

where

the set of objects U' was limited here to numbers 1–42 setting the successive days of measurements during which full information about the value of the selected attributes could be taken into account;

the set C_3 consists of 6 conditional attributes:

p_1– air temperature;

p_2– rainfall;

p_3– water level in Wieprz-Krzna Canal (WKC);

p_4– water level in ditch 12;

p_5– water level in ditch 15;

p_6– ground water level in soil sampling site;

the set D_3 assigns one decision attribute:

p_7 – soil humidity in sampling site.

Thus

$$C_3 = \{p_1, p_2, p_3, p_4, p_5, p_6, p_7\} \quad \text{and} \quad D_3 = \{p_7\}.$$

Measuring ranges of particular attributes $p_1 \ldots p_7$. their corresponding divisions into classes, identifiers of these classes and domain of these attributes are presented in Table 3.

Because in the measuring data of soil humidity no record from class 3 was found, it was accepted that the domain of decision attribute p_7 is only the set $\{1, 2\}$.

TABLE 3. Division classes of measuring ranges of attributes and designating their domain.

Attribute	Measuring range	Division into	Identifier	Domain of attributes
p_1 [°C]	0-20	0- 4.99	1	{1,2,3,4}
		5- 9.99	2	
		10-14.99	3	
		15-20.00	4	
p_2 [mm]	0-90	0-14.99	1	{1,2,3,4,5,6}
		15-29.99	2	
		30-44.99	3	
		45-59.99	4	
		60-74.99	5	
		75-90.00	6	
p_3 [cm]	0-210	0-29.99	1	{1,2,3,4,5,6,7}
		30-59.99	2	
		60-89.99	3	
		90-119.99	4	
		120-149.99	5	
		150-179.99	6	
		180-210.00	7	
p_4, p_5, p_6 [cm]	0-120	0-14.99	1	{1,2,3,4,5,6,7,8}
		15-29.99	2	
		30-44.99	3	
		45-59.99	4	
		60-74.99	5	
		75-89.99	6	
		90-104.99	7	
		105-120.00	8	
p_7 [%]	42-71	42-57.99	2	{1,2}
		58-71.00	1	

Let us notice that classification D_3^* induced by decision attribute p_7 consists of two subsets U': X_1 – for excessive humidity of soil and X_2 – for good humidity of soil. For verification to what extent decision classes X_1 and X_2 can fully or roughly be characterized by subset B of the set of all conditional attributes C_3 and for defining the degree of completeness of our knowledge about X_1 and X_2 using the conditional attributes of B we compute the coefficients $\mu_B(X_1)$ and $\mu_B(X_2)$ of *accuracy of the approximation* of X_1 and X_2 by B, respectively. Calculation results of these coefficients as well as coefficients $\mu_B(D_3^*)$ and $\gamma_B(D_3^*)$ *of the accuracy and quality of classification* D_3^* by B, respectively, are placed in Table 4 (cf. [5]). Any subset B of set C_3 obtained by removing from it conditional attributes $p_{i_1}, \dots p_{i_n} (i_k \in \{1, \dots, 6\})$ is here identified by a sequence of numbers $(i_1, \dots, i_n)$.

TABLE 4. Measures the accuracy of approximation by subset B by C_3.

B Identifier	$\mu_B(X_1)$	$\mu_B(X_2)$	$\mu_B(D_3^*)$	$\gamma_B(D_3^*)$
	1.0000	1.0000	1.0000	1.0000
(1)	1.0000	1.0000	1.0000	1.0000
(2)	1.0000	1.0000	1.0000	1.0000
(3)	0.7692	0.7273	0.7500	0.8571
(4)	0.8400	0.8095	0.8261	0.9048
(5)	1.0000	1.0000	1.0000	1.0000
(6)	1.0000	1.0000	1.0000	1.0000
(1,2)	0.7692	0.7273	0.7500	0.8571
(1,3)	0.7308	0.6957	0.7143	0.8333
(1,4)	0.8400	0.8095	0.8261	0.9048
(1,5)	1.0000	1.0000	1.0000	1.0000
(1,6)	1.0000	1.0000	1.0000	1.0000
(2,3)	0.7407	0.6818	0.7143	0.8333
(2,4)	0.8400	0.8095	0.8261	0.9048
(2,5)	0.9167	0.9000	0.9091	0.9524
(2,6)	0.9167	0.9000	0.9091	0.9524
(3,4)	0.5000	0.4444	0.4737	0.6429
(3,5)	0.6071	0.5600	0.5849	0.7381
(3,6)	0.7407	0.6818	0.7143	0.8333
(4,5)	0.7308	0.6957	0.7143	0.8333
(4,6)	0.7308	0.6957	0.7143	0.8333
(5,6)	1.0000	1.0000	1.0000	1.0000
(1,2,3)	0.5455	0.3750	0.4737	0.6429
(1,2,4)	0.5862	0.5200	0.5556	0.7143
(1,2,5)	0.6786	0.6087	0.6471	0.7857
(1,2,6)	0.7407	0.6818	0.7143	0.8333
(1,3,4)	0.4516	0.3929	0.4237	0.5952
(1,3,5)	0.4828	0.4643	0.4737	0.6429
(1,3,6)	0.6552	0.5652	0.6154	0.7619
(1,4,5)	0.6667	0.6250	0.6471	0.7857
(1,4,6)	0.7308	0.6957	0.7143	0.8333
(1,5,6)	0.9167	0.9000	0.9091	0.9524
(2,3,4)	0.4545	0.3333	0.4000	0.5714
(2,3,5)	0.5172	0.4815	0.5000	0.6667
(2,3,6)	0.6552	0.5652	0.6154	0.7619
(2,4,5)	0.5862	0.5200	0.5556	0.7143
(2,4,6)	0.7037	0.6522	0.6800	0.8095
(2,5,6)	0.7692	0.7273	0.7500	0.8571
(3,4,5)	0.4242	0.3214	0.3770	0.5476

TABLE 4. Cont.

B Identifier	$\mu_B(X_1)$	$\mu_B(X_2)$	$\mu_B(D_3^*)$	$\gamma_B(D_3^*)$
(3,4,6)	0.4688	0.3704	0.4237	0.5952
(3,5,6)	0.6071	0.5600	0.5849	0.7381
(4,5,6)	0.5714	0.5385	0.5556	0.7143
(1,2,3,4)	0.3611	0.2069	0.2923	0.4524
(1,2,3,5)	0.3514	0.1724	0.2727	0.4286
(1,2,3,6)	0.5000	0.2500	0.4000	0.5714
(1,2,4,5)	0.4375	0.3571	0.4000	0.5714
(1,2,4,6)	0.4688	0.3704	0.4237	0.5952
(1,2,5,6)	0.6000	0.5000	0.5556	0.7143
(1,3,4,5)	0.2647	0.2424	0.2537	0.4048
(1,3,4,6)	0.4118	0.2857	0.3548	0.5238
(1,3,5,6)	0.4118	0.2857	0.3548	0.5238
(1,4,5,6)	0.1143	0.1842	0.1507	0.2619
(2,3,4,5)	0.2778	0.1875	0.2353	0.3810
(2,3,4,6)	0.3590	0.1071	0.2537	0.4048
(2,3,5,6)	0.4667	0.4286	0.4483	0.6190
(2,4,5,6)	0.2703	0.1563	0.2174	0.3571
(3,4,5,6)	0.3611	0.2069	0.2923	0.4524
(2,3,4,5,6)	0.2143	0.0000	0.1200	0.2143
(1,3,4,5,6)	0.0000	0.0714	0.0370	0.0714
(1,2,4,5,6)	0.0000	0.0000	0.0000	0.0000
(1,2,3,5,6)	0.3171	0.0345	0.2000	0.3333
(1,2,3,4,6)	0.3171	0.0345	0.2000	0.3333
(1,2,3,4,5)	0.0000	0.0714	0.0370	0.0714

Values of coefficients $\gamma_B(D_3^*)$ or $\mu_B(D_3^*)$ given in Table 4 can be used for determination of *the core* of C_3 and *reducts* of C_3 with respect to classification D_3^*. Reduction of conditional attributes has, as it is well known, a great practical significance. We can reduce superfluous conditional attributes from DT_3 using Table 4. Because the considered *decision table* DT_3 *is consistent* (see Table 4: $\gamma_{C_3}(D_3^*) = \mu_{C_3}(D_3^*) = 1$, hence $C_3 \to D_3$), *the core* of C_3 as well as the *reducts* of C_3 with respect to D_3^* (see [6]), we can calculate, using two measures to describe inexactness of approximate classifications: *the accuracy* and *the quality of classification* D_3^*. For calculation of D_3-Core of C_3 we can use the following property:

> these and only these attributes $p \in C_3$, for which $\gamma_{C_3-\{p\}}(D_3^*) < 1$ (or $\mu_{C_3-\{p\}}(D_3^*) < 1$) belong to the set.

It is apparent, that the set $\{p_3, p_4\}$ is D_3-Core of C_3. However, to ascertain that for $B \subseteq C_3$, B is D_3-reduct of C_3 let us avail ourselves of the criterium according to which B fulfills the conditions:
$\gamma_B(D_3^*) = 1$ (or $\mu_B(D_3^*) = 1$) and for every $p \in B$ $\gamma_{B-\{p\}}(D_3^*) < 1$ (or $\mu_{B-\{p\}}(D_3^*) < 1$).

Hence D_3-reducts of C_3 are the following sets:

$$\begin{aligned} B_{(2)} &= \{p_1, p_3, p_4, p_5, p_6\} , \\ B_{(1,5)} &= \{p_2, p_3, p_4, p_5\} , \\ B_{(1,6)} &= \{p_2, p_3, p_4, p_5\} , \\ B_{(5,6)} &= \{p_1, p_2, p_3, p_4\} . \end{aligned}$$

Let us note additionally, that the set $B = \{p_2, p_3, p_4\}$ on high quality coefficient ($\gamma_B(D_3^*) = 0.9524$) is *the minimal set*. It can be found by the use of similar method as the minimal sets in Słowiński paper [7].

If, as the set of condition attributes we accept only the core $\{p_3, p_4\}$, i.e. the water level in Wieprz-Krzna Canal and ditch 12, then the lower approximation of X_1 and X_2 sets may be graphically shown, as in Fig. 1

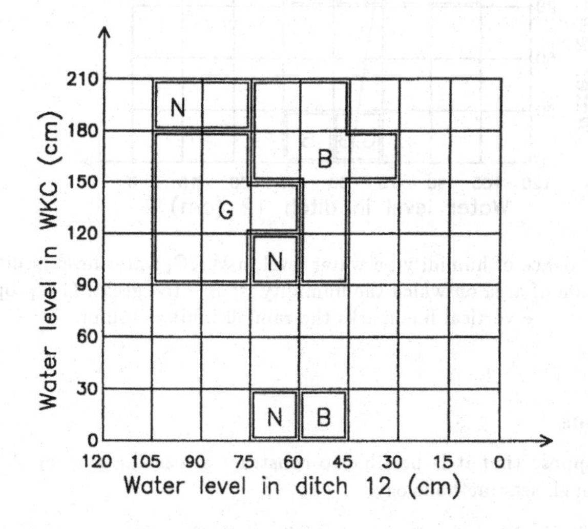

Figure 1. Dependence of humidity on water level in WKC and ditch 12. Location of areas B, G, N, where: G area of lower approximation of X_1 (appropriate humidity), B area of approximation of X_2 (too great humidity), N – boundary area.

As it follows from the distribution of the lower approximation of X_1 and X_2 sets for the preservation of appropriate humidity in soil it is enough to maintain the water level in ditch 12, 75cm below the soil surface.

If we accept now the set $\{p_2, p_3, p_4\}$ as set of condition attributes, we enrich thus the previous model, the quality of approximation will increase and the distribution of lower approximation of X_1 and X_2 sets will be formed as in Fig. 2.

One can see that in order to maintain appropriate humidity in soil it is enough to retain the water level in ditch 12, 75cm below the soil surface, if water level in Wieprz-Krzna

Canal is lower than 180cm. When the water level in Wieprz-Krzna Canal is higher, the water level in ditch 12 should be below 90cm.

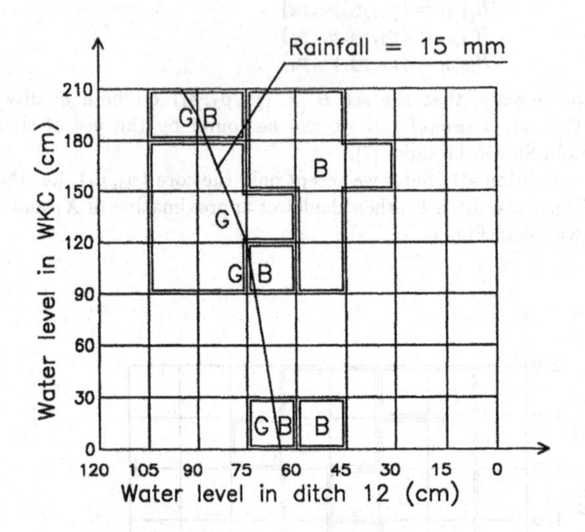

Figure 2. Dependence of humidity on water level in WKC, water level in ditch 12, and rainfall. Location of area on which the humidity is: B – too great, G – proper, where the vertical line marks the rainfall limit = 15mm.

4. Conclusions

1. Authors suppose that it is possible to construct a warning system of flood danger based on rough sets methodology.
2. Authors believe that an attempt at application of rough sets methodology in the control of water-air relation on a polder is noteworthy.

References

[1] Dejas, D., Reinhard, A., Stawski, B., and Weber, T. (1987) *'Optimization of Polder Systems'* (in Polish), Report of Institute of Land and Forest Improvement, Wrocław

[2] Dejas, D., Reinhard, A., Stawski, B., and Weber, T. (1988) *'Optimization of Polder Systems'* (in Polish), Report of Institute of Land and Forest Improvement, Wrocław

[3] Pawlak, Z. (1990) *Rough Sets. Theoretical Aspects of Reasoning about Data*, Kluwer, Dordrecht ,1991.

[4] Reinhard, A., Stawski, B., and Weber, T. (1989) 'Application of Rough Sets in Study of Water Outflow from River Basin' *Bull. Pol. Ac. Sc. Techn. Sc.*, vol. 37, No 1-2, 97-103.

[5] Reinhard, A., Stawski, B., Szwast, W., and Weber, T. (1989) 'An Attempt to Use Rough Set Theory for the Control of Water-Air Relation in a Given Polder', *Bull. Pol. Ac. Sc. Techn. Sc.*, vol. 37, No 5-6, 339-349.

[6] Skowron, A., and Rauszer, C. (1991) *'The Discernibility Matrices and Functions in Information systems'*, Research Report 1/91. Institute of Computer Science, Technical University of Warsaw.

[7] Słowiński, K. (1990) *'Application of Rough Set Theory Analysis of Duodenal Ulcer Treatment by Highly Selective Vagatomy and the Accute Pancreatitis by Peritoneum Lavage'* (in Polish), Work for the Doctor habilitowany Degree, Poznań.

[8] Szajda, J. (1980) *'Working out a Method of Irrigation Prognosis in Conditions of Shallow Ground Water Level'* (in Polish), Institute of Grassland Improvement, Plant of Natural Basis of Land Improvement, Lublin Division, Lublin.

Part I
Chapter 12

USE OF "ROUGH SETS" METHOD TO DRAW PREMONITORY FACTORS FOR EARTHQUAKES BY EMPHASING GAS GEOCHEMISTRY: THE CASE OF A LOW SEISMIC ACTIVITY CONTEXT, IN BELGIUM

Jacques TEGHEM Jean-Marie CHARLET
Faculté Polytechnique de Mons
9, rue de Houdain
7000 Mons, Belgium

Abstract. The "Rough sets" method has been applied to earthquake prediction using the gas geochemistry. The field of application concerns the Mons basin (Belgium) with various geological environment, a geothermal system and a rather low seismic activity in 1987 two seismic sequences with events of magnitude upper than 1,5-2 have been recorded. A data base includes the radon concentrations in soils for eight points of measurement in different geological environments, with different climatological parameters and seismic activity. The "Rough sets" method has allowed to discriminate the sites with a particular sensitivity to a seismic event.

1. The problem

There is a relation between seismic risk and gases concentration in soils or ground waters.

An abundant specialized literature exists on the subject (see an overview in Chi-Yu King,[4]; Charlet and al.[2], and a discussion has been often carried on the reasons of the relation.

Among the gases more concentrated in the earth crust than in the atmosphere (terrestrial gases), radon is particularly interesting because it is easy to detect, due to its radioactivity properties.

Radon 222 is a decay product of uranium with a half live of 3,8 days. Radon escapes from the geological formations by a process named emanation controlled to a large degree by the distribution of the stress conditions or the tectonic events (active fault in relation with seismic zone,...). For earthquake prediction one can use radon in soils or radon in underground waters.

From the scientific literature about the subject one can extract the following points [2,4,5]:

- many authors draw attention to a variation (often an increasing) of radon concentration and other terrestrial gases as a premonitory factor of seismic events,

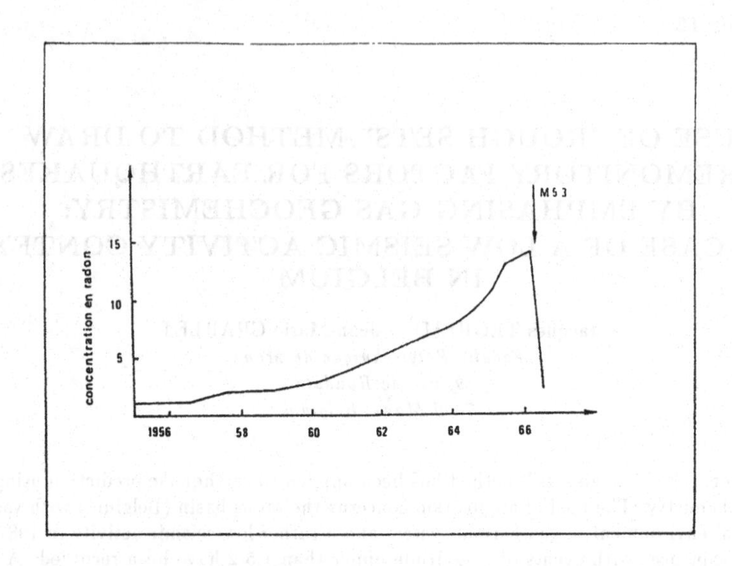

Figure 1. Variation of radon concentration in the Tashkent basin [11]

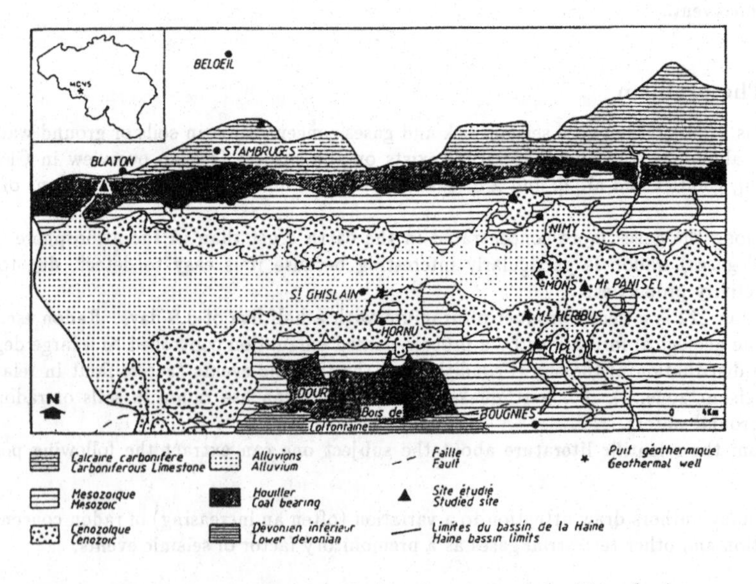

Figure 2. Radon stations in the geological environment of the Mons basin

- the study of the variations of the radon concentration in the Tashkent basin [11] has shown a typical form with a long-term anomalies, several years earlier and a short-term anomalies characterized by a greater increase of the radon concentration in ground waters during the days or the hours preceeding the earthquake, first slowly and more suddenly just before (fig.1),
- there doesnt't exist any correlation between the level of the gas emanation and the amplitude of the seism on the Richter scale,
- its distribution around the epicenter of the earthquake is not uniformly localised, depending on the geological situation of the region and of the nature of the superficial formations; for a well-defined region some points are more relevant than others to measure the emanations,
- the variations of these emanations are probably a better indicator that their absolute values.

Moreover, some climatic parameters certainly influence the level of the gas emanation; but, till now, it is not clearly established what are the effects of these parameters. Some statistical analysis existing in the literature seem to prove that, for instance, the radon emanations increase with the humidity level and decrease when the atmospheric pressure level is increasing [6].

Nevertheless, the reality is probably not so simple: the multivariate model to explain the relationships between climatic parameters, nature of the underground and gas emanation is still to discover.

To study the behaviour of a radon in natural environment, the geology laboratory of the FPMs undertook a survey in the Mons basin, a region characterized by various geological formations, a geothermal system and a seismic active zone of rather low intensity (to magnitude 3).

Of course Belgium is not a very seismic active zone but it can be noted that some authors [1] reported radon anomalies with amplitude of about 20 % above background and durations of 3-8 days before some small earthquakes of magnitude 3-3,5. Moreover the Mons basin has been the subject of numerous studies because of the scientific interest of the geological series which can be found there. Presently geodynamic studies have been undertaken in collaboration with the Royal Observatory of Belgium in the frame of a common scientific project (relations between microseismic activity, tectonic events and gas geochemistry).

2. The data base, a radon survey in the Mons basin (fig. 2)

The Mons basin forms an area of strong subsidence limited by important tectonic accidents like the North-Artois-Shear-Zone and the Variscan front. It is therefore characterized by a tectonic instability which is denoted by some seismic activity (intensity 2,5 - 4,5 MSK).

The carboniferous limestone which outcrops in the north of the basin is karstified in the deep zone. It is the seat of an important geothermic system worked in the central part of the basin (Saint-Ghislain, ...) by drillings some thousand meters deep. Hot and rich inert gas springs (Beloeil - Stambruges) are probably in relation with this geothermic system.

Moreover the geological formations are diversified and sometimes contain uraniferous beds which can constitute local or regional radon sources.

Since 1983 the geological laboratory of the Faculté Polytechnique de Mons (so-called GEFA) set up a system of radon emanation measuring stations in the Mons basin (with the contribution of a so-called project PRIME).

The measures are made in soils with a particular type of detectors - so called "boukoal"- using activated coal and originally developped by the laboratory GEFA (3).

Several sites are chosen to cover the various geological environments in the specified region and several detectors are placed in each site.

The sites are separated in two sets and the measures are kept every week, alternatively for each set, so that measures at one site are available every two weeks, with a difference of one week between the two sets of sites. In the same time, several climatic factors have been registered.

In 1987 two seismic activity periods have been recorded by the Royal Observatory of Belgium : the so-called seismic sequence of Dour covers the period from January to May 87, the one of Charleroi from September to December 1987. So 1987, is a year of seismic activity in comparison with 1985-1986 which one can consider as a background relating to seismic activity.

Logically, the geology laboratory has been interested to use this data base to analyse if some factors are able to characterise the periods of seismic activity.

3. The available knowledge table

The initial information system [7]

$$S =< U; Q = C \cup D;\ V, f >$$

given by the geology laboratory contains the following data :

. U : the set of objects

There are 155 objects in the system. Each one corresponds to the data related to one week, so that the universe covers the measures made during three years (1985 till 1987).

. C : the set of condition attribute

- There are 15 condition attributes.
  The eight first correspond to 8 different sites : C_1 and C_2 belong to the first set of sites, C_3 till C_8 belong to the second one.
  At each site, several "boukoal" detectors are present and a mean of measured rates of radon emanation is calculated on the different detectors.

An assumption of normal distribution $N(m, \sigma)$ is made for each site and five classes are introduced :

$$-\infty, m-\sigma[;]m-\sigma,\ m-\frac{\sigma}{3}[;]m-\frac{\sigma}{3},\ m+\frac{\sigma}{3}[;]m+\frac{\sigma}{3},\ m+\sigma[;]m+\sigma,+\infty,$$

numbered from 1 to 5, respectively.

The descriptor $f(x, q)$ of an object x for such an attribute q takes thus its value in the domain $Vq = \{1, 2, 3, 4, 5\}$, indicating a very low, low, average, high, very high level of radon emanation for the week x on the site corresponding to q.

<u>Remark</u>

As for technical reasons (see section 1), the measures on each site are only available every two weeks, the data are repeated for two successive weeks to obtain a complete information table (see table 1), alternatively for sites C_1, C_2 and C_3 till C_8.

– The conditional attributes C_9 till C_{13} correspond to five climatic factors, respectively :

C_9 : the atmospheric pressure
C_{10} : the sun period
C_{11} : the air temperature
C_{12} : the relative humidity
C_{13} : the rainfall.

The data for an object x is the mean value of an attribute on the week x; a similar assumption of normal distribution is made and again five classes are defined.

For an attribute $q \varepsilon \{C_9, C_{10}, C_{11}, C_{12}, C_{13}\}$ the descriptor $f(x,q)$ of an object x is again a value of $Vq = \{1,2,3,4,5\}$ indicating the level of these climatic attributes during the week x.

– The two last attributes C_{14} and C_{15} indicate if there exists, or not, some frost, at the ground level and two centimeters below this level respectively;

so $\quad f(x,q) = 1 \quad$ case of frost
$\quad\quad = 0 \quad$ otherwise

for $q \varepsilon \{C_{14}, C_{15}\}$

<u>Remarks</u>

(i) It is sufficient to keep these data at the day before the measures of the "boukoal" are taken.

(ii) Probably, it must be interesting to measure the possible frost more deeply below the ground level, but this information is not available for these periods.

. <u>D : the decision attributes</u>

Clearly, there is an unique decision attribute related to the risk level of seismic activity. The geology laboratory proposed to distinguish only two classes for this attribute q:
$f(x,q) = 1$ if the registered magnitude on the Richter scale is less or equal to 1.5,
$f(x,q) = 2$ otherwise.

This limit of magnitude 1.5 corresponds to the perception of some troubles by the human body and it is the reason of this choice.

Nevertheless, this limit introduce a large dissymetry between the two equivalence classes $D^{(1)}$ and $D^{(2)}$ of relation D : effectively (see table 1), the two D–elementary sets (see [7]) contains respectively 147 and ... 8 objects respectively:

$D^{(1)} = \mid \{x \varepsilon U \; f(x,D) = 1\} \mid = 147$
$D^{(2)} = \mid \{x \varepsilon U \; f(x,D) = 1\} \mid = 8$

4. The analysis of the information system by the rough set approach

A major difficulty to apply to this information system a classical statistical method of discriminant analysis - for instance like the one described in [10] - is the very low number of objects present in class $D^{(2)}$: the sample of objects included in this class is not representative enough in the statistical meaning.

For this reason, we have for the first time experimented a rough set approach [7], applying the first version of the software proposed by R. Słowiński and J. Stefanowski [9].

The information table is given in annex 1 ; all the 155 objects are atoms.

4.1.

The main characteristics of the information system described by table 1 is that the quality of the classification corresponding to the set C of all conditional attributes is equal to one :

$$\gamma_C(D) = 1,$$

where $D = \{D^{(1)}, D^{(2)}\}$ is the partition of U defined by the decision attribute D.

The decision table associated to the information system is thus consistent.

4.2.

The CORE(C) is empty, because RED(C) is a large and diversified family of reducts of C.

Here is the sample of 10 possible reducts of C, those containing 4 or 5 attributes :

Table 1. Sample of reducts

$R_1 = \{1,2,4,5\}$
$R_2 = \{1,2,4,6\}$
$R_3 = \{1,2,4,12\}$
$R_4 = \{2,4,5,9\}$
$R_5 = \{1,2,6,8\}$
$R_6 = \{1,2,5,6\}$
$R_7 = \{1,2,6,9\}$
$R_8 = \{1,2,3,6,10\}$
$R_9 = \{1,2,3,6,11\}$
$R_{10} = \{1,2,3,6,12\}$

We observe that :

- there are always at least three attributes corresponding to sites (attributes C_1 till C_8); the attributes related to climatic factors play a less important role, and this is not really a surprise; for this reason, in the following, we will only draw our attention to the atrributes corresponding to sites,
- in the set of ten interesting reducts, the sites appear with the following frequency :

Table 2.

Sites	Frequency in the ten reducts
2	1
1	0.9
6	0.7
4	0.4
3	0.3
5	0.3
8	0.1

4.3.

It is interesting to determine the quality of the classification using some subsets of attributes.
First we consider singletons :

Table 3.

Attribute	1	2	6	3	4	8	5	7
Quality of approximation	0.56	0.45	0.31	0.23	0.18	0.15	0.12	0

We derive from tables 2 and 3 that the two more interesting sites are 1 and 2, and after the sites 6, 4 and 3.

Then, we successively examined several subsets of attributes, with two or three attributes, to analyse the corresponding quality of classification (see table 4).

Table 4.

Subset P	Quality γ_P of the approximation
{1,2}	0.83
{2,6}	0.74
{2,4}	0.75
{1,4}	0.72
{1,6}	0.69
{2,3}	0.69
{1,3}	0.66
{4,6}	0.58
{3,6}	0.53
{3,4}	0.52
{1,4,6}	0.96
{2,4,6}	0.96
{1,2,6}	0.92
{1,2,4}	0.90
{1,2,3}	0.87
{2,3,4}	0.84
{2,3,6}	0.83
{3,4,6}	0.82
{1,3,4}	0.81
{1,3,6}	0.80

4.4

For each reduct, it is possible to obtain a set of deterministic rules; in each case, the number of rules is greater or equal to 25.
The minimum of rules are obtained respectively with :
. reduct R_1 : 25 rules, with 6 rules for class $D^{(2)}$
. reduct R_5 : 26 rules, with 7 rules for class $D^{(2)}$
. reduct R_2 : 27 rules, with 7 rules for class $D^{(2)}$

The deterministic rules corresponding to these three reducts are given in annex 2.

We remark that the nine first rules, using only attributes 1 and 2, are common to the three situations.

4.5.

Clearly the attributes 1 and 2 are of major importance for the analysis.

Moreover we observe that the objects x classified in class $D^{(2)}$ always correspond to low values of attribute C_1 and high values of C_2 ; effectively

$f(x, C_1) \in \{1,2\} \qquad \forall\, x \in D^{(2)}$
$f(x, C_2) \in \{3,4,5\} \qquad \forall\, x \in D^{(2)}$

This observation appears clearly from table 5, giving the distribution of the 155 objects - the 147 one of $D^{(1)}$, the 8 of $D^{(2)}$- in the {1,2} - elementary sets :

Table 5.

C_1 \ C_2	1	2	3	4	5
1	2	8	4	4 / 2	1 / 1
2	10	14	3 / 1	10 / 4	4
3	6	14	4	8	2
4	0	8	2	12	4
5	5	6	6	4	8

Nevertheless, there exist some objects x, classified in $D^{(1)}$, with the same characteristics.

<u>Remark</u>

Similary, we have mean values for attribute C_3 :

$f(x, C_3) \in \{2,3,4\} \qquad \forall\, x \in D^{(2)}$

4.6.

A final interesting observation may be derived by a precise observation of the objects classified in the six particular elementary sets defined by the descriptors.

$\{C_1 = 1 \text{ for } i = 1,2;\ C_2 = j \text{ for } j = 3,4,5\}$

Effectively almost all the 34 objects - with only very few of exceptions - classified in these six elementary sets are either objects - i.e. observations at a precise week - with seismic activity (i.e. belonging to class $D^{(2)}$) or ... observations located in time, just before or just after these weeks; so almost all these observations are coming from the two periods with an anormal high level seismic activity.

This clearly appears from table 6, in which the numbers refered to the objects of the information system (see annex 1).

Table 6.

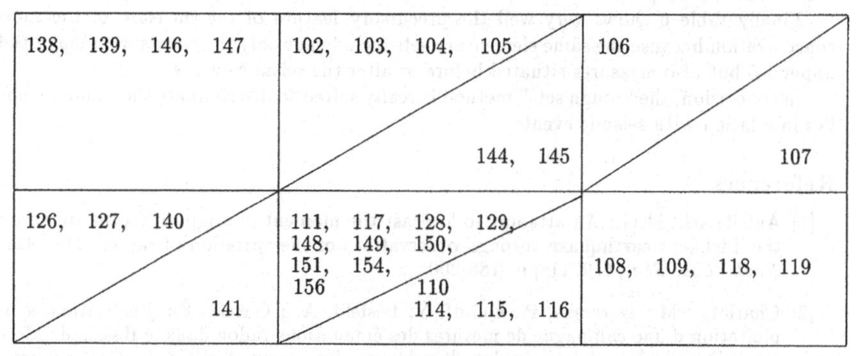

138, 139, 146, 147	102, 103, 104, 105 / 144, 145	106 / 107
126, 127, 140 / 141	111, 117, 128, 129, 148, 149, 150, 151, 154, 156 / 110 114, 115, 116	108, 109, 118, 119

So, we may suppose that the norm used to define the decision classes are not appropriate.

A further analysis is needed : new norms of the decision attribute must be tested, defining, for instance, the limit between classes $D^{(1)}$ and $D^{(2)}$ as the value 1.2 or 1.3 on the Richter scale instead of 1.5. With such modification, we will probably have a better quality of approximation using the {1,2} - elementary sets.

5. Conclusions

The application of the "rough sets" method shows that the attributes 1 and 2 are the most important attributes for earthquake predictions in the Mons basin. These attributes concern two radon points of measurement situated in the Northern part of the Mons basin where the Paleozoic basement outcrops with the carboniferous limestone and the shales of the Namurian. Uraniferous beds occur in the black shales facies of the Upper Visean -lower Namurian (superficial radon source, Blaton, C_1). Otherwise the "Beloeil - Stambruges" site C_2 concerns the environment of a spring in relation with the geothermal system of St Ghislain characterized by a deeper radon source. Besides the geological formations of the Northern part of the Mons basin are affected by transversal and longitudinal faults.

On the other hand, among the sites with a very low or nul value of the quality of approximation, it may be noted the Nimy site (attribute 7) situated in the tertiary sands.

So it is quite logic that the best sites sensitive to seismic prediction are situated in the Northern region.

However, the interpretation introduces a question. Usually a seismic risk lead to an increase of the radon concentration and thus the low radon concentration of the Blaton site with the possibility of a seismic risk could appear abnormal. However, a paper of Monnin and Seidel 1988 [7] about a theoretical study of the radon emission before an important geophysical event shows that a decrease of the radon emission is possible following the situation of the point of measure.

According to Monnin and Siedel, radon in soils does not originate from deep zones. Its variations of concentration depend upon the motion of other gases from deeper origin that transport it. One may point out that Blaton is a site with an uraniferous anomaly situated near the ground surface (at about five meter deep), whereas Stambruges site is connected with spring waters in relation with the geothermal system of the Mons basin and so the radon source is deeper.

Finally table 6 shows very well the precursory feature of the measure of the radon concentration because the same elementary sets include not only the events with magnitude upper 1.5 but also measures situated before or after the seismic events.

In conclusion, the "rough sets" method is really suited to discriminate the radon anomalies in relation with seismic events.

References

[1] Antsilevich, M.G.: An attempt to forecast the moment of origin of recent tremors of the Tashkent earthquake through observations of the variation of radon. *Izv. Acad. Nauk. Uzb. SSR*, (1971), pp. 188-200

[2] Charlet, J.M.; Doremus, P., Quinif, Y; Losfeld, A. ; Garcia, F.: Réalisation et exploitation d'une campagne de mesures des émanations radon dans le Bassin de Mons, essai d'application é la prévision des risques sismiques. *Ann. Soc. Géol. Belg.*, t. 112, 2, (1989), pp. 381-392.

[3] Charlet, J.M.; Doremus, P.; Kotzmann-Routier, V. ; Leclercq, F. ; Vandycke, S.: *International colloquium on gas geochemistry*, field session (radon in the Mons basin), Mons, 3-6 October 1990, 38 p.

[4] King, C.Y.: Gas geochemistry applied to earthquake prediction : an overview. *J. of Geophysical Research*, 91 (B12) (1986), pp. 12269-12281.

[5] King, C.Y.: Radon monitoring for earthquake prediction in China. *Earthquake Predict. Res*, 3 (1985), pp. 47-68.

[6] Klusman, R. ; Webster, J.: Preliminary analysis of meteorological and seasonal influences in crustal gas emission relevant to earthquake prediction. *Bull. Seismol. Soc. Am.* (1981), pp. 211-222

[7] Monnin, M. ; Seidel, J.L. Sur une hypothétique émission intense de radon avant un évènement géophysique majeur : une analyse thórique. C.R. Acad. Sc. (1988), 307, pp. 1363-1368.

[8] Pawlak Z. Rough Sets. Kluwer Academic Publishers, Theory and decision library, Series D (1991)

[9] Słowiński R., Stefanowski J. RoughDAS and RoughClass software implementation of the rough sets approach. Chapter III.8 in this volumme (1992).

[10] Teghem J. ; Benjelloun M. Some experiments to compare rough set theory and ordinal statistical methods. Chapter II.4 in this volume (1992).

[11] Ulomov, V.I. ; Mavashev, B.Z. A precursor of a strong tectonic earthquake. Dokl. Akad. Sc. USSR, Earth Sci. Sect. Engl. transl. 176, (1968), pp. 9-11.

Annex 1. The information system

Atr. / Nr ob.	1	2	3	4	5	6	7	8	9	10	11	12	13	14	15	D_r
1	5	1	5	4	3	5	5	5	5	2	1	4	1	1	1	1
2	5	2	5	4	3	5	5	5	1	1	2	5	2	1	1	1
3	5	2	5	5	3	5	5	4	3	1	1	5	5	2	2	1
4	5	1	5	5	3	5	5	4	4	4	1	3	1	2	2	1
5	5	1	5	4	3	3	5	5	1	5	1	2	3	1	1	1
6	4	2	5	4	3	3	5	5	5	5	1	2	1	1	1	1
7	4	2	5	5	4	4	5	5	5	2	2	4	1	1	2	1
8	5	3	5	5	4	4	5	5	4	3	3	5	2	1	2	1
9	5	3	5	5	3	4	5	5	5	3	2	4	1	1	2	1
10	5	4	5	5	3	4	5	5	2	3	1	4	2	1	2	1
11	5	4	5	5	3	5	5	5	1	2	2	4	5	1	2	1
12	5	3	5	5	3	5	5	5	3	4	4	3	4	2	2	1
13	5	3	4	4	2	5	4	3	1	3	3	4	5	2	2	1
14	5	1	4	4	2	5	4	3	5	4	3	4	5	2	2	1
15	5	1	5	2	1	4	3	4	3	4	3	1	1	1	2	1
16	5	2	5	2	1	4	3	4	3	2	3	3	3	2	2	1
17	5	2	5	2	3	5	3	4	1	1	3	5	4	2	2	1
18	3	1	5	2	3	5	3	4	2	3	4	3	5	2	2	1
19	3	1	5	1	3	5	3	4	1	3	4	4	5	2	2	1
20	3	1	5	1	3	5	3	4	3	5	5	1	1	2	2	1
21	3	1	3	1	2	5	1	2	2	5	5	1	1	2	2	1
22	2	2	3	1	2	5	1	2	2	3	4	2	5	2	2	1
23	2	2	3	2	3	5	2	2	3	3	4	2	3	2	2	1
24	4	2	3	2	3	5	2	2	3	2	4	4	4	2	2	1
25	4	2	4	2	2	5	1	2	4	5	5	2	1	2	2	1
26	2	1	4	2	2	5	1	2	4	4	5	2	1	2	2	1
27	2	1	2	1	3	3	1	2	4	5	5	2	1	2	2	1
28	3	2	2	1	3	3	1	2	4	4	5	4	2	2	2	1
29	3	2	2	2	3	3	1	2	1	3	5	4	5	2	2	1
30	3	2	2	2	3	3	1	2	1	4	4	2	4	2	2	1
31	3	2	3	2	4	4	1	2	3	5	5	4	3	2	2	1
32	3	1	3	2	4	4	1	2	3	4	5	3	1	2	2	1
33	3	1	3	3	1	3	2	2	4	5	4	1	1	2	2	1
34	3	2	3	3	1	3	2	2	3	3	4	3	5	2	2	1
35	3	2	2	1	3	1	1	2	5	5	4	2	1	2	2	1
36	1	2	2	1	3	1	1	2	4	2	4	4	1	2	2	1
37	1	2	2	1	2	1	1	1	4	2	4	5	1	2	2	1
38	2	1	2	1	2	1	1	1	4	5	5	1	2	2	2	1
39	2	1	1	2	3	1	1	1	3	3	4	2	4	2	2	1
40	2	1	1	2	3	1	1	1	5	2	3	5	1	2	2	1
41	2	1	2	3	2	2	2	1	5	5	2	4	1	1	2	1
42	4	3	2	3	2	2	2	1	4	2	2	4	1	2	2	1
43	4	3	3	3	4	4	3	1	1	2	3	2	3	2	2	1
44	5	5	3	3	4	4	3	1	3	3	2	2	1	1	2	1
45	5	5	4	3	4	2	4	3	4	1	1	2	1	1	1	1
46	5	5	4	3	4	2	4	3	1	2	1	5	1	1	2	1
47	5	5	5	4	4	3	5	4	3	1	4	2	3	2	2	1
48	4	5	5	4	4	3	5	4	4	1	2	5	5	2	2	1
49	4	5	3	5	4	5	5	2	4	1	3	4	2	2	2	1
50	5	5	3	5	4	5	5	2	1	2	3	4	3	2	2	1
51	5	5	4	4	4	4	5	2	2	1	1	3	2	1	1	1
52	5	5	4	4	4	4	5	2	2	1	1	4	5	1	2	1
53	.5	5	4	5	2	5	5	4	2	1	2	2	5	2	2	1
54	5	3	4	5	2	5	5	4	2	2	2	4	5	2	2	1
55	5	3	4	3	3	2	3	5	1	2	1	2	2	1	2	1
56	4	4	4	3	3	2	3	5	2	2	1	4	1	1	1	1
57	4	4	3	4	2	1	4	5	5	2	1	2	1	1	1	1
58	5	4	3	4	2	1	4	5	1	2	1	1	1	1	1	1

59	5	4	5	4	3	1	4	4	4	5	1	1	1	1	1	1
60	4	4	5	4	3	1	4	4	2	2	1	4	4	2	2	1
61	4	4	4	5	1	2	4	5	4	4	2	1	1	1	1	1
62	4	4	4	5	1	2	4	5	3	2	3	2	1	2	2	1
63	4	4	1	4	1	2	3	2	1	3	2	2	5	2	2	1
64	4	2	1	4	1	2	3	2	2	4	2	2	5	1	2	1
65	4	2	3	1	1	2	3	4	2	1	2	5	2	2	2	1
66	3	4	3	1	1	2	3	4	1	2	3	3	4	2	2	1
67	3	4	2	2	3	4	4	3	1	4	3	1	5	2	2	1
68	4	4	2	2	3	4	4	3	4	5	3	1	1	2	2	1
69	4	4	2	2	1	2	2	1	2	3	4	1	1	2	2	1
70	3	5	2	2	1	2	2	1	2	5	4	1	1	2	2	1
71	3	5	2	2	3	3	2	2	3	5	5	1	2	2	2	1
72	3	3	2	2	3	3	2	2	4	5	4	1	1	2	2	1
73	3	3	2	2	4	2	2	2	2	2	4	4	5	2	2	1
74	3	4	2	2	4	2	2	2	3	5	4	2	5	2	2	1
75	3	4	3	2	1	3	2	4	3	5	5	3	1	2	2	1
76	3	2	3	2	1	3	2	4	3	4	5	3	5	2	2	1
77	3	2	2	1	1	2	1	3	4	5	5	1	1	2	2	1
78	2	1	2	1	1	2	1	3	3	3	4	2	5	2	2	1
79	2	1	2	2	3	2	1	2	4	4	5	3	1	2	2	1
80	3	2	2	2	3	2	1	2	2	3	5	2	1	2	2	1
81	3	2	1	2	2	2	1	1	3	5	5	1	1	2	2	1
82	2	1	1	2	2	2	1	1	4	5	5	1	1	2	2	1
83	2	1	1	1	2	1	1	1	3	2	5	3	1	2	2	1
84	1	1	1	1	2	1	1	1	3	3	4	2	1	2	2	1
85	1	1	1	3	3	2	1	2	1	4	4	2	3	2	2	1
86	2	2	1	3	3	2	1	2	3	3	4	3	3	2	2	1
87	2	2	2	3	3	1	2	2	4	5	3	2	1	2	2	1
88	2	2	2	3	3	1	2	2	3	1	3	5	5	2	2	1
89	2	2	3	5	1	1	2	1	4	4	4	4	1	2	2	1
90	4	2	3	5	1	1	2	1	5	3	4	5	1	2	2	1
91	4	2	3	2	3	3	2	1	5	2	4	4	1	2	2	1
92	3	2	3	2	3	3	2	1	4	3	4	4	1	2	2	1
93	3	2	2	4	4	3	4	1	1	1	3	4	5	2	2	1
94	1	2	2	4	4	3	4	1	4	3	3	4	3	2	2	1
95	1	2	3	4	2	2	3	3	5	1	3	5	2	2	2	1
96	4	4	3	4	2	2	3	3	3	2	3	5	1	2	2	1
97	4	4	3	5	4	4	3	1	3	1	3	5	5	2	2	1
98	3	2	3	5	4	4	3	1	5	2	3	4	3	2	2	1
99	3	2	3	5	3	4	4	3	5	2	2	5	1	2	2	1
100	4	5	3	5	3	4	4	3	4	1	3	3	2	2	2	1
101	4	5	3	5	3	5	5	4	2	2	2	5	5	2	2	1
102	1	4	3	5	3	5	5	4	3	3	1	5	5	1	2	1
103	1	4	3	3	3	5	5	3	2	1	3	5	5	2	2	1
104	1	4	3	3	3	5	5	3	3	2	2	3	5	1	1	1
105	1	4	3	2	2	1	4	4	3	1	1	2	1	1	1	1
106	1	5	3	2	2	1	4	4	5	1	1	5	1	1	1	1
107	1	5	4	3	3	2	4	2	2	1	1	5	1	1	1	2
108	2	5	4	3	3	2	4	2	4	1	1	5	1	2	2	1
109	2	5	3	5	2	5	5	2	1	2	2	5	4	1	2	1
110	2	4	3	5	2	5	5	2	3	2	1	5	4	1	2	2
111	2	4	3	4	3	2	4	5	4	4	1	3	1	1	1	1
112	1	2	3	4	3	2	4	5	3	3	1	3	5	1	1	1
113	1	2	2	2	2	3	3	3	4	3	1	3	1	1	1	1
114	2	4	2	2	2	3	3	3	2	2	2	4	2	1	2	2
115	2	4	2	1	2	2	4	4	1	1	3	5	5	2	2	2
116	2	4	2	1	2	2	4	4	4	2	2	1	5	2	2	2

117	2	4	2	1	3	3	3	4	2	4	3	2	4	2	2	1
118	2	5	2	1	3	3	3	4	5	2	3	3	1	2	2	1
119	2	5	1	1	1	1	1	1	5	4	3	2	1	2	2	1
120	3	4	2	2	1	1	1	1	4	5	4	1	1	2	2	1
121	3	4	2	2	2	1	2	2	5	4	3	2	2	2	2	1
122	3	3	2	2	2	1	2	2	1	3	3	1	4	2	2	1
123	3	3	3	4	3	3	3	4	3	3	3	2	4	2	2	1
124	4	4	3	4	3	3	3	4	3	5	4	1	1	2	2	1
125	4	4	2	2	4	4	3	3	3	3	4	3	4	2	2	1
126	2	3	2	2	4	4	3	3	2	4	4	2	5	2	2	1
127	2	3	2	3	4	4	3	5	3	2	3	4	4	2	2	1
128	2	4	2	3	4	4	3	5	4	2	4	4	5	2	2	1
129	2	4	2	1	5	3	3	4	4	5	5	2	2	2	2	1
130	2	2	2	1	5	3	3	4	2	5	5	1	1	2	2	1
131	2	2	1	1	5	2	1	2	3	5	5	1	1	2	2	1
132	1	2	1	1	5	2	1	2	2	2	4	4	5	2	2	1
133	1	2	2	3	5	3	3	3	3	1	4	4	5	2	2	1
134	2	2	2	3	5	3	3	3	3	4	4	3	3	2	2	1
135	2	2	2	1	4	3	2	2	3	4	5	1	1	2	2	1
136	2	2	2	1	4	3	2	2	3	3	5	2	1	2	2	1
137	2	2	1	2	5	3	3	2	1	1	4	5	5	2	2	1
138	1	3	1	2	5	3	3	2	3	3	5	4	1	2	2	1
139	1	3	2	2	5	3	2	2	4	3	4	4	4	2	2	1
140	2	3	2	2	5	3	2	2	4	3	4	3	1	2	2	1
141	2	3	2	1	4	3	1	1	2	2	5	4	1	2	2	2
142	2	2	2	1	4	3	1	1	5	4	3	3	1	2	2	1
143	2	2	2	3	5	3	2	2	1	1	4	4	5	2	2	1
144	1	4	2	3	5	3	2	2	1	2	3	4	5	2	2	2
145	1	4	3	5	5	3	3	4	2	3	4	3	4	2	2	2
146	1	3	3	5	5	3	3	4	3	1	3	5	2	2	2	1
147	1	3	3	4	3	3	4	4	5	3	3	5	1	2	2	1
148	2	4	3	4	3	3	4	4	2	1	2	5	1	2	2	1
149	2	4	3	5	5	3	5	4	4	2	3	4	5	2	2	1
150	2	4	3	5	5	3	5	4	1	1	2	4	5	2	2	1
151	2	4	4	4	2	1	5	5	5	4	1	5	3	1	2	1
152	3	4	4	4	2	1	5	5	4	4	1	2	1	1	1	1
153	3	4	4	3	4	2	5	5	1	1	1	5	1	2	2	1
154	2	4	4	3	4	2	5	5	5	2	2	5	1	1	2	1
155	2	4	3	5	4	1	5	4	5	1	3	3	1	2	2	1

Annex 2. Decision rules corresponding to three reducts

Reduct $R_1 = \{1,2,4,5\}$

1.	C1 = 5 ⟶	d = 1			
2.	C1 = 4 ⟶	d = 1			
3.	C1 = 3 ⟶	d = 1			
4.	C1 = 2	C2 = 2 ⟶	d = 1		
5.	C1 = 2	C2 = 1 ⟶	d = 1		
6.	C1 = 1	C2 = 2 ⟶	d = 1		
7.	C1 = 1	C2 = 1 ⟶	d = 1		
8.	C1 = 2	C2 = 5 ⟶	d = 1		
9.	C1 = 1	C2 = 3 ⟶	d = 1		
10.	C1 = 1	C2 = 4	C5 = 3 ⟶	d = 1	
11.	C1 = 1	C2 = 4	C5 = 2 ⟶	d = 1	
12.	C1 = 1	C2 = 5	C5 = 2 ⟶	d = 1	
13.	C1 = 2	C2 = 4	C5 = 3 ⟶	d = 1	
14.	C1 = 2	C2 = 4	C5 = 4 ⟶	d = 1	
15.	C1 = 2	C2 = 4	C5 = 5 ⟶	d = 1	
16.	C1 = 2	C2 = 3	C5 = 5 ⟶	d = 1	
17.	C1 = 2	C2 = 3	C4 = 2	C5 = 4 ⟶	d = 1
18.	C1 = 2	C2 = 3	C4 = 3	C5 = 4 ⟶	d = 1
19.	C1 = 2	C2 = 4	C4 = 4	C5 = 2 ⟶	d = 1
20.	C1 = 1	C2 = 5	C5 = 3 ⟶	d = 2	
21.	C1 = 1	C2 = 4	C5 = 5 ⟶	d = 2	
22.	C1 = 2	C2 = 4	C4 = 5	C5 = 2 ⟶	d = 2
23.	C1 = 2	C2 = 4	C4 = 2	C5 = 2 ⟶	d = 2
24.	C1 = 2	C2 = 4	C4 = 1	C5 = 2 ⟶	d = 2
25.	C1 = 2	C2 = 3	C4 = 1	C5 = 4 ⟶	d = 2

Reduct $R_2 = \{1,2,4,6\}$

1.	C1 = 5 ⟶	d = 1			
2.	C1 = 4 ⟶	d = 1			
3.	C1 = 3 ⟶	d = 1			
4.	C1 = 2	C2 = 2 ⟶	d = 1		
5.	C1 = 2	C2 = 1 ⟶	d = 1		
6.	C1 = 1	C2 = 2 ⟶	d = 1		
7.	C1 = 1	C2 = 1 ⟶	d = 1		
8.	C1 = 2	C2 = 5 ⟶	d = 1		
9.	C1 = 1	C2 = 3 ⟶	d = 1		
10.	C1 = 1	C2 = 4	C4 = 2 ⟶	d = 1	
11.	C1 = 1	C2 = 5	C4 = 2 ⟶	d = 1	
12.	C1 = 2	C2 = 4	C4 = 4 ⟶	d = 1	
13.	C1 = 2	C2 = 3	C4 = 2 ⟶	d = 1	
14.	C1 = 2	C2 = 3	C4 = 3 ⟶	d = 1	
15.	C1 = 2	C2 = 4	C4 = 3 ⟶	d = 1	
16.	C1 = 1	C2 = 4	C4 = 5	C6 = 5 ⟶	d = 1
17.	C1 = 1	C2 = 4	C4 = 3	C6 = 5 ⟶	d = 1
18.	C1 = 2	C2 = 4	C4 = 1	C6 = 3 ⟶	d = 1
19.	C1 = 2	C2 = 4	C4 = 5	C6 = 3 ⟶	d = 1
20.	C1 = 2	C2 = 4	C4 = 5	C6 = 1 ⟶	d = 1
21.	C1 = 1	C2 = 5	C4 = 3 ⟶	d = 2	
22.	C1 = 2	C2 = 4	C4 = 2 ⟶	d = 2	
23.	C1 = 2	C2 = 3	C4 = 1 ⟶	d = 2	
24.	C1 = 2	C2 = 4	C4 = 5	C6 = 5 ⟶	d = 2
25.	C1 = 2	C2 = 4	C4 = 1	C6 = 2 ⟶	d = 2
26.	C1 = 1	C2 = 4	C4 = 3	C6 = 3 ⟶	d = 2
27.	C1 = 1	C2 = 4	C4 = 5	C6 = 3 ⟶	d = 2

Reduct $R_5 = \{1, 2, 6, 8\}$

1.	C1 = 5 ⟶	d = 1			
2.	C1 = 4 ⟶	d = 1			
3.	C1 = 3 ⟶	d = 1			
4.	C1 = 2	C2 = 2 ⟶	d = 1		
5.	C1 = 2	C2 = 1 ⟶	d = 1		
6.	C1 = 1	C2 = 2 ⟶	d = 1		
7.	C1 = 1	C2 = 1 ⟶	d = 1		
8.	C1 = 2	C2 = 5 ⟶	d = 1		
9.	C1 = 1	C2 = 3 ⟶	d = 1		
10.	C1 = 1	C2 = 4	C8 = 3 ⟶	d = 1	
11.	C1 = 1	C2 = 5	C8 = 4 ⟶	d = 1	
12.	C1 = 2	C2 = 4	C8 = 5 ⟶	d = 1	
13.	C1 = 2	C2 = 3	C8 = 3 ⟶	d = 1	
14.	C1 = 2	C2 = 3	C8 = 5 ⟶	d = 1	
15.	C1 = 2	C2 = 3	C8 = 2 ⟶	d = 1	
16.	C1 = 1	C2 = 4	C6 = 5	C8 = 4 ⟶	d = 1
17.	C1 = 2	C2 = 4	C6 = 1	C8 = 4 ⟶	d = 1
18.	C1 = 2	C2 = 4	C6 = 3	C8 = 4 ⟶	d = 1
19.	C1 = 2	C2 = 4	C6 = 1	C8 = 4 ⟶	d = 1
20.	C1 = 1	C2 = 5	C8 = 2 ⟶	d = 2	
21.	C1 = 2	C2 = 4	C8 = 2 ⟶	d = 2	
22.	C1 = 2	C2 = 4	C8 = 3 ⟶	d = 2	
23.	C1 = 2	C2 = 3	C8 = 1 ⟶	d = 2	
24.	C1 = 1	C2 = 4	C8 = 2 ⟶	d = 2	
25.	C1 = 2	C2 = 4	C6 = 2	C8 = 4 ⟶	d = 2
26.	C1 = 1	C2 = 4	C6 = 3	C8 = 4 ⟶	d = 2

Part I
Chapter 13

ROUGH SETS AND SOME ASPECTS OF LOGIC SYNTHESIS

Tadeusz LUBA
Warsaw University of Technology
Institute of Telecommunications
00-447 Warsaw, Poland

Janusz RYBNIK
Hewlett-Packard Polska
00-447 Warsaw
Poland

Abstract. This paper is dedicated to two seemingly different problems. The first one concerns information systems theory and the second one is connected to logic synthesis methods. The common aspect in considering these problems together is the important task of the economic representation of data in information systems and as well as in logic systems. An efficient algorithm to solve the task of attributes/arguments reduction as well as functional decomposition of decision/truth tables is presented. In the latter case a new technique is suggested, which decomposes the original decision/ truth table into an equivalent set of subtables. Using manipulations based on both rough sets and Boolean algebra theory, the decision table is reduced and decomposed so as to get an efficient implementation.

1. Introduction

The aim of the paper is to investigate the connections between currently researched theories: the theory of multiple-valued Boolean functions and the theory of rough sets, particularly its use in information systems and logic synthesis. Both of these theories are the subject of intensive research due to the importance of their practical applications. In the case of multi-valued functions this involves the logic synthesis of VLSI circuits, in the case of rough sets: the analysis and synthesis of information systems.

The two issues, until now researched independently of one another, have resulted in a series of computational methods, algorithms and their computer implementations, which possess so many similarities that it is worthwhile to investigate and apply their common realizations.

An important problem in the practical application of information systems is the reduction of knowledge. This problem has been investigated from a number of points of view: one of them is whether the whole set of attributes is always necessary to define a given partition of an universe and the other concerns the simplification of decision tables, namely the reduction of condition attributes in a decision table.

A similar problem arises in logic synthesis where circuits performance can be presented as truth tables which are in fact decision tables with two valued attributes, where condition attributes are in fact input variables, and decision ones are to represent output variables of the circuit. In the practical application of Boolean algebra the key problem is to represent Boolean functions by formulas which are as simple as possible. One approach to this simplification is to minimize the number of variables appearing in truth table explicitly.

The paper begins with an overview of basic notions related in information systems and rough sets. Then we discuss relations between multiple-valued logic and decision table systems with respect to rough set model of data. Particularly it is shown that elimination of input variables can be easily obtained using standard attribute reduction process used in decision tables. Finally, benefits arising from the logic decomposition are presented.

2. Information systems and rough sets

An information system is a pair $\mathcal{A} = (U, A)$, where

U – is a nonempty, finite set called the universe and

A – is a nonempty, finite set of attributes i.e.each element $a \in A$ is a total function from U into V_a, where V_a is called the domain of a.

With every subset of attributes $B \subseteq A$, a binary relation $IND(B)$, called B – indiscernibility relation, is defined as follows:

$$IND(B) = \{(x, y) \in U^2 : \text{for every } a \in B,\ a(x) = a(y)\}$$

As $IND(B)$ is an equivalence relation and

$$IND(B) = \bigcap_{a \in B} IND(a)$$

the family of all equivalence classes of $IND(B)$ is usually denoted by $U/IND(B)$.

Some subsets (categories) of objects in an information system cannot be expressed exactly by employing available attributes but they can be roughly defined.

If $\mathcal{A} = (U, A)$ is an information system, $B \subseteq A$, and $X \subseteq U$ then the sets

$$\{x \in U : [x]_B \subseteq X\} \quad \text{and} \quad \{x \in U : [x]_B \cap X \neq \emptyset\}$$

where $[x]_B$ denotes the equivalence class of $IND(B)$ including x, are called B–lower and B–upper approximation of X in $\mathcal{A}$. The lower and upper approximations will be denoted by $\underline{B}X$ and $\overline{B}X$, respectively.

The set $\underline{B}X$ is the set of all elements of U which can be with certainty classified as elements of X, in the knowledge represented by attributes B. Set $\overline{B}X$ is the set of elements of U which can be possibly classified as elements of X, employing knowledge represented by attributes from B.

We shall also employ the following denotation:

$$POS_B(X) = \underline{B}X$$

and will refer to $POS_B(X)$ as B–positive region of X.

The positive region $POS_B(X)$ or the lower approximation of X is the collection of those objects which can be classified with full certainty as members of the set X, using classification given by $IND(B)$.

A special class of information systems, which is of great importance in many applications is decision table DT.

The decision tables are information systems with the set of attributes divided into two disjoint sets C and D, called respectively condition and decision attributes i.e. $DT = (U, A, C, D)$.

By C–positive region of D, denoted $POS_{IND(C)}(IND(D))$ or $POS_C(D)$ for simplicity we understand the set

$$POS_C(D) = \bigcup_{x \in U/D} \underline{C}X$$

We say that $c \in C$ is D–dispensable in C, if

$$POS_{IND(C)}(IND(D)) = POS_{IND(C-\{c\})}(IND(D))$$

otherwise c is D–indispensable in C.
In other words c is relatively dispensable in C and c is relatively indispensable in C, respectively.

If every c in C is D–indispensable, we will say that C is D–dependent (or C is independent with respect to D).

The set $S \subseteq C$ will be called a D–reduct of C, if and only if S is D–independent subset of C and $POS_S(D) = POS_C(D)$.

The set of all D–indispensable attributes in C will be called the D–core of C, and will be denoted as $CORE_D(C)$.

$$CORE_D(C) = \bigcap RED_D(C)$$

where $RED_D(C)$ is the family of all D–reducts of C.

A set D of decision attributes in a DT system $T = (U, A, C, D)$ depends on a set C of condition attributes in T, in symbols $C \Longrightarrow D$, if $IND(C) \subseteq IND(D)$.

With every $x \in U$ we associate a function $d_x : A \longrightarrow V$, such that $d_x(a) = a(x)$, for every $a \in C \cup D$; the function d_x will be called a decision rule (in T), and x will be referred to as a label of the decision rule d_x.

If d_x is a decision rule, then the restriction of d_x to C, denoted $d_x|C$, and the restriction of d_x to D, denoted $d_x|D$ will be called conditions and decisions (actions) of d_x, respectively.

The decision rule is consistent (in T) if for every $y \neq x$, $d_x|C = d_y|C$ implies $d_x|D = d_y|D$; otherwise the decision rule is inconsistent. A decision table is consistent if all its decision rules are consistent; otherwise the decision table is inconsistent.

The following equivalence is important property that establishes the relationship among consistency, dependency and positive region of decision attributes in a decision table:

a) decision table $T = (U, A, C, D)$ is consistent
b) D depends on C i.e. $C \Longrightarrow D$
c) $POS_C(D) = U$.

3. Relations between multiple–valued logic and decision tables with respect to rough set model of data

Let x_i be a multiple–valued variable, and $C_i = \{0, 1, ..., c_i - 1\}$ be a set of values it may assume. A generalized Multiple–Valued Boolean function with n input, m output variables is defined as a mapping:

$$F(x_1, ..., x_n) : C_1 \times C_2 \times \cdots \times C_n \longrightarrow D^m,$$

where $D = \{0, 1, *\}$ represents the binary value of the function (0 or 1) [1)]. The value $*$ (don't care) at one of the outputs means that the value is unspecified, and a value of 0 or 1 will be accepted to realize this part of the function.

Every element of the domain $C_1 \times C_2 \times \cdots \times C_n$ is called a minterm. A listing of minterms with the value of the function is called a truth table. Truth tables do not include minterms with the function value not specified for all outputs. Set of minterms for which the function value is unspecified is called DC–set (Don't Care–set). Functions with nonempty DC-set are called partially defined.

It is worth to note that truth tables are in fact decision tables, where condition attributes represent input variables, and decision ones are to represent output variables of the circuit intended to implement the function.

Therefore truth tables may be viewed as Decision Tables $T = (M, A, X, Y)$ where

M — is a nonempty, finite set of objects,
A — is a finite set of variables (arguments); $A = X \cup Y$, where X is a set of input variables and Y is a set of output variables, $X \cap Y = \emptyset$, moreover
a — is a function assigning a value of variable for every object m i.e.

$$a : M \longrightarrow V_a,$$

where V_a is a domain (set of possible values) of variable a. According to physical implementations we assume that the set of values of output variables $V_y = \{0, 1, *\}$.

Applying the notion of a decision rule in the case of truth tables we shall refer to $d_x|X$, and to $d_x|Y$ as input and output vectors, respectively. In this form the

---

1) The assumption of binary values of outputs implies from the structure of commonly used PLAs i.e.multi–valued input, two–valued output PLAs [17], [19], [20], [21].

value of potential outputs are specified for each possible combination of the inputs. According to the definition of Boolean function an input vector will be called a minterm.

In general, any pair of minterms in a table specification of Multiple–Valued Boolean (MVB) function may have identical values for some number of input variables. Therefore an analogous indiscernibility relation as in information systems can be introduced. This relation, denoted IND, is associated with any subset of input variables as follows:

Let $B \subseteq X$, $m_1, m_2 \in M$,

$$(m_1, m_2) \in IND(B) \text{ if and only if } x(m_1) = x(m_2) \text{ for all } x \in B.$$

This means, that $(m_1, m_2) \in IND(B)$ if values of arguments belonging to B are identical for both m_1 and m_2. Minterms m_1 and m_2 are said indiscernible by arguments from B. The indiscernibility relation is an equivalence relation on M and

$$IND(B) = \bigcap_{x \in B} IND(x). \tag{1}$$

Thus, the relation IND partitions M into equivalence classes $M/IND(B)$. Such partitions are of primary importance in logic synthesis [10], [11]. To simplify, we shall denote partition $M/IND(B)$ by $P(B)$ and call such a partition as an input partition generated by set B. Then, the equivalent formula to those of (1) may be written as

$$P(B) = \prod_{x \in B} P(x)$$

where $\prod$ denotes the product of partitions.

Two output vectors, $\mathbf{y_1}$ and $\mathbf{y_2}$, are said to be consistent if their corresponding entries are the same whenever they are both specified i.e. $\forall i \in \{n+1, ..., n+m\}$

$$(\mathbf{y_1}_i = \mathbf{y_2}_i) \vee (\mathbf{y_1}_i = *) \vee (\mathbf{y_2}_i = *).$$

The consistency relation on output vectors is denoted as $\mathbf{y_1} \sim \mathbf{y_2}$.

In general, any pair of minterms in a logic specification table may have consistent output values for some number of output variables. Thus, the relation called output–consistency relation and denoted as CON, can be associated with any subset B of output variables. The output–consistency relation is defined as follows:

Let $B \subseteq Y$, p, $q \in M$,

$$p,\ q \in CON(B) \text{ iff } y(p) \sim y(q) \text{ for every } y \in B.$$

where $a_1 \sim a_2$ if a_1, a_2 are the same whenever are both specified.

A set of minterms constitutes a consistent class, if every pair of minterms in the set is consistent. Those classes that are not subsets of any other output–consistent class are called Maximal Consistent Classes ($MCCs$).

Clearly, the consistency relation is not an equivalence relation on M. Hence, it "partitions" M into non–disjoint subsets; but for a given CON relation there is a unique collection of maximal output–consistent classes of minterms. Therefore, we can use the same notation for output consistency subset, i.e. $P_F(B)$, where index F is intended to distinguish input (IND) and output (CON) relation. When $B = Y$, then we will denote the CON relation simply as P_F. Because of non–disjointness of blocks of P_F, relation CON will be called a rough–partition (r–partition).

Conventions used in denoting r–partitions and their typical operators will be the same as in case of partitions i.e. an r–partition on a set M may be viewed as a collection of non–disjoint subsets of M, where the set union is M. Thus r–partition concepts are simple extensions of partition algebra [5], with which reader familiarity is assumed.

Especially relation *less or equal to* holds between two r–partitions Π_1 and Π_2 ($\Pi_1 \leq \Pi_2$) iff for every block of Π_1 in short denoted by $B_i(\Pi_1)$ there exists a $B_j(\Pi_2)$ such that $B_i(\Pi_1) \subseteq B_j(\Pi_2)$.

If the Π_1 and Π_2 are partitions, this definition reduces to the conventional ordering relation between two partitions.

This points out the main difference between decision tables and incompletely specified Boolean functions. While the equivalence classes of partitions in decision table systems consist of disjoint subsets, the subsets of consistent minterms may be overlapping. This is the reason for generalizing the typical partition description which in the case of Boolean functions proves to be not sufficient in depicting truth tables.

To present a Boolean function F, i.e., functional dependence between outputs Y and inputs X, usually described by formula $Y = F(X)$, the table specification should be consistent.

A logic specification table is consistent iff for every pair of row vectors $\mathbf{r}_1 = (\mathbf{x}_1, \mathbf{y}_1)$, $\mathbf{r}_2 = (\mathbf{x}_2, \mathbf{y}_2)$, $\mathbf{x}_1 = \mathbf{x}_2$ implies $\mathbf{y}_1 \sim \mathbf{y}_2$ (i.e. for every m_1, m_2, $m_1 = m_2$ implies $F(m_1) \sim F(m_2)$).

Example 1

Consider partially defined, multiple–valued Boolean function F shown in Table 1.

In the example:

$M = \{1, \ldots, 10\}$,

$X = \{x_1, \ldots, x_6\}$, $Y = \{y_1, y_2, y_3, y_4\}$,

and $V_{x_1} = \{0, 1\}$, $V_{x_5} = \{0, 1, 2, 3\}$. The IND relations for $B = B_1$ and $B = B_2$, where $B_1 = \{x_1\}$ and $B_2 = \{x_2, x_3\}$ are as follows:

$P(B_1) = \{\{1,6\}, \{2,3,4,5,7,8,9,10\}\}$,

$P(B_2) = \{\{1,3,4\}, \{2\}, \{5\}, \{6,9\}, \{7,8,10\}\}$.

Table 1

	x_1	x_2	x_3	x_4	x_5	x_6	y_1	y_2	y_3	y_4
1	0	1	0	1	0	0	0	0	0	*
2	1	0	0	0	1	3	0	0	1	*
3	1	1	0	2	2	3	1	1	1	0
4	1	1	0	2	3	3	*	0	1	1
5	1	1	1	0	2	3	1	0	0	0
6	0	0	2	0	2	3	*	0	0	*
7	1	1	2	0	2	2	0	1	*	0
8	1	1	2	0	2	3	1	*	1	1
9	1	0	2	2	1	3	1	0	*	1
10	1	1	2	2	3	2	0	1	*	1

Proceeding in the same way for the output–consistency relation CON, we obtain the following r–partitions:

$$P_F(y_1) = \{\{1,2,4,6,7,10\}, \{3,4,5,6,8,9\}\}$$
$$P_F = \{\{1,6\}, \{3\}, \{2,4\}, \{5,6\}, \{6,9\}, \{4,8,9\}, \{7\}, \{10\}\}$$

4. Elimination of input variables

In this section the process of detection and elimination of redundant variables will be described using the concepts of decision systems, however appropriate simplifications caused by functional dependency features as well as the generalization to the case of rough partition will be efficiently applied.

An argument $x \in X$ is called dispensable in a logic specification of function F iff $P(X - \{x\}) \leq P_F$, otherwise i.e. $P(X - \{x\}) \not\leq P_F$, an argument is called indispensable (i.e., an essential variable [9]).

The meaning of an indispensable variable is similar to that of a core attribute i.e. these are the most important variables. In other words, no one indispensable variable can be removed without destroying the consistency of the function specification. Thus, the set of all indispensable arguments will be called the core of X and will be denoted as $CORE(X)$.

In order to find the core set of arguments we have to eliminate an input variable and then to verify whether the corresponding partition inequality holds. A key theorem will be stated below to make this procedure more efficient.

The concept of core plays an important role in reducing the computational complexity relevant to the process of arguments reduction.

Example 2.

Let us characterize indispensability of arguments in the specification of function F from Example 1. Since $P(X - \{x_1\}) \leq P_F$, x_1 is dispensable in function F. In contradiction, x_3 is indispensable argument, because $P(X - \{x_3\}) \not\leq P_F$ and this fact can be observed by inconsistency of the table specification after eliminating x_3.

As the reduction of arguments is of primary importance, in this section we reformulate this problem to apply more useful tools which are efficiently used in switching theory [11]. First of all we reformulate fundamental notions.

A set $B = \{b_1, \ldots, b_k\} \subseteq X$ is called a minimal dependence set (i.e. reduct) of a Boolean function F iff $P(B) \leq P_F$, and there is no proper subset B' of B such that $P(B') \leq P_F$.

It is evident that an indispensable input variable of function F is an argument of every minimal dependence set of F.

Example 3.

The minimal dependence sets for function of table 1 are as follows:

1) $\{x_1, x_3, x_5, x_6\}$,

2) $\{x_2, x_3, x_5, x_6\}$,

From this analysis it follows that, in the case of solution 1), dispensable input variables are x_2, x_4, which means that these variables are superfluous in the definition of logic function described by Table 1, i.e.: $P(x_1, x_3, x_5, x_6) \leq P_F$, and no one proper subset of $B = \{x_1, x_3, x_5, x_6\}$ satisfies this condition. Hence, Table 1 can be replaced by more compact form shown in Table 2.

Table 2

	x_1	x_3	x_5	x_6	y_1	y_2	y_3	y_4
1	0	0	0	0	0	0	0	*
2	1	0	1	3	0	0	1	*
3	1	0	2	3	1	1	1	0
4	1	0	3	3	*	0	1	1
5	1	1	2	3	1	0	0	0
6	0	2	2	3	*	0	0	*
7	1	2	2	2	0	1	*	0
8	1	2	2	3	1	*	1	1
9	1	2	1	3	1	0	*	1
10	1	2	3	2	0	1	*	1

Now we introduce two basic notions, namely discernibility matrix and discernibility function, which will help us to construct an efficient algorithm for argument reduction process.

Let F be a multiple-valued Boolean function and $X = \{x_1, ..., x_n\}$, $M = \{m_1, ..., m_t\}$. By C_{pq}, where p, q are minterms of M, such that $F(p) \not\sim F(q)$, we denote a set of input variables, called discernibility set, defined as follows:

$$C_{pq} = \{x \in X : d_p(x) \neq d_q(x) \text{ for } p, q = 1, \ldots, t \text{ and } p < q\}.$$

where $d_m(x)$ denotes a decision rule restricted to the set X.

A discernibility function f_F for a function F is a Boolean function of n variables $x_1, \ldots x_n$ defined by the conjunction of all expressions $\vee(C_{pq})$, where $\vee(C_{pq})$ is the disjunction of all elements of C_{pq}, $1 \leq p < q \leq t$.

We will describe the collection $C = \{C_1, ..., C_r\}$ of all C_{pq} sets in the form of the binary matrix M for which an element $m_{ij} (i = 1, \ldots r = CARD(C), j = 1, \ldots, t = CARD(X))$ is defined as follows:

$$m_{ij} = \begin{cases} 1 & \text{if } x_j \in C_i \\ 0 & \text{otherwise} \end{cases}$$

Thus, the M matrix is a 0–1 matrix determined by the C_{pq} sets. Our goal is to select an optimal set L of arguments corresponding to columns of M. Here a "column covering " L means that every row of M contains a "1" in some column which appears in L. More precisely, a column cover of binary matrix is defined as a set L of columns such that for every i :

$$\sum_{j \in L} m_{ij} \geq 1.$$

Covers L of M are in one-to-one correspondence with the reduced subsets of arguments (i.e. reducts).

One can easily observe a strong relation between the notion of a reduct of function F ($RED(X)$) and prime implicant of the monotonic Boolean function f_F, namely we have the following equivalence:

$$\{x_{i_1}, \ldots, x_{i_p}\} \in RED(X) \text{ iff } x_{i_1} \wedge \cdots \wedge x_{i_p} \text{ is a prime implicant of } f_F.$$

This remark implies that in a computation of a reduct set of a given F, one can apply known algorithms for computing all prime implicants for a given monotonic Boolean function.

An interesting approach is based on the fact that the unate complementation is intimately related to the concept of a column cover of the binary matrix [1].

Theorem 1 [1]: *Each row i of $\overline{M}$, the binary matrix complement of M, corresponds to a column cover L of M, where $j \in L$ if and only if $\overline{M}_{ij} = 1$.*

The rows of $\overline{M}$ include the set of all minimal column covers of M. If $\overline{M}$ was minimal with respect to containment, then $\overline{M}$ would precisely represent the set of all minimal column covers of M.

Let each column of M corresponds to conjunction factor of F_M, which is defined by the disjunction of all M_i, where M_i is the conjunction of negative literals $\overline{x}_j$ corresponding to $m_{ij} = 1$.

To obtain discernibility function in the minimal disjunctive normal form (DNF) we apply the fast complementation algorithm of unate Boolean functions adopted from ESPRESSO [1].

The fast complementation algorithm for monotonously decreasing function F_M is based on the Shannon expansion of F_M, for simplicity denoted by F

$$F = x_j F_{x_j} + \overline{x}_j F_{\overline{x}_j} \tag{2}$$

where F_{x_j}, $F_{\overline{x}_j}$ are cofactors of F with respect to splitting variable x_j i.e. the results of substituting **1** and **0** for x_j in F.

Hence applying the property of unateness i.e. $F_{x_j} \leq F_{\overline{x}_j}$, F can be expressed as

$$F = \overline{x}_j F_{\overline{x}_j} + F_{x_j} = F_{\overline{x}_j}(\overline{x}_j + F_{x_j}) \tag{3}$$

Thus by complementing (3) we obtain a simplified formula:

$$\overline{F} = x_j \overline{F}_{x_j} + \overline{F}_{\overline{x}_j}$$

which is the key to a fast recursive complementation process.

In order to find an efficient algorithm to complement a function we will again represent F as binary matrix $M(F)$. Let there be an one–to–one correspondence between columns of M and variables of F. Let each row of $M(F)$ correspond to a product term of M. Then

$$M_{ij}(F) = \begin{cases} 1, & \text{if in } i\text{–term there is variable of column } j, \\ 0, & \text{otherwise.} \end{cases}$$

The matrix $M(F)$ will be directly used in complementation algorithm.

We illustrate the unate complementation algorithm with a following example.

Let $F = \overline{x}_1\overline{x}_2\overline{x}_4 + \overline{x}_3\overline{x}_4 + \overline{x}_1\overline{x}_2 + \overline{x}_1\overline{x}_4$. Hence:

$$M(F) = \begin{bmatrix} 1 & 1 & 0 & 1 \\ 0 & 0 & 1 & 1 \\ 1 & 1 & 0 & 0 \\ 1 & 0 & 0 & 1 \end{bmatrix}$$

In order to identify the splitting variables we choose them among the shortest terms in F. Here we select the second term, yielding variables x_3 and x_4. Since the variable that appears most often in the other terms of F is x_4, we decide to choose this one.

Now we compute the cofactors of F with respect to the variable x_4:

$$F_{\overline{x}_4} = \overline{x}_1\overline{x}_2 + \overline{x}_3 + \overline{x}_1\overline{x}_2 + \overline{x}_1 , \qquad F_{x_4} = \overline{x}_1\overline{x}_2$$

$$M(F_{\overline{x}_4}) = \begin{bmatrix} 1 & 1 & 0 & 0 \\ 0 & 0 & 1 & 0 \\ 1 & 1 & 0 & 0 \\ 1 & 0 & 0 & 0 \end{bmatrix} \qquad M(F_{x_4}) = [1 \;\; 1 \;\; 0 \;\; 0]$$

The cofactor with respect to $\overline{x}_j$ is obtained by setting up the jth column to 0, and the cofactor with respect to x_j is obtained by excluding all the rows for which the jth element is equal to 1.

In each branch of the recursion we examine the possibilities for complementation using easily computable special cases:

a) There is a row of all 0's in $M(F)$ (empty conjunction is equal **1**)

b) $M(F)$ is empty (empty disjunction is equal **0**)

c) $M(F)$ has only one row (apply De Morgan Law to the unique term).

In our example the complement of F_{x_4} is : $\overline{F}_{x_4} = x_2 + x_1$

$$M(\overline{F}_{x_4}) = \begin{bmatrix} 0 & 1 & 0 & 0 \\ 1 & 0 & 0 & 0 \end{bmatrix}$$

$F_{\overline{x}_4}$ must be processed further and yields:

$$F_{\overline{x}_4\overline{x}_1} = \mathbf{1} \qquad F_{\overline{x}_4 x_1} = \overline{x}_3$$

$$M(F_{\overline{x}_4\overline{x}_1}) = \begin{bmatrix} 0 & 1 & 0 & 0 \\ 0 & 0 & 1 & 0 \\ 0 & 1 & 0 & 0 \\ 0 & 0 & 0 & 0 \end{bmatrix} \qquad M(F_{\overline{x}_4 x_1}) = [0 \;\; 0 \;\; 1 \;\; 0]$$

and hence $\overline{F}_{\overline{x}_4\overline{x}_1} = \mathbf{0}, \overline{F}_{\overline{x}_4 x_1} = x_3$.

$$M(\overline{F}_{\overline{x}_4\overline{x}_1}) = \emptyset \qquad M(\overline{F}_{\overline{x}_4 x_1}) = [0 \;\; 0 \;\; 1 \;\; 0]$$

By merging these results we obtain

$$\overline{F}_{\overline{x}_4} = x_1\overline{F}_{\overline{x}_4 x_1} + \overline{F}_{\overline{x}_4\overline{x}_1} = x_1 x_3 + \mathbf{0} = x_1 x_3$$

$$M(\overline{F}_{\overline{x}_4}) = [1 \;\; 0 \;\; 1 \;\; 0]$$

and finally

$$\overline{F} = x_4\overline{F}_{x_4} + \overline{F}_{\overline{x}_4} = x_4(x_2 + x_1) + x_1 x_3 = x_2 x_4 + x_1 x_4 + x_1 x_3 .$$

We interpret the final expression as the set of reducts: $\{x_2, x_4\}$, $\{x_1, x_4\}$, $\{x_1, x_3\}$. This agrees with the result that can be obtained using discernibility function proposed in [22]. For the discernibility matrix $M(F)$ the appropriate function f_M is as follows:

$$f_M = (x_1 + x_2 + x_4)(x_3 + x_4)(x_1 + x_2)(x_1 + x_4)$$

Hence, performing the multiplication and applying absorption law we obtain the same set of reducts.

5. Logic decomposition

It is sometimes the case that a set of Boolean functions cannot be made to fit into any single module designated for implementation. The only solution is to decompose the problem in such a way that the requirements can be met by a network of two or more devices each implementing a part of the functions.

A similar problem arises in information systems where functionally dependent data can be projected out of a given DT – decision table and complete DT can always be recovered by means of the joining operation. This resolution is important to the system designer, for it allows him design freedom. He may or may not decide to break down a file into components, depending on possible storage savings or other considerations. It is important to him to be aware of the logical possibilities that functional dependency offers. It would be very interesting to elaborate a method of breaking down a DT into components, depending on possible storage savings. To solve this problem we adopt the decomposition technique that has been developed specially for logic functions.

Basically the need for decomposition arises very naturally in the case of functional dependencies. The meaning of the dependency relation is as follows: holding the condition $C \Longrightarrow D$ assures that if a pair of objects cannot be distinguished by means of attributes belonging to set C, then it cannot be distinguished by attributes from set D, in other words values of attributes from D are determined by values of attributes from set C.

The intuitive meaning of this concept appears from Table 3, where the condition attributes are $x_1, \ldots, x_6$ and decision ones are y_1, y_2.

The values of decision attributes are explicitely indicated by the tuples of condition attributes i.e. we write $D = F(C)$, which means D functionally depends on C. We see, for example, that if we take object number 1, then we have the tuple of decision attributes equal 10, similarly object number 4 implies decisions 01.

Functionally dependent data (such as attributes D) can be projected out of a given file and complete file can always be recovered by means of the joining operation.

To show the implication of the functional dependencies for data base administration consider a functional dependence, F, given in the form of a truth table specification. The set Y of outputs can be partitioned into two disjoint subsets such that the input supports of the obtained components are X_h and X_g, respectively. These may be easily obtained because each output usually depends on a different set of variables, whose cardinality is smaller than for the primary X set. Thus, we can group the outputs into separate sets, to obtain the minimal input support sets X_h and X_g.

Table 3

	x_1	x_2	x_3	x_4	x_5	x_6	y_1	y_2
1	1	1	0	1	0	0	1	0
2	1	1	0	0	0	0	1	0
3	2	2	1	1	2	1	0	1
4	1	0	0	1	2	1	0	1
5	1	0	0	2	0	1	0	0
6	2	2	1	3	3	0	0	1
7	2	2	1	2	0	1	1	0
8	1	1	0	0	0	1	0	0
9	1	0	0	3	3	0	0	1
10	0	2	1	3	3	0	0	1

Example 4.

Consider again the multiple–output function of Table 1. The dependence sets of input variables for every single–output function are as follows:

y_1 : $\{x_2, x_3, x_6\}$, $\{x_3, x_4, x_6\}$, $\{x_3, x_5, x_6\}$, $\{x_2, x_4, x_6\}$, $\{x_4, x_5, x_6\}$

y_2 : $\{x_1, x_3, x_5\}$, $\{x_2, x_3, x_5\}$, $\{x_3, x_5, x_6\}$, $\{x_4, x_5, x_6\}$

y_3 : $\{x_1, x_3\}$

y_4 : $\{x_3, x_5, x_6\}$

Therefore, we can determine an optimal two–block decomposition (in this case called parallel [11]), $G = \{y_1, y_4\}$ and $H = \{y_2, y_3\}$ with the input support sets $X_g = \{x_3, x_5, x_6\}$ and $X_h = \{x_1, x_3, x_5\}$, respectively.

A more complex structure is in hierarchical decomposition in which the global description is broken down sequentially into smaller and smaller subtables and at each step of the process the data associated with the attributes being resolved should be regenerable from the several data collections defined. We shall explain this problem using the truth table model of data.

Let F be a multiple–valued function representing functional dependency $Y = F(X)$, where X is the set of multiple–valued input variables (condition attributes) and Y is the set of binary output variables (decision attributes). Let A, B be the subsets of X such that $X = A \cup B$ and $A \cap B = \emptyset$.

We state the decomposition problem as the following question: when can the functional dependency $Y = F(X)$ be derived in two steps

a) $G(B) = g$,

b) $Y = H(A, g)$

where g is some fictitious, auxiliary attribute.

Example 5.

Consider the DT–system given in Table 3, where $x_1, \ldots, x_6$ are condition attributes and y_1, y_2 are decision ones. As $IND(C) \subseteq IND(D)$, D functionally depends on C i.e. all decisions are uniquely determined by conditions. Now we will consider a possibility of deriving such functional dependency in hierarchical way using the tables which specify dependencies: $g = G(x_4, x_5, x_6)$ and $F = H(x_1, x_2, x_3, g)$, where g is the fictitious decision attribute playing the role of the auxiliary condition attribute. It can be easily verified that, for instance, the tuple of attribute values $1, 1, 0, 1, 0, 0$ indicates the auxiliary decision $g = 0$. Thus joining this result with the values of attributes x_1, x_2, x_3, we have:

$$H(x_1, x_2, x_3, g) = (y_1, y_2).$$

It is the same result as in the table specifying functional dependency F. Verifying other valid objects of table F we conclude that as for all $u \in U$ described by attributes $x_1, \ldots, x_6$

$$H(x_1, x_2, x_3, G(x_4, x_5, x_6)) = F(x_1, \ldots, x_6),$$

so it is possible to regenerate the global table from the subtables each describing a different subset of attributes.

In the above example we used a two–valued fictitious attribute. As the data tables are usually stored in computer memory we are usually interested in fictitious attributes with minimum number of values. The minimum number of values of fictitious attribute g sufficient to represent function F in the form $F = H(A, G(B))$ is equal to

$$r(A) = \Gamma(log_2 \gamma(P(A)|P_F)),$$

where $\gamma(\Pi)$ denotes the number of elements in the largest block of partition Π, $\Gamma(x)$ is the smallest integer equal to or larger to x and $P(A)|P_F$ denotes the quotient partision.

We say that there is a simple hierarchical decomposition of F iff

$$F = H(A, G(B, C)) = H(A, g)$$

where G and H denote functional dependencies: $G(B, C) = g$ and $H(A, g) = Y$. If in addition, $C = \emptyset$, then H is called a simple disjoint decomposition of F.

In other words we try to find a function H depending on the variables of the set A as well on the outputs of a function G depending on the set $B \cup C$. The outputs of the function H are identical with the function values of F.

The following theorem states the sufficient condition for hierarchical decomposition.

Theorem 2: *Functions G and H represent a hierarchical decomposition of function F i.e. $F = H(A, G(B, C))$ iff there exists a partition $\Pi_G \geq P(B \cup C)$ such that*

$$P(A) \cdot \Pi_G \leq P_F \tag{4}$$

where all the partitions are over the set of objects and the number of values of component G is equal to $L(\Pi_G)$, where $L(\Pi)$ denotes the number of blocks of partition Π.

In the theorem partition Π_G represents component G, and the product of partitions $P(A)$ and Π_G corresponds to H. The truth tables as well as decision tables of the resulting components can be easily obtained from these partitions.

Example 6.

Let us decompose the function F of Table 3, where the characteristic partition is as follows:

$$P_F = \{\{1,2,7\}, \{3,4,6,9,10\}, \{5,8\}\}$$

For $A = \{x_1, x_2, x_3\}$, $B = \{x_4, x_5, x_6\}$, we obtain

$$P(A) = \{\{1,2,8\}, \{3,6,7\}, \{4,5,9\}, \{10\}\}$$
$$P(B) = \{\{1\}, \{3,4\}, \{5,7\}, \{6,9,10\}, \{2\}, \{8\}\}$$
$$\Pi_G = \{\{1,2,3,4,6,9,10\}, \{5,7,8\}\}$$

It can be easily verified that since $P(A) \cdot \Pi_G \leq P_F$, function F is decomposable as $F = H(x_1, x_2, x_3, G(x_4, x_5, x_6))$, where G is one–output function of three variables.

The truth tables of components G and H can be obtained from partitions $P(A)$, Π_G, and P_F. Encoding the blocks of Π_G respectively as 0 and 1, we immediately obtain the truth table of function G; it is presented in Table 4. The truth table of function H can be derived by reencoding input vectors of F using an intermediate variable g. The truth table obtained in this way is shown in Table 5.

Table 4

x_4	x_5	x_6	g
1	0	0	0
1	2	1	0
2	0	1	1
3	3	0	0
0	0	0	0
0	0	1	1

Table 5

x_1	x_2	x_3	g	y_1	y_2
1	1	0	0	1	0
1	1	0	1	0	0
2	2	1	0	0	1
2	2	1	1	1	0
1	0	0	0	0	1
1	0	0	1	0	0
0	2	1	0	0	1

The gain of the decomposition implies from the fact that two components (i.e. tables G and H) generally require less memory space than non-decomposed table. If we express the truth table's relative size as $S = a\sum b_i$, where a – the number of objects, b_i – the number of bits needed to represent variable x_i, we can compare the common size of the decomposition components with the size of the primary (non-decomposed table). In the above example the size S_F of primary table is $10 \cdot 12 = 120$ units and the size of decomposed tables is $6 \cdot 6 + 8 \cdot 6 = 84$ units i.e. 70% of the S_F.

The main task of decomposition process is to find a subset of inputs for component G that hierarchically connected with component H will implement function F i.e. to find $P_G = P(B \cup C)$, such that there exists $\Pi_G \geq P_G$ that satisfies condition (4) in Theorem 2. To solve this problem, consider a subset of primary inputs, $D = B \cup C$, and an m–block partition $P(D) = (B_1; B_2; \ldots; B_m)$ generated by this subset.

A relation of compatibility of partition blocks will be used to verify whether or not partition $P(D)$ is suitable for hierarchical decomposition.

Two blocks B_i, $B_j \in P(D)$ are compatible iff partition Π'_G obtained from partition $P(D)$ by merging blocks B_i and B_j into a single block B'_{ij} satisfies condition (4) in Theorem 2, i.e., iff

$$P(A) \cdot \Pi'_G \leq P_F$$

A subset of n partition blocks, $\mathcal{B} = \{B_{i_1}, B_{i_2}, \ldots, B_{i_n}\}$, where $B_{i_j} \in P(D)$, is a class of compatible blocks for partition $P(D)$ iff all blocks in $\mathcal{B}$ are pairwise compatible.

A compatible class is called Maximal Compatible Class (MCC) iff it cannot be contained in any other compatible class.

If we can form suitable MCCs of blocks, we can merge them into a singular block to obtain partition Π_G such that

$$P(A) \cdot \Pi_G \leq P_F \tag{5}$$

where partition Π_G represents function G, and the product of partitions $P(A)$ and Π_G corresponds to function H. The truth table description of these functions can be easily obtained from the partitions.

Example 7.

Let for function F described by the following partitions on $M = \{1, \ldots, 15\}$

$P_1 = \{\{1, \ldots, 7\}, \{8, \ldots, 15\}\}$

$P_2 = \{\{1, 2, 3, 13, 14, 15\}, \{4, \ldots, 12\}\}$

$P_3 = \{\{1, 2, 3, 7, 8, 9, 13, 14, 15\}, \{4, 5, 6, 10, 11, 12\}\}$

$P_4 = \{\{1, 4, 5, 7, 8, 10, 13\}, \{2, 3, 6, 9, 11, 12, 14, 15\}\}$

$P_5 = \{\{1, 3, 4, 6, \ldots, 10, 12, 15\}, \{2, 5, 11, 13, 14\}\}$

$P_F = \{\{1, 9, 14\}, \{5, 7, 8, 13\}, \{2, 6, 12\}, \{4, 11\}, \{3, 10, 15\}\}$

sets A and B are as follows: $A = \{x_3, x_4\}$, $B = \{x_1, x_2, x_5\}$. Then for the quotient partition

$$P(A)|P_F = \{\{1\}, \{7,8,13\}; \{9,14\}, \{2\}, \{3,15\}; \{10\}, \{5\}, \{4\}; \{6,12\}, \{11\}\}$$

and if the blocks of

$$P_G = \{\{1,3\}, \{2\}, \{4,6,7\}, \{5\}, \{8,9,10,12\}, \{11\}, \{13,14\}, \{15\}\}$$

are denoted $B_1, \ldots, B_8$, we obtain the corresponding MCCs

$$MCC1 = \{B_4, B_6, B_7\},$$
$$MCC2 = \{B_1, B_4, B_6, B_8\},$$
$$MCC3 = \{B_2, B_4, B_6\},$$
$$MCC4 = \{B_3, B_8\},$$
$$MCC5 = \{B_3, B_7\},$$
$$MCC6 = \{B_2, B_3\},$$
$$MCC7 = \{B_5, B_7\}.$$

To obtain Π_G the minimum cover of compatible classes must be found i.e. the set $MCC = \{MCC_{i_1}, \ldots, MCC_{i_t}\}$ with minimum cardinality such that $MCC_{i_1} \cup \cdots \cup MCC_{i_t} = \{B_1, \ldots, B_s\}$. For the above obtained MCCs one of feasible minimal covers is $\{\{B_1, B_4, B_6, B_8\}, \{B_2, B_3\}, \{B_5, B_7\}\}$.

Thus $\Pi_G = \{\{1,3,5,11,15\}, \{2,4,6,7\}, \{8,9,10,12,13,14\}\}$.
It is easy to verify that $P(A) \cdot \Pi_G \leq P_F$, and therefore

$$F = H(x_3, x_4, G(x_1, x_2, x_5)).$$

7. Conclusions

Designing of logic circuits is driven by tremendous demand of industry for more powerfull design tools. This paper has demonstrated that advanced algorithms applied in rough set theory can apparently become extremely useful in solving problems typical for logic designing [1], [3], [13], [23]. From this implies the idea of mutual realization of CAD system which could be useful for both these problems. Such a system has been implemented by the authors in a case of attributes/arguments reduction and the results we have obtained so far fully confirm the efficiency of the applied algorithms [12].

The logic decomposition algorithms presented in this paper have been also appied in supporting the logic synthesis system PLATO [11].

References

[1] Brayton R.K.,Hachtel G.D.,McMullen C.T., Sangiovanni-Vincentelli A. (1984) *Logic Minimization Algorithms for VLSI Synthesis.* Kluwer Academic Publ.

[2] Brown F.M., (1990). *Boolean Reasoning. The Logic of Boolean Equations.* Kluwer Academic Publ.

[3] Fang K., Wójcik A.S., (1988). Modular Decomposition of Combinational Multiple–Valued Circuits.*IEEE Trans.on Computers*, vol **37**,No.10, 1293-1301

[4] Grzymała-Busse J.W. and Pawlak Z.,(1984).On Some Subset of the Partition Set.*Fundamenta Informaticae* **VII.3**, 483-488.

[5] Hartmanis J. and Stearns R.E.,(1966). *Algebraic Structure Theory of Sequential Machines.* Prentice–Hall.

[6] Hurley R.B.,(1983). *Decision Tables in Software Engineering.* Van Nostrand Reinhold Company, New York.

[7] Jasiński K., Luba T., Kalinowski J.,(1989). Parallel Decomposition in Logic Synthesis. *Proc.15th European Solid-State Circuits Conf.*, 113-116.

[8] Jasiński K., Łuba T., Kalinowski J.,(1991). CAD Tools for PLD Implementation of ASICs. *Proc. of Second Eurochip Workshop on VLSI Design Training,* Grenoble, 225-230.

[9] Łuba T., (1986). A Uniform Method of Boolean Function Decomposition. *Rozprawy Elektrotechniczne (Journal of the Polish Academy of Science)*, No.**4**, 1041-1054.

[10] Łuba T., Kalinowski J., Jasiński K., Kraśniewski A., (1991). Combining Serial Decomposition with Topological Partitioning for Effective Multi-Level PLA Implementations. *In Logic and Architecture Synthesis*, P.Michel and G.Saucier (Editors), Elsevier Science Publishers B.V. (North-Holland).

[11] Łuba T., Kalinowski J., Jasiński K., (1991). PLATO: A CAD Tool for Logic Synthesis Based on Decomposition. *Proc. of European Conference on Design Automation*, 65–69.

[12] Łuba T., Janowski J., Rybnik J., (1991). Relations Between Multiple–Valued Logic and Decision Logic with Respect to Rough Set Theory Semantic. *Research Report, Institute of Teclecommunications.*No.**39**.

[13] Łuba T., Markowski M.A., Zbierzchowski B., (1992). Logic Decomposition for Programmable Gate Arrays. *Proc. of Euro-ASIC'92*, Paris (to appear).

[14] McCluskey E.J., (1986). *Logic Design Principles. With Emphasis on Testable Semicustom Circuits.* Prentice–Hall.

[15] Pawlak Z., (1983).*Informations Systems. Theoretical Foundations.*(in Polish) WNT.

[16] Pawlak Z., (1991). *Rough Sets. Theoretical Aspects of Reasoning about Data.* Kluwer Academic Publishers.

[17] Rudell R.L, Sangiovanni–Vincentelli A.,(1987). Multiple–Valued Minimization for PLA Optimization. *IEEE Transactions on Computer–Aided Design.*

[18] Rybnik J.,(1990). Minimizations of Partially Defined Switching Functions Using Rough Sets Theory. *Manuscript.*

[19] Sasao T.,(1989). On the Optimal Design of MultipleValued PLA's. *IEEE Trans. on Computers*, Vol.**38**, No.4, 582–592.

[20] Sasao T.,(1988). Multiple–Valued Logic and Optimization of Programmable Logic Arrays. *Computer.*

[21] Sasao T.,(1984). Input Variable Assignment and Output Phase Optimization of PLA's. *IEEE Transactions on Computers.*

[22] Skowron A. and Rauszer C., (1991). The Discernibility Matrices and Functions in Information Systems. *ICS Report., Institute of Computer Science.***1**.

[23] Varma D., Trachtenberg E.A.,(1989). Design Automation Tools for Efficient Implementation of Logic Functions by Decomposition. *IEEE Trans. on CAD* **8**, No.8, 901–916.

Part II

COMPARISON WITH RELATED METHODOLOGIES

Part II
Chapter 1

PUTTING ROUGH SETS AND FUZZY SETS TOGETHER *

Didier DUBOIS Henri PRADE
Institut de Recherche en Informatique de Toulouse - C.N.R.S.
Université Paul Sabatier
31062 Toulouse, France

Abstract. In this paper we argue that fuzzy sets and rough sets aim to different purposes and that it is more natural to try to combine the two models of uncertainty (vagueness for fuzzy sets and coarseness for rough sets) in order to get a more accurate account of imperfect information. First, the upper and lower approximations of a fuzzy set are defined, when the universe of discourse of a fuzzy sets is coarsened by means of an equivalence relation. We then come close to Caianiello's C-calculus. Shafer's concept of coarsened belief functions also belongs to the same line of thought and is reviewed here. Another idea is to turn the equivalence relation relation into a fuzzy similarity relation, for a more expressive modeling of coarseness. New results on the representation of similarity relations by means of a fuzzy partition of fuzzy clusters of more or less indiscernible points are surveyed. The properties of upper and lower approximations of fuzzy sets by similarity relations are thoroughly studied. Lastly the potential usefulness of the fuzzy rough set notions for logical inference in the presence of both fuzzy predicates and graded indiscernibility is indicated. Especially fuzzy rough sets may provide a nice semantic background for modal logic involving fuzzy modalities and/or fuzzy sentences.

1. Introduction

The contemporary concern about knowledge representation and information systems has put forward useful extensions of elementary set theory such as fuzzy sets (Zadeh (1965)) and rough sets (Pawlak (1982)), among others. In this paper we pursue an investigation around these two notions in order to lay bare their respective specificity, instead of turning them into rival theories (e.g. Pawlak (1985)). Basically, rough sets embody the idea of indiscernibility between objects in a set, while fuzzy sets model the ill-definition of the boundary of a sub-class of this set. Rough sets are a calculus of partitions, while fuzzy sets are a continuous generalization of set-characteristic functions. Marrying both notions leads to consider rough approximations of fuzzy sets, but also approximations of sets by means of (fuzzy) similarity relations or fuzzy partitions. These hybrid notions come up in a natural way when a linguistic category, denoting a set of objects, must be approximated in terms of already existing labels, or when the indiscernibility relation between objects no longer obeys the ideal laws of equivalence and is a matter of degree. Moreover, the attempt to mix up vagueness and approximation leads us to bring together past works developed independently of rough sets, and often before them, but based on the same ideas.

* This paper draws from and continues a previous article by the authors, entitled "Rough fuzzy sets and fuzzy rough sets", that appeared in the Int.J.of General Systems in 1990.

The first section contains basic definitions of rough sets and fuzzy sets and points out the contrast between the intended purposes of the two notions. Section two defines upper and lower approximations of fuzzy sets and belief functions, and bridges the gap with Caianiello's C-calculus (Caianiello (1973)). It is indicated that under a different terminology, Shafer (1976) has used the same model as Pawlak in the theory of evidence, i.e. a calculus of partitions.Section three generalizes rough sets by weakening the concept of equivalence into similarity (Zadeh (1971)), as suggested in a previous paper (Fariñas del Cerro and Prade (1986)). The rough set idea then comes close to well-known concerns in mathematical taxonomy and approximation theory. This section equivalently builds fuzzy rough sets from fuzzy partitions. Relevant results on fuzzy partitions defined as fuzzy quotient sets of similarity relations are recalled. Basic properties of fuzzy rough sets are investigated. A last technical section briefly recalls some links between rough sets and modal logic, and reviews some works that have introduced fuzzy sets in these constructs. In the conclusion, some research directions are surveyed.

2. Rough Sets and Fuzzy Sets : Two Different Topics

2.1. ROUGH SETS

Let X be a set, and R be an equivalence relation on X (i.e. reflexive, symmetric and transitive). Let X/R denote the quotient set of equivalence classes, which form a partition in X. X/R is a coarsened version of X ; elementary parts of X/R are usually coarser than the ones of X, and denoted X_1, $X_2, \ldots, X_i, \ldots$. The cardinality of X/R is thus generally smaller than that of X (except if R is the equality on X). An equivalence relation is the simplest model one can think of to represent the fact that, in X, it is not possible to distinguish some elements from others. x R y then means : x is too close (or too similar) to y so that both elements are indiscernible.

Examples

1. Measurement scale $X = [0,2.5]$ is a human size scale between 0 and 2.5 meters, that allows for infinite precision. In practice, only millimeters can be measured, i.e. X/R is a set of adjacent intervals, whose representatives are of the form n/1000 with $0 \leq n \leq 2500$, n integer. x R (n/1000) means that x can be rounded by n/1000, x R y means that x and y are rounded by the same number of millimeters. In usual communication between individuals on this matter, the implicit representation of this scale is even coarser : only centimeters (or inches) make sense. Indiscernibility also occurs in decision theory, when the respective ratings of two potential decisions are too close to ensure that one of these decisions is strictly preferred to the other (Roy (1985)).
2. Information system X is a set of item identifiers (objects), $\mathcal{A}$ is a set of attributes a, V_a the set $\{a(x) \mid x \in X\}$ of attribute values for attribute a. The equivalence relation R is defined by x R y if and only if $\forall a \in \mathcal{A}$, $a(x) = a(y)$. $[x]_R$ denotes the class of objects which have the same description as x in terms of attributes in $\mathcal{A}$. This example is given by Pawlak (1984).
3. Image processing X is a rectangle screen, i.e. a Cartesian product $[0,a] \times [0,b]$, X/R is a discretization grid into pixels, $[(x,y)]_R$ being the pixel that contains a point (x,y) in X. This is the 2-dimensional version of example 1.

Let S be a subset of X. The main question addressed by rough sets (Pawlak (1982)) is : how to represent S by means of X/R ? Denote $[x]_R$ the equivalence class of $x \in X$. A *rough set* is a pair of subsets $R^*(S)$ and $R_*(S)$ of X/R that approach as close as possible S from outside and inside respectively :

$$R^*(S) = \{[x]_R \mid [x]_R \cap S \neq \emptyset, x \in X\} \quad (1)$$

$$R_*(S) = \{[x]_R \mid [x]_R \subseteq S, x \in X\} \quad (2)$$

$R^*(S)$ (resp. : $R_*(S)$) is called the upper (resp. : lower) approximation of S by R. $R^*(S)$ contains $R_*(S)$. When $R^*(S) \neq R_*(S)$, it means that due to the indiscernibility of elements in X, S cannot be perfectly described. More precisely, the set difference $R^*(S) - R_*(S)$ is a rough (imprecise) description of the boundary of S by means of "granules" of X/R.

These notions are actually older than Pawlak's paper. They were already introduced by Shafer (1976) in his book (chapter 6), where coarsenings and refinements of a frame of discernment are introduced. A frame of discernment is a set of alternatives perceived as distinct answers to a question. Coarsening a frame of discernment X comes down to clustering elements and build a partition. Refinement is the converse operation, i.e. distinguishing sub-alternatives corresponding to single elements. In other words X/R is a coarsening of X, and X is a refinement of X/R. Following Shafer, let us denote ω the mapping that, for any subset of X/R, computes its refinement in X. Namely if $X_i \in X/R$,

$$\omega(X_i) = \{x \mid X_i \text{ is the name of the equivalence class } [x]_R\} \quad (3)$$

and for $A \subset X/R$,

$$\omega(A) = \bigcup_{X_i \in A} \omega(X_i). \quad (4)$$

It is important to distinguish between X_i, an element of X/R, and $\omega(X_i)$, a subset of X. $R^*(S)$ and $R_*(S)$ are respectively called outer and inner reductions by Shafer (1976). Viewing X_i as the name of an equivalence class, $\omega(X_i)$ can be viewed as the extension of X_i, and will be termed so in the following, by analogy with logic.

2.2. FUZZY SETS

A fuzzy set (Zadeh (1965)) F of X is defined by a mapping $\mu_F : X \to L$ where L is an ordered set of membership values (often a complete lattice, at least) and $\mu_F(x)$ is the degree of membership of x in F. L = [0,1] generally ; L = {0,1} in the case of usual sets. Allowing for partial membership intends to account for the ill-definition of the extension of the predicate named F.

Examples (continued)

1. Measurement scale : Rounding off sizes to millimeters does not always allow to distinguish between people whose size is close to within one millimeter, i.e. if S = {1.71815, 1.71816}, $R^*(S) = [1.718]_R$ but $R^*(S) = \emptyset$. Contrastedly the set of *tall* sizes is fuzzy because (especially when expressed in millimeters) some sizes are compatible with *tall* to a degree that may be different from total compatibility and total incompatibility. Generally the more refined the

scale, the more likely linguistically relevant subsets will be fuzzy. Vagueness lies in the subset F denoted by "tall" while indiscernibility is a property of the referential itself, as perceived by some observer, not of its subsets. The model of "tall" can be more or less refined by modification of the set L of membership values.

2. Information systems : Fuzziness in information systems often lies in the formulation of queries that describe subsets of relevant objects by a flexible specification of admissible attribute values.Then the set of retrieved objects is fuzzy (e.g. Tahani (1977)). Moreover the presence of ill-described objects in a data base can be expressed in terms of possibility distributions on the set V_a of attribute values ; these possibility distributions are modelled by fuzzy sets, following Zadeh (1978). For instance, all we know about x is that its size a(x) is "about 2 meters", with the underlying assumption that a(x) is in [1.8,2.2] and the most plausible value is 2. If a user is interested in "tall objects", the response of the information system may consist of two (fuzzy) sets of relevant objects : the set of *certainly* relevant objects, and the (larger) set of *possibly* relevant objects (Prade and Testemale (1984)). These two sets are nested ; the corresponding pair is called a *twofold* fuzzy set (Dubois and Prade (1987)). Its existence is due to the presence of (vague) incomplete information in the data base, but not to the indiscernibility of objects. The idea of indiscernibility in fuzzy data bases is at the heart of the approach proposed by Buckles and Petry (1982) where a fuzzy relation models, on each attribute domain, to what extent two attribute values are interchangeable ; see (Prade and Testemale (1987)) for a comparative discussion between this latter approach and vague incomplete information data bases.
3. Image processing : given a subset S of the screen defined by the contour of an object, various approximations $R^*(S)$ and $R_*(S)$ of this subset can be obtained by modifying the graininess of the discretized picture. However generally, objects on a screen appear rather like fuzzy sets of the screen, due to grey levels. The object can be made more or less fuzzy by acting on the number of grey levels, i.e. contrast modification. Number of allowed levels of grey n_1 and number of pixels n_2 in the screen are almost unrelated parameters (only $n_1 \leq n_2$ must hold in the picture).

It should be clear from the above examples that rough sets and fuzzy sets are not meant to play the same role in knowledge representation problems. As a consequence they are not rival theories but capture two distinct aspects of imperfection in knowledge : indiscernibility and vagueness, that may be simultaneously present in a given application. It is then natural to *combine* these notions, rather than compare them from a formal point of view.

Of course, it may be tempting to identify the boundary of a rough set as containing borderline elements and then decree that the upper and lower approximations of a set S can be viewed respectively as the support (i.e., elements with positive membership) and the core (i.e., elements with complete membership) of a fuzzy set F(S) defined on X/R. Then, as suggested by Pawlak (1985), elements in the boundary can have membership value .5. This idea is further extended by Wong and Ziarko (1985) (see also Pawlak et al. (1988)) who suggest to use the conditional probability $P(S \mid \omega(X_i))$ to evaluate the degree of membership $\mu_{F(S)}(X_i)$. However these views can be but partially in agreement with fuzzy set theory since they impose severe restrictions on the choice of fuzzy set-theoretic connectives (see Wygralak (1989) for the three-valued logic approach, and Wong and Ziarko (1985) for the probabilistic view).

3. Upper and Lower Approximations of Generalized Sets

3.1. APPROXIMATIONS OF FUZZY SETS

Let X be a set, R be an equivalence relation on X and F be a fuzzy set in X. The upper and lower approximations $R^*(F)$ and $R_*(F)$ of a fuzzy set F by R are fuzzy sets of X/R with membership functions defined by

$$\mu_{R^*(F)}(X_i) = \sup\{\mu_F(x) \mid \omega(X_i) = [x]_R\} \tag{5}$$

$$\mu_{R_*(F)}(X_i) = \inf\{\mu_F(x) \mid \omega(X_i) = [x]_R\} \tag{6}$$

where $\mu_{R^*(F)}(X_i)$ (resp. : $\mu_{R_*(F)}(X_i)$) is the degree of membership of X_i in $R^*(F)$ (resp. : $R_*(F)$). $(R^*(F), R_*(F))$ is called a rough fuzzy set. These expressions derive from possibility theory (Zadeh (1978), Dubois and Prade (1988)) ; (5) (resp. : (6)) is the degree of possibility (resp. : necessity) of the fuzzy event F, based on the (crisp) possibility distribution defined from the characteristic function of $\omega(X_i)$. To see it, note that the fuzzy extensions $\omega(R^*(F))$ and $\omega(R_*(F))$ can be defined via the extension principle (Zadeh (1965)), as :

$$\mu_{\omega(R^*(F))}(x) = \mu_{R^*(F)}(X_i) = \Pi_i(F), \forall x \in \omega(X_i) \tag{7}$$

$$\mu_{\omega(R_*(F))}(x) = \mu_{R_*(F)}(X_i) = N_i(F), \forall x \in \omega(X_i) \tag{8}$$

where Π_i (resp. N_i) is the possibility (resp. necessity) measure whose distribution is $\mu_{\omega(X_i)}$ hereafter denoted π_i, i.e. $\Pi_i(F) = \sup_x \min(\pi_i(x), \mu_F(x))$; $N_i(F) = \inf_x \max(1 - \pi_i(x), \mu_F(x))$ (see Dubois and Prade (1988) for instance). Note that π_i is a crisp possibility distribution, i.e. $\pi_i(x) \in \{0,1\}$, and $x \in \omega(X_i)$ is equivalent to $\omega(X_i) = [x]_R$ and to $\pi_i(x) = 1$; this is why (5) and (6) are the same as (7) and (8).

The extension of $\omega(A)$ of a fuzzy set A of X/R is defined by

$$\mu_{\omega(A)}(x) = \mu_A(X_i) \text{ if } x \in X_i$$

$\omega(A)$ is a fuzzy set with constant membership on the equivalence classes of R. These fuzzy sets are unaltered by the equivalence relation, as shown now :

<u>Proposition 1</u> : A fuzzy set H on X is equal to its upper or lower approximations if and only if H is constant on the equivalence classes of R.

<u>Proof</u> : Let $x \in X_i$ and $\mu_H(x) = h_i$, $\forall x \in X_i$. Then

$$\begin{aligned} \mu_{\omega(R^*(H))}(x) &= \mu_{R^*(H)}(X_i) \\ &= \sup_{x \in X_i} \mu_H(x) = h_i = \mu_H(x) \\ &= \inf_{x \in X_i} \mu_H(x) = \mu_{\omega(R_*(H))}(x). \end{aligned}$$

Conversely, if $\omega(R^*(H)) = H$ then $\forall x_0 \in X_i$, $\mu_H(x) = \sup_{x \in X_i} \mu_H(x_0) = h_i$. And the same for $R_*(H)$. Q.E.D.

The following properties of upper and lower approximations still hold for rough fuzzy sets :

Proposition 2

$$\omega(R_*(F)) \subseteq F \subseteq \omega(R^*(F))$$
$$R^*(F \cup G) = R^*(F) \cup R^*(G)$$
$$R_*(F \cap G) = R_*(F) \cap R_*(G)$$
$$R^*(F \cap G) \subseteq R^*(F) \cap R^*(G)$$
$$R_*(F \cup G) \supseteq R_*(F) \cup R_*(G)$$
$$R^*(\bar{F}) = \overline{R_*(F)}$$
$$R^*(\omega(R^*(F))) = R_*(\omega(R^*(F))) = R^*(F)$$
$$R_*(\omega(R_*(F))) = R^*(\omega(R_*(F))) = R_*(F)$$

where the fuzzy set union, intersection, complementation and inclusion are defined by (Zadeh (1965)) :

$$\mu_{F \cup G}(x) = \max(\mu_F(x), \mu_G(x))$$
$$\mu_{F \cap G}(x) = \min(\mu_F(x), \mu_G(x))$$
$$\mu_{\bar{F}}(x) = 1 - \mu_F(x).$$
$$F \subseteq G \Leftrightarrow \forall x, \mu_F(x) \leq \mu_G(x)$$

Proof : The reason why the two last equalities hold is that if H is a fuzzy set with constant membership on equivalence classes (as are $\omega(R^*(F))$ and $\omega(R_*(F))$), then $R^*(H) = R_*(H) = H$, and Proposition 1 applies. Q.E.D.

Letting $m_i = \inf_{x \in X_i} \mu_F(x)$ and $M_i = \sup_{x \in X_i} \mu_F(x)$ for any fuzzy set F in X, then it is easy to check that

$$\mu_{\omega(R^*(F))}(x) = \Sigma_{i=1,n} M_i \cdot \mu_{\omega(X_i)}(x) = \sup\{\mu_F(y) \mid x \mathrel{R} y\}$$
$$\mu_{\omega(R_*(F))}(x) = \Sigma_{i=1,n} m_i \cdot \mu_{\omega(X_i)}(x) = \inf\{\mu_F(y) \mid x \mathrel{R} y\}.$$

3.2 THE LINK WITH C-CALCULUS

If we apply (7) and (8) to a fuzzy set of the real line such as "tall" in example 1, we obtain for $\mu_{\omega(R^*(F))}$ and $\mu_{\omega(R_*(F))}$ piecewise constant functions that are used in integration theory to bracket the integral of μ_F by means of Darboux sums. Moreover the two last equations of Section 3.1 are basic equations of C-calculus (Caianiello (1973), (1987)). A composite set or C-set is a triple (χ,m,M) where $\chi = \{X_1, \ldots, X_n\}$ correspond to a partition of X, and m, M are mappings $X \to [0,1]$ such that

$$\forall x \in X, m(x) = \sum m_i \cdot \mu_{\omega(X_i)}(x) = m_i \text{ if } x \in \omega(X_i) \quad (9)$$

$$\forall x \in X, M(x) = \sum M_i \cdot \mu_{\omega(X_i)}(x) = M_i \text{ if } x \in \omega(X_i) \quad (10)$$

where $0 \le m_i \le M_i \le 1$, $\forall i = 1,n$. Every function $f : X \to [0,1]$ defines a composite set, letting $m_i = \inf\{f(x) \mid x \in \omega(X_i)\}$ and $M_i = \sup\{f(x) \mid x \in \omega(X_i)\}$. Clearly these equations are (5) and (6) where $X/R = \chi$, and $f = \mu_F$ is the membership function of a fuzzy set. The links between fuzzy sets and C-calculus were pointed out in the first paper on C-calculus (Caianiello (1973)) and more recently by Caianiello and Ventre (1984) and by Gisolfi (1992). But it is clear that a C-set is nothing but a rough fuzzy set, i.e., a more general and, by the way, earlier notion than a rough set.

A basic operation in C-calculus is C-product (Caianiello and Ventre (1985)). Given two C-sets (χ,m,M) and (χ',m',M'), a C-set is obtained, say (χ'',m'',M'') such that $\chi'' = \{\omega(X_i) \cap \omega(X'_j) \mid X_i \in \chi, X'_j \in \chi'\}$, $m'' = \min(m,m')$, $M'' = \max(M,M')$. However the two latter relations, expressing a fuzzy set intersection and union respectively, do not give exact result for the rough fuzzy set, defined using χ'' ; indeed, if m and M are defined from f as well as m' and M', only the following inequalities hold

$$\begin{aligned} \min(m_i,m'_j) &\le \inf\{f(x) \mid x \in \omega(X_i) \cap \omega(X'_j)\} \\ &\le \sup\{f(x) \mid x \in \omega(X_i) \cap \omega(X'_j)\} \le \max(M_i,M'_j) \end{aligned} \quad (11)$$

It is interesting to notice that the main applications of C-calculus are in image processing, i.e. in the setting of example 3. However C-product of rough sets make sense in example 2. If R_a and R_b denote the equivalence relations defined on X by attributes a and b respectively, the refined equivalence relation R_{ab} defined on X by both attributes correspond to the partition obtained by intersecting the equivalence classes for attribute a and attribute b, i.e. a C-product. In terms of rough sets, (11) writes (Pawlak (1984)) :

$$R_{a*}(S) \cap R_{b*}(S) \subseteq R_{ab*}(S) \subseteq R_{ab}{}^*(S) \subseteq R_a{}^*(S) \cup R_b{}^*(S) \quad (12)$$

3.3. APPROXIMATIONS OF RANDOM SETS

In Shafer (1976)'s book, chap. 6, preliminary results are given about coarsening and refinement of belief functions. The author assumes that a belief function is defined on X/R and computes its refinement on X. Conversely, given a belief function on X, Shafer computes its coarsening on X/R. This work is pursued in Shafer et al. (1987). A belief function on X is defined by a finite set $\mathcal{F}$ of non-empty subsets to which positive masses p(S), $S \in \mathcal{F}$ are allocated, so that $\sum_S p(S) = 1$. When $\mathcal{F}$ contains only a family of nested sets, the belief function is called consonant and is a necessity measure in the sense of possibility theory (Dubois and Prade (1988)) i.e. the random set $(\mathcal{F},p)$ is equivalent to a fuzzy set F with

$$\forall x \in X, \mu_F(x) = \Sigma_{x \in S}\, p(S) \tag{13}$$

μ_F is called the contour function of $(\mathcal{F},p)$ by Shafer (1976), even when $(\mathcal{F},p)$ is not consonant. In the consonant case, the equivalence between $(\mathcal{F},p)$ and F via (13) comes from the fact that $\mathcal{F}$ is the set of level-cuts of F, i.e. $\mathcal{F} = \{F_\alpha \mid \alpha \in (0,1]\}$ with $F_\alpha = \{x \mid \mu_F(x) \geq \alpha\}$; the 1-cut, called the core of F, is not empty because $\mathcal{F}$ does not contain the empty set.

Coarsening X using an equivalence relation R leads to introduce upper and lower approximations of $(\mathcal{F},p)$, say $(R^*(\mathcal{F}),p^*)$, $(R_*(\mathcal{F}),p_*)$ on X/R, with

$$R^*(\mathcal{F}) = \{R^*(S) \mid S \in \mathcal{F}\} \; ; \; p^*(A) = \Sigma\, \{p(S) \mid A = R^*(S)\} \tag{14}$$

$$R_*(\mathcal{F}) = \{R_*(S) \mid S \in \mathcal{F}\} \; ; \; p_*(A) = \Sigma\, \{p(S) \mid A = R_*(S)\} \tag{15}$$

(14-15) generalize (1-2) into what can be called "rough random sets".

The consistency of the rough random sets and the rough fuzzy sets (i.e. the equivalence between (14-15) and (5-6)) is easily achieved if we invert (13). Starting from a fuzzy set F on a finite set X, letting $\alpha_1 = 1 \geq \alpha_2 \geq \ldots \geq \alpha_k$ be the set of positive membership grades, the random set equivalent to F is defined by (Dubois and Prade (1988))

$$\mathcal{F}_F = \{F_{\alpha_i}, i = 1,k\}$$

$$p_F(F_{\alpha_i}) = \alpha_i - \alpha_{i+1}$$

with $\alpha_{k+1} = 0$, by convention. We moreover need the following :

<u>Lemma 1</u> : $\forall \alpha \in (0,1]$, $R^*(F)_\alpha = R^*(F_\alpha)$ and $R_*(F)_\alpha = R_*(F_\alpha)$.

<u>Proof</u> : See Dubois and Prade (1990).

Now the following result becomes easy (Dubois and Prade (1990)) :

<u>Proposition 3</u> : $R^*(F)$ is equivalent to $(R^*(\mathcal{F}_F),p_{F^*})$; $R_*(F)$ is equivalent to $(R_*(\mathcal{F}_F),p_{F_*})$

For instance $R^*(F) \Leftrightarrow (\mathcal{F}_{R^*(F)},p_{R^*(F)}) \Leftrightarrow (R^*(\mathcal{F}_F),p_F^*)$ using the lemma. Indeed $\mathcal{F}_{R^*(F)} = \{R^*(F)_{\alpha_i} \mid i = 1,k\} = \{R^*(F_{\alpha_i}) \mid i = 1,k\}$. As for the masses, $p_{F^*}(R^*(F_{\alpha_i})) = p_{R^*(F)}((R^*(F)_{\alpha_i}) = p_F(F_{\alpha_i}) = \alpha_i - \alpha_{i+1}$. Now if $R^*(F_{\alpha_i}) = R^*(F_{\alpha_{i+1}})$ for some i, then $\mathcal{F}_{R^*(F)}$ would contain possibly non-distinct sets. Allowing for non-distinct focal sets is a matter of convention and does not affect the equation (13) that produces the membership function.

3.4. APPROXIMATIONS OF BELIEF FUNCTIONS

Another way of proceeding is to project the belief and plausibility functions from X to X/R directly. Let Bel and Pl be two set-functions defined on 2^X, built from $(\mathcal{F},p)$, as follows :

$$Bel(S) = \Sigma_{T:T\subseteq S}\, p(T) \text{ (belief function)}$$

$$Pl(S) = \Sigma_{T:T\cap S\neq\emptyset}\, p(T) \text{ (plausibility function).}$$

Then define Bel_R and Pl_R on $2^{X/R}$ as $Bel_R(A) = Bel(\omega(A))$; $Pl_R(A) = Pl(\omega(A))$, $\forall A \subseteq X/R$. Let Bel^* and Bel_* on $2^{X/R}$ derive from $(R^*(\mathcal{F}),p^*)$ and $(R_*(\mathcal{F}),p_*)$ respectively. It is easy to see that $Bel_R = Bel^*$. Indeed,

$$\forall A \in 2^{X/R}, Bel^*(A) = \Sigma_{B:B\subseteq A}\, p^*(B) = \Sigma_{S:R^*(S)\subseteq A}\, p(S) = \Sigma_{S:S\subseteq\omega(A)}\, p(S) = Bel(\omega(A))$$

since $R^*(S) \subseteq A$ is equivalent to $S \subseteq \omega(R_*(\omega(A))) = \omega(A)$. Similarly, $Pl^* = Pl_R$.

The effect of coarsening X by means of R can be analyzed on X itself. Namely, noticing that the belief function is defined by means of the set inclusion, the presence of R leads to four possible definitions of inclusion of a subset T in a subset S, when only their upper and lower approximations are discerned : $R_*(T) \subseteq R_*(S)$; $R^*(T) \subseteq R_*(S)$, $R_*(T) \subseteq R^*(S)$, $R^*(T) \subseteq R^*(S)$, respectively denoted ${}_*\!\subseteq_*$, ${}^*\!\subseteq_*$, ${}_*\!\subseteq^*$, ${}^*\!\subseteq^*$. Their strength compares as follows

$$T \,{}^*\!\subseteq_*\, S \text{ implies } T \,{}_*\!\subseteq_*\, S \text{ which implies } T \,{}_*\!\subseteq^*\, S \quad (16)$$

$$T \,{}^*\!\subseteq_*\, S \text{ implies } T \,{}^*\!\subseteq^*\, S \text{ which implies } T \,{}_*\!\subseteq^*\, S \quad (17)$$

These concepts, that may be called "rough implications", can be compared with the usual implication. Namely ${}^*\!\subseteq_*$ is stronger than $\subseteq$, and $\subseteq$ is stronger than the three others. ${}_*\!\subseteq_*$ and ${}^*\!\subseteq^*$ do not compare. Based on these implications four definitions of a rough belief function are obtained on 2^X

$$Bel^*_*(S) = \Sigma_{T\,{}^*\!\subseteq_*S}\, p(T)$$

$$Bel_{**}(S) = \Sigma_{T\,{}_*\!\subseteq_*S}\, p(T)$$

$$Bel^{**}(S) = \Sigma_{T\,{}^*\!\subseteq^*S}\, p(T)$$

$$Bel_*^*(S) = \Sigma_{T\,{}_*\!\subseteq^*S}\, p(T)$$

Notice the following equivalences : $T \,{}^*\!\subseteq_*\, S \Leftrightarrow T \subseteq \omega(R_*(S))$ and $T \,{}^*\!\subseteq^*\, S \Leftrightarrow T \subseteq \omega(R^*(S))$. They can be justified by the identities $R^*(\omega(R_*(S))) = R_*(S)$, $R^*(\omega(R^*(S)) = R^*(S)$ and the inclusion $T \subseteq \omega(R^*(T))$. As a consequence

$$Bel^*_*(S) = Bel^*(R_*(S)) = Bel(\omega(R_*(S))) \quad ; \quad Bel^{**}(S) = Bel^*(R^*(S)) = Bel(\omega(R^*(S))) \quad (18)$$

i.e. Bel^*_* and Bel^{**} derive from the belief function $Bel^* = Bel_R$ with underlying random set $(R^*(\mathcal{F}),p^*)$.

Dubois and Prade (1985) study the concept of upper and lower belief function induced by a multiple-valued mapping. If Ω and Ω' are two sets, Γ a mapping from Ω to $2^{\Omega'} - \emptyset$, Bel a belief function on Ω, and A' a subset of Ω', two set functions Ubel and Lbel can be defined on Ω' as follows

$$\mathrm{Ubel}(A') = \mathrm{Bel}(\{\omega \in \Omega \mid \Gamma(\omega) \cap A' \neq \emptyset\}) \text{ (upper belief function)}$$

$$\mathrm{Lbel}(A') = \mathrm{Bel}(\{\omega \in \Omega \mid \Gamma(\omega) \subseteq A'\}) \text{ (lower belief function)}$$

It is nothing than iterating Dempster (1967)'s construction of belief functions from probability spaces. It can be proved that Lbel ≤ Ubel and that Lbel is still a belief function while Ubel is not, generally. As a consequence, it is easy to figure out that

- Bel^*_* is the lower belief function induced by Bel^* from X/R to X by means of the multiple-valued mapping ω that associates to each X_i the equivalence class $[x]_R$ such that $\omega(X_i) = [x]_R$
- Bel^{**} is the upper belief function induced by Bel^* in the same way.

This is obvious noticing that $R^*(S) = \{X_i \mid \omega(X_i) \cap S \neq \emptyset\}$ and $R_*(S) = \{X_i \mid \omega(X_i) \subseteq S\}$. Moreover the set-function Bel^*_* is a belief function on 2^X, while Bel^{**} is not in spite of the misleading appearance of (18). The random set equivalent to Bel^*_* has focal elements $\{\omega(R^*(S)) \mid S \in \mathcal{F}\}$, and the mass allocated to $\omega(R^*(S))$ is $p^*(R^*(S))$. Bel^*_* derives from the extension of $(R^*(\mathcal{F}), p^*)$. To see it just remember the identity between Bel_R and Bel^* on X/R, that is based on the equivalence between $T \subseteq \omega(R_*(S))$ and $\omega(R^*(T)) \subseteq S$, so that

$$\begin{aligned} Bel^*_*(S) &= \Sigma_{T \subseteq \omega(R_*(S))}\, p(T) \qquad \text{from (18)} \\ &= \Sigma_{\omega(R^*(T)) \subseteq S}\, p(T) \\ &= \Sigma_{\omega(A) \subseteq S}\, p^*(A). \end{aligned}$$

These properties of Bel^*_* have been studied by Shafer et al. (1987), where Bel^*_* is called the "coarsening" of Bel, moreover Shafer et al. (1987) notice that

$$Bel^*_*(S) = \max\{Bel(T) \mid T \subseteq S, T = \omega(A) \text{ for some } A \subseteq X/R\} \tag{19}$$

a relation that is obvious from (18). Similarly

$$Bel^{**}(S) = \min\{Bel(T) \mid S \subseteq T, T = \omega(A) \text{ for some } A \subseteq X/R\} \tag{20}$$

since $\omega(R^*(S))$ is the smallest subset containing S and being the extension of a subset of X/R. But Bel^{**} is generally not a belief function on 2^X.

A similar analysis can be carried out for the lower approximation of $(\mathcal{F},p)$, and it can be checked that $Bel_*^*(S) = Bel_*(R^*(S))$ and $Bel_{**}(S) = Bel_*(R_*(S))$. See Dubois and Prade (1990) for details. Especially Bel_{**} is still a belief function.

Now remembering the relationship between the various "rough" inclusions we obtain the following inequalities :

$$Bel^{*}{}_{*}(S) \leq Bel(S) \leq Bel^{**}(S)$$
$$Bel_{*}{}^{*}(S) \geq Bel_{**}(S) \geq Bel(S)$$

The first one is also clear from (19-20) which indicate a tight bracketing. However there is no inequalities between Bel** and Bel_{**}, generally. The interest of the latter inequality lies in the fact that Bel_{**} is a belief function that may act as an upper bound to Bel, so that Bel can be bracketted by two belief functions namely $Bel^{*}{}_{*}$ (from below) and Bel_{**} (from above) that correspond to the upper and lower approximations of $(\mathcal{F},p)$ and are a generalization of rough fuzzy sets. This result extends Lemmas 11 and 12 of Shafer et al. (1987) that only consider $Bel^{*}{}_{*}$ as an approximation of Bel on X due to R. He never considers Bel_{**}. Of course dual results can be obtained for the plausibility function since $Bel(S) = 1 - Pl(\bar{S})$.

4. Approximation of Sets with Graded Similarity Relations

4.1. FUZZY SIMILARITY RELATIONS AND FUZZY EQUIVALENCE CLASSES

Another extension of rough sets consists in equipping X with a proximity relation, i.e. a fuzzy set R on X^2 such that $\mu_R(x,x) = 1$ (reflexivity), $\mu_R(x,y) = \mu_R(y,x)$ (symmetry) and a *-transitivity property of the form

$$\mu_R(x,z) \geq \mu_R(x,y) * \mu_R(y,z) \quad (21)$$

for some operation * satisfying $a * b \leq \min(a,b)$. Zadeh (1971) has introduced such relations. They generalize equivalence relations in the sense that the core of R, i.e. $\{(x,y) \mid \mu_R(x,y) = 1\}$ is an equivalence relation. Examples of admissible transitivity axioms are obtained for * = min (Zadeh's similarity relations), * = product, * = Tm (a Tm b = max(0, a + b – 1)), see Bezdek and Harris (1978). Particularly $1 - \mu_R$ is an ultrametric distance for * = min, and satisfies the usual triangle inequality for * = Tm. Trillas and Valverde (1984), (1985) have assumed * is a triangular norm (Schweizer and Sklar (1983)), i.e a non-decreasing semi-group of the unit interval with unity 1, that subsumes the three basic operations. We thus get close to the numerous works devoted to approximation in metric-like spaces, a brief survey of which is in Fariñas del Cerro and Prade (1986).

A simple way to get a fuzzy relation satisfying (21) with * = Tm is to start with n classical equivalence relations $R_1, \ldots, R_n$ on X and to define (Bezdek and Harris (1978))

$$\mu_R(x,y) = \Sigma_{i=1,n}\, \alpha_i\, \mu_{R_i}(x,y) \quad (22)$$

where $\Sigma\alpha_i = 1$ and $\alpha_i > 0$, $\forall i$. When the R_i are nested ($R_1 \subseteq R_2 \subseteq \ldots \subseteq R_n$), R satisfies (21) with * = min. When X has cardinality more than 3, Tm-transitivity is more general than convex decomposability into equivalence classes. However the problem of finding a criterion to determine if a Tm-transitive proximity decomposes into (22) is still unsolved, except for min-transitive ones for which the R_i's are the (nested) level cuts of R, i.e. $\{\{R_\lambda \mid x R_\lambda y \Leftrightarrow \mu_R(x,y) \geq \lambda\} \mid \lambda \in [0,1]\}$ (Zadeh (1971)). More insight into this problem has been recently provided by Kitainik (1991).

Another interesting problem is to define the counterpart of equivalence classes for *-transitive proximity relations. In fact that is a basic problem in taxonomy. Following Zadeh (1971) we can define the fuzzy class $[x]_R$ of elements close to x by

$$\mu_{[x]_R}(y) = \mu_R(x,y) \ \forall y \in X \tag{23}$$

This definition coincides with the one of usual equivalence classes when R is a non-fuzzy relation. Höhle (1988) has proposed a definition of what should be a fuzzy equivalence class X_i by means of three axioms :

i) μ_{X_i} is normalized, i.e. $\exists x, \mu_{X_i}(x) = 1$

ii) $\mu_{X_i}(x) * \mu_R(x,y) \leq \mu_{X_i}(y)$

iii) $\mu_{X_i}(x) * \mu_{X_i}(y) \leq \mu_R(x,y)$

i) correspond to the requirement that an equivalence class be not empty ; ii) requires that elements in the neighborhood of y should be in the equivalence class of y. Lastly, iii) states that R should contain the Cartesian product (here in the sense of *) of any equivalence class by itself ; in other words any two elements in X_i are related via R. Note that axiom ii) can be expressed under the form

$$X_i \otimes R \subseteq X_i$$

where $\mu_{X_i \otimes R}(y) = \sup_x \mu_{X_i}(x) * \mu_R(x,y)$ acts as a matrix-vector product. Due to reflexivity of R, the property $X_i \subseteq X_i \otimes R$ always hold so that ii) expresses that X_i is an eigen fuzzy set of R (Sanchez (1978) in the algebra (max, *)). The following facts are easy to establish.

Lemma 2 : A fuzzy set $[x]_R$ as in (23) is a fuzzy equivalence class.

Proof : Reflexivity of R ensures that $[x]_R$ is normalized. ii) and iii) both reduce to the *-transitivity, provided that we account for the symmetry of R.
Q.E.D.

Lemma 3 : For any normal fuzzy set F, $F \otimes R$ is an eigenvector of R.

Proof : $(F \otimes R) \otimes R = F \otimes (R \otimes R) = F \otimes R$ due to transitivity and reflexivity of R. Q.E.D.

Lemma 4 : If X_i and X_j are two equivalence classes such that $\mu_{X_i}(x) = \mu_{X_j}(x) = 1$ for some $x \in X$, then $X_i = X_j$.

Proof (inspired by Höhle (1988)) : Let $z \in X$

$$\mu_{X_j}(z) = \mu_{X_j}(z) * \sup_{x \in X} \mu_{X_i}(x) * \mu_{X_i}(x) \quad \text{(since } X_i \text{ is normalized)}$$
$$= \sup_{x \in X} \mu_{X_j}(z) * \mu_{X_i}(x) * \mu_{X_i}(x)$$
$$\leq \sup_{x \in X} \mu_R(x,z) * \mu_{X_i}(x) \quad \text{(due to axiom iii)}$$
$$= \mu_{X_i}(z)$$

since X_i is an eigen fuzzy set ; by symmetry we conclude that $X_i = X_j$. Q.E.D.

Particularly, if $X_i \subseteq X_j$ then $X_i = X_j$. In other words the set of fuzzy equivalence classes cannot be ordered via fuzzy set inclusion. So far we have proved that the set of fuzzy equivalence classes contain the fuzzy sets $\{[x]_R \mid x \in X\}$ (all members of this family are not necessarily distinct) and no fuzzy subsets of these fuzzy sets. We conclude by the last result :

Lemma 5 : Assume that F is a normalized fuzzy set such that $\nexists\, x \in X, F \subseteq [x]_R$. Then F is not a fuzzy equivalence class of R.

Proof : By assumption, $\forall x, \exists y, \mu_R(x,y) < \mu_F(y)$. Let x such that $\mu_F(x) = 1$. Then for some y, $\mu_R(x,y) < \mu_F(y) = \mu_F(y) * \mu_F(x)$. Hence axiom iii) is violated. Q.E.D.

What has been proved is that a fuzzy equivalence class can neither be strictly included in some $[x]_R$, but at the same time should be included in some $[x]_R$. To conclude

Proposition 4 : The set of fuzzy equivalence classes in the sense of Höhle is the set $\{[x]_R, x \in X\}$, for any $*$-transitive similarity relation.

If $X_1 \ldots X_n$ are the distinct fuzzy equivalence classes, we even have that R is the fuzzy union of the fuzzy Cartesian products $X_i \times X_i$ in the sense of the triangular norm $*$. Indeed

$$\mu_{\bigcup_{i=1,n} X_i \times X_i}(x,y) = \max_{i=1,n} \mu_{X_i}(x) * \mu_{X_i}(y)$$
$$= \max_{z \in X} \mu_R(x,z) * \mu_R(y,z) \quad \text{since } \forall i, \exists z, X_i = [z]_R$$
$$= \mu_R(x,y)$$

through symmetry and $*$-transitivity. Note that Höhle (1988) works in a more general setting where reflexivity of R is changed into weaker properties, equality in X is itself changed into a special similarity relation, and the unit interval is changed into a more abstract algebraic structure.

Since the notion of fuzzy quotient set is well-defined, the fuzzy equivalence classes $[x]_R$ allow for an extension of rough sets to account for graded indistinguishibility. But the set X/R is now a collection of fuzzy sets that makes a "fuzzy partition" of X. However note that if the core R_1 of R is a very fine equivalence relation, i.e. such that $x \neq y \Rightarrow [x]_{R_1} \neq [y]_{R_1}$, then $[x]_R \neq [y]_R$ as well so that $X/R = \{[x]_R \mid x \in X\}$ contains as many elements as X. This problem motivated Höhle's investigations. However, the specialization of his results to the unit interval and *-transitive fuzzy relations turns out to exactly correspond to Zadeh's original proposal for fuzzy equivalence classes.

4.2. FUZZY PARTITIONS

A more direct way of obtaining a fuzzy coarsening is to start with a family Φ of normal fuzzy sets of X, say $F_1, F_2, \ldots, F_n$, with $n < |X|$ generally. Φ is supposed to cover X sufficiently, i.e.

$$\inf_x \max_{i=1,n} \mu_{F_i}(x) > 0. \tag{24}$$

Moreover, a disjointness property between the F_i's can be requested, e.g.

$$\forall i,j,\ \sup_x \min(\mu_{F_i}(x), \mu_{F_j}(x)) < 1 \tag{25}$$

This is the weakest possible view of a fuzzy partition ; in the literature (e.g. Bezdek (1981)) a stronger definition is often adopted i.e.

$$\Sigma_{i=1,n}\, \mu_{F_i}(x) = 1, \forall x \in X.$$

However it is not requested here. $F_1, \ldots, F_n$ play the role of fuzzy equivalence classes of a similarity relation, e.g. $\Phi = \{[x]_R \mid x \in X\}$. This situation often occurs when a measurement scale X is shared into a few linguistic categories, e.g. [0,2.5] m. shared into fuzzy sets which mean "*short*", "*medium-sized*", "*tall*", "*very tall*",.... In that case the term set $\Phi = \{F_1, \ldots, F_n\}$ plays the role of the quotient set X/R. The problem of deriving a fuzzy partition from a fuzzy similarity relation is solved by (23). Indeed X/R satisfies (24) and (25) : the fuzzy union of the $[x]_R$ is exactly X : (24) holds and its left-hand side is 1. Moreover if $[x]_R \neq [y]_R$ then $\nexists z, \mu_R(x,z) = \mu_R(y,z) = 1$ hence (25) holds too.

The converse problem i.e. given a family of fuzzy sets on X that represents clusters of similar elements, find the underlying relation has been solved by Valverde (1985). Namely given a triangular norm (Schweizer and Sklar (1983)) $*$ on [0,1], let $\Rightarrow$ be a multiple-valued implication that derives from $*$ by residuation, i.e.

$$a \Rightarrow b = \sup\{x \in [0,1] \mid a * x \leq b\}$$

Note that $a \Rightarrow b = 1$ as soon as $a \leq b$, and $1 \Rightarrow b = b$. Let $F_1,\ldots, F_n$ be fuzzy sets on X. Then the fuzzy relation R defined by

$$\mu_R(x,y) = \min_{i=1,n} (\max(\mu_{F_i}(x), \mu_{F_i}(y)) \Rightarrow \min(\mu_{F_i}(x), \mu_{F_i}(y))) \tag{26}$$

is a max-$*$ transitive similarity relation. Moreover if the family of fuzzy sets is $X/R = \{[x]_R \mid x \in X\}$, then (26) applied to this family produces R again.

However if we start with any family $\{F_1, F_2, \ldots, F_n\}$ of fuzzy sets on X, the relation R is defined only up to the choice of the transitivity property (via the choice of $*$), and $X/R \neq \{F_1, F_2, \ldots, F_n\}$ generally. Hence the problem is to select the operation $*$ that produces a fuzzy quotient set X/R that is as close as possible to $\{F_1, F_2, \ldots, F_n\}$. See López de Mántaras and Valverde (1988) on this topic.

The application of (26) to the 3 basic kinds of transitive similarity relation gives :

– max-min transitivity

$$\mu_R(x,y) = \min_{i:\mu_{F_i}(x) \neq \mu_{F_i}(y)} \min(\mu_{F_i}(x), \mu_{F_i}(y))$$
$$= 1 \text{ otherwise}$$

– max-product transitivity (Ovchinnikov (1982))

$$\mu_R(x,y) = \min_{i=1,n} \min\left(\frac{\mu_{F_i}(x)}{\mu_{F_i}(y)}, \frac{\mu_{F_i}(y)}{\mu_{F_i}(x)}\right)$$

– max-linear transitivity

$$\mu_R(x,y) = \min_{i=1,n} 1 - |\mu_{F_i}(x) - \mu_{F_i}(y)|$$

The basic reason why we recover R when synthetizing a $*$-transitive fuzzy relation from its fuzzy quotient set using (26) is that it comes down to solve the relational equation (with symmetric and reflexive relations R) derived from the transitivity property. This property can be viewed as acknowledging R as a solution to $R \otimes T \subseteq R$. The equation $R \otimes T \subseteq R$ is equivalent to $T \subseteq R \otimes\!\!\rightarrow R$ where $\otimes$ is the max-$*$ matrix composition and $\otimes\!\!\rightarrow$ is the min-$\Rightarrow$ matrix composition, after results by Sanchez (1976) and Pedrycz (1985). Moreover R is also a solution to $T \otimes R \subseteq R$, so that $T \subseteq (R \otimes\!\!\rightarrow R)^{-1}$ (where for a fuzzy relation S, $\mu_{S^{-1}}(x,y) = \mu_S(y,x)$). As a consequence

$$T \subseteq T^* = (R \otimes\!\!\rightarrow R)^{-1} \cap (R \otimes\!\!\rightarrow R)$$

where $\cap$ translates into min, i.e.

$$\mu_{T^*}(x,y) = \min(\min_z (\mu_R(x,z) \Rightarrow \mu_R(z,y)), \min_{z'} (\mu_R(y,z') \Rightarrow \mu_R(z',x)))$$
$$= \min_z(\min(\mu_R(x,z) \Rightarrow \mu_R(y,z), \mu_R(y,z) \Rightarrow \mu_R(x,z)))$$

(because, R is symmetric, and due to the properties of min).

$$= \min_z(\max(\mu_R(x,z), \mu_R(y,z)) \Rightarrow \min(\mu_R(y,z), \mu_R(x,z)))$$

since $\min(a \Rightarrow b, c \Rightarrow b) = \max(a,c) \Rightarrow b$

$\min(a \Rightarrow b, a \Rightarrow c) = a \Rightarrow \min(b,c)$ (see Di Nola et al. (1989))

By construction $R \subseteq T^*$. Letting $z = x$ in the expression of μ_{T^*}, leads to $T^* \subseteq R$, due to reflexivity and symmetry of R. (26) is thus completely justified by the identity $T^* = R$.

4.3. FUZZY ROUGH SETS

Given a fuzzy partition Φ on X,counterparts of (1-2) and (5-6) in this setting allow a description of any fuzzy set F by means of the term set Φ, under the form of an upper and a lower approximation $\Phi^*(F)$ and $\Phi_*(F)$ as follows

$$M_i \triangleq \mu_{\Phi^*(F)}(F_i) = \sup_x \mu_{F_i}(x) * \mu_F(x) \quad (27)$$

$$m_i \triangleq \mu_{\Phi_*(F)}(F_i) = \inf_x \mu_{F_i}(x) \to \mu_F(x) \quad (28)$$

where $a \to b = 1 - a * (1 - b)$ is called an S-implication (Trillas and Valverde (1985)). Clearly, these equations also generalize the basic definitions of C-calculus. (27) was first proposed with $*$ = min by Willaeys and Malvache (1981) in order to approximately represent a fuzzy set in X on a coarser referential of fuzzy sets of X. (27) and (28) are nothing but the degrees of possibility and necessity of the fuzzy event F, in the sense of Zadeh when $*$ = min (Zadeh (1978), Dubois and Prade (1988)). M_i being the degree of possible membership of F_i in F, and m_i the corresponding degree of certain membership.This definition makes sense even when F is not fuzzy. The pair $(\Phi_*(F), \Phi^*(F))$ can be called a fuzzy rough set. When F_i is crisp, it particularizes to the rough fuzzy set of Section 2.1.

Assume now that Φ is a fuzzy quotient set that derives from a fuzzy relation R via (23).The knowledge of R is anyway sufficient to define the "extension" of fuzzy rough sets in the sense of (3-4). Given a subset S of X, (7) and (8) generalize into

$$\forall x \in X, \mu_{\omega(R^*(S))}(x) = \sup\{\mu_R(x,y) \mid y \in S\} \quad (29)$$

$$\forall x \in X, \mu_{\omega(R_*(S))}(x) = \inf\{1 - \mu_R(x,y) \mid y \notin S\} \quad (30)$$

Moreover (29-30) in turn generalize in accordance with possibility theory into

$$\forall x \in X, \mu_{\omega(R^*(F))}(x) = \sup_y \mu_F(y) * \mu_R(x,y) \quad (31)$$

$$\forall x \in X, \mu_{\omega(R_*(F))}(x) = \inf_y \mu_R(x,y) \to \mu_F(y) \quad (32)$$

which define upper and lower approximations of a fuzzy set F through a similarity relation R. Again $a \to b = 1 - a * (1 - b)$ in (32). (31) and (32) have been proposed by Fariñas del Cerro and Prade (1986) with $*$ = minimum. The equivalence between (27-28) and (31-32), when $\Phi = X/R$, is almost self-evident, since the index i in (27-28) is just replaced by the element x in (31-32), based on (23) ; moreover, if x and x' generate the same fuzzy equivalence class, their degrees of membership in (31-32) coincide. Besides, (31) and (32) represent the distorsion of a fuzzy set F due to the indiscernibility relation on X, as if looking at F with blurring glasses.

(27-28) suggest a new approach to linguistic approximation. Assume $X/R = \{F_1, F_2, \ldots, F_n\}$ is a term set that partition X into linguistically meaningful fuzzy subsets. The problem of linguistic approximation is one of finding the best term that may qualify a given unnamed fuzzy set F. This problem is commonly encountered in fuzzy set-based software that accept linguistic inputs, model them by fuzzy sets, and must ouput responses in a linguistic format again. Then the fuzzy sets computed by the program must be named by means of some user- oriented vocabulary. See for instance Novak (1987),... If I_* and I^* denote the set of indices such that

$$m_i = \mu_{R_*(F)}(F_i) > 0 \text{ and } M_i = \mu_{R^*(F)}(F_i) > 0$$

respectively, F can be bracketed by ORing the F_i's in I_* and in I^* (which contains I_*). The M_i's and m_i's may be used to define linguistic modifiers for the selected terms in the term set. Namely, F means "$L_{i1}\ F_{i1}$ or $L_{i2}\ F_{i1}$ or... $L_{ip}\ F_{ip}$" where p is the cardinality of I^*, and L_i is a modifier of F_i. Rougly speaking, the following modifiers could be chosen :

if $M_i \leq 0.5$	L_i means "not very possibly"
if $m_i = 0$, $M_i \geq 0.5$	L_i means "possibly"
if $m_i \geq 0.5$	L_i means "rather certainly"
if $m_i = 1$	L_i means "certainly" (and can be omitted)

This approach to linguistic approximation is quite different from the one of Bonissone (1979) and followers, based on minimizing a distance between F and the fuzzy sets produced from the F_i's by means of unary modifiers and logical connectives. Here we view linguistic approximation as a problem of rough classification in the sense of Pawlak et al. (1986). Our procedure looks more robust than the ones based on a nearest-neighbour classification process. But it cannot precisely name fuzzy sets included in the F_i's, by construction.

Nakamura ((1988), (1989), (1991a, b)), and Nakamura and Gao (1991) have proposed another view of fuzzy rough sets. They consider the upper and lower approximations of fuzzy sets in the sense of a min-transitive fuzzy relation. These upper and lower approximations are viewed as a family of rough fuzzy sets in the sense of Section 2.1, induced by the level-cuts of the fuzzy relation. Namely if R is a fuzzy similarity relation, let $R_\alpha = \{(x,y), \mu_R(x,y) \geq \alpha\}$. When R is min-transitive, R_α is an equivalence relation for all $\alpha \in [0,1]$. Moreover we have the representation theorem (Zadeh (1971)) :

$$\mu_R(x,y) = \sup\{\alpha \mid x\ R_\alpha\ y\}.$$

The upper and lower approximations of a fuzzy set F with respect to a min-transitive fuzzy relation is then defined by Nakamura as the sets of fuzzy sets $\mathcal{R}^*(F)$ and $\mathcal{R}_*(F)$ such that

$$\mathcal{R}^*(F) = \{R^*(\alpha,F), \alpha \in [0,1]\}$$
$$\mathcal{R}_*(F) = \{R_*(\alpha,F), \alpha \in [0,1]\}$$

where $(R^*(\alpha,F), R_*(\alpha,F))$ is a rough fuzzy set with respect to R_α, i.e.

$$\mu_{\omega(R^*(\alpha,F))}(x) = \sup\{\mu_F(y),\ y \in [x]_{R_\alpha}\}$$
$$\mu_{\omega(R_*(\alpha,F))}(x) = \inf\{\mu_F(y),\ y \in [x]_{R_\alpha}\}.$$

It is easy to verify that $\mathcal{R}^*(F)$ and $\mathcal{R}_*(F)$ are nested families of fuzzy sets. More specifically

$$\alpha > \beta \text{ entails } R^*(\alpha,F) \subseteq R^*(\beta,F) \text{ and } R_*(\beta,F) \subseteq R_*(\alpha,F).$$

Note that the nestedness property works opposite ways in $\mathcal{R}^*(F)$ and $\mathcal{R}_*(F)$ since when α increases, R_α becomes finer, and the upper (resp. : lower) approximations become smaller (resp. : larger).

As pointed out in an earlier paper (Dubois and Prade (1990)). Nakamura's approach can be related to ours (when $*$ = min) through the notion of median. Namely $R^*(F)$ and $R_*(F)$ are medians of the families $\mathcal{R}^*(F)$ and $\mathcal{R}_*(F)$ in the sense of Sugeno's integral, i.e.

$$\mu_{\omega(R^*(F))}(x) = \sup_\alpha \min(\alpha, \mu_{\omega(R^*(\alpha,F))}(x)) \tag{33}$$
$$\mu_{\omega(R_*(F))}(x) = \inf_\alpha \max(1-\alpha, \mu_{\omega(R_*(\alpha,F))}(x)). \tag{34}$$

The first equality holds because

$$\sup_\alpha \min(\alpha, \sup_y \min(\mu_F(y), \mu_{R_\alpha}(x,y)) = \sup_y \min(\mu_F(y), \sup_\alpha \min(\alpha, \mu_{R_\alpha}(x,y)))$$

using Zadeh's representation theorem. The other equality obtains letting $\mathcal{R}_*(F) = \overline{\mathcal{R}^*(\overline{F})}$. Note that the synthesis process is not the same in (33) and (34) because $\mu_{\omega(R^*(\alpha,F))}(x)$ is decreasing with α while $\mu_{\omega(R_*(\alpha,F))}(x)$ is increasing, especially $\sup_\alpha \min(\alpha, \mu_{\omega(R_*(\alpha,F))}(x)) = \mu_{\omega(R_*(1,F))}(x)$, i.e. the rough fuzzy set with respect to the finest equivalence relation obtained from R. To see that (33) is a Sugeno integral, it is enough to notice that the set function g on X × X such that $g(R_\alpha) = \mu_{\omega(R^*(\alpha,F))}(x)$ is indeed a fuzzy measure since $R_\beta \subseteq R_\alpha$ entails $g(R_\beta) \leq g(R_\alpha)$. The other expression is also a fuzzy integral noticing that

$$\mu_{\omega(R_*(F))}(x) = \inf_\beta \max(\beta, \mu_{\omega(R_*(1-\beta,F))}(x))$$
$$= \sup_\beta \min(\beta, \mu_{\omega(R_*(1-\beta,F))}(x))$$

Again $g((R)_\beta) = \mu_{\omega(R_*(1-\beta,F))}(x)$ is a fuzzy measure in the sense of Sugeno.

4.4. PROPERTIES OF FUZZY ROUGH SETS

Basic properties of rough sets can be extended to fuzzy rough sets. Using classical pointwise definitions of union, intersection and complementation, by maximum, minimum, and $1 - \cdot$ respectively, and the usual definition of fuzzy set inclusion, the following properties hold on the term set $\Phi = X/R$

$$R_*(F \cup G) \supseteq R_*(F) \cup R_*(G)\ ;\ R^*(F \cup G) = R^*(F) \cup R^*(G) \tag{35}$$

$$R^*(F \cap G) \subseteq R^*(F) \cap R^*(G)\ ;\ R_*(F \cap G) = R_*(F) \cap R_*(G) \tag{36}$$

$$R^*(\bar{F}) = \overline{R_*(F)} \tag{37}$$

where inclusion is in the sense of fuzzy set theory. They are another way of stating basic relations between N(F), N(G), N(F $\cup$ G), N(F $\cap$ G) on the one hand, and Π(F), Π(G), Π(F $\cup$ G), Π(F $\cap$ G), on the other hand, in possibility theory.

The following properties also hold for fuzzy rough sets

$$\omega(R_*(F)) \subseteq F \subseteq \omega\,(R^*(F))$$
$$R_{**}(F) = R_*(F)\ ;\ R^{**}(F) = R^*(F)$$

where $R_{**}(F)$ is short for $R_*(\omega(R_*(F))$; $R^{**}(F) = R^*(\omega(R^*(F))$. The latter equalities take advantage of the $*$-transitivity of R, i.e.

$$\begin{aligned}
\mu_{\omega(R^{**}(F))}(x) &= \sup_y \mu_{\omega(R^*(F))}(y) * \mu_R(x,y) \\
&= \sup_y (\sup_z \mu_F(z) * \mu_R(z,y)) * \mu_R(x,y) \\
&= \sup_{y,z} \mu_F(z) * \mu_R(y,z) * \mu_R(x,y) \\
&= \sup_z \mu_F(z) * \sup_y \mu_R(x,y) * \mu_R(y,z) \\
&= \sup_z \mu_F(z) * \mu_R(z,x) \quad \text{since } R \otimes R = R \\
&= \mu_{\omega(R^*(F))}(x).
\end{aligned}$$

However the equalities $R^*(R_*(F)) = R_*(F)$ and $R_*(R^*(F)) = R^*(F)$ that hold for rough fuzzy sets no longer hold for all kinds of fuzzy rough sets.

<u>Proposition 5</u> : For continuous triangular norm $*$ and a symmetric, $*$-transitive fuzzy relation, the upper approximation of the lower approximation satisfies the following properties :

–) if $*$ = minimum : $\mu_{\omega(R_*(F))} \leq \mu_{\omega(R^*(R_*(F)))} \leq \max(0.5, \mu_{\omega(R_*(F))})$

–) if $*$ = product : $\mu_{\omega(R^*(R_*(F)))} = \mu_{\omega(R_*(F))}$ if $\mu_{\omega(R_*(F))}(x) \geq 1/2$

$$\mu_{\omega(R_*(F))} \leq \mu_{\omega(R^*(R_*(F)))} \leq \frac{1}{4(1 - \mu_{\omega(R_*(F))}(x))} \quad \text{otherwise}$$

–) if $* = Tm$ then $R^*(R_*(F)) = R_*(F)$

–) if $Tm \le *$ then $\mu_\omega(R_*(F)) \le \mu_\omega(R^*(R_*(F))) \le \frac{1+\mu_\omega(R_*(F))}{2}$

Proof : $R_*(F) \subseteq R^*(R_*(F))$ is obvious. Let us check the upper bounds

$$\mu_\omega(R^*(R_*(F))(x) = \sup_y [\inf_z 1 - \mu_R(y,z) * (1 - \mu_F(z))] * \mu_R(x,y).$$

from (32). This expression can be upper bounded using transitivity, changing $\mu_R(y,z)$ into $\mu_R(y,x) * \mu_R(x,z)$

$$\begin{aligned}\mu_\omega(R^*(R_*(F))(x) &\le \sup_y [\inf_z 1 - \mu_R(y,x) * \mu_R(x,z) * (1 - \mu_F(z))] * \mu_R(x,y) \\ &= \sup_y (1 - \mu_R(y,x) * \mu_\omega(R^*(\bar{F}))(x)) * \mu_R(x,y) \\ &= \sup_y (\mu_R(x,y) \to \mu_\omega(R_*(F))(x)) * \mu_R(x,y) \\ &\le \sup_{a\in[0,1]} (a \to \mu_\omega(R_*(F))(x)) * a.\end{aligned}$$

This quantity is not easy to compute in the general case.

If $* \ge Tm$ then $(1 - a * b) * a \le (1 - \max(0, a + b - 1)) * a \le \min(1, 2 - a - b, a)$. This quantity reaches its maximum for $a = 1 - b/2$, and we thus obtain

$$\mu_\omega(R^*(R_*(F))(x) \le 1 - (1/2) \cdot \mu_\omega(R^*(\bar{F}))(x) = \frac{1+\mu_\omega(R_*(F))(x)}{2}.$$

For the three basic t-norms, we obtain more precise results :

<u>* = minimum</u> : then we have to compute

$$\begin{aligned}\sup_{a\in[0,1]} \min[\max(1 - a, b), a] &= \max[\sup_a \min(1 - a, a), \sup_a \min(a,b)] \\ &= \max(1/2, b). \text{ Hence the result.}\end{aligned}$$

<u>* = product</u> : we have to compute

$$\sup_{a\in[0,1]} (1 - a + ab)a = \begin{cases} \frac{1}{4(1-b)} & \text{if } b \le 1/2 \\ b & \text{otherwise}\end{cases}$$

Hence the result.

<u>* = Tm</u> : we have to compute

$$\sup_{a\in[0,1]} \max(0, \min(1, 1 - a + b) + a - 1) = \sup_{a\in[0,1]} \max(0, \min(a,b)) = b.$$

Hence the result. Q.E.D.

Dual inequalities hold for $R_*(R^*(F))$.

Proposition 6 : For a continuous triangular norm $*$ and a symmetric, $*$-transitive relation, the lower approximation of the upper approximation satisfies the following properties

–) if $*$ = minimum : $\mu_\omega(R^*(F)) \geq \mu_\omega(R_*(R^*(F))) \geq \min(0.5, \mu_\omega(R^*(F)))$

–) if $*$ = product : $\mu_\omega(R^*(F)) = \mu_\omega(R_*(R^*(F)))$ if $\mu_\omega(R^*(F)) \leq 1/2$

$$\mu_\omega(R^*(F)) \geq \mu_\omega(R_*(R^*(F))) \geq 1 - 1/4 \cdot \frac{1}{\mu_\omega(R^*(F))(x))} \quad \text{otherwise}$$

–) if $*$ = Tm then $R_*(R^*(F)) = R^*(F)$

–) if $* \geq$ Tm then $\mu_\omega(R^*(F)) \geq \mu_\omega(R_*(R^*(F))) \geq 1/2 \cdot \mu_\omega(R^*(F))$.

Proof : $R_*(R^*(F)) = R_*(\overline{R_*(\bar{F})})$. Hence it is enough to use the bounds for $R_*(\bar{F})$. Q.E.D.

The definition of $R_*(F)$ as in (32) is based on so-called S-implications, that derive from triangular norms through the classical definition of material implication $\varphi \to \psi = \neg\varphi \vee \psi = \neg(\varphi \wedge \neg\psi)$; hence $a \to b = 1 - (a * (1 - b))$. There is another definition of multivalued implication, based on the deduction theorem, i.e. φ ; $\psi \vdash \xi \Leftrightarrow \varphi \vdash \psi \to \xi$, which translates in the multiple-valued framework by $a * b \leq c$ if and only if $a \leq b \Rightarrow c$, hence $b \Rightarrow c = \sup\{x \mid x * b \leq c\}$. These are called R-implications (R means residuation) by Trillas and Valverde (1985) and encountered in Section 4.2. These two families were laid bare by these authors as well as the authors of this paper (Dubois and Prade (1984a, b)). The expression (32) suggests another definition of the lower approximation, based on the residuated implications

$$\mu_\omega(R_*(F))(x) = \inf_y \mu_R(x,y) \Rightarrow \mu_F(y).$$

The following properties of residuated implications based on continuous triangular norms will be useful :

Lemma 5

i) $((a * b) \Rightarrow c) * a \leq b \Rightarrow c$

ii) $a \Rightarrow (b \Rightarrow c) = (a * b) \Rightarrow c$

iii) $a \Rightarrow b * c * a \geq b * c$.

Proof : Let us show i)

$$\begin{aligned} ((a * b) \Rightarrow c) * a &= (\sup\{x \mid (a * b) * x \leq c\}) * a \\ &= \sup\{a * x \mid (a * x) * b \leq c\} \text{ ; let } y = a * x \\ &\leq \sup\{y, y * b \leq c\} = b \Rightarrow c. \qquad (\text{since } \{a * x \mid x \in [0,1]\} \subseteq [0,1]). \end{aligned}$$

As for ii) first notice that due to i), $(a * b) \Rightarrow c \in \{x \mid x * a \leq b \Rightarrow c\}$; hence $(a * b) \Rightarrow c \leq \sup\{x \mid x * a \leq b \Rightarrow c\} = a \Rightarrow (b \Rightarrow c)$. The converse inequality can be established as follows :

$$a \Rightarrow (b \Rightarrow c) = \sup\{x \mid a * x \leq \sup\{y \mid b * y \leq c\}\}$$
$$\leq \sup\{x \mid a * b * x \leq \sup\{y * b \mid b * y \leq c\}\} \qquad \text{(since } s \leq t \text{ entails } s * u \leq t * u)$$
$$= \sup\{x \mid a * b * x \leq c\} = (a * b) \Rightarrow c$$

Lastly, $a \Rightarrow b * c * a = \sup\{x \mid x * a \leq b * c * a\} \geq \sup\{x \mid x \leq b * c\} = b * c$, since $x \leq b * c$ entails $x * a \leq b * c * a$, which proves iii). Q.E.D.

<u>Proposition 7</u> : The lower approximations $R_*(F)$ based on a R-implication satisfy the following properties

i) $R_*(F \cap G) = R_*(F) \cap R_*(G)$
ii) $R_*(F \cup G) \supseteq R_*(F) \cup R_*(G)$
iii) $R_*(F) \subseteq F$
iv) $R_{**}(F) = R_*(F)$
v) $R^*(R_*(F)) = R_*(F)$
vi) $R_*(R^*(F)) = R^*(F)$.

<u>Proof</u> : i) is a simple consequence of $\Rightarrow$ being monotically increasing in its second argument ; ii) is as obvious ; iii) obtains since $\inf_y \mu_R(x,y) \Rightarrow \mu_F(y) \leq (\mu_R(x,x) \Rightarrow \mu_F(x)) = \mu_F(x)$ because R is reflexive.

iv) $$\mu_{\omega(R_{**}(F))}(x) = \inf_y \mu_R(x,y) \Rightarrow (\inf_z (\mu_R(y,z) \Rightarrow \mu_F(z)))$$
$$= \inf_{y,z} \mu_R(x,y) \Rightarrow (\mu_R(y,z) \Rightarrow \mu_F(z))$$
$$= \inf_z (\sup_y (\mu_R(x,y) * \mu_R(y,z)) \Rightarrow \mu_F(z)) \qquad \text{using ii) in lemma 5}$$
$$= \inf_z \mu_R(x,z) \Rightarrow \mu_F(z) = \mu_{\omega(R_*(F))}(x)$$

v) $$\mu_{\omega(R^*(R_*(F)))}(x) = \sup_y \mu_R(x,y) * (\inf_z (\mu_R(y,z) \Rightarrow \mu_F(z))$$
$$\leq \sup_y \inf_z \mu_R(x,y) * ((\mu_R(y,x) * \mu_R(x,z)) \Rightarrow \mu_F(z))$$
due to $*$-transitivity of R ;
$$\leq \sup_y \inf_z \mu_R(x,z) \Rightarrow \mu_F(z) \qquad \text{due to i) in lemma 5, and symmetry of R}$$
$$= \mu_{\omega(R_*(F))}(x) ;$$

hence $R^*(R_*(F)) \subseteq R_*(F)$. But $R_*(F) \subseteq R^*(R_*(F))$ is also obvious (by letting x = y and taking advantage of the reflexivity of R in the expression of $\mu_{\omega(R^*(R_*(F)))}(x)$). Hence the result.

vi) $\mu_{\omega(R_*(R^*(F)))}(x) = \inf_y (\mu_R(x,y) \Rightarrow (\sup_z \mu_R(y,z) * \mu_F(z)))$

$\geq \inf_y \sup_z \mu_R(x,y) \Rightarrow (\mu_R(y,x) * \mu_R(x,z) * \mu_F(z))$

using the $*$-transitivity of R

$\geq \inf_y \sup_z \mu_R(x,z) * \mu_F(z)$ due to symmetry of R and iii) in lemma 5

$= \mu_{\omega(R^*(F))}(x)$.

Hence $R^*(F) \subseteq R_*(R^*(F))$. Hence the equality $R^*(F) = R_*(R^*(F))$, since $R^*(F) \supseteq R_*(R^*(F))$ is obvious by letting x = y in the expression of $\mu_{\omega(R_*(R^*(F)))}(x)$. Q.E.D.

Note that all the properties pertaining to the lower approximation must be computed independently of the properties of the upper approximation because when the lower approximation is based on a residuated implication, we do not have $R^*(\bar{F}) = \overline{R_*(\bar{F})}$ any longer. To keep this property means to modify the definition of the upper image by means of a non-commutative conjunction defined, symmetrically to S-implication, as $a \curlywedge b = 1 - a \Rightarrow (1 - b)$ given a R-implication $\Rightarrow$. The study of the upper approximations based on non-commutative conjunctions is left for further research. However, remembering that the non-commutative conjunction can also be built by residuation from the S-implication (i.e. $a \curlywedge b = \sup\{x \mid a \rightarrow x \leq b\}$ for a large class of t-norms (e.g. Dubois and Prade (1984b), Fodor (1991)), there is clearly an interesting investigation to envisage along this research line.

Note that Nakamura's approach has the advantage of preserving all properties of rough sets since for rough fuzzy sets everything is preserved indeed (see Section 3.1). Then for all $\alpha \in [0,1]$, $(R^*(\alpha,F),R_*(\alpha,F))$ verifies all the properties enjoyed by rough fuzzy sets, and this can be written in a more compact way in terms of $(\mathcal{R}^*(F),\mathcal{R}_*(F))$. However Nakamura's approach has the disadvantage of being very heavy, since a fuzzy rough set in his sense may correspond to an infinity of rough fuzzy sets, indexed by α. On the contrary, our approach leads to a more compact description of fuzzy rough sets, at the expense of losing some properties.

4.5. FUZZY ROUGH SETS AND MODAL LOGICS

There have been several works by Orlowska (1984), Fariñas del Cerro and Orlowska (1985) that relate rough sets to modal logic. The basic idea is to interpret a rough set in terms of the two modalities L (necessary) and M (possible). Namely if p is a proposition whose meaning is defined via the subset $S \subseteq X$, Lp and Mp then correspond to $R_*(S)$ and $R^*(S)$ respectively. R plays the role of the accessibility relation on X in order to equip modal logic with the usual semantics (Chellas (1980)). A modal logic in the usual sense is thus defined at the syntactic level. At the semantic level, a model is viewed as (X, R, m) where X is a set of objects, R an equivalence relation on X and m is the meaning function. m defines for each formula p the set of objects m(p) for which p is true. Satisfiability is defined in the usual sense, especially for Lp and Mp :

x satisfies p	if and only if	$x \in m(p)$
x satisfies Lp	if and only if	$\forall y \in [x]_R, y \in m(p)$, i.e. $[x]_R \subseteq m(p)$.

x satisfies Mp if and only if $\exists\, y \in [x]_R,\ y \in m(p)$, i.e. $[x]_R \cap m(p) \neq \emptyset$.

Clearly this can be written $m(Lp) = R_*(p)$, $m(Mp) = R^*(p)$. Since R is an equivalence relation the corresponding modal logic is the S5 system (Orlowska (1984), Nakamura (1991a)). The axioms of this system are embodied in the properties of rough sets with respect to set-theoretic operations and iteration of lower and upper approximation operations.

Fuzzy rough sets and rough fuzzy sets offer a nice opportunity for developping a meaningful modal-like logic involving fuzzy modalities (induced by a fuzzy accessibility relation) acting on fuzzy propositions (whose meanings are fuzzy sets of X). Some attempts have been made in the past along this line. Nakamura and Gao (1991) have pointed out that since fuzzy rough sets (viewed as an indexed family of rough fuzzy sets) satisfy all properties of rough sets, it is possible to develop a S5-like modal fuzzy logic ; this logic is presented in (Nakamura (1991a)). The same author considers the case when the fuzzy relation is only symmetric and reflexive in (Nakamura (1989)). He basically focuses on indexed modalities, L_α denoting the necessity modality associated with the level-cut relation R_α (see also Nakamura (1991b)).

In a different context Ruspini (1991) envisages a fuzzy logic where fuzziness comes from indiscernibility between possible worlds. Namely, m(p) is not attainable, only fuzzy upper and lower approximations $R^*(m(p))$ and $R_*(m(p))$ make sense. R is defined as a $*$-transitive similarity relation, and (Lp,Mp) correspond again to the fuzzy rough sets $R^*(m(p))$ and $R_*(m(p))$. Ruspini focuses on the notion of rough deduction ; namely when a proposition p implies the proposition q to degree α, it means that $\forall\, x \in m(p)$, $\exists\, y \in m(q)$ which is α-similar to x, i.e. such that $\mu_R(x,y) \geq \alpha$. In other words the degree of implication of q by p is the value

$$I(q \mid p) = \inf_{y \in m(p)} \sup_{x \in m(q)} \mu_R(x,y)$$
$$= \inf_{y \in m(p)} \mu_{R^*(m(q))}(y).$$

This degree is a degree of inclusion of m(p) in the upper approximation of m(q) obtained by "stretching" the set of objects x satisfying q in a suitable way. A companion degree of consistence is defined by

$$C(q \mid p) = \sup_{y \in m(p)} \mu_{R^*(m(q))}(y).$$

Denoting $I(q \mid x) = \mu_{R^*(m(q))}(x)$, Ruspini (1991) considers another implication degree, here denoted $I_S(q \mid p)$ constructed from stretching both p and q and taking into account a piece of incomplete evidence s under the form of a subset S of possible worlds, one of which is the true one :

$$I_S(q \mid p) = \inf_{x \in S} I(p \mid x) \Rightarrow I(q \mid x)$$

where $\Rightarrow$ is a residuated implication associated to the triangular norm $*$. In fact Ruspini calls "conditional necessity distribution" Nec(q | p) any lower bound to this quantity and "conditional possibility distribution" any upper bound.

Note that what is computed is a degree of inclusion of the restriction to S of the upper approximation $R^*(m(p))$ into $R^*(m(q))$. It generalizes the classical notion of deducing q from p, given evidence S since letting R be the equality relation

$$I_S(q \mid p) = 1 \quad \text{if and only if } S \subseteq m(\neg p \vee q)$$
$$\text{if and only if } S \cap m(p) \subseteq m(q).$$

A transitivity property is obtained, namely

$$I(q \mid x) \geq I_S(q \mid p) * I(p \mid x)$$

which is due to the property $a * (a \Rightarrow b) \leq b$ that characterizes R-implications. Ruspini (1991) considers the above inequality as an inference rule that is the basis of Zadeh's generalized modus ponens. The other inequality

$$C(q \mid x) \geq I_S(q \mid p) * C(p \mid x)$$

is also proved by Ruspini (1991).

The above view of fuzzy logic considers fuzziness as a by-product of graded indistinguishability. Namely any subset A of a set X of possible worlds is perceived as a fuzzy set $R^*(A)$ due to the similarity relation R on X. This is well in accordance with Orlowska and Pawlak's intuitions. However Ruspini never considers the lower approximation $R_*(A)$ which is another fuzzy set included in $R^*(A)$. It is the pair $(R_*(A), R^*(A))$ that should be regarded as the result of blurring A by means of R. This fuzzy rough set involves both fuzziness (grades of membership) and imprecision (interval-valued membership grades). Particularly, there are three companion entailment degrees to $I_S(q \mid p)$, changing upper approximations into lower approximations, as done in Section 3.4, corresponding to so-called "rough implications". Ruspini's fuzzy logic considers that given evidence s whose set of models is $m(s) = S$, and indistinguishability R, necessity and possibility degrees of any proposition p Nec(p) and Pos(p) are defined as

$$\mathrm{Nec}(p) \leq I(p \mid s)$$
$$\mathrm{Pos}(p) \geq C(p \mid s).$$

These notions of possibility and necessity differ from the ones proposed by Zadeh (1978) and embedded in the authors' possibilistic logic (Dubois and Prade (1991)). In possibilistic logic all elements of X can be distinguished, but the available evidence s is fuzzy, thus inducing an *ordering* relation on X, encoded by the possibility distribution $\pi = \mu_{m(s)}$. The preferred worlds x are such that $\pi(x)$ is maximal. The possibility and necessity of a proposition p are then defined by

$$N(p) = \inf_{x \notin m(p)} 1 - \pi(x)$$
$$\Pi(p) = \sup_{x \in m(p)} \pi(x)$$

which are again degrees of inclusion of m(s) in m(p) and consistence of m(s) and m(p) respectively. (Nec,Pos) and (N,Π) do not have the same properties generally. Namely it is possible to prove (Ruspini (1991))

$$I(p \wedge q \mid s) \leq \min(I(p \mid s), I(q \mid s))$$
$$I(p \vee q \mid s) \geq \max(I(p \mid s), I(q \mid s))$$
$$C(p \wedge q \mid s) \leq \min(C(p \mid s), C(q \mid s))$$

$$C(p \vee q \mid s) = \max(C(p \mid s), C(q \mid s)).$$

Moreover $I(\neg p \mid s) \neq 1 - C(p \mid s)$. On the contrary, $N(p) = 1 - \Pi(\neg p)$ and $N(p \wedge q) = \min(N(p),N(q))$, an equality which fails in Ruspini's approach. But the equality would hold if $I(p \mid s)$ were based on the inclusion of S in the lower approximation of m(p), instead.

To conclude, Ruspini's approach to fuzzy logic comes close to attempts at devising modal logics of rough sets by Orlowska and to extensions as done by Nakamura, although Ruspini does not try to propose a modal logic system per se. Moreover Ruspini's fuzzy logic contrasts with possibilistic logic in the sense that the former is a logic of similarity while the other is a logic of preference. It would be fruitful to put them together by keeping the indistinguishability relation, but allowing for fuzzy evidence under the form of a possibility distribution ranking the possible worlds in terms of preference. Then the necessity-like index could be generalized into

$$I(p \mid s) = \inf_x \pi(x) \rightarrow \mu_{R^*(m(p))}(x)$$

with $\pi = \mu_{m(s)}$, using an S-implication. Clearly when R is the equality on X, $I(p \mid s)$ reduces to the degree of necessity of possibilistic logic. Similarly,

$$C(p \mid s) = \sup_x \pi(x) * \mu_{R^*(m(p))}(x)$$

generalizes both the consistence index and the possibility measure. Ruspini's notion of conditional possibility and necessity does not coincide to the similar notion in possibilistic logic. Namely when R reduces to an equality, $I_S(q \mid p) = N(\neg p \vee q)$ where N is a {0,1}-valued necessity measure defined from $\pi(x) = 1$ if $x \in S$ and $\pi(x) = 0$ otherwise. On the contrary, in possibilistic logic $N(q \mid p)$ is defined as

$$\begin{aligned} N(q \mid p) &= N(\neg p \vee q) \text{ if } N(\neg p \vee q) > N(\neg p \vee \neg q) \\ &= 0 \text{ otherwise} \end{aligned}$$

i.e. $N(q \mid p)$ expresses to what extent q is a more plausible conclusion than $\neg q$ in the case when p is true, given evidence s. The conditional possibility is $\Pi(q \mid p) = 1 - N(\neg q \mid p)$ while Ruspini views it as an upper bound on $I_S(q \mid p)$. When S is fuzzy, the following extension can be proposed

$$I_S(q \mid p) = \inf_x \pi(x) \rightarrow (I(p \mid x) \Rightarrow I(q \mid x)).$$

where $\pi = \mu_S$ and $\rightarrow$ is a S-implication. Note that we must use an S-implication to combine π with the other expression in order to encompass $N(\neg p \vee q)$ as a particular case of the above expression, when R is an equality. Indeed $N(p) = \inf_{x \notin m(p)} 1 - \pi(x) \neq \inf_x \pi(x) \Rightarrow \mu_{m(p)}(x)$, generally.

5. Conclusion

This paper has shown that the idea of a rough set can be combined with fuzzy sets in a fruitful way. It enables several independent approaches to approximation models to be unified. Further research is needed on the following aspects :

Modal logic approaches

It is obvious that fuzzy rough set offer a good opportunity to relate and/or put together fuzzy sets and modal logic, a task that has been considered in the past from various perspectives (see Dubois and Prade (1980) for a survey of early attempts) especially fuzzy accessibility relations. Prade (1984) and Dubois et al. (1988) have considered incomplete information systems for which a modal formalization is due to Lipski, and have extended it to fuzzy incomplete information systems (indistinguishability between attribute values is discussed in this context in (Prade and Testemale (1987)). More recently, Fariñas del Cerro and Herzig (1991) have related possibilistic logic to the modal conditional systems of Lewis (1973). Fuzzy rough sets offer a tool for graded extensions of the S5 system as pointed out by Nakamura (1991a, b). Here the difficulty depends about where the fuzzy component lies. Deriving a modal logic of rough fuzzy sets looks difficult because one must start from the syntax of a multiple-valued logic where conjunction and disjunction translate into max and min. To the authors knowledge, such a fuzzy counterpart of propositional calculus does not exist. Especially, this is not Lukasiewicz logic, but a multiple-valued logic based on a Kleene algebra. If the corresponding syntax can be worked out in harmony with fuzzy set-based semantics then the modalities induced by an equivalence relation will correspond to those of S5. The converse task, i.e. adding fuzzy modalities deriving from a similarity relation to propositional calculus has been considered by Nakamura (1991b) using a multi-modal calculus. Doing it in the style of fuzzy rough sets defined in this paper looks less straightforward because all properties of the S5 modalities do not hold, and there is a degree of freedom as to the choice of the connective family that defines the lower approximations.

Handling several similarity relations

The full rough sets theory handles several indiscernibility relations that correspond to, for instance, as many attributes in the framework of information systems. The way equivalence relations may combine has been the topic of algebraic studies, and these results have been cast in the modal logic DAL of Fariñas del Cerro and Orlowska (1985). An algebraic study of the combination modes of *-transitive fuzzy relations is certainly an important topic.

Modeling independence in a fuzzy setting

Lastly, Pawlak (1984) as well as Shafer et al. (1987) and Pearl and Verma (1987) have studied concepts of independence and redundancy between partitions, each with different terminologies. Especially Pawlak's notion of independence is weaker than the one of Shafer et al. (1987). It might be useful to pursue the unification work along this line, and generalize it with fuzzy partitions, viewed as fuzzy quotient sets of similarity relations.

References

Bezdek, J.C. (1981) Pattern Recognition with Fuzzy Objective Function Algorithms, Plenum Press, New York.

Bezdek, J.C. and Harris, J.D. (1978) 'Fuzzy partitions and relations — An axiomatic basis for clustering', Fuzzy Sets and Systems 1, 112-127.

Bonissone, P.P. (1979) 'A pattern recognition approach to the problem of linguistic approximation in system analysis', in Proc. IEEE Inter. Conf. on Cybernetics and Society, Denver, pp. 793-798.

Buckles, B.P. and Petry, F.E. (1982) 'A fuzzy representation of data for relational databases', Fuzzy Sets and Systems 7, 213-226.

Caianiello, E.R. (1973) 'A calculus for hierarchical systems', in Proc. 1st Inter. Conf. on Pattern Recognition, Washington, D.C., pp. 1-5.

Caianiello, E.R. (1987) 'C-calculus : an overview', in E.R. Caianiello and M.A. Aizerman (eds.), Topics in the General Theory of Structures, D. Reidel Dordrecht, pp. 163-173.

Caianiello, E.R. and Ventre, A.G.S. (1984) 'C-calculus and fuzzy sets', in Proc. 1st Napoli Meeting on the Mathematics of Fuzzy Systems, Università degli Studi di Napoli, pp. 29-33.

Caianiello, E.R. and Ventre, A.G.S. (1985) 'A model for C-calculus', Int. J. of General Systems 11, 153-161.

Chellas, B.F. (1980) Modal Logic : An Introduction, Cambridge University Press, Cambridge, UK.

Dempster, A.P. (1967) 'Upper and lower probabilities induced by a multiple-valued mapping', Annals of Mathematical Statistics 38, 325-339.

Di Nola, A., Sessa, A., Pedrycz, W. and Sanchez, E. (1989) Fuzzy Relation Equations and their Applications to Knowledge Engineering, Kluwer Academic Publ., Dordrecht, The Netherland.

Dubois, D. and Prade, H. (1980) Fuzzy Sets and Systems : Theory and Applications, Academic Press, New York.

Dubois, D. and Prade, H. (1984a) 'Fuzzy logics and the generalized modus ponens revisited', Int. J. of Cybernetics and Systems 15, 293-331.

Dubois, D. and Prade, H. (1984b) 'A theorem on implication functions defined from triangular norms', Stochastica 8(3), 267-279.

Dubois, D. and Prade, H. (1985) 'Evidence measures based on fuzzy information', Automatica 21, 547-562.

Dubois, D. and Prade, H. (1987) 'Twofold fuzzy sets and rough sets — Some issues in knowledge representation', Fuzzy Sets and Systems 23, 3-18.

Dubois, D. and Prade, H. (with the collaboration of Farreny, H., Martin-Clouaire, R and Testemale, C.) (1988) Possibility Theory : An Approach to Computerized Processing of Uncertainty, Plenum Press, New York.

Dubois, D. and Prade, H. (1990) 'Rough fuzzy sets and fuzzy rough sets', Int. J. of General Systems 17, 191-200.

Dubois, D. and Prade, H. (1991) 'Possibilistic logic, preferential models, non-monotonicity and related issues', in Proc. of the 12th Inter. Joint Conf. on Artificial Intelligence (IJCAI-91), Sydney, Australia, Aug. 24-30, pp. 419-424.

Dubois, D., Prade, H. and Testemale, C. (1988) 'In search of a modal system for possibility theory', in Proc. of the Conf. on Artificial Intelligence, Munich, Germany, Aug. 1-5, pp. 501-506.

Fariñas del Cerro, L. and Herzig, A. (1991) 'A modal analysis of possibility theory', in Proc. of the Inter. Workshop on Fundamentals of Artificial Intelligence (FAIR'91), Smolenice Castle, Czechoslovakia, Sept. 8-12, 1991, Ph. Jorrand and J. Kelemen (eds.), Springer Verlag, Berlin, pp. 11-18.

Fariñas del Cerro, L. and Orlowska, E. (1985) 'DAL — A logic for data analysis', Theoretical Computer Science 36, 251-264.

Fariñas del Cerro, L. and Prade, H. (1986) 'Rough sets, twofold fuzzy sets and modal logic — Fuzziness in indiscernibility and partial information', in A. Di Nola and A.G.S. Ventre (eds.), The Mathematics of Fuzzy Systems, Verlag TÜV Rheinland, Köln, pp. 103-120.

Fodor, J.C. (1991) 'On fuzzy implication operators', Fuzzy Sets and Systems 42, 293-300.

Gisolfi, A. (1992) 'An algebraic fuzzy structure for approximate reasoning', Fuzzy Sets and Systems 45, 37-43.

Höhle, U. (1988) 'Quotients with respect to similarity relations', Fuzzy Sets and Systems 27, 31-44.

Kitainik, L. (1991) 'Notes on convex decomposition of bold fuzzy equivalence relations', in BUSEFAL n° 48 (IRIT, Univ. P. Sabatier, Toulouse, France), 27-35.

Lewis, D. (1973) Counterfactuals, Blackwell, Oxford.

López de Mántaras, R. and Valverde, L. (1988) 'New results in fuzzy clustering based on the concept of indistinguishability relation', IEEE Trans. on Pattern Analysis and Machine Intelligence 10, 754-757.

Nakamura, A. (1988) 'Fuzzy rough sets', Note on Multiple-Valued Logic in Japan 9(8), 1-8.

Nakamura, A. (1989) 'On a KTB-modal fuzzy logic', Tech. Report n° C-31, Dept. of Applied Mathematics, Hiroshima University, Japan.

Nakamura, A. (1991a) 'Topological soft algebra for the S5 modal fuzzy logic', in Proc. of the 21st Inter. Symp. on Multiple-Valued Logic, Victoria, B.C., pp. 80-84.

Nakamura, A. (1991b) 'On a logic based on fuzzy modalities', Report MCS-10, Dept. of Computer Science, Meiji University, Japan.

Nakamura, A. and Gao, J.M. (1991) 'A logic for fuzzy data analysis', Fuzzy Sets and Systems 39, 127-132.

Novak, V. (1987) 'Automatic generation of verbal comments on results of mathematical modelling', in E. Sanchez and L.A. Zadeh (eds.), Approximate Reasoning in Intelligent Systems, Decision and Control, Pergamon Press, Oxford, U.K., pp. 55-68.

Orlowska, E. (1984) 'Modal logics in the theory of information systems', Zeitschrift für Mathematische Logik und Grundlagen der Mathematik 30(3), 213-222.

Ovchinnikov, S.V. (1982) 'On fuzzy relational systems', in Proc. 2d World Conf. on Mathematics at the Service of Man, Las Palmas, Spain, pp. 566-569.

Pawlak, Z. (1982) 'Rough sets', Int. J. of Computer and Information Science 11, 341-356.

Pawlak, Z. (1984) 'Rough classification', Int. J. of Man-Machine Studies 20, 469-485.

Pawlak, Z. (1985) 'Rough sets and fuzzy sets', Fuzzy Sets and Systems 17, 99-102.

Pawlak, Z., Slowinski, K. and Slowinski, R. (1986) 'Rough classification of patients after highly selective vagotomy for duodenal ulcer', Int. J. of Man-Machine Studies 24, 413-433.

Pawlak, Z., Wong, S.M.K. and Ziarko, W. (1988) 'Rough sets : probabilistic versus deterministic approach', Int. J. of Man-Machine Studies 29, 81-95.

Pearl, J. and Verma, T. (1987) 'The logic of representing dependencies by directed graphs', in Proc. 6th AAAI National Conf. on Artificial Intelligence, Seattle, pp. 374-379.

Pedrycz, W. (1985) 'On generalized fuzzy relational equations and their applications', J. of Mathematical Analysis and Applications 107, 520-536.

Prade, H. (1984) 'Lipski's approach to incomplete information data bases restated and generalized in the setting of Zadeh's possibility theory', Information Systems 9(1), 27-42.

Prade, H. and Testemale, C. (1984) 'Generalizing database relational algebra for the treatment of uncertain/imprecice information and vague queries', Information Sciences 34, 27-42.

Prade, H. and Testemale, C. (1987) 'Fuzzy relational databases : representational issues and reduction using similarity measures', J. of the American Society for Information Science 38(2), 118-126.

Roy, B. (1985) Méthodologie Multicritère d'Aide à la Décision, Editions Economica, Paris.

Ruspini, E.H. (1991) 'On the semantics of fuzzy logic', Int. J. of Approximate Reasoning 5, 45-88.

Sanchez, E. (1976) 'Resolution of composite fuzzy relation equations', Information and Control, 30 38-48

Sanchez, E. (1978) 'Resolution of eigen fuzzy sets equations', Fuzzy Sets and Systems 1(1), 69-74.

Schweizer, B. and Sklar, A. (1983) Probabilistic Metric Spaces, North-Holland, Amsterdam.

Shafer, G. (1976) A Mathematical Theory of Evidence, Princeton Univ. Press, Princeton, NJ.

Shafer, G., Shennoy, P. and Mellouli, K. (1987) 'Qualitative Markov trees', Int. J. Approximate Reasoning 1, 349-400.

Sugeno, M. (1977) 'Fuzzy measures and fuzzy integrals : a survey', in M.M. Gupta, G.N. Saridis and B.R. Gaines (eds.), Fuzzy Automated and Decision Processes, North-Holland, Amsterdam, pp. 89-102.

Tahani, V. (1977) 'A conceptual framework for fuzzy query processing — A step towards very intelligent database systems', Information Processing and Management 13, 289-303.

Trillas, E. and Valverde, L. (1984) 'An investigation into indisguishability operators', in H.J. Skala, S. Termini and E. Trillas (eds.), Aspects of Vagueness, D. Reidel Dordrecht, pp. 231-256.

Trillas, E. and Valverde, L. (1985) 'On implication and indistinguishability in the setting of fuzzy logic', in J. Kacprzyk and R.R. Yager (eds.), Management Decision Support Systems using Fuzzy Sets and Possibility Theory, Interdisciplinary Systems Research, Vol. 83, Verlag TÜV Rheinland, Köln, pp. 198-212.

Valverde, L. (1985) 'On the structure of F-indistinguishability operators', Fuzzy Sets and Systems 17, 313-328.

Willaeys, D. and Malvache, N. (1981) 'The use of fuzzy sets for the treatment of fuzzy information by computer', Fuzzy Sets and Systems 5, 323-328.

Wong, S.M.K. and Ziarko, W. (1985) 'A probabilistic model of approximate classificaton and decision rules with uncertainties and inductive learning', Technical Report CS 85-23, University of Regina, Saskatchewan.

Wygralak, M. (1989) 'Rough sets and fuzzy sets : some remarks on interrelations', Fuzzy sets and Systems 29, 241-243.

Zadeh, L.A. (1965) 'Fuzzy sets', Information and Control 8, 338-353.

Zadeh, L.A. (1971) 'Similarity relations and fuzzy orderings', Information Sciences 3, 177-200.

Zadeh, L.A. (1978) 'Fuzzy sets as a basis for a theory of possibility', Fuzzy Sets and Systems 1, 3-28.

Part II
Chapter 2

APPLICATIONS OF FUZZY-ROUGH CLASSIFICATIONS TO LOGICS

Akira NAKAMURA
Department of Computer Science
Meiji University
Tama-ku Kawasaki, 214, Japan

Abstract. From a point of view of handling imperfect knowledge like uncertainty, vagueness, imprecision, etc., a concept of fuzzy-rough classifications is introduced. This is a notion defined as a kind of modification of the rough classifications. Two logics based on the fuzzy-rough classifications are proposed, and various properties are examined as compared with the indiscernibility relations. Also, decision procedures for the proposed logics are given in the tableau style which is useful in automated reasoning.

1. Introduction

In ancient times the logic was built as the science of reasoning on human knowledge, and the establishment of symbolic logic in the early years of the 20th century suggests the possibility of mechanical processing of this reasoning. Further, rapid progress of computer in these days enable us to realize this aim. In such a stream, the researches on knowledge have been one of the central topics in artificial intelligence. Originally, the logic has two aspects, namely, the formal one (syntax) and the material one (semantics), and these two are strongly related each other. In this reason, the problem of knowledge representation must be argued from the point of view of logic.

In recent years a variety of formalisms have been developed which address several aspects of handling imperfect knowledge like uncertainty, vagueness, imprecision, incompleteness, and partial inconsistency. This is a reason that the formal system must be discussed from the meaning. In [11] Pawlak introduced the notion of *rough sets* whose issue is reasoning from imprecise data. In [12], he provided the basic theory and results about the rough sets from a point of view of its possible application to artificial intelligence. The idea of the rough sets consists in approximations of a set by a pair of sets called the *lower* and the *upper* approximations. In fact these approximations are *interior* and *closure* operations in a certain topology generated by available data about elements of the set. In [10], Orlowska discussed theoretical foundations of knowledge representation from a point of view of the above rough sets modeling indiscernibility of knowledge. Indiscernibility relation is a binary relation that identifies objects which have the same descriptions with respect to a set of attributes of objects. For example, the persons who were born at the same time are not distinguishable with respect to "age".

Meanwhile, the theory of fuzzy set formalized by Zadeh [13] is a well-established and active field, with numerous applications in such areas as artificial intelligence, pattern recognition, circuit theory, control theory, etc.. Further, in [14] the *similarity* relation in the fuzzy theory has been introduced as a generalization of the equivalence relation. Along this line, in [7] the present author developed a logic for fuzzy data analysis which was built by extending indiscernibility relations to the fuzzy case. Further, in [8], starting the fact that the equivalence relation is the base of model for the S5-modal system the author proposed the S5-modal fuzzy logic and examined some relationships of this logic to a topological soft algebra.

As a continuous work of previous papers [7] - [9], in the first half (the Section 2 to the Section 4) of this paper we propose a logic based on fuzzy-rough classifications and examine various properties of this logic. Roughly speaking, fuzzy-rough classifications mean a kind of degree of the equivalence relation, which correspond to the fuzzy degree of S5-modality. We introduce the family of modal operations $\{[\lambda]\}_{\lambda \in [0,1]}$. This operation $[\lambda]$ is also considered as a kind of modification of $\underline{\text{ind}}(A)$ based on indiscernibility relation. Also, it is related to the quantitative modal logic in [3]. And by making use of the technique based on the tableau method the paper gives a decision procedure for this logic. Next, in the second half (the Section 5 to the Section 6) we propose a fuzzy logic based on the above fuzzy modalities. This logic is obtained from [9] by generalizing the quantifiers to fuzzy-rough classifications. And we will show a decision procedure which is obtained by extending the results of the first part.

Finally, relationships of imprecise monadic predicate logic to S5-modal fuzzy logic are discussed. This is a similar relation to that between the ordinary monadic predicate logic and the usual S5 modal logic. Further, remarks about an axiomatization of the logic L_1 in the first part and a relationship of these modal operations to rough quantifiers are given with references [4] and [6]. Also, some problems of practical techniques of a tableau method for the proposed logics are suggested as a future view.

2. Preliminaries and basic definitions

Let us consider a system $S = < OB, AT, \{VAL_a\}_{a \in AT}, f >$. This system is usually called the *information system*, and is the same as that denoted by $S = <U, Q, V, f>$. That is, OB, AT, VAL_a and f are defined as follows:

OB is a set (not necessarily finite) of object,

AT is a set (not necessarily finite) of attributes,

VAL_a is a set of values of attribute a for each AT, and VAL is the union of all the set VAL_a,

f is a mapping from OB× AT into VAL.

We define a binary relation R on the set OB as follows:

$(o_i, o_j) \in R$ iff $f(o_i, a) = f(o_j, a)$ for each $a \in AT$.

The relation R on the set OB is referred to as *indiscernibility* with respect to attributes from the set AT ([10] and [11]). It is easily shown that R is an equivalence relation on the set OB. Speaking more generally, let R be an equivalence relation defined on X. Let us denote the equivalence class of x in X in the sense of R by $[x]_R$. Given a subset S of X and an equivalence relation R, a *lower* approximation $R_*(S)$ and an *upper* one $R^*(S)$ of S with respect to R are defined as follows:

Definition 2.1. Let S be a subset of a given set X and R be an equivalence relation defined on X. $R_*(S)$ and $R^*(S)$ are defined as follows:

$R_*(S) = \{x \in X \mid [x]_R \subseteq S\}$, $R^*(S) = \{x \in X \mid [x]_R \cap S \neq \phi\}$. □

Now, let us begin to discuss the fuzzy case of the previous definition. Fuzzy sets are defined by their membership function μ. Let S and T be fuzzy sets. The membership functions of $S \cap T$, $S \cup T$ and $\overline{S}$ are defined as follows:

$\mu_{S \cap T}(x) = \min(\mu_S(x), \mu_T(x))$,

$\mu_{S \cup T}(x) = \max(\mu_S(x), \mu_T(x))$,

$\mu_{\overline{S}}(x) = 1 - \mu_S(x)$.

A fuzzy relation R is defined as a fuzzy collection of ordered pairs. In this paper, our attention is focused on a special case of the fuzzy relation. That is, we employ the notion of similarity relation due to Zadeh [14]. The similarity relation is essentially a generalization of the notion of equivalence, and it serves as a very useful concept for the logical approach to fuzzy data analysis.

Definition 2.2. A similarity relation R in X is fuzzy relation in X which is:

(a) reflexive: $\mu_R(x,x) = 1$ for all x in dom R,

(b) symmetric: $\mu_R(x,y) = \mu_R(y,x)$ for all x,y in dom R,

(c) transitive: $\mu_R(x,z) \geq \vee_y(\mu_R(x,y) \wedge \mu_R(y,z))$ for all x,y, z in dom R. □

We generalize here Definition 2.1 to the fuzzy case. To this end, we introduce a new notion denoted by $R(\lambda)$.

Definition 2.3. For a similarity relation R and $\lambda \in [0,1]$, $R(\lambda)$ is defined as follows:

$$R(\lambda) = \{(x,y) \mid \mu_R(x, y) \geq \lambda\}.$$ □

It is well known that $R(\lambda)$ is the usual (non-fuzzy) equivalence relation ([14]). Also, for $\alpha \leq \beta$ $R(\beta) \subseteq R(\alpha)$ ([14]). Note that $\forall x([x]_{R(\alpha)} \subseteq [x]_{R(\beta)})$ or $\forall x([x]_{R(\beta)} \subseteq [x]_{R(\alpha)})$.

Definition 2.4. For a subset S of X, $\lambda \in [0,1]$, and a similarity relation R $R_*(\lambda)S$ and $R^*(\lambda)S$ are defined as follows:

$$R_*(\lambda)S = \{x \in X \mid [x]_{R(\lambda)} \subseteq S\}$$

$$R^*(\lambda)S = \{x \in X \mid [x]_{R(\lambda)} \cap S \neq \phi\}.$$ □

The above definition is obtained from Definition 2.3 of [7] by modification.

3. Logic of fuzzy modalities

In this section, we describe the syntax and semantics of a logic L_1 based on fuzzy modalities.

Consider three classes of symbols:

(i) Propositional variables: p, q, ..., p_1, p_2,... .

(ii) Propositional operations: $\neg$ (negation), $\wedge$ (conjunction), $\vee$ (disjunction).

(iii) Family of modal operations: $\{[\lambda]\}_{\lambda \in [0,1]}$.

Then, the set FOR of well-formed formulas is the least set defined inductively as follows:

p, q, ... $\in$ FOR and if F, G $\in$ FOR then $\neg$F, F$\wedge$G, F$\vee$G, $[\lambda]$F $\in$ FOR.

Also, $\langle\lambda\rangle$F is defined as $\neg[\lambda]\neg$F , and there is a case that parentheses of consecutive modal operations are omitted, e.g., $\langle\alpha\rangle([\beta]F)$ is written as $\langle\alpha\rangle[\beta]F$. The modal operations are denoted by making use of the small Greek letter.

By a *model*, we mean a system M = < OB, R, μ,v >where

OB is a nonempty set of objects (data items),

R is the similarity relation on OB,

μ is a membership function,

v is a valuation function such that for a propositional variable p, v(p) is a subset in OB, and

$v(\neg F) = \overline{v(F)}$, $v(F \wedge G) = v(F) \cap v(G)$, $v(F \vee G) = v(F) \cup v(G)$,
$v([\lambda]F) = R_*(\lambda)v(F)$, $v(<\lambda>F) = R^*(\lambda)v(F)$.

Note that the above $v([\lambda]F)$ and $v(<\lambda>F)$ are well defined by the abbreviation $<\lambda>$ of $\neg[\lambda]\neg$ Let us call $[\lambda]$ as a *fuzzy-modal operation.* The model is an abstract form of information system which is described in Section 2.

We now examine some of properties of this logic. For two well-formed formulas F and G, we write $F \leq G$ iff $v(F) \subseteq v(G)$ holds for every model. In this case, $F \leq G$ is said to be *valid*. In the usual way, $F = G$ is defined as $F \leq G$ & $G \leq F$. Also, a well-formed formula G is said to be *valid* iff $G = I$ is valid where I is defined as $\neg F \vee F$.

Proposition 3.1. We have the following expressions:

(1) $[\alpha]F \leq [\beta]F \leq F \leq <\gamma>F \leq <\delta>F$ for $\alpha \leq \beta$ and $\gamma \geq \delta$.

(2) $[\alpha]F \vee [\beta]G \leq [\gamma](F \vee G)$ for $\alpha \leq \gamma$ and $\beta \leq \gamma$.

(3) $<\alpha>(F \vee G) = <\alpha>F \vee <\alpha>G$.

(4) $[\alpha](F \wedge G) = [\alpha]F \wedge [\alpha]G$.

(5) $<\gamma>(F \wedge G) \leq <\alpha>F \wedge <\beta>G$ for $\alpha \leq \gamma$ and $\beta \leq \gamma$.

(6) $[\alpha](\neg F \vee G) \leq \neg[\alpha]F \vee [\alpha]G$.

Proof.

We prove $[\alpha]F \leq [\beta]F$ of (1), (2), (3), and (6) only. The other expressions are similarly probable.

(1) $R(\beta) \subseteq R(\alpha)$ for $\alpha \leq \beta$. Hence, for a given x in OB $x \in [\alpha]F \Rightarrow [x]_{R(\alpha)} \subseteq F$, and also $[x]_{R(\alpha)} \supseteq [x]_{R(\beta)}$. Thus, we get $[x]_{R(\beta)} \subseteq F$.

Therefore, we have $x \in [\beta]F$, and also $[\alpha]F \leq [\beta]F$.

(2) is provable as follows: Without loss of generality, we can assume $\alpha \leq \beta$. From (1) we have $[\alpha]F \leq [\beta]F$ and $[\beta](F \vee G) \leq [\gamma](F \vee G)$. Thus, it is sufficient to prove $[\beta]F \vee [\beta]G \leq [\beta](F \vee G)$. But this is easy from $[\beta]F \leq [\beta](F \vee G)$ and $[\beta]G \leq [\beta](F \vee G)$ which follow from $F \leq F \vee G$ and $G \leq F \vee G$.

(3) $x \in <\alpha>(F \vee G) \Leftrightarrow \exists y(x \in [y]_{R(\alpha)}$ & $(y \in F \vee y \in G))$

$\Leftrightarrow \exists y(x \in [y]_{R(\alpha)} \,\&\, x \in F) \vee \exists y(x \in [y]_{R(\alpha)} \,\&\, y \in G)$

$\Leftrightarrow x \in \langle\alpha\rangle F \vee x \in \langle\alpha\rangle G.$

(6) Generally, we have $(\neg F \vee G) \wedge F \leq G$.

Thus, we get $[\alpha](\neg F \vee G) \wedge [\alpha]F \leq [\alpha]G$.

Therefore, we get $\neg([\alpha](\neg F \vee G) \wedge [\alpha]F) \vee [\alpha]G = I$. That is, we get $\neg[\alpha](\neg F \vee G) \vee \neg[\alpha]F \vee [\alpha]G = I$.

Thus, $[\alpha](\neg F \vee G) \leq \neg[\alpha]F \vee [\alpha]G$. □

Proposition 3.2. We have the following expressions:

(7) $\langle\alpha\rangle\langle\beta\rangle F = \langle\beta\rangle\langle\alpha\rangle F = \langle\alpha\rangle F$ for $\alpha \leq \beta$.

(8) $[\alpha][\beta]F = [\beta][\alpha]F = [\alpha]F$ for $\alpha \leq \beta$.

(9) $[\alpha]\langle\beta\rangle F = \langle\beta\rangle F$ for $\alpha \geq \beta$.

(10) $\langle\alpha\rangle[\beta]F = [\beta]F$ for $\alpha \geq \beta$.

(11) $\langle\alpha\rangle(F \wedge [\beta]G) = \langle\alpha\rangle F \wedge \langle\alpha\rangle[\beta]G$ for $\alpha \geq \beta$.

(12) $[\alpha](F \vee \langle\beta\rangle G) = [\alpha]F \vee [\alpha]\langle\beta\rangle G$ for $\alpha \geq \beta$.

(13) $\langle\alpha\rangle(F \wedge \langle\beta\rangle G) = \langle\alpha\rangle F \wedge \langle\alpha\rangle\langle\beta\rangle G$ for $\alpha \geq \beta$.

(14) $[\alpha](F \vee [\beta]G) = [\alpha]F \vee [\alpha][\beta]G$ for $\alpha \geq \beta$.

Proof.

We prove (7), (9), (11), and (13) only. Others are similarly provable.

(7) $x \in \langle\alpha\rangle(\langle\beta\rangle F) \Leftrightarrow [x]_{R(\alpha)} \cap \langle\beta\rangle F \neq \phi$

$\Leftrightarrow \exists y(y \in [x]_{R(\alpha)} \,\&\, y \in \langle\beta\rangle F)$

$\Leftrightarrow \exists y(y \in [x]_{R(\alpha)} \,\&\, ([y]_{R(\beta)} \cap F \neq \phi))$

$\Leftrightarrow \exists y(y \in [x]_{R(\alpha)} \,\&\, \exists z(z \in [y]_{R(\beta)} \,\&\, z \in F))$

$\Leftrightarrow \exists z(z \in [x]_{R(\alpha)} \,\&\, z \in F)$

$\Leftrightarrow x \in \langle\alpha\rangle F.$

Thus, we get $\langle\alpha\rangle(\langle\beta\rangle F) = \langle\alpha\rangle F$. $\langle\beta\rangle(\langle\alpha\rangle F) = \langle\alpha\rangle F$ is similarly provable.

(9) $x \in [\alpha]\langle\beta\rangle F \Leftrightarrow [x]_{R(\alpha)} \subseteq \langle\beta\rangle F$

$\Leftrightarrow \forall y(y \in [x]_{R(\alpha)}$ implies $y \in \langle\beta\rangle F)$

$\Leftrightarrow \forall y(y \in [x]_{R(\alpha)}$ implies $\exists z(z \in [y]_{R(\beta)} \,\&\, z \in F))$

$\Leftrightarrow \ \exists z(z \in [x]_{R(\beta)} \ \& \ z \in F)$

$\Leftrightarrow \ x \in <\beta>F.$

(11) $x \in <\alpha>(F \wedge [\beta]G)$

$\Leftrightarrow \ \exists y(y \in [x]_{R(\alpha)} \ \& \ y \in (F \wedge [\beta]G))$

$\Leftrightarrow \ \exists y(y \in [x]_{R(\alpha)} \ \& \ y \in F \ \& \forall z(z \in [y]_{R(\beta)} \text{ implies } z \in G))$

$\Leftrightarrow \ \exists y(y \in [x]_{R(\alpha)} \ \& \ y \in F)$

$\& \ \exists y(y \in [x]_{R(\alpha)} \ \& \ \forall z(z \in [y]_{R(\beta)} \text{ implies } z \in G))$

(for $\Leftarrow$, we take the same y for $\exists y(.......) \& \exists y(....)$. This is possible since $[y_1]_{R(\alpha)} = [y_2]_{R(\alpha)}$ follows from $y_1 \in [x]_{R(\alpha)} \ \& \ y_2 \in [x]_{R(\alpha)}$.)

$\Leftrightarrow \ x \in <\alpha>F \ \& \ x \in <\alpha>[\beta]G$

$\Leftrightarrow \ x \in <\alpha>F \wedge <\alpha>[\beta]G.$

(13) $x \in <\alpha>(F \wedge <\beta>G)$

$\Leftrightarrow \ \exists y(y \in [x]_{R(\alpha)} \ \& \ y \in (F \wedge <\beta>G))$

$\Leftrightarrow \ \exists y(y \in [x]_{R(\alpha)} \ \& \ y \in F \ \& \ y \in <\beta>G)$

$\Leftrightarrow \ \exists y(y \in [x]_{R(\alpha)} \ \& \ y \in F \ \& \ \exists z(z \in [y]_{R(\beta)} \ \& \ z \in G)$

$\Leftrightarrow \ \exists y(y \in [x]_{R(\alpha)} \ \& \ y \in F) \ \& \ \exists y(y \in [x]_{R(\alpha)} \ \& \exists z \ (z \in [y]_{R(\beta)} \ \& \ z \in G))$

(for $\Leftarrow$, we take the same y for $\exists y(.......) \& \exists y(....)$. This is possible since $[y_1]_{R(\alpha)} = [y_2]_{R(\alpha)}$ follows from $y_1 \in [x]_{R(\alpha)} \ \& \ y_2 \in [x]_{R(\alpha)}$.)

$\Leftrightarrow \ x \in <\alpha>F \wedge <\alpha><\beta>G).$ □

Note that (7) and (8) do not hold for $\alpha > \beta$ and (9), (10), (11), (12), (13), and (14) do not hold for $\alpha < \beta$.

Also, note that for $\alpha = \beta$ (1) - (14) show the corresponding expressions of S5-modal logic.

4. Decision procedure

Let us give a decision procedure to solve whether a given well-formed formula of this logic is valid. This procedure is given in a modified way of the usual one for S5-modal logic.

The idea and diagram applied to solve the decision problem for validity of well-formed formula follow from [5]. To test the validity of a given formula F, a counter-model to F is to be searched;

if such a counter-model exists then F is not valid; otherwise F is valid. By the diagram, a rectangle means a *world* of the usual model of modal logic.

Now, let us describe the procedure below.

Procedure

Beginning: assign the initial value 0 to F in the first rectangle. In the following, G, F_1 and F_2 are subformulas of F.

1) Rules for propositional operations

$\neg 1$: If $\neg$G has the assignment 1 in a rectangle, then assign 0 to G in the same rectangle.

$\neg 0$: If $\neg$G has the assignment 0 in a rectangle, then assign 1 to G in the same rectangle.

$\wedge 1$: If $F_1 \wedge F_2$ has the assignment 1 in a rectangle, then assign 1 to F_1 and F_2 in the same rectangle.

$\wedge 0$: If $F_1 \wedge F_2$ has the assignment 0 in a rectangle, then introduce two alternatives of the rectangle, and
(i) assign 0 to F_1 on one alternative, and
(ii) assign 0 to F_2 in another alternative.

$\vee 1$: If $F_1 \vee F_2$ has the assignment 1 in a rectangle, then introduce two alternatives of the rectangle, and
(i) assign 1 to F_1 in one alternative,
(ii) assign 1 to F_2 in another alternative.

$\vee 0$: If $F_1 \vee F_2$ has the assignment 0 in a rectangle, then assign 0 to F_1 and F_2 in the same rectangle.

2) Rules for modal operations

[ ]1: If a subformula of the form $[\alpha]$G has the assignment 1 in a rectangle r, then any rectangle r' such that $(r, r') \in R(\alpha)$ put G into r' and assign 1 to G in r'.

[ ]0: If a subformula of the form $[\alpha]$G has the assignment 0 in a rectangle r and G has no the assignment 0 in r, then
(i) start out a new rectangle r', and
(ii) put G into r' where $(r, r') \in R(\alpha)$,
(iii) assign 0 to G in r'. □

Since $\langle\lambda\rangle$ is the same as $\neg[\lambda]\neg$, it is not necessary to give rules for $\langle\lambda\rangle$. Let r_1 and r_2 be two rectangles. For the proof of the procedure, if $(r_1, r_2) \in R(\alpha)$ then r_2 is called *R(α)*-

accessible from r_1; if there is a relation $R(\alpha)$ such that $(r_1, r_2) \in R(\alpha)$ then r_2 is called *accessible* from r_1.

We say that a rectangle r is *closed* iff one of the following conditions is satisfied:

1) some subformula in r has two assignments 1 and 0,
2) there is a closed rectangle r' such that r' is accessible from r and r' is closed in the sense of 1),
3) all alternatives of r are closed.

Consider a sequence of non-closed rectangles $r_1, r_2, ..., r_n, ...$ where r_{i+1} is obtained from r_i by applying the rule [ ]0. We call the sequence a *path*. In a path, r_j is said to be *contained* in r_i iff every subformula in r_j also appears in r_i and the same assignment as in r_i (i<j).

By the following reasons, there must be merely finitely many different assignments to all the formulas in all rectangle in a path:

i) only subformulas of the formula F occur in any rectangle, but there are only finitely many subformulas of the F,
ii) different sets of the subformulas in all the rectangles of the path is less than 2^n (n is the number of all the subformulas).

By the above procedure, we produce successively rectangles in *tree form*. In this tree, a rectangle at a leaf is called a *leaf rectangle*.

Termination of the procedure: An application of rules to rectangles terminates when one of the following conditions is satisfied:

i) a rectangle remains unaffected by any rule,
ii) some r_j is contained in some r_i (in this case, the corresponding path goes into a cycle and r_j terminates),
iii) a rectangle is closed.

The procedure terminates when applications of rules to all rectangles terminate. Also, the root rectangle is called *closed* iff its every leaf rectangle is closed.

From the facts described above, it is known that our procedure is always terminated.

We have the following lemma:

Lemma 4.1. If the rectangle for F is closed, then F is valid.

Proof.

This lemma is provable in the usual way which is similar to that in [5].

Roughly speaking, the proof is as follows:

We construct a well-formed formula from subformulas in a rectangle by the following stipulation: if a subformula F_i is assigned to 0 then we take F_i, if a subformula F_j is assigned 1 then we take $\neg F_j$. After that, we consider a disjunction D_t of these F_i's and $\neg F_j$'s. Since the rectangle for F is closed, D_t constructed from the last rectangle is valid. Let a rectangle k be obtained from a rectangle h by one application of some rule. And, let D_h and D_k be well-formed formulas constructed from the rectangle h and k, respectively. Let a rectangle k be accessible from a rectangle h. Then, it is shown by the same technique described in [5] that if D_k in the rectangle k is valid then D_h in the rectangle h is also valid. This is as follows:

This is easily shown for the case that h and k are the same rectangle (also, for the case of two alternatives in $\wedge 0$ and $\vee 1$ if two alternatives are valid then the original one is valid.)

Let the rectangle k be obtained from the rectangle h by the rule [ ]0 of the procedure, and D_k be constructed from $\neg F_1 \vee \ldots \vee \neg F_s \vee G$. Then, D_h is of the form $\neg[\alpha_1]F_1 \vee \ldots \vee \neg[\alpha_s]F_s \vee [\alpha]G$, where $\alpha_1, \ldots, \alpha_s \leq \alpha$. In this time, we get the validity of $[\alpha](\neg F_1 \vee \ldots \vee \neg F_s \vee G)$ from the validity of $\neg F_1 \vee \ldots \vee \neg F_s \vee G$, i.e., $[\alpha]\neg(F_1 \wedge \ldots \wedge F_s) \vee G)$. Thus, it is known from (4) and (6) of Proposition 3.1 that if $\neg F_1 \vee \ldots \vee \neg F_s \vee G$ is valid then $\neg([\alpha]F_1 \wedge \ldots \wedge [\alpha]F_s) \vee [\alpha]G$ is valid. Thus, from (1) of Proposition 3.1 we know that $\neg([\alpha_1]F_1 \wedge \ldots \wedge [\alpha_s]F_s) \vee [\alpha]G$ is valid, i.e., $\neg[\alpha_1]F_1 \vee \ldots \vee \neg[\alpha_1]F_1 \vee [\alpha]G$.

By repeating this process, we know that the well-formed formula F in the first rectangle is valid. Because F is assign to 0. □

Lemma 4.2. If F is valid, then the rectangle for F is closed.

Proof.

Let us assume that F is not closed. Then, there is at least a rectangle which is not closed or goes into a cycle. In this case, every propositional variable appeared in the given F are assigned to 0 or 1. Using these assignments and the construction of rectangles by rules , we last can build a counter-model for the first well-formed formula F.

Thus, F is not valid. Therefore, we get this lemma. □

Theorem 4.3. The procedure mentioned above is the decision procedure for the logic L_1 of fuzzy modalities.

Proof.

From the discussion mentioned above, for the decision procedure it is sufficient to check whether or not the first rectangle is closed. And it is easily known from the <u>Termination of the procedure</u> that this process ends after finite steps.

Therefore, we get this theorem. □

5. Modal operation on fuzzy logic

In the above sections, we considered propositional variables p's for which the valuation v(p) is a subset of OB. Now, we extend it to a case that v(p) goes to a *fuzzy* subset of OB. This is also consider as an extension of the logic proposed in [8]. We shall define the formal language of this logic L_2 in the usual way. The symbols and the definition of FOR of L_2 are the same to L_1. Also, the definition of *model* for L_2 is the same as L_1, where Definition 2.4 is modified as follows:

Definition 5.1. For a fuzzy subset S of X, $\lambda \in [0,1]$, and a similarity relation R $R_*(\lambda)S$ and $R^*(\lambda)S$ are defined as follows:

$$\mu_{R^*(\lambda)S}(x) = \sup_{R(x,x')\geq\lambda} \mu_S(x')$$

$$\mu_{R_*(\lambda)S}(x) = \inf_{R(x,x')\geq\lambda} \mu_S(x')$$

□

When a model M is given, a fuzzy value (denoted by M(A)) of a wff A by M is determined. It will be noticed here that the logic L_2 has no tautologies, namely, wff's which are always true (the fuzzy value 1) in all models. This, however, is no obstacle to define validity by making use of the notion of logical consequence. To this end, we use a meta-symbol $\rightarrow$ between wff's A and B. And we consider an expression $A \rightarrow B$. Then, B is called a *logical consequence* of A iff $M(A) \leq M(B)$ for every model M. When B is a logical consequence of A, we use the notation $\models A \rightarrow B$.

6. Decision procedure

We consider the decision problem for our logic L_2, i.e., to decide whether or not for given well-formed formulas A and B $\models A \rightarrow B$. We use a method similar to that of monadic predicate logic. This is a convenient to consider relationships between L_2 and the monadic predicate logic. To this end, hereafter we treat $\mu_A(x)$'s as well-formed formulas defined before and denote them A(x)'s .

By a *sequent* we mean an expression of the form $\Gamma \rightarrow \Delta$, where Γ, Δ are sequences of wff's. Usually Γ, Δ are finite but it allows the possibility of being infinite. The sequent $\Gamma \rightarrow \Delta$ is *valid* iff for every modal M $\min\{M(A): A\in\Gamma\} \leq \max\{M(B): B\in D\}$. If Γ

$=\{A_1,\ldots,A_n\}$ and $\Delta = \{B_1,\ldots,B_m\}$ then $\Gamma \rightarrow \Delta$ is valid iff $B_1 \vee \ldots \vee B_m$ is a logical consequence of $A_1 \wedge \ldots \wedge A_n$.

Let p, q are arbitrary propositional variables. Also, let Γ, Δ, and Θ be finite sequences of zero or more wff's. Then, the procedure consists of the (R1), (R2), and (Ax), which are in a style of tableau method given by the following rules:

(R1)

$(\vee\rightarrow)$

$$\frac{\Gamma,\ A(a) \vee B(b), \Delta \rightarrow \Theta}{\Gamma,\ A(a), \Delta \rightarrow \Theta\ ;\quad \Gamma,\ B(b), \Delta \rightarrow \Theta}$$

$(\rightarrow\vee)$

$$\frac{\Theta \rightarrow \Gamma,\ A(a) \vee B(b), \Delta}{\Theta \rightarrow \Gamma,\ A(a), B(b), \Delta}$$

$(\wedge\rightarrow)$

$$\frac{\Gamma,\ A(a) \wedge B(b), \Delta \rightarrow \Theta}{\Gamma,\ A(a), B(b), \Delta \rightarrow \Theta}$$

$(\rightarrow\wedge)$

$$\frac{\Theta \rightarrow \Gamma,\ A(a) \wedge B(b), \Delta}{\Theta \rightarrow \Gamma,\ A(a), \Delta\ ;\quad \Theta \rightarrow \Gamma,\ B(b), \Delta}$$

$([\ ]\rightarrow)$

$$\frac{\Gamma, [\alpha]A(x), \Delta \rightarrow \Theta}{\Gamma, A(a), [\alpha]A(x), \Delta \rightarrow \Theta}$$

,where a is an arbitrary object satisfying $R(x,a) \geq \alpha$.

$(\rightarrow[\ ])$

$$\frac{\Theta \rightarrow \Gamma,\ [\alpha]A(x), \Delta}{\Theta \rightarrow \Gamma,\ A(a), \Delta}$$

,where a is a new object satisfying $R(x, a) \geq \alpha$.

$$\frac{\Gamma, <\alpha>A(x), \Delta \to \Theta}{\Gamma, A(a), \Delta \to \Theta} \quad (< > \to)$$

,where a is a new aobject satisfying $R(x,a) \geq \alpha$.

$$\frac{\Theta \to \Gamma, <\alpha>A(x), \Delta}{\Theta \to \Gamma, A(a), <\alpha>A(x), \Delta} \quad (\to < >)$$

,where a is an arbtrary object satisfying $R(x, a) \geq \alpha$.

(R2)

Let Γ and Δ be sequences of wff's. $\neg(A(x) \wedge B(y))$, $\neg(A(x) \vee B(y))$, $\neg[\alpha]A(x)$, $\neg<\alpha>A(x)$ which appear in Γ or Δ are replaced by $\neg A(x) \vee \neg B(y)$, $\neg A(x) \wedge \neg B(y)$, $<\alpha>\neg A(x)$, $[\alpha]\neg B(x)$, respectively and vice versa.

By repeated applications of the above rules to a given sequent $F(a) \to G(a)$, we get a tree. Then, we define a "closed" sequent $\Gamma \to \Delta$, in a similar way to that of L_1 as follows:

$\Gamma \to \Delta$ satisfies either

(Ax)

(i) there exist atomic formulas p(a), q(b) such that

$p(a) \in \Gamma, \neg p(a) \in \Gamma$ and $q(b) \in \Delta, \neg q(b) \in \Delta$,

(ii) there exists an atomic formula p(c) such that

$p(c) \in \Gamma$, $p(c) \in \Delta$ or $\neg p(c) \in \Gamma$, $\neg p(c) \in \Delta$.

Further, F is defined to be *closed* iff every leaf of the tree for F is closed. We make a correspondence of a sequent to a rectangle in the Section 5. The following notice is given here:

A new object o_j for $\Gamma_j \to \Delta_j$ by the rule $(\to[\,])$ can be replaced by an object o in $\Gamma_i \to \Delta_i$ which has been already introduced if the result is equivalent to $\Gamma_i \to \Delta_i$ with respect to the validity. In this case, the rule $(\to[\,])$ is not applied to this $\Gamma_j \to \Delta_j$. The situation is the same for $(< > \to)$.

<u>Termination of the procedure</u>: An application of rules to a sequent terminates when one the following conditions is satisfies:,

(i) a sequent remain unaffected (or not applied in the situation mentioned above) by any rule,

(ii) a sequent is closed.

The procedure terminates when applications of rules to all sequents terminate.

Note that a number of introduced objects is finite which depends to the number of propositional variables appearing in the firstly given sequent $F(a) \rightarrow G(a)$. Because a number of permutations of order of fuzzy values for atomic formulas is finite. Thus, it is known that our procedure is always terminated.

Now, we prove the following lemmas:

Lemma 6.1. If $A \rightarrow B$ is closed, then $A \rightarrow B$ is valid.

Proof.

It is easily known from the definition that a closed sequent is valid. Also, It is also known that if a sequent (two sequents for alternatives) of the conclusion is valid then a sequent of the premise is valid. This fact is shown as follows:

Since $A \rightarrow B$ is closed, every sequent at the leaves is of the form (i) or (ii) in (Ax). Thus, it is easily known that these are valid. Therefore, it is sufficient to show that the validity is preserved in the reverse direction of the rules (R1). Here, we prove it for the rule $(\rightarrow\wedge)$ and $(\rightarrow[\ ])$. For other rules, it is shown similarly.

$(\rightarrow\wedge)$:

Let us assume that $\Theta \rightarrow \Gamma,\ A(a), \Delta$ and $\Theta \rightarrow \Gamma,\ B(b), \Delta$ are valid. The, for every model M we have the following inequalities $\min\{M(W) : W \in \Theta\} \leq \max\{M(U) : U \in \Gamma \cup \{A(a)\} \cup \Delta\}$ and $\min\{M(W) : W \in \Theta\} \leq \max\{M(U) : U \in \Gamma \cup \{B(b)\} \cup \Delta\}$. Therefore , for every model we have $\min\{M(W) : W \in \Theta\} \leq \max\{M(U) : U \in \Gamma \cup \{A(a) \wedge B(b)\} \cup \Delta\}$.

$(\rightarrow[\ ])$:

Let us assume that $\Theta \rightarrow \Gamma,\ [\alpha]A(x), \Delta$ is not valid. Then, there is a model M such that $\min\{M(W) : W \in \Theta\} > \max\{M(U) : U \in \Gamma \cup \{[\alpha]A(x)\} \cup \Delta.\}$ Let c be an individual which satisfies $v([\alpha]A(x)) = v(A(c))$. By making this model M, we define a model M' as follows: The free individual variable a does not occur in the conclusion. Hence, we can construct a model M' in which every v' is the same to that of M except $v'(A(a)) = v'(A(c))$. Thus, we get $\min\{M'(W) : W \in \Theta\} > \max\{M'(U) : U \in \Gamma \cup \{A(a)\} \cup \Delta\}$.

Thus, we have this case.

Therefore, we have this lemma. □

Lemma 6.2. If $A \rightarrow B$ is valid, then $A \rightarrow B$ is closed.

Proof.

Let $A \to B$ be not closed. From the assumption, there is a model by which there exists at least one sequent of every leaf for $A \to B$ which is not valid. Hence, by going upward preserving this non-validity we arrive the root. Hence, we know that $A \to B$ is not valid. □

Theorem 6.3. The procedure given above is a decision procedure for the logic L_2.

Proof.

From the discussion mentioned above, for the decision procedure of $A \to B$ it is sufficient to check whether or not $A \to B$ is closed. And it is easily known from the Termination of the procedure that this process ends after finite steps.

Therefore, we get this theorem. □

The procedure in the proof of Theorem 6.3 gives a solution of the decision problem of this logic. It is mentioned here that the technique is strongly related to automated reasoning.

7. Relationships of the imprecise monadic predicate logic to the S5-modal fuzzy logic

Let us consider relations of the imprecise monadic predicate logic to the S5-modal fuzzy logic. The imprecise monadic predicate logic $L^{\#}$ is that proposed in [9], and it is obtained by considering only one operation [0] instead of the family $\{[\lambda]\}_{\lambda \in [0,1]}$. Now, in the decision procedure of L_2 we wrote $A(x)$ for $\mu_A(x)$. By considering this $A(x)$ as a wff of monadic predicates, the proposed logic L_2 is an extension of the classical monadic predicate logic to the fuzzy case. It is well-known that the classical monadic predicate logic is strongly related to the S5-modal logic. That is, it is possible to interpret the modal operations M, N as the quantifiers $\exists$, $\forall$, respectively. We have the similar relation between $L^{\#}$ and the S5-modal fuzzy logic. Here, the S5-modal fuzzy logic is that introduced in [8] and is defined as follows:

Fuzzy sets are defined in the usual way by their membership functions μ. For fuzzy sets S and T, $S \cap T$, $S \cup T$, $\overline{S}$ are also defined in the usual way. Further, a similarity relation R in X is a fuzzy relation in X which satisfies:

(a) reflexive: $\mu_R(x,x) = 1$ for all x in dom R,

(b) symmetric: $\mu_R(x,y) = \mu_R(y,x)$, for all x,y in Dom R,

(c) transitive: $\mu_R(x,z) \geq \bigvee_y(\mu_R(x,y) \wedge \mu_R(y,z))$ for all x,y, z in dom R.

By making use of fuzzy sets and the similarity relation R, we define a function μ° over $[0,1] \times X$.

$$\mu^{\circ}_{S \cap T}(\tau, x) = \min(\mu_S(\tau, x), \mu_T(\tau, x)),$$

$$\mu^{\circ}_{S \cup T}(\tau, x) = \max(\mu_S(\tau, x), \mu_T(\tau, x)),$$

$$\mu^{\circ}_{\overline{S}}(\tau, x) = 1 - \mu^{\circ}_S(\tau, x),$$

$$\mu^{\circ}_{R^*(S)}(\tau, x) = \sup_{\mu^{\circ}_R(x,x') \geq \tau} \mu^{\circ}_S(\tau, x'),$$

$$\mu^{\circ}_{R_*(S)}(\tau, x) = \inf_{\mu^{\circ}_R(x,x') \geq \tau} \mu^{\circ}_S(\tau, x'),$$

, where $\mu^{\circ}_S(\tau, x) = \mu_S(x)$ if S does not contain R* or R∗ .

Then, the S5-modal fuzzy logic means the following system:

Consider three classes of symbols: propositional variables p, q,..., propositional operators ¬, ∨, ∧, and modal operations < >, [ ]. The set of well-formed formulas is defined in the usual way. Further, semantics of this logic is defined as follows: First, by a model we mean a system M(OB,[0,1],R, μ,v) where OB is a non-empty set of objects, R is the similarity relation on OB, μ is a membership function such that for o∈ OB and a fuzzy set A in OB, $\mu_A(o) \in [0,1]$, v is a valuation function such that for propositional variable p, v(p) is a fuzzy set in OB, and

$$v(\neg F) = \overline{v(F)},\ v(F \vee G) = v(F) \cup v(G),\ v(F \wedge G) = v(F) \cap v(G),$$

$$v([\ ]F) = R_* v(F),\ v(<>F) = R^*(v(F)).$$

For two wff's F and G, we say $F \leq G$ iff $\mu^{\circ}_{v(F)}(\tau, x) \leq \mu^{\circ}_{v(G)}(\tau, x)$ for every model and every $\tau \in [0, 1]$.

Then, in [7] the following equalities and inequalities were proved:

$$[\ ](F \vee G) \geq [\ ](F) \vee [\ ](G), \qquad <>(F \vee G) = <>(F) \vee <>(G),$$

$$[\ ](F \wedge G) = [\ ](F) \wedge [\ ](G), \qquad <>(F \wedge G) \leq <>(F) \wedge <>(G),$$

$$[\ ](F \vee [\ ](G)) = [\ ](F) \vee [\ ](G), \qquad <>(F \wedge <>(G)) = <>(F) \wedge <>(G),$$

$$[\ ](F \vee <>(G)) = [\ ](F) \vee <>(G), \qquad <>(F \wedge [\ ](G)) = <>(F) \wedge [\ ](G),$$

From the above-mentioned formulas, it is known that the modal operations <0>, [0] can be interpreted as the quantifies ∃ , ∀ , respectively. Let $A^{\#}$ be a wff of $L^{\#}$ which is obtained from a wff of the S5-modal fuzzy logic by the correspondence of ∃x,∀x to <0>, [0], respectively. Then, it is known that $A^{\#} \rightarrow B^{\#}$ is valid in $L^{\#}$ iff $A \rightarrow B$ is valid in the S5-modal fuzzy logic. This can be shown by making use of the following facts:

(1) The modal operations of S5-modal fuzzy logic are interpreted as those of the usual S5 modal logic by taking 0 as a level value τ for fuzzy relations.

(2) The above value 0 does not violate the discussion of system.

Remarks

This paper proposed two logics on fuzzy modalities and gave decision procedures in the style of tableau method. From a point of view of the automated reasoning, it seems to useful since these procedures are practically applicable. As one of future problems, it is desirable to give practical techniques, e.g. programming which efficiently works. Also, this paper has a close relationship to [1] and [2]. Further, as mentioned in Introduction $[\lambda]$ can be considered as a special case of ind(A) of Orlowska [10], in which A, B, ..., of ind(A), ind(B), ... are partially ordered but λ's are totally ordered; namely, in a sense that in a correspondence of A, B,... to α, β ,... $\alpha \cup \beta$ is α or β. In this paper, we considered the logics from a side of semantics. As mentioned previously, its syntax (formal system) must be discussed. It is an interesting open problem to axiomatize the logics L_1 and L_2. The family of modal operations $\{[\lambda]\}_{\lambda \in [0,1]}$ of this paper has the *local agreement* proposed in [4]. That is, for any object $x \in X$ and for the corresponding two equivalence classes to two relations $R(\alpha)$ and $R(\beta)$ the following local agreement holds: $[x]_{R(\alpha)} \subseteq [x]_{R(\beta)}$ or $[x]_{R(\beta)} \subseteq [x]_{R(\alpha)}$. An axiomatic system for the logic based on local agreement has been given in [4]. Thus, the technique of [4] seems to be useful our axiomatization problem for L_1. On the other hand, we pick up some of valid formulas such as (1)-(14) of Proposition 3.1 and 3.2, which have been used in the proof of Lemma 4.1 and consider them as axioms. Then, it will be possible to axiomatize the logic L_1. And also in the same way it will be possible to axiomatize the *logic of indiscernability relation* by Orlowska. Further, the family of modal operations $\{[\lambda]\}_{\lambda \in [0,1]}$ is considered as a kind of *fuzzy-rough quantifier* in the monadic predicates. In this consideration, [6] and [10] is also strongly related to this paper. The details of these problems will be discussed in further papers.

Rferences

[1] D. Dubois and H. Prade: Rough fuzzy sets and fuzzy rough sets, *Proc. of Internat. Conf. Fuzzy Sets in Informatics* , Moscow, (Sept. 1988), 20-23.

[2] L. Farinas del Cerro and H. Prade: Rough sets, twofold fuzzy sets and modal logics. Fuzziness in indiscernibility and partial information, in: A.Di Nola and A.G. Ventre (ed.), **The Math. of Fuzzy Systems**, Verlag TÜV Rheinland, Köln, (1986), 103-120.

[3] L. Farinas del Cerro and Andreas Herzig: A modal analysis of possibility theory, *Lecture Notes in Computer Science*, (Symbolic and Quantitative Approaches to Uncertainty), 548, 58-62.

[4] G. Gargov: Two completeness theorems in the logic for data analysis, *ICS PAS Reports, 581, Polish Academy of Sciences*, (1986).

[5] G.E. Hughes and M. Cresswell: **An introduction to modal logic**, Methuen, London, 1968

[6] M. Krynicki and H-P. Tuschik: An axiomatization of the logic with the rough quantifier, *J. of Symbolic Logic*, 56, (1991), 608-617.

[7] A. Nakamura and J-M. Gao: A logic for fuzzy data analysis, *Fuzzy Sets and Systems* , 39, (1991), 127-132.

[8] A. Nakamura: Topological soft algebra for the S5-modal fuzzy logic, *Proc. of the 21st ISMVL, May 26-29, 1991, Victoria* , 80-84.

[9] A. Nakamura: A logic of imprecise monadic predicates and its relation to the S5-modal fuzzy logic, *Lecture Notes in Computer Science*, (Symbolic and Quantitative Approaches to Uncertainty), 548, 254-261.

[10] E. Orlowska: Logic of indiscernibility relation, *Bulletin of the Polish Academy of Sciences, Mathematics*, (1985), 475-485.

[11] Z. Pawlak: Rough sets, *International J. of Information and Computer Sciences* , 11, (1982), 345-356.

[12] Z. Pawlak: **Rough Sets, Theoretical Aspects of Reasoning about Data**, Kluwer Academic Publishers, Dordrecht, 1991.

[13] L.A. Zadeh: Fuzzy sets, *Information and Control* , 8, (1965), 338-353

[14] L.A. Zadeh: Similarity relations and fuzzy orderings, *Information Sciences*, 3, (1971), 177-200.

Part II
Chapter 3

COMPARISON OF THE ROUGH SETS APPROACH AND PROBABILISTIC DATA ANALYSIS TECHNIQUES ON A COMMON SET OF MEDICAL DATA

Ewa KRUSIŃSKA
Institute of Computer Science
University of Wrocław
51-151 Wrocław, Poland

Ankica BABIC *
Faculty of Electrical and Computer Engineering
University of Ljubljana, Slovenia

Roman SLOWIŃSKI
Jerzy STEFANOWSKI
Institute of Computing Science
Technical University of Poznań
60-965 Poznań, Poland

Abstract. The paper presents a comparison study of the rough sets approach and probabilistic techniques, in particular, discriminant analysis and probabilistic inductive learning, to data analysis on a common set of medical data. This study completes the comparison done in [9], by taking into account, in addition to the location model of discriminant analysis, the linear Fisherian discrimination and Bayesian tree classifiers derived via inductive learning approach. A general discussion on similarities and differences among compared methods is given. Particular attention is paid to data reduction and creation of decision rules. The outcomes of a computational experiment on the common set of data are described and discussed.

1. Introduction

Very often data recording certain experience concern a set of objects (examples, individuals, observations etc.) described by a set of multi-valued attributes (features, tests, characteristics, variables etc.). The set of objects is usually classified into a disjoint family of classes, e.g. according to the expert's experience and knowledge in the considered field. It is also typical that attributes, in particular in medicine, are both discrete and continuous. Data of this type can be represented in a structure of an information system.

One of main problems in analysis of data coming from experience is reduction of all superfluous attributes. As a result of this reduction the most significant attributes for the classification are determined, i.e. attributes which distinguish in the best way between classes of classification. Another essential problem refers to creation of decision rules which represent not only relationships between the description of objects by attributes and their

* on leave in the Department of Medical Informatics, Linköping University, Sweden

assignment to particular classes, but can be also used to the classification of new objects.

The approaches which can be used to solve above problems can come from different fields of data analysis. One of the recent approaches is based on the rough sets theory (Pawlak [16]). It has been proved to be a useful tool for analysis of several real information systems. However, one can also refer to other methods from statistics, pattern recognition or machine learning. Within the framework of multivariate statistical approaches, the practical problems of selection of attributes and derivation of decision rules are studied by the discriminant analysis. A preliminary theoretical study on comparison of rough sets theory and discriminant analysis was carried out in [20]. Then, a set of medical data [9] has been used to compare rough sets analysis with the location model. The location model is one of the methods of discriminant analysis to be used for mixed continuous and discrete attributes. This is a cost-effective method which applies a collection of linear discriminant rules derived for subsamples of the data in relation to the values of discrete variables (attributes). Other methods of discriminant analysis, as logistic discrimination reached as a solution of a non-linear system and, especially, non-parametric kernel methods requiring density estimation, are more time consuming.

In the present paper, we extend our comparison to other probabilistic techniques, i.e. linear Fisherian discrimination and inductive learning. In linear discrimination, we apply a linear classification rule, similarly to the location model, but only one for the whole set of data. Probabilistic inductive learning results in creation of a tree classifier, so it provides a kind of data division into branches. Moreover, as inductive learning is a sequential technique (on the contrary to statistical discriminant analysis), it can be sometimes judged as resembling more to the 'human' way of medical decision making [13].

The aim of the paper is to compare the rough sets approach, the location model method, linear Fisherian discrimination and probabilistic inductive learning on the common data set. The analysed information system comes from medicine and concerns patients suffering from duodenal ulcer treated by highly selective vagotomy.

In the comparison particular attention is paid to the following problems: selection of the most significant attributes, creation of decision rules or tree classifiers and its verification. The verification is performed by reclassification of the data set on the 'leaving-one-out' basis.

The discussion of results obtained by different approaches may be useful to draw practical conclusions for prospective data analysis.

2. Considered data analysis methods

2.1. The rough sets approach

The rough sets approach considered in this paper is based on the methodology developed and used in several applications, e.g. medical (cf. [2],[17],[18],[19]) or technical ones (cf. [14],[15]). This methodology allows to analyse information systems with both discrete and continuous attributes. Discrete (qualitative) attributes are handled naturally in the analysis because the rough sets theory is dealing mainly with information systems where domains of attributes are finite sets of rather low cardinality. Continuous (quantitative) attributes are handled in the analysis after application of some norms dividing the original domains of those attributes into subintervals and assignment of qualitative codes to these subintervals.

The rough sets methodology consists in : approximation of classes, calculation of accuracies of approximations and quality of classification, searching for minimal subsets of attributes ensuring a satisfactory quality of classification, reduction of non-significant attributes, derivation of a decision algorithm from the reduced information system. The decision algorithm obtained using the rough sets methodology is composed of deterministic (certain) and non-deterministic (possible) rules (cf. [2]) which represent important relationships between the description of objects and their assignment to particular classes. The algorithm may be also used for classification of new objects (cf [21]). This possibility will be emphasized in the present paper in the part concerning verification of the decision algorithm. Let us notice that all decision rules are not equally important or reliable. Some rules are built using information about greater number of objects than other rules. In the worst case, a rule may refer to a single object in the information system. This difference in the importance of derived rules can be described by an additional parameter for each rule. This parameter, called 'strength' of the rule, is expressed by the number of objects in the information system supporting the considered decision rule. It has particular interpretation for non-deterministic rules. In these rules decisions are not uniquely determined by conditions, so parameters describe each possible assignment. For example, if one of possible decisions is supported by a significantly greater number of objects than others, one can conclude that this decision is the most possible.

When classifying a new object it may happen that there is no decision rule consistent with the description of the classified object by the condition attributes. In this case the idea of, so called, 'nearest' rules can be employed. The nearest rules are rules which are close to the description of the considered object in the sense of a chosen certain measure. Different measures were proposed in [21] but in this paper nearest rules are obtained using a simplified distance measure (cf [20]). According to it nearest rules are rules differing from the description of the classified object on a position of one attribute only when the value of this difference is not the greatest possible. The nearest rules are used by the following procedure for approximate assignment of the new object to a class of decisions. The k nearest rules are considered . Each of them is described by the number of objects supporting the assignment to a class, i.e. its strength. Then, supports for all classes are summed up and the object is assigned to the class with the greatest number of supporting objects. If there are several classes with the greatest score, then the new object is not classified at all. Moreover, if it is not possible to find the nearest rules because of restrictions for the allowed difference, the new object is also not classified.

2.2. Linear Fisherian discrimination

Discrimination problem is to distinguish between two (or generally more) distinct classes (called also groups) on the basis of values of attributes under consideration. The attributes involved in the classification process should have a high discriminatory power. The classical solution of this problem was given by Fisher (for overview see : [11]). Let us assume that each object considered is described by vector $y^T = (y_1, \ldots, y_p)$ of p attributes. We assume that the distribution of y is multivariate normal with the mean vector μ_i $(i = 1, 2, \ldots, g)$; g – number of considered classes) different for each class, and a common covariance matrix $\sum$. The 'a priori' probability that y belongs to the i–th class is described by p_i . The classification rules are obtained by determining an 'a posteriori' probability for y using the Bayes theorem. The rules are optimal only when assumption on normality and homogeneity of covariance matrices is fulfilled. The decision algorithm is based on comparison of discriminant scores :

$$S_i(y) = \mu_i^T \sum^{-1} y_i - \frac{1}{2}\mu_i^T \sum^{-1} \mu_i + \log p_i \tag{1}$$

or 'aposteriori' probabilities given by the formula

$$Pr(i \mid y) = \frac{exp\{S_i\}}{\sum_{j=1}^{g} \exp\{S_j\}} \qquad i = 1, 2, \ldots, g; \tag{2}$$

The object is assigned to the class with the greatest 'a posteriori' probability.

In practice, parameters μ and $\sum$ are unknown and should be estimated by class means and within-class sample covariance matrix, respectively. 'A priori' probabilities p_i can be set, e.g. as proportional to the number of objects in classes under consideration or as equal. This gives, in fact, two different classification rules and leads to different results.

The rules (1) and (2) are based on a given number of attributes. It is obvious, however, that not all of them possess considerable discriminatory ability. The selection, prior to classification is necessary. Many various criteria may be used. One applied commonly is the Wilks Λ statistic defined as a ratio of two determinants

$$\Lambda = \frac{\mid G \mid}{\mid G + H \mid} \tag{3}$$

where G is a within-class adjusted squares and products matrix, H is a between-class adjusted squares and products matrix, $G+H$ builds the total adjusted squares and products matrix.

The Wilks Λ, which measures the relation between within-class and total variability has a clear interpretation. It is equal to 0 when there is a complete discrimination and 1 if there is no discriminatory ability in the set of considered attributes. Thus, it can be used in exploratory manner (without testing). However, in our study we applied the stepwise search with significance testing of the attribute introduced or dropped out of the discrimination set. The same distributional technique was studied in [9] in relation to the location model. In the case of the linear Fisherian discrimination the significance test is based on the normality assumption. The significance of the variable introduced to the discrimination set as well dropped out of discrimination set is tested using Snedecor F statistic. The reader is referred to [11] for detailed formulae. Moreover, the significance of the current discrimination subset can be tested using another F statistics called 'overall F' [23].

The statistical studies have proved [11] that the linear discrimination is quite robust against violation of normality or homogeneity assumptions. This was a reason, that in the present paper we study this classical kind of discriminative approach on the set of mixed continuous and discrete attributes taking additionally into account the impact of changes in a priori probabilities to find classification results. In spite of robustness of the linear function one should stress that from the point of view of statistical correctness it is possible to use only binary and discrete ordinal attributes in the function. Discrete nominal attributes with more than two possible stages should be decoded to the set of binary attributes.

2.3. Location model

The location model [7,8,10] is a specific procedure used for discrimination with both discrete and continuous variables. The method has been already compared with the rough sets theory in [9]. The procedure is a natural extension of classical linear discrimination. The objects are divided into subsamples (cells) in correspondence to values of binary (discrete) attributes, e.g. having q binary attributes we get 2^q subsamples. For each subsample linear discriminant rules are calculated and used for discrimination. The additional extension in relation to classical linear discrimination is the different method of estimation of group and cell means and covariance matrices. Moreover, the probabilities that an object from the m-th cell belongs to the i-th class, i.e. p_{im} are estimated by the interactive scaling procedure [5] instead of taking them as equal or substituting them by fraction of objects occurrence in the m-th cell of the i-th class. The reduction of the number of attributes in the location model is also based on the matrices H and G as for linear Fisherian discrimination, but taking into account the location model cell structure. The generalized discriminatory measure is then calculated as $T^2 = tr(HG^{-1})$. Under assumption about normality of continuous attributes in the model, the transformed T^2 has Snedecor F distribution [8]. The distributional approach can be used (as in [9]) to determine the best subset. Additionally, one should stress that discrete (nominal) attributes have to be binarized before selection and classification as for the classical discrimination.

2.4. Probabilistic inductive learning

The inductive learning algorithm which is considered in this paper (its software implementation is called Assistant Professional) is based on Quinlan's ID3 algorithm [1]. A principle of induction is used to learn rules or to obtain a classifier, which should be complete and consistent with the given set of learning objects [4]. Knowledge is represented in the form of a binary tree where tree nodes correspond to attribute values and tree leaves to classes. The most informative attribute is placed in the tree root, whilst less informative attributes follow in lower nodes. An optimal classifier is obtained for the minimal complexity and for the maximal possible accuracy corresponding to it. Complexity can be measured by the number of leaves [4]. The pruning or truncation of classification reduces the complexity of the induced classifier. It practically means pruning the unreliable parts of the decision tree, and from viewpoint of the data set it is a kind of selection of attributes.

Pruning is done when building the classifier, according to the informativity of each attribute, at each tree node. The function measuring informativity is based on the entropy of the set of objects [13]. According to this measure, an attribute chosen for splitting a sample at a tree node is the one that gives the greatest decrease of entropy. If that set is characterized by l values of a given attribute and objects belong to t classes, then the number of objects in the i-th class ($i = 1,\ldots,t$) with j value of the considered attribute can be denoted by $n(i,j)$. The probability that object with the attribute value j, belongs to the i-th class is denoted by $p(i,j)$. The probability that the object belongs to the i-th class is denoted by $p(i)$. The decrease of entropy is measured as the difference between expected entropy and the entropy before splitting of the sample at the current node and is denoted by :

$$-\sum_{i,j}(p(i,j) * \ln p(i,j)) + 2\sum_{i}(p(i) + \ln p(i)) \tag{4}$$

For continuous attributes the prior discretization is done before entering the inductive procedure. In the medical application, the reference values intervals can be taken for this purpose as it was done in the rough sets approach. Moreover, during building a tree, an additional binarization of all attributes can be performed to normalize informativity of them with respect to the number of their values. Any attribute is binarized by finding such a division of the set of its possible values into two subsets, which maximizes attribute's informativity [1].

Estimation of reliability of the tree is performed at each node, i.e. for each attribute during the process of the tree building. Parameters which are used for it are the following : attribute suitability, class frequency and node weight. Attribute suitability is calculated as $(Inf(a_j)/E) * W_r$, where E is entropy of a learning set in the current node (see formula 4), W_r is the weight of the learning set in node r. This weight, also called the minimal weight threshold, determines the required minimum relative sum of objects at the current node of the tree sufficient to continue the generation of a subtree. The weight of an object itself determines its importance for the classification problems, and is usually equal to 1. $Inf(a_j)$, or the informativity of attribute a_j , is the decrease of the entropy of the learning set after introducing attribute a_j [1]. This parameter relates the attribute success $(Inf(a_j)/E)$ with the confidence into its success (W_r). Class frequency, which is calculated as $p(i) * 100\%$, where $p(i)$ is the probability that an object belongs to the majority class i in the learning set of node r. It shows the share of the majority class in the learning set. Node weight is calculated as $(W_r/W_o) * 100\%$ where W_r is the weight of the learning set in the node r and W_o is the weight of the learning set in the root node. The parameter is used to stop over-splitting of the learning set when the relative frequency shows that the subset of objects is too small.

Pruning can be performed in accordance to informativity using pruning factor (PF), which minimizes an information gain required from the currently selected attributes. The gain depends on the relative informativity content of the attribute in relation to the sum of weights of examples in the current node. The values of the informativity pruning factors may be between 0 and 9.9, where 0 corresponds to no pruning at all. The recommended value is usually $PF = 3$ [12].

Classification of a new objects results in the prediction of its class membership using the tree classifier. Classification can be also used to test the accuracy of the tree, which can be done in different manners. In this study, two methods are applied [12]. The first one, 'tree and Bayesian' uses the decision tree to calculate a priori probabilities for the Bayesian method and then classifies objects with Bayesian methods considering only attributes which do not appear in the corresponding branch of the tree. The other considered method, 'Bayesian and tree', uses a decision tree for classification, taking into account Bayesian probabilities already calculated in leaves.

3. Data analysis results

All methods considered in this study were applied to the analysis of the information system concerning patients suffering from duodenal ulcer treated by highly selective vagotomy (HSV). It will be shortly called HSV information system. A detailed description of this system may be found in [10]. In brief, the HSV information system is composed of 122 patients treated by HSV described by 11 attributes (variables). The set of all patients is classified into 4 classes according to long term results of the operation. The cardinalities of each classes are : class 1 (excellent result) - 81 patients, class 2 (good result) - 19 patients, class 3 (satisfactory result) - 8 patients, class 4 (unsatisfactory result) - 14 patients. Attributes describing patients are qualitative (attribute 1 - sex and attribute 4 - complication of ulcer) and quantitative (they concern anamnesis or laboratory tests).

In this study, similarly to [2], we have decided to consider also the case of aggregated classes 1+2 (denoted by I) and 3+4 (denoted by II). This aggregation was done because classes 1 and 2 correspond both to good results of treatment and classes 3 and 4 correspond to bad results.

3.1. Results of the rough sets analysis

The case of four classes was considered first. Proceeding in the usual way of the rough sets methodology, we created approximations of each particular class by the set of all attributes, and the calculated accuracy of each approximation. The results are presented in Table 1. The quality of whole classification is equal 0.97.

Table 1. The rough sets approach.
Accuracy of approximation of each class by attributes
from set Q in the HSV information system.

Class X_i	Number of patiens card (X_i)	Lower approx. X_i card$(\underline{Q}X_i)$	Upper approx. X_i card$(\overline{Q}X_i)$	Accuracy $\mu_Q(X_i)$
1	81	79	83	0.95
2	19	19	20	0.9
3	8	8	8	1.0
4	14	13	15	0.87

Then, the only minimal subset of attributes consisting of 9 attributes $\{2,3,4,5,6,7,9,10,11\}$ was found. Because it has a high quality (0.97) we looked for more reduced subset of attributes ensuring a satisfactory quality of classification. As a result of this searching we found subset $H = \{3,4,6,9,10\}$ giving the quality 0.68.

The same procedure was applied to the analysis of the two class case. We found one minimal subset composed of 7 attributes $\{3,4,5,6,7,10,11\}$ (quality of classification equal to 0.97) and then the reduced subset $F = \{3,4,6,10,11\}$ giving a good quality of classification equal to 0.84.

Obtained reduced subsets of attributes were used to reduce the information systems (i.e system with 4 classes and system with 2 classes). Then, decision algorithms were derived from both reduced information systems for both cases. As it was possible to obtain more than one possible decision algorithm, the criterion of the lowest number of descriptors in the decision algorithm was used to choose one of them. The algorithms are presented in Table 2 (four classes) and Table 3 (two classes).

The verification of these decision algorithms was performed by means of the reclassification of the old sample (the HSV information system) using the 'leaving-one-out method', which is typical for statistical approaches. It relies on reclassification of each particular patient using the decision algorithm derived from the information system which does not

Table 2. The rough sets approach. Decision algorithm derived from the reduced HSV information system. Four class case.

Condition attributes					Class	Condition attributes					Class
3	4	6	9	10		3	4	6	9	10	
0	3				⇒ 1	1	2	1	0		⇒ 2
1	1		2		⇒ 1	2	0	2	1	2	⇒ 2
1	0		2		⇒ 1	0	0				⇒ 3
2	1		1		⇒ 1	1	3	0			⇒ 3
2	2		2		⇒ 1	1	3	2		1	⇒ 3
2	1		2		⇒ 1	1	1	2		2	⇒ 3
2	3		0		⇒ 1	1	0	0	1	1	⇒ 3
2	0	1	2		⇒ 1	2	0	0		0	⇒ 4
1	1	0	1		⇒ 1	1	3	2		2	⇒ 4
1	1	2	1		⇒ 1	1	2	1		0	⇒ 4
2	4	2	0		⇒ 1	2	0	2	2	2	⇒ 4
0	1	2	0		⇒ 1	1	0	2	0	1	⇒ 4
1	0	1	1		⇒ 1	2	4	2	2	1	⇒ 4
2	1	0	0		⇒ 1	2	1	1	0	1	⇒ 4
2	0	2	2	0	⇒ 1	2	2	1	0	1	⇒ 1/2
1	0	0	0	0	⇒ 1	2	3	1	1	1	⇒ 1/2
2	0	1	0	2	⇒ 1	2	0	2	2	1	⇒ 1/2/3
1	1	2	0	1	⇒ 1	2	0	2	0	1	⇒ 1/2
2	0	2	0	2	⇒ 1	1	0	2	1	1	⇒ 1/2
2	0	1	1	2	⇒ 1	2	0	1	0	1	⇒ 1/4
2	0	0	1	1	⇒ 1	1	0	0	0	1	⇒ 1/4
2	0	0	0	1	⇒ 1	2	2	0	0	1	⇒ 1/4
	4	0			⇒ 2	2	0	0	1	2	⇒ 1/4
	3	1	2		⇒ 2	1	0	2	1	2	⇒ 1/2
	3	1	0		⇒ 2	1	1	1	1	2	⇒ 1/2
0	1	1	0		⇒ 2	2	0	1	1	1	⇒ 1/2
1	2	1	2		⇒ 2	2	0	2	1	1	⇒ 2/4
2	2	1	1		⇒ 2						

Table 3. The rough sets approach. Decision algorithm derived from the reduced HSV information system. Two class case.

Condition attributes					Class	Condition attributes					Class
3	4	6	10	11		3	4	6	10	11	
2	3				⇒ 1	2	0	2	0	1	⇒ 1
0	1				⇒ 1	1	0	0	0	0	⇒ 1
0	3				⇒ 1	1	0	2	1	1	⇒ 1
1	1	0			⇒ 1	2	0	0	1	0	⇒ 1
2	1	0			⇒ 1	2	0	1	2	0	⇒ 1
2	2	1			⇒ 1	1	2	1	1	0	⇒ 1
1	0	1			⇒ 1	2	0	1	1	0	⇒ 1/2
1	3	1			⇒ 1	1	0	0	1	0	⇒ 1/2
1	1	1			⇒ 1	2	2	0	1	0	⇒ 1/2
2	4	0			⇒ 1	2	0	2	1	1	⇒ 1/2
2	1	2			⇒ 1	2	0	0	2	2	⇒ 1/2
2	0	1		1	⇒ 1	0	0				⇒ 2
1	0	0		3	⇒ 1	1	3	0			⇒ 2
2	0	1		3	⇒ 1	1	3	2			⇒ 2
1	0	0		1	⇒ 1	1	0	2		0	⇒ 2
1	1	2		1	⇒ 1	2	4	2		3	⇒ 2
2	0	2		0	⇒ 1	1	1	2		2	⇒ 2
2	4	2		0	⇒ 1	2	1	1		0	⇒ 2
1	1	2		0	⇒ 1	1	0	0		2	⇒ 2
2	0	2		2	⇒ 1	2	0	0	0	0	⇒ 2
1	2	1		3	⇒ 1	2	0	2	2	3	⇒ 2
2	1	1		1	⇒ 1	1	2	1	0	0	⇒ 2
1	0	2		2	⇒ 1	1	0	2	2	1	⇒ 2
2	0	1		2	⇒ 1						
2	0	0		1	⇒ 1						
2	2	0		1	⇒ 1						
2	1	1		2	⇒ 1						
2	0	1	0	0	⇒ 1						
2	0	2	1	3	⇒ 1						

contain the patient considered (i.e. the information system was decreased by the considered patient and contained only 121 others).

The reclassification was performed twice: first using only certain, i.e. deterministic rules, then using also non-deterministic rules (with the analysis of possible assignments) and the procedure for finding nearest rules. In the latter reclassification $k = 5$ nearest rules were analysed and the patient was assigned to the most supported class. This approach is a bit different form one considered in [9] where only first nearest rule was analysed.

The results of reclassification are presented in Table 6 (four classes) and Table 7 (two classes).

All necessary calculation were performed with the RoughDAS software [6].

3.2. Discriminant analysis results

As the attribute 4 is a discrete nominal one, it was transformed into a set of 4 binary attributes, e.g. 4–1 is equal to 1 when the complication of ulcer no. 1 was present and 0, otherwise. The same transformation was made in [9] for the location model.

For the linear Fisherian discrimination two methods of estimation of 'a priori' probabilities were analysed : probabilities proportional to the number of patients in classes under consideration or equal to this number.

Subsets selected for linear discrimination are given in Table 4 (four class case) and Table 5 (two class case). The results of reclassification are summarized in Table 6 and 7. The best results obtained via location model method [9] are also given in Tables 4,5,6 and 7.

Table 4. Subsets selected by different methods. Four classes.

method	subset
Rough sets approach	{3,4,6,9,10}
Location model – A	{4–3,4–4,6,9,10,11}
– B	{4–3,4–4,6}
linear discrimination	{2,3,4–1,4–2,5,8,9}
inductive learning	{1,2,3,4,5,6,7,11}

Table 5. Subsets selected by different methods. Two classes.

method	subset
Rough sets approach	{3,4,6,10,11}
Location model – C	{2,3,4–1,4–4,7,8,10}
– D	{4–1,4–4,7,8,10}
linear discrimination	{1,2,3,5,8,11}
inductive learning	{2,3,4,5,6,7,8,9,10}

3.3. Inductive learning analysis

The inductive learning algorithms are represented by a tree structure. The tree structure depends on a pruning factor. Although, it is known that optimal results are obtained with the pruning factor equal to 3, we have also checked how the increase of this factor influences the complexity of tree and classification results.

Table 6. Results of reclassification of the HSV information system (numbers of correctly classified objects) for the four class case.

method and subset	Correct classification in class 1	2	3	4	Whole system
Rough sets approach					
deterministic rules (*)	48	2	0	3	53
first nearest rule (**)	71	10	2	6	89
k nearest rules (**)	79	4	0	3	86
Location model					
subset B	51	2	0	5	58
Linear discrimination					
equal probabilities					
all attributes	30	6	3	2	41
selected subset	35	6	3	3	47
proportional prob.					
all attributes	69	1	0	0	70
selected subset	71	1	0	0	72
Inductive learning					
tree and Bayesian	68	1	1	1	71
Bayesian and tree	78	0	0	0	78

(*) – classification on the basis of deterministic rules only as performed in [9]
(**) – classification with non-deterministic rules and 'nearest' rules

Table 7. Results of reclassification of the HSV information system (numbers of correctly classified objects) for the two class problem

method and subset	Correct classification in class I	II	Whole system
Rough sets approach			
deterministic rules (*)	75	6	81
first nearest rule (**)	88	10	98
k nearest rules (**)	99	6	105
Location model			
subset C	64	12	76
Linear discrimination			
equal probabilities			
all attributes	60	2	62
selected subset	59	10	69
proportional prob.			
all attributes	94	0	94
selected subset	98	0	98
Inductive learning			
tree and Bayesian	92	2	94
Bayesian and tree	99	0	99

(*) – classification on the basis of deterministic rules only as performed in [9]
(**) - classification with non-deterministic rules and 'nearest' rules

For the four class case, we have obtained a tree based on attributes $\{1,2,3,4,5,6,7,11\}$. It is presented in Fig. 1. Then, we have found that increasing the pruning factor PF we can reach simpler tree structures (less attributes) but branches leading to the group 2 and 3 are cut off from the tree and classification is unreliable. The tree presented in Fig. 1 is derived from the whole data set but for verification calculations the leaving-one-out method was applied as in the rough sets and discriminant procedures. The results are given in Table 6.

For two class case, the tree (Fig 2.) is based on attributes $\{2,3,4,5,6,7,8,9,10\}$. The classification results are given in Table 7. Further increase of the pruning factor gives an unreliable tree using only attribute 4 and classifying all the patients to group I.

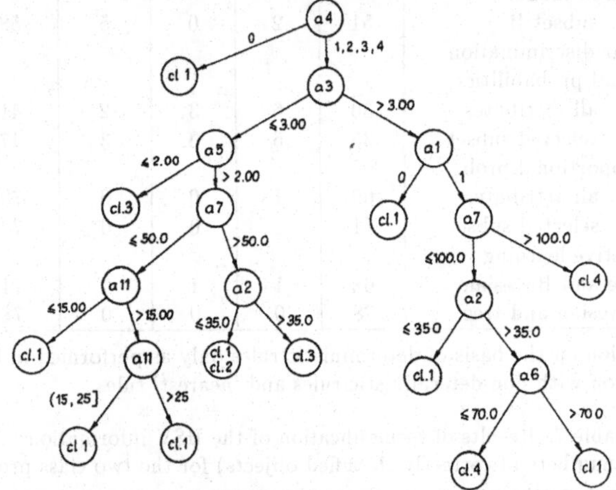

Figure 1: The tree structure for four class case.

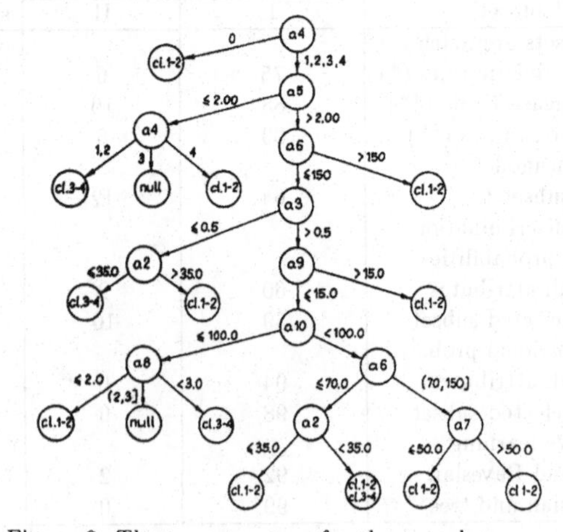

Figure 2: The tree structure for the two class case.

4. Discussion

Comparative evaluation and discussion concerns two aspects of the analysis of the HSV information system, i.e. reduction of attributes and classification ability.

The first aspect refers to comparison of subsets of significant attributes (see tables 4 and 5) selected by the four methods. These subsets can replace the whole set of attributes and reduce in this way the information system. Reduction of attributes is important in many practical application, e.g. in medicine, some unnecessary but harmful or expensive tests may be avoided in this way. The comparison of subsets chosen for four and two classes, respectively, by the same method shows that they are consistent only in the case of the rough sets approach. Indeed, subsets $\{3,4,6,9,10\}$ and $\{3,4,6,10,11\}$ are nearly the same. It is an interesting and useful feature from the practical point of view because the aggregation was done to join the classes having similar meaning. Subsets chosen by other methods differ significantly.

For discriminant techniques, this difference can be explained in such a way that in the measure of a discriminative power the within-class variability is related to the total variability. As in the case of aggregated classes two mean values are aggregated to calculate the mean of an aggregated class, the dispersion between objects and a joint mean can be different than for means obtained separately. This influences the discriminative power of attributes under study.

In inductive learning, selection of attributes results from pruning of trees. So, in this method there is no preliminary step of data reduction performed before building decision rules as in rough sets or discrimination techniques. It can also be noticed that attribute selection by pruning trees has been relatively ineffective since 8 and 9 attributes (four and two class case, respectively) remained in the structure.

Reclassification with 'leaving-one-out' shows that the best results (see Tables 6 and 7) are obtained by means of the rough sets approach. The second best is inductive learning and classical linear discrimination with proportional a priori probabilities.

Considering results of the rough sets approach (the highest results of reclassification for both cases) it can be noticed that analysis of nearest rules and non-deterministic rules, instead of using deterministic rules only, has considerably increased the score of reclassification (in the case of four classes the number of correctly classified objects has increased from 53 to 89). Moreover, in the case of two classes the analysis of k nearest rules was slightly better than taking into account the first rule only as it was done in [9]. In the case of four classes, global results were similar. It should be noticed, however, that objects correctly classified by means of the analysis of k nearest rules mainly belong to the first class. It results from the fact that the first class has much greater cardinality than other classes in information system and, in consequence, majority of analysed nearest rules concern this class. On the other hand, analysis of the first nearest rule only, while giving worse results of global reclassification, enabled classification of objects from other classes than the first one (see [9]).

It is also interesting to note that in four class case, linear discrimination performs better than location model when 'a priori' probabilities are set proportional to the number of patients in classes. For equal 'a priori' probabilities the global reclassification score is worse but classes 2–4 are classified better. The advantage of linear discrimination over the location model is not surprising as for the location model the sample size is relatively small and yet divided by two binary attributes (4–3, 4–4). Thus, because of many subparts, and therefore poor estimates, we cannot expect better results for many binary attributes. The location model performs better in the two class case. Its global reclassification score is smaller than for linear discrimination and proportional a priori probabilities but on the other hand kind of linear discrimination misclassifies almost all patients from class II to class I. In addition, one should stress that linear discrimination with leaving-one-out performs better for subsets selected than for the entire set of attributes. This effect called 'peak effect' is caused by noise reduction in the case of selection.

In inductive learning, results of reclassification are similar to linear discrimination, both for 'Bayesian and tree' and 'tree and Bayesian' techniques. This means that the classification is biased again towards the greatest class. Despite of good global results, it is impossible to classify patients from smaller classes in a reliable way (the worst results among all the methods).

Taking into account all compared results, one can conclude that in the case of unbalanced class sizes and many discrete attributes in information system the linear discrimination could be recommended over location model (although assumptions of normality and homogeneity are not fulfilled). The location model could be better for larger sample sizes and relatively lower number of discrete attributes. Inductive learning is also ineffective for unbalanced class sizes, the rough sets approach is generally better than other methods in that case. A more general conclusion is that the rough sets approach is dealing particularly well with analysis of information systems coming from human experience, where qualitative attributes exist naturally. Moreover, the rough sets approach makes no assumption about data as probabilistic methods do, e.g. normality of distributions, similar cardinality or large sizes of classes, and it does not use their typical operators of data aggregation (e.g. mean value) in the course of the analysis. The structure of final classifiers (decision rules) is also different from probabilistic methods. The most similar is the tree representation in probabilistic inductive learning but it could not be converted to decision rules because the same attribute can be used in binary tree several times down to it and its different values can have different 'decision power'.

References

[1] Cestnik B., Kononenko I., Bratko I., ASSISTANT86, a knowledge–elicitation tool for sophisticated users. In Bratko I., Lavrac N. (eds), *Progress in Machine Learning*, Sigma Press, Wilmshow (1987) (distributed by Wiley), 31–45.

[2] Fibak J., Pawlak Z., Słowiński K., Słowiński K., Rough sets based decision algorithm for treatment of duodenal ulcer by HSV. *Bull. PAS, Biological Series* **34**, 10–12 (1986), 227–246.

[3] Gale J., Marsden P., Diagnosis process not conduct. In Sheldon M., Brook J., Rector A. (eds), *Decision Making in General Practice*, Macmillan (1985).

[4] Garns M., Lavrac N., Review of five empirical learning systems within a proposed schemata, In Bratko I., Lavrac N. (eds) *Progress in Machine Learning*, Sigma Press, Wilmshow (1987) (distributed by Wiley), 46–66.

[5] Haberman S.J., Log linear fit for contingency tables. Algorithm AS51. *Applied Statistics* **21** (1972), 218–225.

[6] Gruszecki G., Słowiński R., Stefanowski J., *RoughDAS – Rough Sets Based Data Analysis Software – user's manual.* APRO S.A. Warszawa (1990).

[7] Krusińska E., New procedure for selection of variables in location model for mixed variable discrimination. *Biometrical Journal* **31** (1989), 511–523.

[8] Krusińska E., Linear and quadratic classification rule in the location model. A comparison for heterogeneous data. *Biocybernetics and Biomedical Engineering* **9** (1989), 103–113.

[9] Krusińska E., Słowiński R., Stefanowski J., Discriminat versus rough sets approach to vague data analysis. *Applied Stochastic Models and Data Analysis* **8**, (1992) (to appear).

[10] Krzanowski W.J., Discrimination and classification using both binary and continuous variables. *J.Am.Statistic Assoc.* **70** (1975), 782–790.

[11] Lachenburch P.A., *Discriminant Analysis*, Hafner Press, New York (1975).

[12] Man J., Kononenko B., Cestnik I., Bratko I., *Assistant Proffesional. User's manual.* Institute Jozef Stefan, Ljubliana (1988).

[13] Niblet T., Construction of decision trees in noisy domains, in Bratko I., Lavrac N. (eds) *Progress in Machine Learning*, Sigma Press, Wilmshow (1987) (distributed by Wiley).

[14] Nowicki R., Słowiński R., Stefanowski J., Rough sets analysis of diagnostic capacity of vibroacoustic symptoms, *Computers and Mathematics with Applications* (1992) (to appear).

[15] Nowicki R., Słowiński R., Stefanowski J., Evaluation of vibroacoustic symptoms by means of the rough sets theory, *Computer in Industry* (1992) (submitted).

[16] Pawlak Z., Rough sets. *International Journal of Information and Computer Sciences* **11**, 5 (1982), 341–356.

[17] Pawlak Z., Słowiński K., Słowiński R., Rough classification of patients after highly selective vagotomy for duodenal ulcer. *International Journal of Man-Machine Studies* **24** (1986), 413–433.

[18] Słowiński K., Słowiński R., Stefanowski J., Rough sets approach to analysis of data from pertioneal lavage in acute pancreatitis. *Medical Informatics* **13**, 3 (1989), 143–159.

[19] Słowiński K., Słowiński R., Sensitivity analysis of rough classification. *International Journal of Man-Machine Studies* **32** (1990), 693–705.

[20] Stefanowski J., Rough classification and discriminant analysis. Comparative study. (in Polish) *Postępy Cybernetyki* (1991) (to appear).

[21] Stefanowski J., Classification support based on the rough sets theory, *Proceedings of the IIASA Workshop on User-Oriented Methodology and Techniques of Decision Analysis and Support.* Serock, 9–13 Sept. (1991) (to appear).

[22] Watkins C.J. C.H., Combining cross–validation and search. In Bratko I., Lavrac N. (eds) *Progress in Machine Learning.* Sigma Press, Wilmshow 1987 (distributed by Wiley) 79–97.

[23] SPSSX, Statistical Algorithms. SPSS Inc. Chicago (1983) Mc Graw–Hill

Part II
Chapter 4

SOME EXPERIMENTS TO COMPARE ROUGH SETS THEORY AND ORDINAL STATISTICAL METHODS

Jacques TEGHEM Mohammed BENJELLOUN
Faculté Polytechnique de Mons
9, rue de Houdain
B-7000 Mons, Belgium

Abstract. In statistics, classical discriminant analysis methods explain a predefined classification of a set of objects, by using the discrete values taken by these objects on some conditional attributes. Rough sets theory has the same objective. The purpose of the paper is to describe some experiments made to compare both approaches. In a first section, we recall the results obtained by Wong et al. for the comparison of rough sets theory with the Quinlan's method, using the notion of entropy. Section two is devoted to a comparison of rough sets approach with a specific discriminant method, the Elysee method; an application is solved by both methods and similar results are obtained. In the last section, we analyse a real case study concerning a production problem, in a printing company, by both methodologies.

1. Rough sets theory and information theory

In this preliminary section, we want to recall some interesting results of Wong et al. [6] , related to a comparison between rough sets theory and Quinlan's method, using the notion of entropy.

Let us consider a decision table $\langle U, C \cup D, V, f \rangle$. If $D^* = \{D_1, \ldots, D_R\}$ represents the family of all D-elementary sets defined by the equivalence relation $\tilde{D}$, we denote by

$$p_r = \frac{card(D_r)}{card(U)}$$

the relative frequency of objects in class D_r.

From the theory of information, we know that the entropy

$$H_U^D = -\sum_{r=1}^{R} p_r \log_R p_r$$

is a good measure of the uncertainty of the distribution of the objects in D^*.
The entropy verifies some basic properties :

- $H_U^D = 0$ iff $R = 1$, i.e., $U \equiv D_1$
- for R fixed, H_U^D is maximum if $p_r = \frac{1}{R}\, \forall r$
- if $p_r = \frac{1}{R}\, \forall r, H_U^D$ is an increasing function of R

The following property is important for the analysis of a decision table by the Quinlan's method : if the set of objects U is considered as composed of several subsystems P_n, $n = 1, \ldots, N$ - for instance, corresponding to elementary sets of a subset P of the conditional attributes - the entropy H_U^P, measuring the uncertainty of the distribution of the classes D_r in the classes P_n, is expressed by

$$H_U^P = \sum_{n=1}^{N} q_n\, H_{P_n}^D$$

where

$$q_n = \frac{card(P_n)}{card(U)}$$

$$H_{P_n}^D = -\sum_{r=1}^{R} p_{rn} \log_R p_{rn}$$

with

$$p_{rn} = \frac{card(D_r \cap P_n)}{card(P_n)}$$

The Quinlan methods consists to build a decision tree determining progressively the decision rules. At the root of the tree, the attribute $q_1 \in C$ is selected such that

$$H_U^{\{q_1\}} = \min_{q \in C} H_U^{\{q\}},$$

i.e the attribute q_1 for which the uncertainty of the distribution of classes D_r in the $\{q_1\}$-elementary sets, is minimum.

Let $X_i, i = 1, \ldots, I$ be the $\{q_1\}$-elementary sets. If X_i is pure - i.e. if $\exists r | X_i \subseteq D_r$ - then no further branching of X_i is needed.

Otherwise, if X_i is impure - i.e. the objects of X_i belong to more that one class D_r - the process is continued, determining for this class, the attribute $q_2 \in C \setminus \{q_1\}$ such that

$$H_{X_i}^{\{q_2\}} = \min_{q \in C \setminus \{q_1\}} H_{X_i}^{\{q\}};$$

some classes $X_{i,j}$ - which are $\{q_1, q_2\}$-elementary sets - are thus obtained.

If the decision table is consistent, the process ends with no impure terminal classes and deterministic rules are obtained. Otherwise, there exists some impure terminal nodes corresponding to non deterministic rules.

Wong et al. [6] obtained an interesting proposition which gives, in a particular case, a direct relation between Quinlan's method and rough sets theory. More precisely, this proposition lies the value H_U^P of the entropy with the value $\gamma_P(D^*)$ expressing the quality of approximation of D^* by the P-elementary sets :

Proposition [6]

If the objects belonging to each impure class of P^* are uniformly distributed in all the classes D_r, i.e. if

$$card(P_n \cap D_r) = \frac{card(P_n)}{R} \quad \forall r = 1, \ldots, R, \ \forall P_n \text{ impure},$$

then

$$\begin{aligned} H_U^P &= 1 - \alpha_P(D_r) \quad r = 1, \ldots, R \\ &= 1 - \gamma(D^*) \end{aligned}$$

Remark

The assumption of uniform distribution is equivalent to

$$card(B_{n_P}(D_i) \cap D_r) = \frac{card(B_{n_P}(D_i))}{R} \quad \forall r = 1, \ldots, R$$

For this particular case, an important consequence of this proposition is, that the decision rules derived from a subset P of attributes are identical if we use Quinlan's method or rough sets approach !

We illustrate this result, considering the following didactic example :

Example

There are 9 objects - a till i -, 4 conditional attributes - C_1 till C_4 - and one decision attribute D; the information system is given in table 1.

	C_1	C_2	C_3	C_4	D
a	1	1	2	1	-
b	1	1	2	1	-
c	2	1	1	2	-
d	1	1	2	1	+
e	1	1	1	2	+
f	2	1	1	2	+
g	1	2	2	2	+
h	1	2	1	1	+
i	2	2	1	1	+

Table 1. Information system of the example

The atoms of this information system are $\{a, b, d\}; \{c, f\}; \{e\}; \{g\}; \{h\}; \{i\}$.

We consider $P = \{C_1, C_2\}$ and the corresponding partition $P^* = \{P_1, P_2, P_3, P_4\}$ of U : $\{a, b, d, e\}; \{c, f\}; \{g, h\}; \{i\}$.

It is easy to obtain

$$\begin{aligned} H_U^P &= \tfrac{4}{9}\ H_{P_1}^D + \tfrac{2}{9}\ H_{P_2}^D + \tfrac{2}{9}\ H_{P_3}^D + \tfrac{1}{9}\ H_{P_4}^D \\ &= \tfrac{4}{9}\ 1 + \tfrac{2}{9}\ 1 + \tfrac{2}{9}\ 0 + \tfrac{1}{9}\ 0 \\ &= \tfrac{2}{3} \end{aligned}$$

and $\alpha_P(D^+) = \alpha_P(D^-) = \gamma_P(D^*) = \frac{1}{3}$.

Effectively, the assumption of the proposition is verified : the impure classes P_1 and P_2 have an uniform distribution of their elements in classes D^+ and D^-, or equivalently it is true for the P-doubtful region of D^+ and D^-

$$B_{n_P}(D^+) = B_{n_P}(D^-) = \{a, b, c, d, e, f\}$$

As noted in [6] , this result reveals a drawback of the rough sets theory in the common case where the very particular assumption of the proposition is not verified.

Effectively, the discriminatory coefficients $\alpha_P(D_r)$ and $\gamma_P(D^*)$ the quality of approximation D^* by P, only use the informations related to the cardinality of the pure classes of P^* and the cardinality of the doubtful regions $B_{n_P}(D_r)$, without taking into account the internal distribution of the objects into the impure classes of P^* and thus into the doubtful regions $B_{n_P}(D_r)$.

So, rough sets theory is unaware of the degree of impurity of the doubtful regions.

This observation can be explained more precisely using the example.

We now consider the subset $\bar{P} = \{C_3, C_4\}$ of attributes, with the corresponding partition $\bar{P}^* = \{\bar{P}_1, \bar{P}_2, \bar{P}_3, \bar{P}_4\}$:

$$\{a, b, d\}; \{c, e, f\}; \{g\}; \{h, i\}$$

By the rough sets theory, the results obtained with $\bar{P}$ are similar to those obtained with P:

$$\alpha_{\bar{P}}(D^+) = \alpha_{\bar{P}}(D^-) = \gamma_{\bar{P}}(D^*) = \frac{3}{9}$$

$$B_{n_{\bar{P}}}(D^+) = B_{n_{\bar{P}}}(D^-) = \{a, b, c, d, e, f\}$$

So using this approach, there is no reason to choose $\bar{P}$ instead of P. Nevertheless, the impure classes of $\bar{P}^*$ are less impure that those of P^* because

$\bar{P}_1$ contains 2 objects of D^- and 1 object of D^+

$\bar{P}_2$ contains 1 object of D^- and 2 objects of D^+

Thus, in one sense, the approximation of D^* by $\bar{P}$ is better that with P.

<u>Remark</u>

Let us note that the notion of entropy takes this information into account :

$$H_U^{\bar{P}} = 0.61219 < \frac{2}{3}$$

To emphasize the importance of this observation, let us modify the preceeding example, considering now that there are n identical objects b - denoted $b_1, \ldots, b_n$ - all classified in D^-, and n identical objects e - denoted by $e_1, \ldots, e_n$ -, all classified in class D^+.

The classes of P^* and $\bar{P}^*$ and are now respectively :

$$P^* : \{a, b_1, \ldots, b_n, d, e_1, \ldots, e_n\}, \{c, f\}, \{g, h\}, \{i\}$$

$$\bar{P}^* : \{a, b_1, \ldots, b_n, d\}, \{c, e_1, \ldots, e_n, f\}, \{g, h\}, \{i\}$$

Again $\gamma_P(D^*)$ and $\gamma_{\bar{P}}(D^*)$ have the same value, equal to $\frac{3}{2n+7}$, but, in a certain sense, $\bar{P}^*$ is clearly better than P^* to approximate D^*. On one hand, for P, we still have an uniform distribution of the objects included in the impure classes into D^* ; on the other hand, for $\bar{P}$, the impure classes are " almost " pure, with a proportion $\frac{n+1}{n+2}$ objects to the same classes of D^*.

Remark

For this extended example, the entropy values - in function of n - for P and $\bar{P}$ are respectively

$$H_U^P(n) = \frac{6}{9} \; \forall n$$

$$\lim_{n \to \infty} H_U^{\bar{P}}(n) = 0$$

In conclusion of this section, if rough sets theory certainly is an efficient tool to analyse information systems, often more simple and comprehensive that Quinlan's method using the entropy notion, nevertheless the comparison suggests that some improvements can still be effected in the rough sets approach.

Further researchs are necessary to investigate how the distribution of the objects into the doubtful regions can be taken into account.

One way may be to define a simple index of purity of a P-elementary set, equal to one for a pure class, and decreasing with the degree of uncertainty in case of an impure class.

2. Comparison between rough sets theory and ordinal discriminant analysis

In statistics, classical discriminant analysis methods consist, from a sample of observations - or set of objects U - to explain a classification $D^* = \{D_r, \; r = 1, \ldots, R\}$ - defined by some decision attributes - by using the values taken by the observations on some factors C_q - or conditional attributes.

The aim of the rough set theory is thus not so new. Nevertheless, the statistical methods are commonly based on attributes defined on continuous scales [3] .

However, some ordinal statistical methods exist also in the literature; it is for instance the case of the Elysee method [2].

Logically it appears useful to compare these two different methodologies, the statistical approach and the rough sets approach.

2.1. The Elysee method [2]

We shortly recall the principle of this method, using the terminology of the rough sets theory.

The method generates a tree of subsets of objects ; the root of the tree is the set U of all objects. At each level, a dichotomy is defined to separate a node in two parts, like in figure 1.

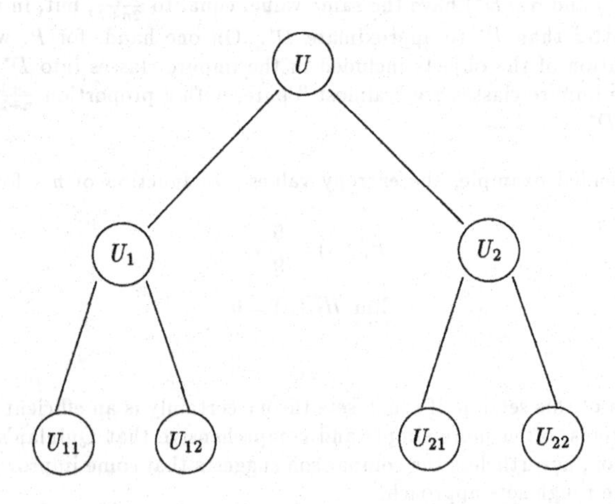

Figure 1. Decision tree of the Elysee method

Let us describe this branching process.

At the root, a conditional attribute $C_{\bar{q}}$ and a value $\bar{v} \in V_{\bar{q}}$ are choosen to define the two subsets by

$$U_1 \equiv U_1(\bar{q}, \bar{v}) = \{x \in U | f(x, \bar{q}) \leq \bar{v}\}$$
$$U_2 \equiv U_2(\bar{q}, \bar{v}) = \{x \in U | f(x, \bar{q}) > \bar{v}\}$$

So we obtain the following table 2, where

$$n_{ri} = card(D_r \cap U_i), \; r = 1, \ldots, R; \; i = 1, 2$$

$$n_{r.} = \sum_{i=1}^{2} n_{ri} \text{ and } n_{.i} = \sum_{r=1}^{R} n_{ri}$$

The objective of this branching process is to obtain two sets U_1 and U_2 which are, regarding the classification D^*, very homogeneous each one but very heterogeneous between them.

	U_1	U_2	
D_1	n_{11}	n_{12}	$n_{1.}$
$\vdots$			
D_r	n_{r1}	n_{r2}	$n_{r.}$
$\vdots$			
D_R	n_{R1}	n_{R2}	$n_{R.}$
	$n_{.1}$	$n_{.2}$	$n_{..} = n$

Table 2. Statistical table resulting from the branching process

A distance $D(U_1, U_2)$ is introduced to measure the slack between the distribution defined by table 2, and the distribution corresponding to statistical independance of the $2R$ subsets $D_r \cap U_i$.

Using a "Benzekri distance" [3] , the Elysee method uses the following coefficient

$$D^2(U_1, U_2) = \sum_{r=1}^{R} \sum_{i=1}^{2} \frac{(n_{ri} - \frac{n_{r.}n_{.i}}{n})^2}{n_{r.}n_{.i}}$$

The variable $nD^2(U_1, U_2)$ is ditributed like a classical distribution χ^2 [3] . To have a normalized coefficient, taking its value on [0,1], it is necessary to consider the coefficient

$$T^2(U_1, U_2) = \frac{nD^2(U_1, U_2)}{\sqrt{R-1}}$$

The pair $(\bar{q}, \bar{v})$ is determined by the criterion

$$\max_{v} \max_{q} \; T^2(U_1(q,v); U_2(q,v))$$

The optimal pair $(\bar{q}, \bar{v})$ corresponds to the dichotomy presenting the larger statistical dependance with the classification D^*.

On each subset U_k, $k = 1, 2$, the branching process is again applied, determining a pair (q, x) possibly different from one subset to another.

The branching process is stopped

- either when a generated subset is a pure class regarding the classification D^*.
- either when the sample of the corresponding table - similar to table 1 - is not representative enough from a statistical point of view. Generally , a branching is not performed for instance, if the cardinality of a generated subset is inferior to a fixed level $\bar{n}$, i.e. if

$$\exists \, (r, i) \mid n_{ri} < \bar{n}.$$

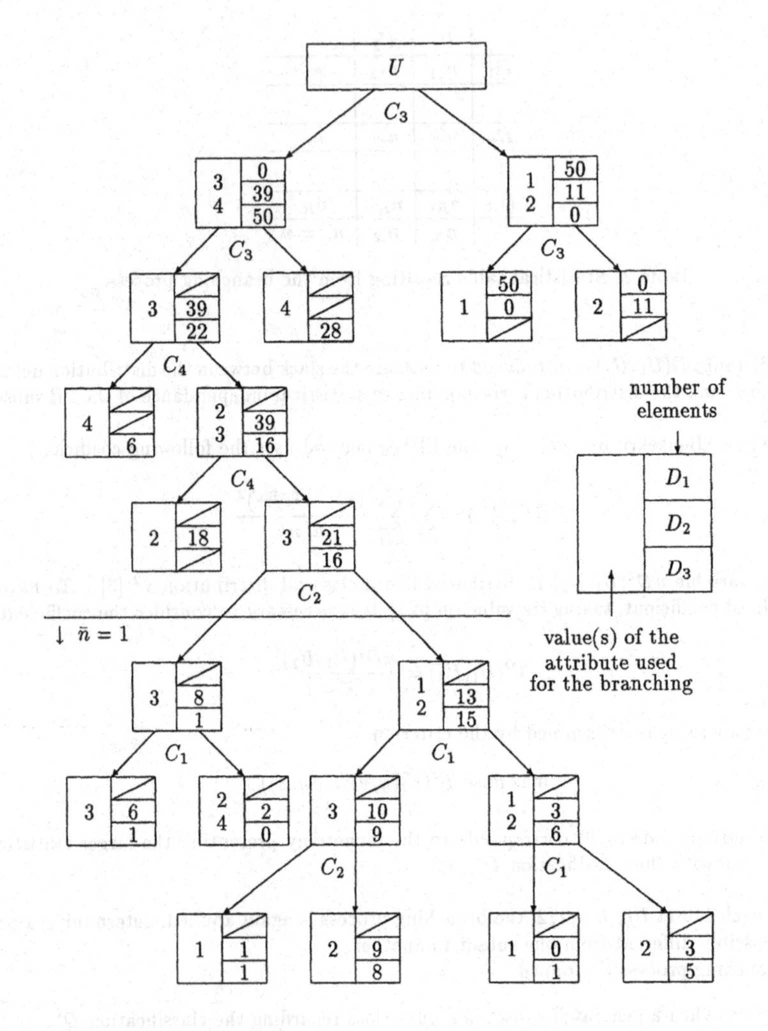

Figure 2. Decision tree obtained by the Elysee method

2.2. Comparison with the rough sets approach based on an example

We have applied the method described in 3.1. and also the rough sets method [4] to a classical application often used in the field of discrete statistical discriminant analysis.

The example concerns a set U of 150 flowers (objects) characterized by four conditional attributes C_q, $q = 1, \ldots, 4$ which describe the length and the breadth of the petal and the sepal respectively. For each of these attributes, the descriptor of an object can take four different values, corresponding to pre-defined classes of length :

$$f(x,q) \in V_q = \{1,2,3,4\} \ \forall x \in U, \ q = 1, \ldots, 4$$

The decision attribute D defines a partition of U in three classes D_r, $r = 1,2,3$, each one of cardinality 50. The corresponding information system is given in annex 1.

2.2.1. Application of the Elysee method

The decision tree generated by this method is described in figure 2. The complete tree corresponds to the case where no lower bound $\bar{n}$ is fixed for the cardinality of a set $(D_r \cap U_i)$, or equivalently $\bar{n} = 1$ (see section 3.1.); otherwise, if $\bar{n} > 1$, the part of the left branch of the tree, included in a box (see figure 2) will not appear.
The classification rules furnished by the terminal nodes of the tree are the following :

- if $\bar{n} > 1$
  There are five deterministic rules
  1 . $C_3 = 1 \rightarrow d = 1$
  2 . $C_3 = 2 \rightarrow d = 2$
  3 . $C_3 = 4 \rightarrow d = 3$
  4 . $C_3 = 3, C_4 = 2 \rightarrow d = 2$
  5 . $C_3 = 3, C_4 = 4 \rightarrow d = 3$
  and one non deterministic rule
  $C_3 = 3,\ C_4 = 3 \rightarrow d = 2$ or 3
- if $\bar{n} = 1$
  In this case, the non-deterministic rule provides three new deterministic rules
  6 . $C_3 = 3, C_4 = 3, C_1 = 1 \rightarrow d = 3$ (see the remark below)
  7 . $C_3 = 3, C_4 = 3, C_2 = 3, C_1 = 2 \rightarrow d = 2$
  8 . $C_3 = 3, C_4 = 3, C_2 = 3, C_1 = 4 \rightarrow d = 2$
  and four new non-deterministic rules
  9 . $C_3 = 3, C_4 = 3, C_2 = 3, C_1 = 3 \rightarrow d = 2$ or 3
  10. $C_3 = 3, C_4 = 3, C_1 = 3, C_2 = 1 \rightarrow d = 2$ or 3
  11. $C_3 = 3, C_4 = 3, C_1 = 3, C_2 = 2 \rightarrow d = 2$ or 3
  12. $C_3 = 3, C_4 = 3, C_1 = 2 \rightarrow d = 2$ or 3 (see the remark below).

Remark

The case $C_3 = 3, C_4 = 3, C_1 = 1, C_2 = 1$ does not correspond to any object, so it is not necessary to discriminate rule 6 by $C_2 = 1$ or 2.

Similarly, the case $C_3 = 3, C_4 = 3, C_1 = 2, C_2 = 1$ does not correspond to any object, so it is not necessary to discriminate rule 12 by $C_2 = 1$ or 2.

2.2.2. Application of the rough sets method

We applied the first version of the software proposed by R. Słowiński and J. Stefanowski [5] . There are 31 atoms only. The quality of the classification corresponding to the set C of the four conditional attributes is

$$\gamma_C(D^*) = 0.773$$

and the quality of each class D_r is

$$\alpha_C(D_1) = 1 \; ; \; \alpha_C(D_2) = 0.48 \; ; \; \alpha_C(D_3) = 0.50$$

All the reductions of C correspond to an inferior value of the quality of the approximation.

Using this set C, we obtain eight deterministic rules and four non-deterministic rules :

1' . $C_3 = 1 \rightarrow d = 1$
2' . $C_4 = 2 \rightarrow d = 2$
3' . $C_3 = 2, C_4 = 3 \rightarrow d = 2$
4' . $C_1 = 4, C_3 = 3, C_4 = 3 \rightarrow d = 2$
5' . $C_1 = 2, C_2 = 3, C_3 = 3, C_4 = 3 \rightarrow d = 2$
6' . $C_4 = 4 \rightarrow d = 3$
7' . $C_3 = 4, C_4 = 3 \rightarrow d = 3$
8' . $C_1 = 1, C_3 = 3, C_4 = 3 \rightarrow d = 3$
9' . $C_1 = 3, C_2 = 3, C_3 = 3, C_4 = 3 \rightarrow d = 2$ or 3
10'. $C_1 = 3, C_2 = 2, C_3 = 3, C_4 = 3 \rightarrow d = 2$ or 3
11'. $C_1 = 2, C_2 = 2, C_3 = 3, C_4 = 3 \rightarrow d = 2$ or 3
12'. $C_1 = 3, C_2 = 1, C_3 = 3, C_4 = 3 \rightarrow d = 2$ or 3

Nevertheless, attributes 3 and 4 play a more important rules; with these two attributes, we still have a quality of $\gamma_{\{3,4\}}(D^*) = 0.753$.

Using only this subset of attributes, we obtain six rules - five deterministic and one non-deterministic - which are exactly the same that those corresponding to the case $\bar{n} > 1$ in the Elysee method (see section 2.2.1).

2.2.3. Comparison of the results

The comparison of both sets of rules, those described in a) and in b) respectively, reveals a certain drawback of each method.

Effectively, the Elysee method as well the software [5] based on the rough sets approach, generate rules on a hierarchy of attributes (a global hierarchy for all the classes in the first method, a specific hierarchy for each class in the second). Sometimes, when a new rule is obtained, some of the preceeding attributes in the hierarchy are superflous to describe this rule.

For instance, by inspection of the information table it is easy to see that some rules can be simplified:

- for the Elysee method, rules 4, 5 and 8 can be simply written

  4 . $C_4 = 2 \rightarrow d = 2$
  5 . $C_4 = 4 \rightarrow d = 3$
  8 . $C_3 = 3, C_4 = 3, C_1 = 4 \rightarrow d = 2$

- for the rough set approach, rules 3' and 7' can be simply written
  3'. $C_3 = 2 \rightarrow d = 2$
  7'. $C_3 = 4 \rightarrow d = 3$

With these simplifications, the two set of rules are now identical !
Remark

Let us note an another drawback of the generation of rules based on a hierarchy of attributes, in both methods.

By examination of the information system (see annex 1), we establish that

$$\{x \in U | f(x,3) = 1\} \equiv \{x \in U | f(x,4) = 1\}$$

So the rule 1 and 1'

$$C_3 = 1 \rightarrow d = 1$$

is equivalent to the rule

$$C_4 = 1 \rightarrow d = 1$$

Nevertheless, this rule is not directly generated by the two methods

To conclude, let us note that it seems difficult to compare both methods from a theoretical point of view: one is based on some statistical assumption, the other on simple logical rules.

Nevertheless, this experimental comparison proves that rough sets theory may be considered like a valid competitive tool to analyse such information system.

3. Analysis of a real case study by both methods

We have carried on the comparison between Elysee and rough sets method, applying both methods on a real application [1] . In this paper, we just emphasize the major lines of the analysis without going in the details.

3.1. The problem concerns a workshop in a large printing company. During the production of a job, a variable quantity of paper is lost. There are two types of lost:

- starting wastes are generated when the job is initialized and the machines are adjusted; sometimes several initializations are necessary during a same job
- working wastes appear also during the process of the job itself.

These wastes are due to several factors difficult to precise. So, it is not easy for the manager of the workshop to predict the volume of wastes for a specific job; for this reason, he forecasts the volume of wastes by over-estimating the production in regard with the exact quantity needed. Sometimes this over-valuation is enough and a short excess quantity is produced; sometimes the real production is inferior to the required quantity and this is of course a more embarrasing situation.

The objective of the analysis is to try to determine the characteristics of a job which have a large influence on the lost of paper.

The information system (see [1]) contains 253 objects (jobs) and 8 conditional attributes (characteristics of a job). No interesting results have been obtained from the initial information system given by the company, neither by the Elysee method, nor by the rough sets method.

Several reasons have been emphasized:

- the domain of some conditional attributes have too much possible values, introducing an important lack of balance between attributes
- Similarly, the number of equivalence classes of the decision attribute is too large
- It appears difficult to treat simultaneously the two types of wastes, because the characteristics of a job explaining both types are possibly different. So a regression analysis has been made to estimate separately the starting and the working wastes.

We only present here the analysis for the first type of wastes, using a modified information system.

The eight conditional attributes are the following, with in brackets, the number of possible values in the definition domain of the attribute:

C_1 : type of rotary (2)
C_2 : type of paper (4)
C_3 : quality of paper (3)
C_4 : breadth of the paper (5)
C_5 : weighth of the paper (5)
C_6 : type of job (4)
C_7 : number of initializations (6)
C_8 : printing importance.

3.2. By the rough sets approach, several experiments have been realized; the results are given in table 3 and 4.

- First, different sets of norms of the decision attribute have been tested (see columns A till H of table 3).
- The core of the system is formed of the eight attributes: the elimination of any attribute decreases the quality of the approximation (see table 3)
- Different subsets of attributes have been compared (see table 4)
  - from the point of view of the quality of the approximation, globally for the three (or four in cases D,E,G) classes D_r and individually for each one
  - from the number of decision rules generated

From these tests, finally,

- the set H of norms has been chosen :

$$\gamma_C(D^*) = 0.82$$

Norm	A	B	C	D	E	F	G	H
Quality	0.71	0.75	0.80	0.68	0.71	0.78	0.72	0.82
Elimination of one attribute	Quality of approximation							
1	0.68	0.72	0.77	0.65	0.68	0.75	0.68	0.78
2	0.52	0.54	0.59	0.54	0.57	0.63	0.61	0.72
3	0.64	0.69	0.74	0.62	0.63	0.73	0.65	0.80
4	0.63	0.66	0.73	0.60	0.60	0.71	0.63	0.76
5	0.55	0.66	0.70	0.64	0.67	0.73	0.66	0.76
6	0.69	0.69	0.76	0.67	0.69	0.78	0.70	0.80
7	0.42	0.43	0.50	0.45	0.46	0.55	0.54	0.59
8	0.42	0.45	0.52	0.44	0.44	0.54	0.52	0.60
Subset of attributes	Quality of approximation							
{2,7,8}	0.37	0.37	0.37	0.43(3)	0.43(5)	0.44(7)	0.45(9)	0.51(11)
{2,3,7,8}	0.45	0.47	0.51(1)	0.49(4)	0.49(6)	0.55(8)	0.50(10)	0.58(12)
{2,4,7,8}	0.43	0.45	0.46(2)	0.49	0.49	0.55	0.49	0.58(13)
{3,7,8}	0.13	0.13	0.19	-	-	0.19	-	0.34

Table 3: Analysis of the quality of approximation

Norm	Subset of attribute	Quality of approximation of each class				Number of rules		
		class 1	class 2	class 3	class 4	non deterministic	deterministic	total
C	(1)	0.32	0.33	0.26	-	21	40	61
	(2)	0.28	0.32	0.19	-	20	40	60
D	(3)	0.36	0.17	0.11	0.00	21	26	47
	(4)	0.47	0.29	0.04	0.04	23	38	61
E	(5)	0.38	0.15	0.02	0.00	21	25	46
	(6)	0.47	0.24	0.05	0.04	23	36	59
F	(7)	0.38	0.15	0.04	-	22	25	47
	(8)	0.47	0.25	0.16	-	21	37	58
G	(9)	0.45	0.02	0.01	0.00	23	22	45
	(10)	0.52	0.06	0.04	0.04	19	39	58
H	(11)	0.47	0.12	0.13	-	15	28	43
	(12)	0.54	0.22	0.27	-	15	36	51
	(13)	0.52	0.18	0.30	-	16	41	57

Table 4: Analysis of the number of rules

- the subset $P = \{2,7,8\}$ of attributes has been selected :

$$\gamma_P(D^*) = 0.51$$

$$\alpha_P(D_1) = 0.47 \; ; \; \alpha_P(D_2) = 0.12 \; ; \; \alpha_P(D_3) = 0.13.$$

The decision algorithm [5] furnishes 28 deterministic and 15 non deterministic rules. There are given in annex 2.

It is not easy to analyse so much rules. So, different ways to visualize these rules have been proposed (see [1]). We just mention here the most simple one, based on the two more discriminant attributes C_8 and C_2. Let us note that some representations with three attributes are proposed in [1] .

Table 5 indicates which classes of D^* (1,2 or 3) are represented by objects taking the different possible values of the two attributes C_8 and C_2: the squares with only one number represent deterministic rules, thoses with several numbers represent non deterministic rules. For instance

$$C_8 = 1(\text{ for any value of} C_2) \rightarrow d = 1$$

$$C_8 = 3, C_2 = 4 \rightarrow d = 1 \text{ or } 2 \text{ or } 3$$

	$C_2 = 1$	$C_2 = 2$	$C_2 = 3$	$C_2 = 4$
$C_8 = 1$	1	1	1	1
$C_8 = 2$	1	1-2	1-2	2
$C_8 = 3$	1	1-2	1-2	1-2-3
$C_8 = 4$	1	1-2-3		1-2-3
$C_8 = 5$		2-3		2-3

Table 5. Vizualisation of the rules based on $P = \{8,2\}$

A clear tendancy appears from the north-west corner till the south-east corner of the rectangle.

If we note $C_8 = i$ and $C_2 = j$, objects of class 1 is essentially present for short values of $i+j$ (less value, more homogeneous squares) and objects of class 3 are only represent in squares with large values of $i + j$. Class 2 appears as an intermediary case.

3.3. The Elysee method has also be applied on this example. The resulting tree is presented in annex 3 ; a node has been considered as terminal as soon as a further separation provides three subsets with less than 10 objects.

No evident statements are derived from this method.

Only two terminal nodes are homogeneous and correspond to deterministic rules :
$C_2 = 1, C_1 = 1 \rightarrow d = 1$
$C_2 = 1, C_1 = 2, C_8 = 1 \text{ or } 2 \rightarrow d = 1$

The tendancy induced by the rough set approach is not provided by this statistical method.

4. Conclusions

From the experiments already made, we are able to conclude that rough sets theory is a good alternative to some ordinal statistical methods.

Section 2 shows that for an easy application - presented as a test problem for ordinal discriminant analysis - rough sets theory furnishs very similar results. Moreover, from the analysis, made in section 3, of a real case study - difficult because no very clear classification can be extracted of the information system -, it appears that a rough set approach may give better results that, for instance, the Elysee method.

The three main advantages of rough sets theory are

- its very clear interpretation for the user
- its independance to any statistical assumptions
- its efficiency and its rapidity

In our opinion, the present drawback of this new approach are (see section 1)

- the non integration of the degree of impurity of the doubtful regions, in the comparison of subsets of attributes
- the use of a hierarchy of attributes to define the decision rules for each class.

Nevertheless, it is important to emphasize that, in our opinion, the main role of the rough sets approach is to analyse a known information system and to explain a given classification, but not really to predict a further classification of new objects, at least in the statistical sense.

Finally, we think that further research must be oriented

- to a more theoretical - and not only experimental - comparison between rough sets methods and statistical methods
- to improve the tool of rough sets theory, essentially to correct the drawbacks defined above.

References

[1] Benjelloun M., Problème de production dans une imprimerie, Travail de fin d'études, Faculté Polytechnique de Mons (1989).

[2] Ellard C. et al., Programme Elysee : présentation et application, *Revue Metra*, (1967) 503–520.

[3] Lebart L. et al., Techniques de la description statistique; méthodes et logiciels pour l'analyse des grands tableaux, *Dunod* (1977).

[4] Pawlak Z., Rough sets, *Kluwer Academic Publishers, Theory and Decision library*, Series D (1991).

[5] Słowiński R.; Stefanowski J., RoughDas and RoughClass software implementations of the rough sets approach, Chapter III.8 in this volume (1992).

[6] Wong et al., Comparison of rough sets and statistical methods in inductive learning, *Int. J. Man-Machines studies*, 24 (1986) 53–72.

Annex 1. Information system of the application

Objects	C_1	C_2	C_3	C_4	D_r	Objects	C_1	C_2	C_3	C_4	D_r
1	2	3	1	1	1	26	2	2	1	1	1
2	1	2	1	1	1	27	2	3	1	1	1
3	1	3	1	1	1	28	2	3	1	1	1
4	1	3	1	1	1	29	2	3	1	1	1
5	2	3	1	1	1	30	1	3	1	1	1
6	2	4	1	1	1	31	1	3	1	1	1
7	1	3	1	1	1	32	2	3	1	1	1
8	2	3	1	1	1	33	2	4	1	1	1
9	1	2	1	1	1	34	2	4	1	1	1
10	1	3	1	1	1	35	1	3	1	1	1
11	2	3	1	1	1	36	2	3	1	1	1
12	1	3	1	1	1	37	2	3	1	1	1
13	1	2	1	1	1	38	1	3	1	1	1
14	1	2	1	1	1	39	1	2	1	1	1
15	2	4	1	1	1	40	2	3	1	1	1
16	2	4	1	1	1	41	2	3	1	1	1
17	2	4	1	1	1	42	1	1	1	1	1
18	2	3	1	1	1	43	1	3	1	1	1
19	2	4	1	1	1	44	2	3	1	1	1
20	2	4	1	1	1	45	2	4	1	1	1
21	2	3	1	1	1	46	1	2	1	1	1
22	2	3	1	1	1	47	2	4	1	1	1
23	1	3	1	1	1	48	1	3	1	1	1
24	2	3	1	1	1	49	2	3	1	1	1
25	1	3	1	1	1	50	2	3	1	1	1

Objects	C_1	C_2	C_3	C_4	D_r	Objects	C_1	C_2	C_3	C_4	D_r
51	4	3	3	3	2	76	3	2	3	3	2
52	3	3	3	3	2	77	3	2	3	3	2
53	3	3	3	3	2	78	3	2	3	3	2
54	2	1	3	2	2	79	3	2	3	3	2
55	3	2	3	3	2	80	2	2	2	2	2
56	2	2	3	2	2	81	2	2	2	2	2
57	3	3	3	3	2	82	2	2	2	2	2
58	1	2	2	2	2	83	2	2	2	2	2
59	3	2	3	2	2	84	3	2	3	3	2
60	2	2	2	3	2	85	2	2	3	3	2
61	2	1	2	2	2	86	3	3	3	3	2
62	2	2	3	3	2	87	3	3	3	3	2
63	3	1	3	2	2	88	3	1	3	2	2
64	3	2	3	3	2	89	2	2	3	2	2
65	2	2	2	2	2	90	2	2	3	2	2
66	3	3	3	3	2	91	2	2	3	2	2
67	2	2	3	3	2	92	3	2	3	3	2
68	2	2	3	2	2	93	2	2	3	2	2
69	3	1	3	3	2	94	2	1	2	2	2
70	2	2	2	2	2	95	2	2	3	2	2
71	2	3	3	3	2	96	2	2	3	2	2
72	3	2	3	2	2	97	2	2	3	2	2
73	3	2	3	3	2	98	3	2	3	2	2
74	3	2	3	2	2	99	2	2	2	2	2
75	3	2	3	2	2	100	2	2	3	2	2

Objects	C_1	C_2	C_3	C_4	D_r	Objects	C_1	C_2	C_3	C_4	D_r
101	3	3	4	4	3	126	4	3	4	3	3
102	2	2	3	3	3	127	3	2	3	3	3
103	4	2	4	4	3	128	3	2	3	3	3
104	3	2	4	3	3	129	3	2	4	4	3
105	3	2	4	4	3	130	4	2	4	3	3
106	4	2	4	4	3	131	4	2	4	3	3
107	1	2	3	3	3	132	4	4	4	3	3
108	4	2	4	3	3	133	3	2	4	4	3
109	3	2	4	3	3	134	3	2	3	3	3
110	4	3	4	4	3	135	3	2	4	3	3
111	3	3	3	3	3	136	4	2	4	4	3
112	3	2	3	3	3	137	3	3	4	4	3
113	3	2	4	4	3	138	3	3	4	3	3
114	2	2	3	3	3	139	3	2	3	3	3
115	2	2	3	4	3	140	3	3	3	4	3
116	3	3	3	4	3	141	3	3	4	4	3
117	3	2	4	3	3	142	3	3	3	4	3
118	4	4	4	4	3	143	2	2	3	3	3
119	4	2	4	4	3	144	3	3	4	4	3
120	3	1	3	3	3	145	3	3	4	4	3
121	3	3	4	4	3	146	3	2	3	4	3
122	2	2	3	3	3	147	3	2	3	3	3
123	4	2	4	3	3	148	3	2	3	3	3
124	3	2	3	3	3	149	3	3	3	4	3
125	3	3	4	4	3	150	2	2	3	3	3

Annex 2. Decision tree obtained by the Elysee method for the case study

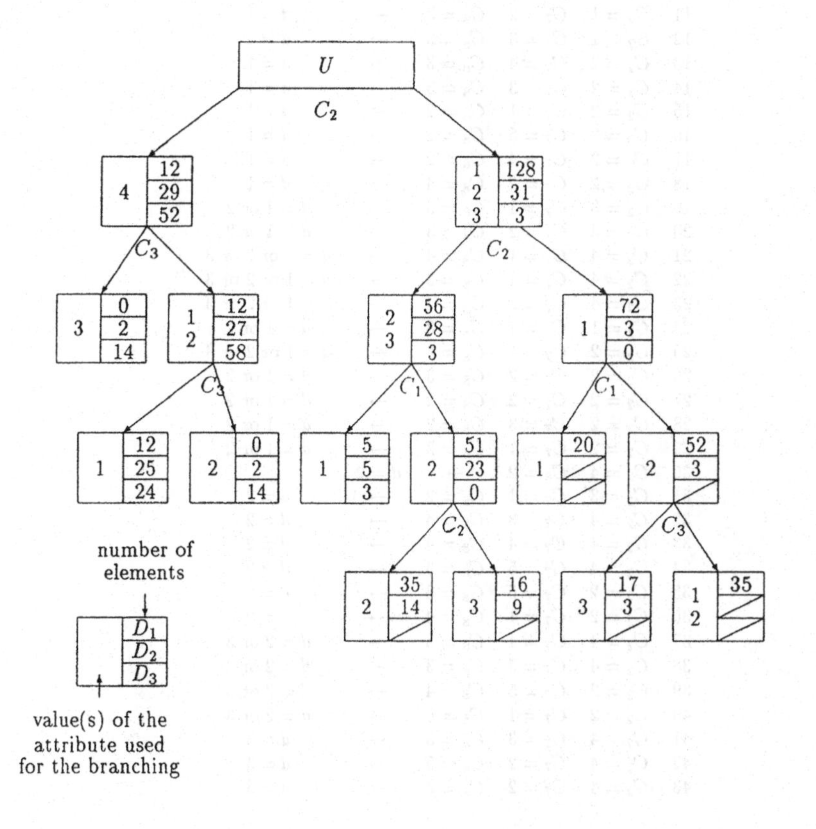

Annex 3. Rules obtained for the case study with the subset $\{2,7,8\}$

1	$C_8=1$	$\rightarrow$	$d=1$		
2	$C_2=1$	$C_8=4$	$\rightarrow$	$d=1$	
3	$C_2=1$	$C_8=2$	$\rightarrow$	$d=1$	
4	$C_2=3$	$C_7=4$	$C_8=2$	$\rightarrow$	$d=1$
5	$C_2=3$	$C_7=5$	$C_8=2$	$\rightarrow$	$d=1$
6	$C_2=3$	$C_7=2$	$C_8=2$	$\rightarrow$	$d=1$
7	$C_2=3$	$C_7=3$	$C_8=3$	$\rightarrow$	$d=1$
8	$C_2=4$	$C_7=4$	$C_8=3$	$\rightarrow$	$d=1$
9	$C_2=1$	$C_7=4$	$C_8=3$	$\rightarrow$	$d=1$
10	$C_2=1$	$C_7=5$	$C_8=3$	$\rightarrow$	$d=1$
11	$C_2=1$	$C_7=2$	$C_8=3$	$\rightarrow$	$d=1$
12	$C_2=1$	$C_7=3$	$C_8=3$	$\rightarrow$	$d=1$
13	$C_2=2$	$C_7=4$	$C_8=3$	$\rightarrow$	$d=1$
14	$C_2=2$	$C_7=3$	$C_8=3$	$\rightarrow$	$d=1$
15	$C_2=2$	$C_7=1$	$C_8=2$	$\rightarrow$	$d=1$
16	$C_2=2$	$C_7=5$	$C_8=2$	$\rightarrow$	$d=1$
17	$C_2=2$	$C_7=4$	$C_8=2$	$\rightarrow$	$d=1$
18	$C_2=2$	$C_7=2$	$C_8=4$	$\rightarrow$	$d=1$
19	$C_2=3$	$C_7=1$	$C_8=3$	$\rightarrow$	$d=1$ or 2
20	$C_2=3$	$C_7=2$	$C_8=3$	$\rightarrow$	$d=1$ or 2
21	$C_2=4$	$C_7=1$	$C_8=4$	$\rightarrow$	$d=1$ or 2 or 3
22	$C_2=4$	$C_7=1$	$C_8=3$	$\rightarrow$	$d=1$ or 2 or 3
23	$C_2=4$	$C_7=2$	$C_8=4$	$\rightarrow$	$d=1$ or 2 or 3
24	$C_2=1$	$C_7=1$	$C_8=3$	$\rightarrow$	$d=1$ or 2
25	$C_2=2$	$C_7=4$	$C_8=4$	$\rightarrow$	$d=1$ or 2 or 3
26	$C_2=2$	$C_7=2$	$C_8=3$	$\rightarrow$	$d=1$ or 2
27	$C_2=2$	$C_7=2$	$C_8=2$	$\rightarrow$	$d=1$ or 2
28	$C_2=2$	$C_7=3$	$C_8=2$	$\rightarrow$	$d=1$ or 2
29	$C_2=2$	$C_7=1$	$C_8=3$	$\rightarrow$	$d=1$ or 2
30	$C_2=4$	$C_8=2$	$\rightarrow$	$d=2$	
31	$C_2=3$	$C_7=1$	$C_8=2$	$\rightarrow$	$d=2$
32	$C_2=4$	$C_7=3$	$C_8=4$	$\rightarrow$	$d=2$
33	$C_2=4$	$C_7=4$	$C_8=4$	$\rightarrow$	$d=2$
34	$C_2=4$	$C_7=5$	$C_8=3$	$\rightarrow$	$d=2$
35	$C_2=2$	$C_7=5$	$C_8=3$	$\rightarrow$	$d=2$
36	$C_2=2$	$C_7=1$	$C_8=4$	$\rightarrow$	$d=2$
37	$C_2=4$	$C_7=1$	$C_8=5$	$\rightarrow$	$d=2$ or 3
38	$C_2=4$	$C_7=3$	$C_8=3$	$\rightarrow$	$d=2$ or 3
39	$C_2=2$	$C_7=5$	$C_8=4$	$\rightarrow$	$d=2$ or 3
40	$C_2=2$	$C_7=1$	$C_8=5$	$\rightarrow$	$d=2$ or 3
41	$C_2=4$	$C_7=3$	$C_8=5$	$\rightarrow$	$d=3$
42	$C_2=4$	$C_7=2$	$C_8=5$	$\rightarrow$	$d=3$
43	$C_2=4$	$C_7=2$	$C_8=3$	$\rightarrow$	$d=3$

Part II
Chapter 5

TOPOLOGICAL AND FUZZY ROUGH SETS

Tsau Y. LIN
Department of Mathematics and Computer Science
San Jose State University
San Jose, California 95192, U.S.A.

Abstract. The approximation theory is studied via rough sets, fuzzy sets and topological spaces (more precisely, Frechet spaces). Rough set theory is a set theory via knowledge bases. This set theory is extended to fuzzy sets and Frechet topological spaces. By these results one can show that the classification preserves the approximation. We also showed that within the approximation theory, fuzzy set and Frechet topology are intrinsically equivalent notions. Finally, we show that even though approximation is a compromised solution, the three theories allow one to draw an exact solution whenever there are adequate approximations. This implies that these three approaches are good approximation theories.

1. Introduction

Given a universe U of discourse, according to Pawlak, a subset X is called a concept and a family of concepts is referred to as knowledge about U. Further, a family of classifications (partitions of equivalence relations) over U is called a knowledge base [1](pp. 14). Pawlak's rough set theory can be viewed as a study of concepts or knowledge via knowledge bases. In this paper, we extend the study to fuzzy concepts (fuzzy subsets) and "continuous" concepts (neighborhood systems [2]).

Rough set theory intrinsically is a study of an equivalence relation, an approximation of equality. Naturally, the rough set theory is most useful in classification or learning by induction. On the other hand, Pawlak also observed that an equivalence relation induces a partition and hence generates a topology for U (called Pawlak topology). Topology is a notion for studying limit and approximation. Pawlak's upper and lower approximation is the closure and the interior of a set. We take Pawlak's two views (classification and approximation) into one frame work. Keeping the classification, we generalize his approximation from sets into fuzzy sets and neighborhood systems (Frechet topology) [3,2]. We can characterize this paper as a rough fuzzy sets/neighborhood systems theory.

By an approximation in U, we mean that an approximation is assigned to each concept (subset):

1. In rough set theory, the approximation is the upper and lower approximation (which is closure and interior in Pawlak topology) of the concept.
2. In topological theory, the approximation is the closure and the interior of the concept.

3. In fuzzy set theory, we interpret a fuzzy concept as an approximation to its real set (the subset in which the value of membership function is one).

Then, a natural question is whether we can import the notion of approximation into the knowledge base, or more precisely, the quotient set U/R? For case (1), this is not very meaningful, because the approximation in U/R is exact. For case (2) and (3) We found positive answers. They can be summarized as

the classification (or learning) preserves the approximation.

Approximation is a compromised answer. Any good approximation theory, should show that an adequate or successive efforts on the approximation can lead to the exact answer. For rough, fuzzy, and topological approximation, we give a criteria when the approximation will become an exact answer. We also show that fuzzy set theory is intrinsically equivalent to topological theory from the perspective of approximation (fuzzy sets have other perspectives, such as probability or possibility).

2. Rough Sets

2.1. Equivalence Relations and Quotient Sets

A binary relation is an equivalence relation iff it is reflexive, symmetric and transitive. For every equivalence relation there is a partition and vice versa [4](pp. 184). Let R be a given equivalence relation over U, there are several notions associated with R:

U: universe of discourse
R: equivalence relation
$P(R,U)$: the partition of equivalence relation over U, or $P(U)$, $P(R)$, P, if R and U is clearly understood.
U/R: the quotient set of equivalence classes.
$[u]$: the equivalence class containing u which is a subset of U and is an element of U/R;
$name([u])$: to emphasize the fact that $[u]$ is treated as element in U/R; often we just use $[u]$ to mean $name([u])$, when the context is clear.
NQ: natural projection (quotient) of U to U/R
$INV.NQ$: inverse image of $NQ(INV.NQ(q)) = [u]$, where $q = name([u])$

The family of all equivalence classes is a set, it is called quotient set and denoted by U/R. There is a natural projection from U to U/R.

$$NQ : U \longrightarrow U/R.$$

defined by $NQ(u) = [u]$ (read as natural quotient), where $[u]$ is the equivalence class containing u. We should note here that $[u]$ has dual roles; it is an element, not a subset, of U/R, but it is a subset of U. In [3], elements in U/R are called names of equivalence classes.

Let us denote the complete inverse image of NQ by

$$INV.NQ(q) = \{u : NQ(u) = q\} = [u]$$

or more generally, for a subset X of U/R

$$INV.NQ(X) = \{u : NQ(u) \text{is in} X\}$$

Note that $INV.NQ(q)$ is an equivalence class and $INV.NQ(X)$ is a union of equivalence classes. It is a R-definable set. There is a nice correspondence between the partition $P(R, U)$ and the quotient set U/R. Every equivalence class of $P(R, U)$ corresponds to an element in U/R. A subfamily of equivalence classes corresponds to a subset of U/R.

Example 1. Let Z be integers. Let R denote the equivalence relation called congruence *modm*. That is,

$$x\ R\ y \text{ if } x - y \text{ is divisible by } m.$$

Let $m = 4$. Then the equivalence classes are

[0] = {... -8, -4, 0, 4, 8...}
[1] = {... -7, -3, 1, 5, 9,..}
[2] = {... -6, -2, 2, 6, 10,.}
[3] = {... -5, -1, 3, 7, 11,.}

In other words, [0], [1], [2], [3] is a partition for the integers Z. The quotient set of this equivalence relation is denoted by Z_m. Z_4 = {[0], [1], [2], [3]}.

Congruence mod m is one of the few early examples of equivalence relations. It has been used as a generalized equality in number theory.

Example 2. Let U be real numbers. Let R be the equivalence relation called congruence mod 1. That is,

$$x\ R\ y \text{ if } x - y \text{ is an integer.}$$

Following are some of the equivalence classes.

[0] = {. . .-2, -1, 0, 1,2 . . .}
[0.5] = {. . .-1.5, -0.5, 0.5, 1,1.5,..}

In general,

$$[r] = \ldots r-2, r-1, r, r+1, r+2, \ldots, \text{ where } 0 \leq r < 1.$$

The family of all equivalence classes is the quotient set U/R. It has a nice geometric representation, namely, the Unit Circle. Each point of the *UnitCircle* represents an equivalence class. One can also represent the *UnitCircle* by complex numbers

$$U/R = \{exp(ir) : 0 \leq r < 1\}$$

This equivalence relation is called congruence mod 1. U/R is a very useful space in approximation. Fourier series, signal processing among others live in this space. The topology in the *UnitCircle* is induced from the topology of real numbers. Such topology is not Pawlak topology.

2.2. Rough Sets

Let U be the universe of discourse. Let $RCol$ be a finite **Col**lection of equivalence **R**elations R over U. In general we will use Pawlak's terminology and notations . An ordered pair

$$K = (U, RCol)$$

is called a **knowledge base over U**. (In most cases, there is only one equivalence relation R in $RCol$, so $K = (U, R)$). A subset X of U is called a concept. For an equivalence relation R, an equivalence class is called R-elementary concept, R- elementary set, R-basic category or R-elementary knowledge (about U in K). The empty set is assumed to be elementary. A set which is a union of elementary sets is called R-definable or R-exact. A finite union is called composed set in U. The set of equivalence classes is the quotient set U/R. There is a neat correspondence between the elementary sets of U and the quotient set U/R. Each elementary set in U corresponds to an element in U/R. Comp(U) is equivalent to the "finite power set" of U/R (The set of all finite subsets in U/R). An R- definable set corresponds to a subset of U/R and vice versa.

Let $SCol$ be a nonempty **SubCol**lection of $RCol$. The intersection of all equivalence relations in $SCol$, denoted by $IND(SCol)$, is an equivalence relation and will be called an indiscernibility relation over $SCol$. The quotient set $U/IND(SCol)$ will be abbreviated by $U/SCol$. Equivalence classes of $IND(SCol)$ are called basic categories (concepts) of knowledge K. A concept X is exact in the knowledge base K if there exists an equivalence relation R in $IND(K)$ such that X is R-exact, where $IND(K)$ is the collection of all possible equivalence relations in K, that is,

$$IND(K) = \{IND(SCol) : \text{for all } SCol\text{'s in } RCol\}.$$

For each X, we associate two subsets, upper and lower approximation:

$$\begin{aligned} L\_APP(X) &= \{u : [u] \text{ is a subset of } X\} \\ U\_APP(X) &= \{u : [u] \text{ and } X \text{ has non-empty intersection}\} \end{aligned}$$

A subset X of U is definable iff $U\_APP(X) = L\_APP(X)$. The lower approximation of X in U is the greatest definable set in U contained in X. The upper approximation of X in U is the least definable set in U containing X.

As Pawlak pointed out that the equivalence classes form a topology for U (it will be called Pawlak topology). So we can rephrase the upper and lower approximations as follows:

$$\begin{aligned} L\_APP(X) &= \text{Interior point of } X \\ &= \text{The largest open set contained in } X \\ U\_APP(X) &= \text{Closure of } X \\ &= \text{The smallest closed set containing X.} \end{aligned}$$

Rough set theory serves two functions: one is a generalization of the equality which leads to classification, the other is the approximation in Pawlak topology. We will keep its classification, but to generalize its topology to neighborhood systems (Frechet topological spaces) and Fuzzy set theory. The topology of the *UnitCircle* in Example 2 above can not be a Pawlak topology of any equivalence relation. Hence a generalization is necessary.

3. Topological Rough Sets

Intuitively neighborhood systems handle the notion of "close to", "analogous to", or "approximate to". Such notion usually is **not** a transitive relation. For example, East LA is "close to" LA, LA is "close to" West LA. However, East LA is not considered to be "close to" West LA. Thus "approximate to" is not an equivalence relation. Therefore Pawlak topology can not handle such general approximations. If the given neighborhood system is pairwise disjoint, then the topology is Pawlak topology (we called it category; see [5,6]). We use neighborhood systems to do such approximation. Rough set theory is very useful in classification (learning). However, some extra structures may be useful in some applications. In [7], Pawlak, Wong and Ziarko introduced the probabilistic framework into the rough set theory. In [3], Dubois introduced fuzzy theory into the rough set theory. In this section, we formalize neighborhood systems into the framework of rough set theory [6, 8, 9, 10, 11, 12]. Topological rough set theory is most suitable to formulate the phenomena of learning rules from information systems that contains the semantics of approximation.

3.1. Frechet Spaces and Topological Spaces

Let U be the universe of discourse and p be an object (element) in U. A neighborhood N of p is a subset of U, which may or may not contain the point p itself. The same notion can be extended to a subset of U. Let p be a subset. A neighborhood $N(P)$ of P is a subset of U which may or may not contain P. However, in all our applications, the neighborhood **does contain** p or P. A neighborhood system of p is a family of neighborhood(s) of p, which satisfies NO further axioms.

Definition 3.1. Each object p in U possesses a neighborhood system is called a Frechet topological space, or briefly an F-space. The totality of the neighborhood system of each point is called a neighborhood system of F-space U.

Note that a neighborhood system in a classical topological space (called T-space) satisfies certain axioms. However, in Frechet space, there is no such requirements. So F-space is more general than topological space. In [2], Sierpiński gave a very concise exposition on the theory of Frechet space; we urge readers to have some glances at it. This author believes that in computer science the notion of Frechet space is more useful than topological space. The extra axioms satisfied by topological spaces are very often irrelevant to "finite" applications in computer science.

Let us examine some examples to illustrate the notion of neighborhood systems.

Example 1. The set of points in the plane is a Frechet space if a neighborhood of a point p is taken to be the interior of an arbitrary circle with center at p. A simplest kind of F-space can be defined by assigning each point of the plane only one such neighborhood.

Example 2. The set of all real valued functions is a F-space, if a neighborhood of $f(x)$ is defined to be the set of all functions $g(x)$ such that for any given positive real number c

$$| f(x) - g(x) | < c \text{ for all } x.$$

Example 3. A topological space U is a set of points, in which every point is assigned a neighborhood system that satisfies certain axioms [13]. So, a classical space is a Frechet space.

Example 4. A metric space U is a set of points (elements), in which there is a real valued function $m(p, q)$ defined on all the pair (p, q) of points such that the function $m(p, q)$ satisfy certain axioms [13]. Intuitively, $m(p1, p2)$ is a real number that is equal to the distance of the points $p1$ and $p2$. A metric space is a classical space, hence is a Frechet space.

A function f from X to Y is said to be continuous at x, if for every neighborhood V of $f(x)$, there exists a neighborhood N of x such that f map N into V. Two F-spaces X and Y is said to be F-homeomorphic if f and its inverse are both continuous. Two neighborhood systems is said to define the same F-space if the identity map is continuous in both directions. Let N_r and M_r be two neighborhood systems for the same point p (or a subset P). Further, for each r there is s such that N_s is a subset of M_r and vice versa, then the two neighborhood systems are said to be equivalent. It follows immediately that two neighborhood systems is equivalent at point p if the identity map is continuous at p in both directions.

Proposition 3.2. Every F-space has a maximal neighborhood system.

Proof: Given a F-space X, we will consider a new F-space NX by considering a new neighborhood system as follows: Let p be a point in X, then any subset which contains a neighborhood of X is a new neighborhood of p in NX. Obviously NX is an F-space. Note the identity map from NX to X is a continuous map, because every neighborhood of X is a neighborhood of NX. Conversely, the identity map from X to NX is continuous. Because every neighborhood V of NX, there is a neighborhood of X which is a subset of V (be definition). This implies that the identity map X to NX is coninuous. So this new neighborhood system define the same F-space as X. This new neighborhood system is the maximal neighborhood system of X.

Proposition 3.3. Every Frechet space generates a classical topological space.

Proof: In classical topological space, the family of all open sets is called topology which determines the space. The open set of Frechet space will be denoted by F-open set. Let FOPEN be the family of F-open sets; FOPEN is not closed under intersection, so by itself can not determine a classical topology. However, we can let FOPEN be a subbase (its finite intersections form a base). Then the resulting topology gives U a classical topological space.

3.2. Approximation by Neighborhoods

The most successful application of neighborhood systems is in numerical analysis. Let us reexamine its procedure more closely from our perspective. Before the approximation begins, a radius of tolerance (error neighborhood or error radius) is usually chosen, in other word, for each real number a neighborhood of tolerance is chosen. Throughout the process of finding an approximate solution, the neighborhood never changes; the only relevant notion of the real line topology is the chosen neighborhood. So a space with such chosen neighborhoods most likely be the correct abstraction to model the intuitive notion of approximation. Such neighborhood could be called an **instance** of the topology; such instance is a Frechet space. So for a given neighborhood system, there are a F-space and a T-space that is constructed in Proposition 3.3. Following Pawlak topology, we define the two approximations by the neighborhood system.

$$
\begin{aligned}
L\_APP(X) &= \text{Interior of } X \text{ (denoted by } In(X)) \\
&= \text{The largest open set contained in } X \\
U\_APP(X) &= \text{Closure of } X \text{ (denoted by } Cl(X)) \\
&= \text{The smallest closed set containing X.}
\end{aligned}
$$

3.3. Topological Rough Set

We have shown earlier, in many information systems, neighborhood systems naturally present in their attribute domains [5, 6, 8, 9, 10, 11, 12]. Here we will simply assume that such a Frechet topology is given in U. Our main theme here is that a **cluster of approximate information will be classified into a cluster of approximate equivalence classes.** For example, a group of patients who have approximately the same symptom, even though they may not be classified into the same medical rule (classification), the resulting medical rules (classifications) must remain close (approximate) to each other.

Let X be an object of U, we will use $N(x)$ to denote a typical neighborhood of x. Let Y be a subset of Frechet space U. A set $N(Y)$ is a neighborhood of Y iff it is the union of $N(y)$'s for all y in Y, where $N(y)$ is a neighborhood of y.

Proposition 3.4. Let U be a Frechet space and R be an equivalence relation. Then there is a natural Frechet topology on the quotient set U/R.

Proof: We will use the Maximal Neighborhood System of U (see 3.2). Let $NQ : U \rightarrow U/R$ be the natural projection. That is, $NQ(u) = [u]$. Let y be a point in U/R. Then the inverse image of y i.e., $X = INV.NQ(y)$ is the union of elementary sets which are mapped into y by the natural projection NQ. Let $N(X)$ be a R-definable neighborhood of X in U, then $NQ(N(X))$ is a neighborhood of y in U/R. For each y in U/R, we define all possible $NQ(N(X))$ as the neighborhood system for y. So U/R is a Frechet space and is called the quotient F-space, or simple quotient space.

Let y be in a neighborhood of x. Assume that the equivalence relation R classifies x and y into two different equivalence classes $[x]$ and $[y]$. Using the quotient F-topology, the two classes $[x]$ and $[y]$ are still close to each other, in fact, $[y]$ is in a neighborhood of $[x]$. In other words, **classification(learning) preserves the approximation.** This is useful in applications.

Example. Let U be a database of patient records in which diseases have been properly identified by its symptoms. In other words, medical rules (knowledge base) have been extracted from this database. Now, suppose a new patient comes and the medial rules can not be applied. Suppose his symptoms is found to be closed to some classified disease, say x. A physician then may apply the approximate medical rule $[x]$ to treat this new patient; this is a healthy way of applying an approximated rule.

Let U be a Frechet space and R an equivalence relation. Then we can give U a new F-topology from the quotient F-space U/R. In this new F-topology each neighborhood is also a union of elementary sets. By abuse of language, we shall say that U has a quotient topology. In fact there is one to one correspondence between the neighborhoods in the quotient topology of U and quotient topology of U/R. We will use NU/R to denote this new quotient F-topology for U. In fact, NU/R is the R-exact F-topology. Its neighborhoods are R-exact neighborhoods. Thus any neighborhood can be approximated by neighborhood of this new R-exact F-topology. Let X be a neighborhood in U, then we have following approximations of X by R-exact neighborhoods.

$$L\_APP(X) = \text{Union of } R\text{-exact neighborhoods contained in } X$$
$$U\_APP(X) = \text{Union of } R\text{-exact neighborhoods contained in the complement of } X$$

If U is a classical topological space, then the family of open sets characterize the topology of U. We can approximate the topology in terms of R-exact topology. (Topology is the family of open sets). Let X be an open set in U.

$$L\_APP(X) = \text{Union of } R\text{-exact open sets contained in } X$$
$$U\_APP(X) = \text{Union of } R\text{-exact open sets contained in the complement of } X$$

Roughly, we approximate the U-topology by the U/R-topology.

4. Fuzzy Rough Sets

A concept may or may not be defined by a given knowledge base K. If not, can it be defined "approximately"? Pawlak's rough set theory answer this question in classical set theory. In this section we develop the rough set theory in fuzzy mathematics. For finite universe, our results are equivalent to [3]. However, we view fuzzy set theory as a field of functional analysis; it extends to infinite universe.

4.1. Fuzzy Sets

The theory of fuzzy sets deals with subsets where the membership function is real valued, not boolean valued. Intuitively the fuzzy subsets have no well defined boundaries in the universe of discourse.

Let U be the universe of discourse. Then a fuzzy set is an ordered pairs:

$$(U, FX)$$

where $FX : U \to [0,1]$ is a function from U to closed unit interval. If both $FX(0)$ and $FX(1)$ are nonempty, we call the fuzzy set **proper fuzzy set**. Note that FX is called a membership function. When context is clear, we will use FX both as the fuzzy set and the membership function. (Mathematically, FX is a function, fuzzy set is only an interpretation.)

A set in classical sense will be called classical set or simply set. Its membership function is a real valued function assuming only real values 0 and 1. An element x is said to be fuzzily belonged to fuzzy set FX if $FX(x) > 0$ and x is said to be absolutely not belong to fuzzy set FX if $FX(x) = 0$.

Let

$$RFX = \{x : FX(x) = 1\}$$

Then RFX is the Real set (classical set) of the fuzzy set FX. We could consider R as an operator which maps a fuzzy set to its real set.

The elements which are absolutely not in FX will be called absolute complement

$$CFX = \{x : FX(x) = 0\}$$

Then CFX is a classical set in which no elements have positive membership values. We could consider C as an operator which maps a fuzzy set to its absolute complement.

Example 1. The set of reals that are very small.

$$\text{Small} = \{x \,\|\, x \mid \textit{is}\ \text{small}\}$$

This set may be defined by a membership function: For some small positive number e

$$\begin{array}{ll} \text{Small}(x) = (e - x)/e & \text{for } \mid x \mid < e \\ \text{Small}(x) = 0 & \text{for } \mid x \mid \geq e. \end{array}$$

or by another membership function

$$\begin{array}{ll} \text{Small}(x) = e/(e + x) & \text{for } \mid x :< e, \\ \text{Small}(x) = 0 & \text{for } \mid x \mid \geq e. \end{array}$$

The point here is that the choice of the membership functions is not unique. In other words, an intuitive real world fuzzy set Small has more than one membership function to represent it.

Example 2. Let $U = \{c\}$ be a singleton. What are the possible fuzzy subsets? How many are there?

By definition, any membership function $M : U \to [0,1]$ is a fuzzy subset. Even though U is a singleton, there are infinitely many membership functions that can be defined on X. The following are some sample examples:

$$\begin{array}{l} M_1(c) = 0 \\ M_2(c) = 1/2 \\ \vdots \\ M_n(c) = 1/n \\ \vdots \end{array}$$

Each $M_n, n = 1, 2, ..$ is a different fuzzy membership functions. So, there are infinitely many different fuzzy subsets. This is quite a contrast to classical set theory. There are only two subsets in U, namely the empty set and whole U (singleton).

These examples illustrate that there are too many functional representations of one intuitive real world fuzzy set. So there is a need to define an equivalence among the representations of an intuitive real world fuzzy set (see Section 5).

4.2. Quasi Classical Sets

Let X be a classical set; in this case the membership function X assumes only two real values, 0 and 1. We would like to consider the membership function

$$c * X : U \longrightarrow [0,1]$$

defined by $(c * X)(u) = c * (X(u))$ for constant c, where $*$ is the multiplication of real numbers. Then $c * X$ is a special type of fuzzy set, we will call it quasi classical set. The meaning of such quasi classical set is that an object x in U is either not in X or the degree (possibility, probability) of its membership is c. We also would like to consider the "union" and "intersection" of quasi classical sets. Let X and Y be two classical sets. Then for two real numbers a, b in [0,1], $a * X$ and $b * Y$ are two quasi classical sets. As in fuzzy set, we can define the union and intersection by MAX and MIN:

$$(a * X \cup a * Y)(x) = MAX(a * X(x), b * Y(x))$$
$$(a * X \cap a * Y)(x) = MIN(a * X(x), b * Y(x))$$

4.3. Fuzzy Rough Sets

The central theme in this section is to express or to approximate a fuzzy set in U by fuzzy sets in the knowledge base, or more precisely, in the quotient set U/R. Let R be an equivalence relation over U. Let $FCol(U/R)$ be the **Co**llection of all Fuzzy sets over U/R. Then the natural projection induces a subfamily of fuzzy sets on U.

$$U \xrightarrow{NQ} U/R \xrightarrow{FX} [0,1]$$

$$SubFCol(U) = \{FX * NQ : FX \text{ is in } FCol(U/R)\}$$

where $*$ is the composition of functions. This subfamily $SubFCol$ is the family of all R-exact fuzzy sets. We would like to have more explicit description of this $SubFCol$ of fuzzy sets on U. We will relate our work to Dubois' [3].

We will borrow some notions from topology of function spaces. Let X be a classical set. Let $FCol(X)$ be the family of all functions from X to [0,1]. If X is infinite set, we will give $FCol(X)$ the pointwise convergence topology [13]. Let f and g be two functions in $FCol(X)$, we define

$$f \leq g \text{ iff } f(x) \leq g(x) \text{ for all } x$$

Then $FCol(X)$ is a poset, in fact, a complete lattice, that is every collection has a greatest lower bound (denoted by **inf**) and least upper bound (denoted by **sup**). If X is a finite set, these properties are obvious. We suggest readers just to proceed with the assumption that the universe is finite universe U; our result is good for infinite universe (in database, the data may grow indefinitely, so we may need infinite universe).

Let the membership function of the equivalence classes (R-elementary concepts) be

$$EC_i : U \longrightarrow [0,\ 1], i = 1, 2, \ldots n.$$

Since EC_i (i-th equivalence class) is a classical set, its membership function assumes 0 and 1 only; it may be referred to as **classical equivalence class.** Let us consider the collection of the **quasi classical equivalence classes** ($c * EC_i$) and all their unions (as fuzzy sets, they are sup in $FCol(U)$); it will be called Complete Quasi Equivalence classes CQE.

Proposition 4.1. $CQE = SubFCol$.

Proof: Let f be a function in CQE, i.e...

$$f = sup\{\ldots, c_i * EC_i, \ldots\}$$

then f induces a function on U/R defined by

$$[f](EC_i) = c_i \text{ for all } i$$

Obviously, $f = [f] * NQ$.

Conversely, let $[f]$ be a function in $FCol(U/R)$, then

$$[f](x) * NQ = [f](EC_i) \text{ for all } x \text{ in } EC_i$$

So, (write $[f](EC_i) = c_i$), we have

$$[f] * NQ = sup\{\ldots, c_i * EC_i, \ldots\}$$

This proves the converse.

If U is a finite universe, CQE can be represented by linear sum of EC_i's. Let f and g be two real valued functions, we define the sum $f + g$ as follows:

$$(f + g)(x) = f(x) + g(x) \text{ for all } x$$

Note that EC_i and EC_j are disjoint (if i not equal to j), so

$$sup\{\ldots, c_i * EC_i, \ldots\} = \ldots + c_i * EC_i + \ldots$$

This proves the assertion. In other words, FX is R-definable iff

$$FX = c_1 * Ec_1 + c_2 * EC_2 + \ldots c_n * EC_n.$$

The R-definable fuzzy set may also be called R-exact. A fuzzy set (concept) is R-undefinable iff it is not R-definable; it may also be called R-inexact.

For each fuzzy set FX, we associate two subsets, upper and lower approximation:

$$U\_APP(FX) = inf\{FY : FX \leq FY \text{ for all } FY \text{ in } CQE\}$$
$$L\_APP(FX) = sup\{FY : FX \geq FY \text{ for all } FY \text{ in } CQE\}$$

where inf and sup is in the poset of function space. This is equivalent to Dubois' [3].

Suppose x fuzzily belongs to real set X, that is, there is FX such that

$$0 < FX(x) \leq 1, \text{ and}$$
$$X = RFX$$

Does $[x]$ fuzzily belongs to $[X]$? Let NQ be the natural projection of U to its quotient set U/R. Then

$$NQ(x) = [x] \text{ and,}$$
$$NQ(X) = [X] = \text{Union of } \{[p] : p \text{ in } X = RFX\}$$

Note that (p is an element in $[X]$)

$$0 < FX(x) \leq U\_APP(FX)[x] \leq 1, \text{ and}$$
$$1 = FX(p) \leq U\_APP(FX)[p] \leq 1$$

So $U\_APP(FX)$ is the fuzzy set such that

$$0 < U\_APP(FX)[x] \leq 1, \text{ and}$$
$$R(U\_APP(FX)) = [X]$$

which says $[x]$ is fuzzily belong to $[X]$ (R is the operator which map a fuzzy set to its real set). So we conclude that **classification preserves fuzzy belonging.**

5. Fuzzy Sets and Neighborhood Systems

Fuzzy mathematics is designed to deal with "what is qualitative and fuzzy" (in contrast to the traditional "what is quantitative and precise") [14]. However, even though Zadeh's fuzzy mathematics [15] is qualitative in its applications, the representation is rather quantitative (numerical). In many applications of fuzzy set theory, very often the actual numerical value of the membership function is not important, the important point is the intuition represented by the numerical value. So we proposed a geometric approach to Zadeh's fuzzy idea. The geometry is not new, it is some form of topological spaces.
For a fuzzy set, we will give the universe U the minimal classical topology so that the membership function is a continuous function [13].

Definition 5.1. Let $\{FX\}$ be a family of fuzzy sets, there is a unique minimal classical topology on U such that all FX are continuous functions. For a given topological space U, let $C(U,[0,1])$ [or simply $C(U)$] be the set of all continuous functions on U.

Obviously, $C(U)$ is the collection of the continuous fuzzy sets.
For most applications of fuzzy sets, there is an intuitive real world fuzzy set, and the membership function is merely one of several possible representations. The Example 1, 2 in Section 3 are illustrations of such situation. Here, we propose a method to identify intrinsic equivalence among different representations (membership functions).

Let h be a homeomorphism (both h and its inverse are continuous map [13])

$$h : [0,1] \rightarrow [0,1]$$

with fixed points set A (i.e,. $h(x) = x$ for x in A). We will call these h relative self homeomorphism on the pair $([0,1], A)$, where A is fixed point set. We will be interest in A which consists of 0 and 1, 0 only, or 1 only.

Definition 5.2. Two Fuzzy membership functions KX and KY are equivalent iff both induce the same topology on U and there is a relative self homeomorphism h on the unit interval [0,1] so that $KX = KY * h$. Such equivalence class will be called a geometric fuzzy set.

For proper fuzzy sets we will be most interested in $A = \{0, 1\}$. The two representations in Example 1 are equivalent. If we choose A to be 1, then all M_n, $n = 1, 2, ..$ are equivalent.

Theorem 5.3. Every proper fuzzy set FX defines a neighborhood system for the real set RFX.

Proof: Let FX be a fuzzy set. Let us consider the real set of FX,

$$RFX = \{x : FX(x) = 1\}.$$

Let r be any real number in [0,1]. Let

$$N_r = \{x : FX(x) \geq 1 - r\}$$

Then the family $\{N_r$; r is a real$\}$ is a neighborhood system for RFX. This proves the theorem.

Corollary. Two equivalent proper fuzzy sets FX and FY can define the same neighborhood system for the common real set RFX.

Proof: First, note that two equivalent proper fuzzy sets have the same real set $RFX = RFY$. Let $FX = FY * h$, then the two neighborhoods

$$N1_r = \{x : FX(x) \geq r\}$$
$$N2_r = \{x : FY(x) \geq h(r)\}$$

are the same.
A neighborhood system N_i, $i = 1, 2, ..$ is said to be monotonic decreasing if N_{i+1} is a subset of N_i for every i.

Theorem 5.4. Let X be a proper classical subset of U with monotonic decreasing neighborhood system N_s. Then there is a fuzzy set FX such that for every N_s there is a real s in [0,1] so that $N_s = (x : FX(x) \geq 1 - s$ and $RFX = X$.

Proof: We assume the universe X is countable, and hence the neighborhood system is countable too. To simplify the exposition we will restrict our proof to countable case. For more general case, one can extend our inductive arguments to transfinite induction.

Let $N_0, N_1, N_2, \ldots$ be a monotonically "decreasing" sequence of neighborhoods for X. We will define a fuzzy subset FX by a sequence of function $f_0, f_1, \ldots, f_{(i-1)}, f_i, \ldots$ as follows:

$$f_0(x) = 1 \text{ for } x \text{ in } X$$
$$f_0(x) = 0 \text{ for } x \text{ in } U\backslash X \text{ (difference set)}$$

Assume $f_{(i-1)}$ is defined, we will define f_i as follows:

$$f_i(x) = 1 \text{ for } x \text{ in } X$$
$$f_i(x) = MAX(f_{(i-1)}, 1 - 1/i) \text{ for } x \text{ in } N_i\backslash X$$
$$f_i(x) = f_{(i-1)}(x) \text{ for } x \text{ in } U\backslash N_i$$

Let $FX(x) = SUP\{f_i(x) : i = 0,1,2,\ldots\}$. Then FX is a fuzzy set and $N_i = \{x : FX(x) \geq (1 - 1/i)\}$. This completes the proof.

6. Exact Information Via Approximations

As a user has accumulated his approximate knowledge (rough sets, fuzzy subsets, neighborhoods), will he eventually possess a sufficient knowledge to describe his universe precisely? In this section we give a positive answer to this question for the three approximate knowledge.

A person has exact information about a specific concept (classical set) if he can specify the concept precisely using his knowledge. A person has the exact information about his universe U if he can specify precisely all possible concepts in U using his knowledge. If a person can describe objects precisely, he has exact information about each object of his universe, and hence he can specify all concepts.

6.1. Exact Information via Rough Sets

Let $RCol$ be a collection of equivalence relations. Let the $IND(RCol)$ be the intersection of all the equivalent relations in $RCol$; it is called indiscernibility relation over $RCol$. When the indiscernibility relation reduces to identity, then each partition is a singleton, so we have

Proposition 6.1. If $IND(RCol)$ is the identity, then

$$X = U\_APP(X) = L\_APP(X)$$

This proposition says that if we have $ADEQUATE$ family of approximations, then the approximation is exact.

6.2. Exact Information via Fuzzy Sets

Now, we will examine the approximation in fuzzy world. A user's knowledge is represented by fuzzy sets on the universe U.

Definition 6.2. Let $F = \{FX_i : i \text{ in the set } I\}$ be a family of fuzzy subsets (knowledge of a user). F is said to be $ADEQUATE$ if the following map is one-to-one

$$Ev : U \longrightarrow HC$$

where HC, the Hilbert cube, is a Cartesian product of copies of unit interval [0,1] and Ev is defined by

$$Ev(x) = \text{Ordered tuple of } \{FX_i(x)\}.$$

If I is a finite set, say $I = \{1,2,3,4\}$, then

$$Ev(x) = (FX_1(x), FX_2(x), FX_3(x), FX_4(x))$$

If I is a countable set, say $I = \{1,2,3,\ldots\}$, then

$$Ev(x) = (FX_1(x), FX_2(x), \ldots).$$

Each x represents an object U, each $FX_i(x)$ is a fuzzy description about x in the FX_i-component. So if Ev is one-to-one, then each object x has a unique description by fuzzy data. In other words, we have an exact identification of each object through fuzzy data and hence we can uniquely identify each concept of U precisely – This is precisely the exact information about U through fuzzy data.

We need to establish easily verifiable conditions for *ADEQUACY*. A family F of fuzzy subsets distinguishes objects if and only if for each pair of distinct objects x and y there is an FX in F such that $FX(x) \neq FX(y)$.

Theorem 6.3. Let U be a universe. Let F be a collection of fuzzy subsets. Then, F is *ADEQUATE* if and only if F distinguishes objects.

Proof: If F is *ADEQUATE*, by definition, Ev is one-to-one map. That is, for every x and y, one of its coordinate, say $FX_i(x) \neq FX_i(y)$. That is F distinguishes objects. If F distinguishes objects, then for each distinct pair of objects there is one of the coordinate $FX_i(x) \neq FX_i(y)$, so $Ev(x) \neq Ev(y)$. That proves Ev is one-to-one.

Theorem 6.4. Let U be a universe. Let F be an *ADEQUATE* collection of fuzzy subsets. Then, one can introduce a metric into U, so that every two distinct objects in the universe U has a positive distance.

Proof: Recall that U and hence F are countable. Let the metric between x and y be denoted by $dist(x,y)$. Then, we can define the

$$dist(x,y) = SUM \text{ of } \{\mid FX_i(x) - X_i(y) \mid /2 * *i\}$$

where $2 * *i$ is the power of 2 (as in Fortran language). By the assumption F being *ADEQUATE*, $\mid FX_i(x) - FX_i(y) \mid$, for some i, is positive. So $dist(x,y)$ is a metric on U. (This is precisely the metric of metric space in mathematics [13]).

Example 1. John has a low quality spectral analyzer, so he can tell from his instruments that certain objects are approximately red (including red) or not. He also has a broken scale which can only tell him an objects is heavy or not. John acquires two fuzzy knowledge about his universe through his instruments, namely, the fuzzy red set FX and the fuzzy heavy set FY. Now, the question is can he tell any two objects in his universe apart by using his fuzzy knowledge FX and FY. The following is his data

$$\begin{array}{ll} FX(x) = 1 & FY(x) = 0.8 \\ FX(y) = 1 & FY(y) = 0.8 \\ FX(z) = 0.5 & FY(z) = 0.8 \\ FX(u) = 0.5 & FY(u) = 0.8 \\ FX(v) = 0 & FY(v) = 0 \\ FX(w) = 0 & FY(w) = 0 \end{array}$$

Certainly, he cannot tell x and y apart because $FX(x) = FX(y) = 1$. $FY(x)$ and $FY(y)$ both near 1. In other words, in the current state of his knowledge, there is no way to distinct x and y. So he can not describe exact information about his universe. However, as time goes on John acquires more instruments, and he has more fuzzy information on the universe U, say fuzzy properties FX, FY, FZ, FW. Let us examine the status of his knowledge.

	FX	FY	FZ	FW
x	1	0.8	0.5	0
y	1	0.8	0	0
z	0.5	0.8	0.5	0
u	0.5	0.8	0	0
v	0	0	0.5	0
w	0	0	0.5	0

John's knowledge is very good now, he can distinguish most of objects, except v and w. If he is lucky enough to acquire one more important fuzzy knowledge, he perhaps can have a clear view about his universe.

Assume that finally he acquires a new fuzzy knowledge about the shape of the object in his universe, namely, he acquired a fuzzy spherical property FV of spherical objects. That is, he has the following FV

$$FV(x) = (\text{Volume of } x)/(\text{smallest sphere containing } x)$$
$$FV(x) = FV(y) = 0.9,\ FV(z) = FV(u) = 0.5,\ FV(v) = 0.0,\ FV(w) = 1$$

	FX	FY	FZ	FW	FV
x	1	0.8	0.5	0	0.9
y	1	0.8	0	0	0.9
z	0.5	0.8	0.5	0	0.5
u	0.5	0.8	0	0	0.5
v	0	0	0.5	0	0.0
w	0	0	0.5	0	1

Now, with FX, FY, FZ, FW and FV, every objects in the universe can be distinguished, namely, each horizontal row of the table above represent an object of the universe. That means John has $ADEQUATE$ knowledge about his universe, in the sense that he can identify individual objects in his universe and specify any concepts of U. In other words, he has the total knowledge of his universe; he can specify any exact concept (classical subset) in U.

6.3. Exact Information via Neighborhood Systems

Now, we will formalize the notion of exact information on the world of neighborhood systems. A user's knowledge is represented by neighborhoods of each object in U. More precisely, a user may not have sufficient knowledge to describe any particular object, but he may be able to specify some approximate objects (via neighborhood). For example, a student may not have sufficient knowledge about the exact value of e, but he knows that it is located somewhere in the interval [2.71, 2.72]). If he gains more and more approximate knowledge about e, for example, he knows now that e is in the intervals [2.718, 2.719], [2.71828, 2.71829],... eventually he may be able to give a precise information about e (the sequence will converges to e). There is no numerical measure about amount of our knowledge in terms of neighborhood systems. However, we can use the terminology of topology to measure our knowledge about the neighborhood systems **qualitatively**.

Definition 6.5. U is called a Tychonoff space, if for every point x and each neighborhood $N(x)$ of x there is a continuous function FX from U to [0,1] such that $f(x) = 1$ and identically zero on the complement of $N(x)$.

Let us quote a theorem from Kelly [13], pp. 118.

Proposition 6.6. In order that U be Tychonoff it is necessary and sufficient that it be homeomorphic to a subspace of Hilbert Cube.

Theorem 6.7. There are $ADEQUATE$ fuzzy subsets in U if and only if U is a Tychonoff space.

Proof: If U is a Tychonoff space, then, by Proposition 6.6, it can be embedded into Hilbert Cube. Then, we use the "coordinate function $FX_i(x)$" as our fuzzy subsets of U. By 6.3, this family of fuzzy subset is $ADEQUATE$. On the other hand if the fuzzy subsets $\{FX_i\}$ is $ADEQUATE$, we can topologize U by demanding that all FX_i's be continuous. By 6.3., it is a subset of Hilbert Cube. Hence by 6.6, U is a Tychonoff space.

Remark: This theorem gives an intrinsic characterization of $ADEQUATE$ fuzzy knowledge about the universe U. Tychonoff is an intrinsic property of U, while Definition of $ADEQUACY$ is an external property of U (represented by functions of U). So we believe neighborhood system is better. We could know that U is a Tychonoff from other reasons without explicit constructing the membership functions.

References

[1] Z. Pawlak. *Rough sets. - Theoretical aspects of reasoning about data.* Kluwer Academic Publishers, 1990.

[2] W. Sierpiński and C. Krieger. *General topology.* University of Toronto Press, 1956.

[3] Didier Dubois and Henri Prade. Rough fuzzy sets and fuzzy rough sets. In *Int. Journal of General Systems*, pages 191–209, 1990.

[4] Donlod F. Stanat and David F. McAllister. *Discrete Mathematics in Computer Science.* Prentice-Hall, Inc., Englewood Cliffs, New Jersey, 1977.

[5] Stella Bairamian. Goal search in relational databases. Master's thesis, California State University, 1989.

[6] T.Y. Lin and S. Bairamian. Neighborhood systems and goal queries. Manuscript, California State University, Northridge, California, 1987.

[7] Z. Pawlak, S.K.M. Wong, and W. Ziarko. Rough sets: Probabilistic versus deterministic approach. In *Int. J. Man-Machine Studies*, pages 81–95, 1988. Vol. 29.

[8] T.Y. Lin. Neighborhood systems and relational database. In *Proceedings of CSC '88*, 1988.

[9] T.Y. Lin. Topological data models and approximate retrieval and reasoning. Annual ACM Conference, 1989.

[10] T.Y. Lin. Neighborhood systems and approximation in database and knowledge base systems. In *Proceedings of the Fourth International Symposium on Methodologies of Intelligent Systems, Poster Session*, 1989.

[11] T.Y. Lin, Q. Liu, K.J. Huang, and W. Chen. Rough sets, neighborhood systems and approximation. In *Fifth International Symposium on Methodologies of Intelligent Systems*, 1990. Selected Papers.

[12] T.Y. Lin, Qing Liu, and K.J. Huang. A model of topological reasoning expert system with application to an expert system for computer-aided diagnosis and treatment in acupuncture and moxibustion. In *International Symposium on Expert Systems and Neural Network Theory and Application*, 1990.

[13] John Kelly. General topology, 1955.

[14] Abraham Kandel. *Fuzzy Mathematical Techniques with Applications.* Addison - Wesley, Reading, Massachusetts, 1986.

[15] L.A. Zadeh. Fuzzy sets. *Information and Control*, 8: pages 338–353, 1965.

[16] T.Y. Lin. Probabilistic measure on aggregation. In *Proceedings of the 6th Annual Computer Security Application Conference*, pages 286–294, Tucson, Arizona, 1990.

[17] T.Y. Lin. "Inference" free multilevel database system. In *Proceedings of the Fourth RADC Database Security Workshop*, Little Compton, Rhode Island, 1991.

Part II
Chapter 6

ON CONVERGENCE OF ROUGH SETS

Lech T. POLKOWSKI
Institute of Mathematics
Warsaw University of Technology
00-650 Warszawa, Poland

Abstract. Various considerations of analysis of knowledge by means of rough sets lead to the following situation: we are given a universe U along with a descending sequence of equivalence relations. We propose metrics for rough sets that make them into a complete metric space; the metric convergence encompasses and generalizes approximative convergence studied by Marek, Pawlak and Marek, Rasiowa.

1. Introduction.

Rough sets as a tool to deal with uncertain knowledge were introduced in [6]. Since their introduction they have become a widely studied topic with many applications. Their formal counterparts have entered logic and Computer Science. Their investigations involve logical, set–theoretical, and topological tools. Our paper arises from taking all these techniques into account: while dealing with some topological properties of rough sets, it is motivated by some recent developments in knowledge analysis and Computer Science.

The interplay between Pawlak's rough sets and Pawlak's topology permits us to discuss rough sets by taking advantages of both set–theoretical and topological, points of view: given a set U, any equivalence relation R induces on U a topology π_R (equivalence classes of R forming an open base for π_R); π_R is peculiar: every π_R–open set is π_R–closed, too, but this is exactly what is necessary for the converse to hold, viz., any topology π on U for which π–Open = π–Closed induces an equivalence relation R_π such that $\pi_{R_\pi} = \pi$ (simply let $x\ R_\pi\ y$ iff x,y have identical sets of π–neighborhoods). We may and we do regard therefore Pawlak's R–lower and R–upper approximations of $X \subset U$ as, resp., the interior and the closure of X in the topology π_R: $\underline{R}(X) = Int_{\pi_R}$, $\overline{R}(X) = Cl_{\pi_R}X$.

Various considerations of knowledge analysis by means of knowledge bases and information systems (both in the sense of [5]) as well as various considerations of computer science (cf.[3],[4]) lead to the following situation: one is given a universe U along with an infinite descending sequence $(R_n)_n$ of equivalence relations. In [3] and [4] an approach to the problem of approximating sets with $R_n'^{\,s}$ via approximating sequences was proposed.

One may regard approximating sequences as a tool to describe a kind of strong convergence: given $X \subset U$, one takes the closure Cl_iX of X in π_{R_i} for each i, and the approximating sequence $(T_i = Cl_iX)_i$ "converges" to X in the sense that $Cl_iX = Cl_iT_i$ for each i, i.e., X and T_i have identical "approximations of order i".

We pursue this line of investigations with the Pawlak duality between rough sets and topology in mind.

We study π–rough sets determined by the topology π obtained by taking equivalence classes of all elements of U with respect to all $R_n'^s$ as an open base for π. As π–rough sets may be expressed as pairs of π–closed sets in U, subject to certain conditions, we apply the classical concept of Hausdorff metric to define two metrics D, $\overline{D}$ on the set of π–rough sets with $D \subseteq \overline{D}$. Roughly speaking, $\overline{D}$ controls the distance between, resp., positive, negative, and uncertain knowledge regions, while D controls the distance between, resp., positive and negative knowledge regions. It turns out that D and $\overline{D}$ make π–rough sets into a complete metric space: any $\overline{D}$–fundamental sequence of π–rough sets D– converges to a π–rough set; moreover approximating sequences of π– rough sets do converge to π–rough sets they approximate: the metric D –convergence encompasses and generalizes approximative convergence.

2. π - rough sets

Let U be a non–empty set and $(R_n)_n$ a sequence of equivalence relations on U with $R_{n+1} \subset R_n$ for each natural n. For $x \in U$ and a number n, we denote by $[x]_n$ the class of x with respect to R_n. Let $\mathbf{T} = \{([x_j]_j)_j : [x_{j+1}]_{j+1} \subset [x_j]_j$ for natural $j\}$ i.e. $\mathbf{T}$ consists of descending sequences of equivalence classes. For $([x_j]_j)_j \in \mathbf{T}$, we will call elements of the intersection $\bigcap_j [x_j]_j$ *indiscernibles.*

We will assume throughout this note that U is *complete with respect to indiscernibles* i.e. $\bigcap_j [x_j]_j \neq \emptyset$ for every $([x_j]_j)_j \in \mathbf{T}$.

We denote by π the topology on U whose base is the set

$$\{[x]_j : x \in U, j \text{ a natural number }\}$$

of all equivalence classes of all relations R_n.

By a *π–rough set* we will understand a pair (P, Q) of sets in U such that P is open in π, Q is closed in π, $P \not\subseteq Q$, and there exists a set $X \subset U$ with the properties that $Int_\pi X = P$ and $Cl_\pi X = Q$, where Int_π, Cl_π denote, respectively, the interior and the closure operators in the topology π.

The following simple and easy to establish fact holds

2.1.Proposition. *A pair (P, Q) of subsets of U such that P is π–open, Q is π–closed, and $P \not\subseteq Q$ is a π– rough set if and only if the set $Q \setminus P$ does not contain an isolated point of U i.e. a point that is simultaneously π–open and π–closed.*

We introduce an equivalence relation r on the set of π–rough points by letting

$$(P_1, Q_1)\, r\, (P_2, Q_2) \text{ if and only if } Q_2 \setminus P_2 = Q_1 \setminus P_1.$$

We denote by $[P, Q]$ the r–equivalence class of (P, Q) and we will call this class *a reduced π–rough set*; the π–rough set $(\emptyset, Q \setminus P)$ will be called *the canonical representative of* $[P, Q]$ and it will be denoted by $[[P, Q]]$.

We will need yet another representation for π–rough sets: for a π– rough set (P, Q), we let $T = U \setminus P$, and we represent (P, Q) as the pair (Q, T) of closed sets. Obviously, a pair (Q, T) of closed sets will represent a π–rough set, viz. $(U \setminus T, Q)$, if and only if,

(i) $U = Q \cup T$,
(ii) $Q \cap T \neq \emptyset$, and
(iii) $Q \cap T$ does not contain an isolated point of U.

3. A metric for π–rough sets

For $x, y \in U$ and a natural number j, we let

$$d_j(x,y) = 0 \text{ if } [x]_j = [y]_j \quad \text{and} \quad d_j(x,y) = 1 \text{ if } [x]_j \neq [y]_j$$

Now, for $x, y \in U$, we let

$$d(x,y) = \sum_{j=1}^{\infty} \frac{1}{10^j} \cdot d_j(x,y).$$

Clearly, d is a compatible pseudometric on (U, π), and $d(x,y) = 0$ if and only if x and y are indiscernibles.

Example 1. The Cantor set **C** is the standard playground for studying topology generated by a descreasing sequence of equivalence relations. Technically, **C** is the product $\{0, 1\}^N$, $N = \{1,2,3,\ldots\}$ i.e. the set of all sequences $(x_1, x_2, \ldots)$, where $x_i = 0$ or $x_i = 1$ for $i = 1, 2, \ldots$. In this case relations R_n' are defined as follows:

$$\text{for } x, y \in \mathbf{C},\ (x, y) \in R_n \text{ if and only if } (x_1, \ldots, x_n) = (y_1, \ldots, y_n).$$

Hence $d_n(x, y) = 0$ if and only if $(x_1, \ldots, x_n) = (y_1, \ldots, y_n)$. If our sequences are, say, $x = (1,0,1,0,1,0,\ldots)$ and $y = (0,1,0,1,0,1,\ldots)$ then the distance $d(x,y)$ between x and y is, according to our formula, equal to $\sum_{j=1}^{\infty} \frac{1}{10^j} \cdot 1 = \frac{1}{9}$.

We employ the pseudometric d to introduce metrics on the set of π–rough sets. To this end, we observe that the representation of a π–rough set via a pair of closed sets permits us to use the concept of the Hausdorff metric (cf.[1] or [2]),viz., we will measure the distance between two π–rough sets by means of the Hausdorff distances between their regions of, resp., positive, negative, and uncertain knowledge.

Formally, given two closed sets A, B in (U, π), we let

$$\delta(A, B) = max\{max\{min\{d(x,y) : y \in B\} : x \in A\}\}, max\{min\{d(x,y) : x \in A\} : y \in B\}\}\}$$

and we define functions $\overline{D}$ and D on pairs of π–rough sets by letting

$$\overline{D}((Q_1, T_1), (Q_2, T_2)) = max\{\delta(Q_1, Q_2)\, \delta(T_1, T_2), \delta(Q_1 \cap T_1, Q_2 \cap T_2)\}$$

and

$$D((Q_1, T_1), (Q_2, T_2)) = max\{\delta(Q_1, Q_2), \delta(T_1, T_2)\}$$

Example 2. Consider the Cantor set **C** again, this time, look at two closed sets in **C**, viz.,

$$A = \{x = (x_j)_j \in \mathbf{C}\colon x_j = 0 \text{ for at most one } j\}$$

and

$$B = \{y = (y_j)_j \in \mathbf{C}\colon y_j = 1 \text{ for at most one } j\}.$$

These A and B are π–closed and we want $\delta(A, B)$, the Hausdorff distance between A and B. So pick $x \in A$; we have two cases:

(a) $x_j = 1$ for every j,

(b) there exists a unique j_0 with $x_j = 0$ for $j = j_0$, $x_j = 1$ for $j \neq j_0$.

We need to find the mininal value of $d(x, y)$ for $y \in B$. In case (a), we have two values of $d(x, y)$:

(a1) for $y \in B$ with $y_1 = 1$, we have $d_j(x,y) = 0$ for $j = 1$, and $d_j(x, y) = 1$ for $j > 1$, so $d(x, y) = \sum_{j=2}^{\infty} \frac{1}{10^j} \cdot d_j(x, y) = \frac{1}{90}$

(a2) for $y \in B$ with $y_1 = 0$, we have $d_j(x, y) = 1$ for each j, so $d(x, y) = \frac{1}{9}$.

The value of $min\{d(x, y) : y \in B\}$ is therefore $\frac{1}{90}$.

Now, the other way round: pick $y \in B$ and establich $min\{d(x, y) : x \in A\}$; clearly, there is nothing to do here but argue by symmetry: $min\{d(x, y) : x \in A\} = \frac{1}{90}$.

So the maximum of the values is $\frac{1}{90}$ i.e. $\delta(A, B) = \frac{1}{90}$.

Our metrics D, $\overline{D}$ do the same, but for pairs or triples of closed sets.

We leave out the proof of the following

3.1.Proposition. *$\overline{D}$, D are metrics on the set of π–rough sets.*

We will need two following, easy to check, properties of D.

3.2 Proposition.

(a) For every n there exists a natural number j_n such that $D((Q_1, T_1), (Q_2, T_2)) < \frac{1}{n}$ implies $[x]_{j_n} \cap Q_1 \neq \emptyset$ if and only if $[x]_{j_n} \cap Q_2 \neq \emptyset$ and $[x]_{j_n} \cap T_1 \neq \emptyset$ if and only if $[x]_{j_n} \cap T_2 \neq \emptyset$ for every $x \in U$; moreover $\lim_n j_n = +\infty$;

(b) For every natural number j there exists $\epsilon_j > 0$ such that $[x]_j \cap Q_1 \neq \emptyset$ if and only if $[x]_j \cap Q_2 \neq \emptyset$ and $[x]_j \cap T_1 \neq \emptyset$ if and only if $[x]_j \cap T_2 \neq \emptyset$ for every $x \in U$ imply $D((Q_1, T_1), (Q_2, T_2)) \leq \epsilon_j$; moreover, $\lim_j \epsilon_j = 0$.

Finally, we observe that D induces a metric D_1 on the set of reduced π–rough sets via the formula $D_1([P_1, Q_1], [P_2, Q_2]) = D([[P_1, Q_1], [[P_2, Q_2]])$.

4. Completeness of the metric space of π–rough sets

We begin with the main property of metrics $\overline{D}$ and D: they make π–rough sets into a ($\overline{D}$, D)–complete metric space.

4.1.Proposition. *Every $\overline{D}$–Cauchy sequence $((Q_n, T_n))_n$ of π–rough sets D–converges to a π– rough set.*

Proof. By Proposition 3.2, there exist two increasing sequences $(j_n)_n$, $(m_n)_n$ of natural numbers with the properties that

(i) $Cl_{j_n}Q_m = Cl_{j_n}Q_{m_n}$ for $m \geq m_n$,

(ii) $Cl_{j_n}T_m = Cl_{j_n}T_{m_n}$ for $m \geq m_n$,

(iii) $Cl_{j_n}(Q_m \cap T_m) = Cl_{j_n}(O_{m_n} \cap T_{m_n})$ for $m \geq m_n$.

where Cl_{j_n} denotes the closure operator in the topology generated by equivalence classes of R_{j_n}.

Consider sets

$$Q^* = \bigcup\left\{\bigcap_j [x_j]_j : ([x_j]_j)_j \in \mathbf{T}\, and\, [x_{j_n}]_{j_n} \cap Q_{m_n} \neq \emptyset\right\}$$

and

$$T^* = \bigcup\left\{\bigcap_j [x_j]_j : ([x_j]_j)_j \in \mathbf{T}\, and\, [x_{j_n}]_{j_n} \cap T_{m_n} \neq \emptyset\right\}.$$

The sets Q^* and T^* being obviously closed, our Proposition will follow from the following claims.

Claim 1. $U = Q^* \cup T^*$.

Suppose, to the contrary, that x is not in $Q^* \cup T^*$ for some $x \in U$.

Let $\overline{x} \in \mathbf{T}$ satisfy $\overline{x}_j = [x]_j$ for each j; then there exist m, n such that $\overline{x}_{j_n} \cap Q_{m_n} = \emptyset = \overline{x}_{j_m} \cap T_{m_n}$. Assume, e.g., that $m \geq n$; then (i) implies $\overline{x}_{j_m} \cap Q_{m_m} = \emptyset$ i.e. x is not in $Q_{m_m} \cup T_{m_m} = U$, a contradiction.

Claim 2. $Q^* \cap T^* \neq \emptyset$.

Pick $x_1 \in Q_{m_1} \cap T_{m_1}$; as, by (iii), $Cl_{j_1}(Q_{m_2} \cap T_{m_2}) = Cl_{j_1}(Q_{m_1} \cap T_{m_1})$, there exists $x_2 \in [x_1]_{j_1}$ with $[x_2]_{j_2} \cap Q_{m_2} \cap T_{m_2} \neq \emptyset$. Continuing by induction, we find a sequence $(x_k)_k$ in U such that $[x_{k+1}]_{k+1} \subset [x_k]_k$ and $[x_k]_{j_k} \cap Q_{m_k} \cap T_{m_k} \neq \emptyset$ for each k. Let $\overline{x}$ be a unique member of $\mathbf{T}$ with $\overline{x}_{j_k} = [x_k]_{j_k}$ for each k. Then $\emptyset \neq \bigcap \overline{x} \subset Q^* \cap T^*$.

Claim 3. For each n,

(iv) $Cl_{j_n}Q^* = Cl_{j_n}Q_{m_n}$,

(v) $Cl_{j_n}T^* = Cl_{j_n}T_{m_n}$.

It suffices to prove (iv); the other proof goes along the same lines. Assume x_0 is in $Cl_{j_n}Q^*$; let y_0 be in $[x_0]_{j_n} \cap Q^*$; by definition of Q^*, $\emptyset \neq [y_0]_{j_n} \cap Q_{m_n} = [x_0]_{j_n} \cap Q_{m_n}$ hence x_0 is in $Cl_{j_n}Q_{m_n}$.

Conversely, assume that x_0 is in $Cl_{j_n}Q_{m_n}$; employing (i), define inductively a sequence $(x_k)_k$ in U such that $[x_{k+1}]_{j_{n+k+1}} \subset [x_k]_{j_{n+k}}$ and $[x_k]_{j_{n+k}} \cap Q_{m_{n+k}} \neq \emptyset$ for each k. Let $\overline{x}$ be a unique number of $\mathbf{T}$ with $\overline{x}_{j_{n+k}} = [x_k]_{j_{n+k}}$ for each k. Then $\emptyset \neq \bigcap \overline{x} \subset Q^* \cap [x_0]_{j_n}$ i.e. x_0 is in $Cl_{j_n}Q^*$.

Claim 4. $Q^* \cap T^*$ does not contain an isolated point of U.

Indeed, if an isolated point of U, say x_0, was in $Q^* \cap T^*$, we would have $[x_0]_{j_n}$ a singleton for some n, hence x_0 would be in $Q_{m_n} \cap T_{m_n}$, a contradiction.
Claim 5. $D((Q_{m_n}, T_{m_n}), (Q^*, T^*))$ converges to 0 as n tends do infinity.
Indeed, this follows from (i), (ii), above, and 3.2.
Claim 6. The sequence $((Q_n, T_n))_n$ D–converges to (Q^*, T^*).
Indeed, this follows from Claim 5 and the fact that $((Q_n, T_n))_n$ is $\overline{D}$–Cauchy hence D–Cauchy.
This concludes the proof.

One can prove along similar lines the following counterpart of Proposition 4.1. for the set of reduced π–rough sets.

4.2.Proposition. *The set of reduced π–sets endowed with the metric D_1 is a complete metric space.*

5. Approximative convergence

The notion of an approximative sequence was studied in [3] and [4]; a sequence $(T_j)_j$ of subsets of U is said to be an *approximating sequence* if there exists a subset X of U with the property that $Cl_jX = T_j$ for each j in which case X is said to be *approximated* by $(T_j)_j$.

We will say that a sequence $((A^1_n, ..., A^k_n))_n$ of k–tuples of π–closed sets *converges approximatively* to a k–tuple $(A^1, ..., A^k)$ of π– closed sets if $Cl_nA^i_n = Cl_nA^i$ for each n and $i \leq k$. Clearly, if $(A^i_n)_n$ is the approximating sequence for A^i for $i \leq k$, then $((A^1_n, ..., A^k_n))_n$ converges approximatively to $(A^1, ..., A^k)$. We have the following proposition which states that the above introduced metric convergence extends the convergence suggested by the notion of an approximating sequence.

5.1.Proposition.

(a) *If $((Q_i, T_i, Q_i \cap T_i))_i$ converges approximatively to $(Q, T, Q \cap T)$, then $((Q_j, T_i))_i$ $\overline{D}$–converges to (Q, T) in the space of π–rough sets,*
(b) *If $((Q_i, T_i))_i$ converges approximatively to (Q, T), then $((Q_i, T_i))_i$ D–converges to (Q, T) in the space of π– rough sets,*
(c) *If $([[P_i, Q_i]])_i$ converges approximatively to $[[P, Q]]$, then $((P_i, Q_i))_i$ D_1–converges to (P, Q) in the space of reduced π–rough sets.*

We assume from now on that the space (U, π) does not contain isolated points i.e. no equivalence class of no relation R_n is a singleton, an assumption very natural from practical point of view.

It is tempting to conjecture that a stronger completeness property takes place i.e. every $\overline{D}$–Cauchy sequence of π–rough sets $\overline{D}$–converges; it is, however, not the case, as simple examples bear out. Still, the idea of an approximating sequence may be refined to provide a special case of sequences for which $\overline{D}$–Cauchy implies $\overline{D}$–convergence. A sequence $((Q_i, T_i))_i$ of π–rough sets will be called *saturated* if $Q_i = Cl_iQ_i$ and $T_i = Cl_iT_i$ for each i; observe, again, that if $(Q_i)_i$ and $(T_i)_i$ are approximating sequences, then $((Q_i, T_i))_i$ is saturated. We have the following proposition, whose proof parallels that of 4.1.

5.2.Proposition. *If $((Q_i, T_i))_i$ is a $\overline{D}$- Cauchy saturated sequence of π-rough sets, then $((Q_i, T_i))_i$ $\overline{D}$-converges to a π-rough set.*

Proof. We need only small addition to the proof of 4.1, viz., we need to prove that $Cl_{j_n}(Q^* \cap T^*) = Cl_{j_n}(Q_{m_n} \cap T_{m_n})$ for each n (the notation is that of 4.1). To this end, observe that the inclusion $Cl_{j_n}(Q_{m_n} \cap T_{m_n}) \subset Cl_{j_n}(Q^* \cap T^*)$ holds for each n by an argument similar to that applied in proving Claims 2,3 in 4.1.
To prove the converse, pick $y \in Cl_{j_n}(Q^* \cap T^*)$ and $x \in [y]_{j_n} \cap Q^* \cap T^*$. We have $[x]_{j_k} \cap Q_{m_k} \neq \emptyset \neq [x]_{j_k} \cap T_{m_k}$ for each k i.e. x is in $Q_{m_k} \cap T_{m_k}$ for each k; in particular, y is in $Cl_{j_n}(Q_{m_n} \cap T_{m_n})$.

6. A Bolzano–Weierstrass — type property

Assume that each relation R_n has finitely many equivalence classes, a case most often met in applications. We have

6.1.Proposition. *Each sequence of π-rough sets contains a D - convergent subsequence.*

Proof. It suffices to define inductively a decreasing sequence $(\mathbf{s}_n)_n$ of increasing sequences of natural numbers such that the given sequence $((Q_i, T_i))_i$ of π-rough sets satisfies
$Cl_n Q_m = Cl_n Q_p$, $Cl_n T_m = Cl_n T_p$, and $Cl_n(Q_m \cap T_m) = Cl_n(Q_p \cap T_p)$ for $m, p \in \mathbf{s}_n$ for each n, and to observe that the sequence $((Q_{\mathbf{s}_n(n)}, T_{\mathbf{s}_n(n)}))_n$ is $\overline{D}$-Cauchy, hence it D-converges by 4.1.

References

[1] Hausdorff F., *Grundzüge der Mengenlehre*, Leipzig (Vien), 1914.

[2] Kuratowski K., *Topology I*, Warszawa, 1966.

[3] Marek W., Pawlak Z., Information systems and rough sets, *Fund.Inform.* **VIII** (1984), 105–115.

[4] Marek W., Rasiowa H., Approximating sets with equivalence relations, *Theoretical Computer Sc.* **48**(1986), 145–152

[5] Pawlak Z., Rough sets: *Theoretical Aspects of Reasoning About Data*, Kluver 1991.

[6] Pawlak Z., Rough sets, *International J.Computer and Information Sciences* **11**(1982), 341–356

Part III

FURTHER DEVELOPMENTS

Part III
Chapter 1

MAINTENANCE OF KNOWLEDGE IN DYNAMIC INFORMATION SYSTEMS

Maria E. ORLOWSKA
Computer Science Department
University of Queensland
St. Lucia, QLD 4072, Australia

Marian W. ORLOWSKI
School of Information Systems
Queensland Univ. of Technology
Brisbane, QLD 4001, Australia

Abstract. In the process of data analysis in Pawlak's information system and in the derivation of decision algorithms from decision tables, the main computational effort is associated with the determination of the reducts. The notion of the reduct plays a major role in the definitions of other fundamental notions of the information systems theory, such as functional dependencies, redundancy etc. The purpose of the presentation is to show that role and demonstrate different ways of computing reducts. The computational complexity of the problem of reducts enumeration is, in the worst case, exponential with respect to the number of attributes. Therefore the only way to achieve its faster execution is by providing an algorithm, with a better constant factor. The algorithm presented computes reducts from the sample set of objects. The corresponding data available for the analysis in real-life applications is subject to some dynamic changes. Thus it would be useful to have a facility to support such a behaviour. The presented algorithm is based on the strategy of the maintenance of the set of reducts - providing their necessary alternations triggered by an expansion of the set of objects. The algorithm is demonstrated in a nontrivial example. The significance of the offered solution is twofold; the avoidance of the repetition of the whole (standard) procedure for the dynamically changing data, and an improvement of the execution time in comparison to the standard methods.

1. Introduction

The relational approach to information systems was initiated by Codd ([3]) and developed by many researchers (see [4] for references). Another, alternative approach to relational information systems was proposed by Pawlak ([7]). The two approaches differ but there are many tangent areas and it is possible for both of them to cross-benefit from their philosophical foundations. It is remarkable how many similarities exist between them, although at first sight it is difficult to spot them.

On the basis of the notions introduced in the Pawlak theory of information systems, one can define and adopt several notions existing in the Codd approach. Consequently, these adopted notions could be used as auxiliary tools to investigate the behaviour of Pawlak's information systems. On the other hand, the concepts and notion specific to Pawlak's Theory can be used to explain, on the set-theoretical platform, a number of issues in Codd's information systems.

The backbone of the Codd theory of databases is the notion of the relation, which is understood as a (finite) subset of the cartesian product of the domains of attributes. The corresponding set of attributes is called the relation schema. The classical Codd approach is based on an identification of properties of relational schemata and subsequently, if necessary, decomposing them into a set of schemata satisfying required quality criteria. The relational algebra operators are then used for manipulations of the relations defined in the decomposed schemata. The properties of the relational schema can be captured by investigating the relation instance(s) assuming that it is a representative sample of an appropriate size. In general, if there is a reassurance that all instances of the relation are satisfying a certain property then it is the property of the relational schema.

The Pawlak approach to information systems concentrates on studying the properties of objects rather than the properties of attributes and, more precisely, the main concern is the classification of objects imposed by equivalence relations which are defined by comparing values of attributes in the set of objects. Formally the Pawlak concept is to investigate the partitioning of the set of objects into equivalence classes, generated by preimages of one point sets of the functions:

$$\mathbf{f_P} : \mathbf{X} \rightarrow \prod_{q \in P} \mathbf{V_q}$$

It is usual practice that, when studying the properties of the equivalence classes of an equivalence relation, one selects a single representative from each class for examination. From now on, the representation of objects shall satisfy this condition. It means that assuming **Q** as a set of all attributes in the universe, the equivalence classes of the indiscernibility relation **IND(Q)** are one-element sets. This assumption can be made without any loss of generality as any manipulations will be carried out on whole equivalence classes, and a further partition of an equivalence class would require an additional attribute(s). This is in contradiction to the assumption. Using notions of the indiscernibility relation as well as the reduct or the dispensable attribute, etc., one can characterise other notions relevant to the data analysis.

The main focus here is on the investigation of the notion of reduct, which plays a similar role as the key in the Codd theory.

2. Reducts, properties and applications

The concept of the reduct is very powerful in information analysis. Extensive studies of it are provided in many sources. In this section we shall examine reducts as a bridge to other important notions of the relational theory.

It will be useful to have a short notation x(**P**) for the set of descriptors related to an object x and a set **P** (⊆ **Q**) of attributes, i.e.:

$$x(\mathbf{P}) = \{f(x,q) \colon q \in \mathbf{P}\}$$

The symbol **X(P)** will be used to denote the set of all x(**P**) for x ∈ **X**.

(2.1) Set **P** ⊆ **Q** is a called a *weak reduct* of **Q** in **X(Q)** (notation **P** ∈ **WRED(Q)**, if **IND(P)** = **IND(Q)**.

The following facts are immediate consequences of the definition

(2.2) If **P** ⊆ **Q** is a reduct of **Q** in **X(Q)** then **P** ∈ **WRED(Q)** in **X(Q)**.

(2.3) **RED(WRED(P) = RED(P)** in **X(Q)**.

All the notions in the Pawlak theory are defined for a fixed finite set of objects, but the universality of their properties can be easily achieved by adding the phrase *for all admissible set of objects X in a given universe of discourse*. (That is, the set **P** is a reduct of **Q** if **P** is a reduct of **Q** for any set **X** of admissible objects in the given universe).

The notion of the functional dependency (FD) of attributes in Pawlak's information systems is defined and studied in a number of papers ([1],[5],[8],[11],[12],[13]). This concept is close to the concept of functional dependency in the Codd theory, although the definition is different. The notion of FD was originally defined in [5] in the following way:

(2.4) $\mathbf{P' \rightarrow P}$ iff $\mathbf{IND(P') \subset IND(P)}$

The characterisation of the FD can be obtained as a generalisation of the notion of a weak reduct:

(2.5) $\mathbf{P' \rightarrow P}$ iff **P'** is a weak reduct of $\mathbf{P' \cup P}$.

It is immediately visible from (2.3) and from the above characterisation that

(2.6) If $\mathbf{P' \rightarrow P}$ then **P'** contains a reduct of $\mathbf{P' \cup P}$.

(2.7) If $\mathbf{P' \rightarrow P}$ and for no proper subset **R** of **Q**, $\mathbf{R \rightarrow P}$, then **P'** is a reduct of $\mathbf{P' \cup P}$.

The notion of reduct characterises the *arity* of functional dependency, that is, the minimum set of attributes from **P'** necessary to distinguish objects if they are distinguished by **P**. More precisely:

(2.8) DEFINITION The functional dependency $\mathbf{P' \rightarrow P}$ is said to be *left-reduced* if **P'** is a reduct of $\mathbf{P' \cup P}$.

The information which could be derived is superfluous and in order to avoid this, there is a need to identify it. A useful description of this phenomenon (known as redundancy) can be provided using an indiscernibility relation:

(2.9) DEFINITION The left-reduced functional dependency $\mathbf{P' \rightarrow P}$, where **P** is one-attribute set, is said to be *redundant* if there is a set **R** not containing **P'** and not containing **P** such that $\mathbf{IND(P') \subseteq IND(R) \subseteq IND(P)}$.

A notion regarded as very helpful in the computation of reducts is **CORE(Q)** of the set of attributes, which consists of indispensable attributes. The knowledge of the **CORE(Q)** is useful indeed, when the strategy of computing reducts is *bottom-up* (see next section). The problem of identifying the core set is an easy one. The computational complexity of the process of computing core attributes is low and, even using the *brute force* approach, it is quadratic with respect to the number of objects in the system. Therefore the knowledge of the core attributes has only *marginal* impact on the overall (expected or worst case) execution time of any algorithm for determining reducts. The purpose of the set **CORE(Q)** is auxiliary in computing reducts and it is beneficial if it is nonempty.

The situation where the core set can be found independently of reducts is when it is possible to detect all functional dependencies in an informal way. Such a situation occurs when the given universe has *familiar* semantics and the functional dependencies are easy to identify; for instance

LECTURER → DEPARTMENT,

SUBJECT YEAR/SEMESTER → LECTURER, etc.,

are obvious functional dependencies in the universe "UNIVERSITY".

The following example describes how different levels of the importance of attributes can be extracted from the knowledge of functional dependencies valid in the system.

EXAMPLE 1. Given the set of functional dependencies on the set of attributes Q = {a,b,c,d,e,f,g,h,i,j}:

ade → f, cd → be, ae → cgi, g → h, dh → j

Consider the following table of appearances of attributes on the left- and right-hand-sides of functional dependencies:

Attribute	LHS	RHS
a	*	
b		*
c	*	*
d	*	
e	*	*
f		*
g	*	*
h	*	*
i		*
j		*

The above table provides more detailed information about the role of attributes than just dispensable and indispensable. The core attributes are those which do not appear on the right hand sides of the functional dependencies. In this example, a and d are core attributes. This table is useful also to recognise the attributes which do not appear in any reduct. These are b, f, i and j - they do not appear on the left hand sides of functional dependencies. The remaining attributes, c, e, g, h may appear in some reducts. The set of reducts for this example consists of two sets {a,d,e} and {a,c,d}; g and h do not participate in any reduct. The dispensable attributes can be classified into categories of different degree of dispensability; for instance, the attribute e is dispensable in the set {acde}, but indispensable in {adefg}.

It is worth noting that the use of functional dependencies in computations is beneficial only if they can be easily detected, otherwise one experiences the same level of difficulty, but shifted into another area.

The computation of reducts is an essential procedure in several applications of rough sets. It could also be an essential part of the manipulation of information systems. A survey of a number of applications ([2],[10], for more see [9]) develops an impression that the computation of reducts is a straightforward process. There are basically two sources of such optimism:

- The (relatively) small size of the set of different objects in the systems considered.
- The examples in the case studies usually have a *sound* semantical real-world meaning and therefore the number of reducts (which actually determine the objects) is small and their computation could indeed be straightforward.

Unfortunately a Universe does not always satisfy the above conditions. In anticipation of a non-semantic Universe (which could be found in some engineering applications) one has to be prepared for the worst case. The common perception is that the problem of finding all the reducts is non-polynomial. This is justified by the fact that the number of reducts in the worst case depends exponentially on the number of attributes and is $\binom{n}{\lfloor n/2 \rfloor}$. An example of a system with such a number of reducts is given below. Note that the entries in the places displayed as empty (for the sake of clarity) must all be different in each column.

	a	b	c	d	e	f
1)	1	1	1			
2)	1	1		1		
3)	1	1			1	
4)	1	1				1
5)	1		1	1		
6)	1		1		1	
7)	1		1			1
8)	1			1	1	
9)	1			1		1
10)	1				1	1
11)		1	1	1		
12)		1	1		1	
13)		1	1			1
14)		1		1	1	
15)		1		1		1
16)		1			1	1
17)			1	1	1	
18)			1	1		1
19)			1		1	1
20)				1	1	1

Therefore, the worst case complexity of any algorithm for the determination of reducts is non polynomial in respect to the number of attributes. This is because all reducts must be identified. In cases where the number of reducts is small, the computational complexity still could be in the NP range as many subsets of the set of attributes have to be tested. Obviously, there are cases that can be resolved with little computational effort, due to a good configuration of data (for instance a large set of core attributes).

An important problem in many applications is the maintenance of knowledge in a dynamic environment. There is a possibility that the current knowledge would have to be altered when a new piece of information is delivered with a new object. Obviously the reducts play an essential formal part of the knowledge. It is remarkable how large alterations could be triggered by the addition of one object. The following example illustrates this phenomenon.

EXAMPLE 2. Consider information system with 8 attributes consisting of one object.

	a	b	c	d	e	f	g	h
1)	1	1	1	1	0	0	0	0

The set of reducts are all comprised of one-attribute subsets of **Q**. Suppose that the new object is added. This is depicted in the next table:

	a	b	c	d	e	f	g	h
1)	1	1	1	1	0	0	0	0
2)	1	1	1	1	1	1	1	1

It is apparent that {e}, {f}, {g} and {h} are the only reducts. The insertion of the following third object makes dramatic changes.

	a	b	c	d	e	f	g	h
1)	1	1	1	1	0	0	0	0
2)	1	1	1	1	1	1	1	1
3)	0	0	0	0	1	1	1	1

None of the one-attribute sets are the reducts any longer. There are 16 reducts formed by the pairs of attributes; one from the set {a,b,c,d} and the second one from the set {e,f,g,h}.

This example shows that the quality and quantity of reducts are *unpredictable* and the set of reducts can vary for different representative objects in the same universe. Also, when the semantics of a real-world universe is involved, the inadequate size and quality of a sample can distort the true knowledge about it. To support this statement consider an example of the universe of entries to students transcripts involving STUDENT_NR, SUBJECT_CODE, YEAR/SEMESTER, GRADE, and LECTURER. On the basis of a review of (even large) numbers of objects related to successful enrolments, one could conclude that {STUDENT_NR, SUBJECT_CODE} is a reduct. However, it is not true when an entry arrives which relates to a student, who has repeated a subject. In such a familiar universe it is possible to handle all the parts of knowledge related to reducts; a similar problem for a more complex and nonfamiliar universe could be difficult to deal with.

Despite that negative side of the problem there is still room for optimism. Pawlak's philosophy of knowledge acquisition and manipulation is based on the examination of the data available. Therefore the assumption can be made that the size of the object's sample (in many cases all possible objects) is allowed to be dealt with and the set of all attributes is small. In such circumstances the NP-completeness of the problem of finding reducts could become irrelevant and the only issue is the evaluation of the computational complexity of the problem, in respect to the size of the sample set of objects with a fixed set of attributes.

One observation which is vital in the evaluation of the number of reducts generated by sample data is as follows:

(2.10) LEMMA If $\mathbf{P} \subseteq \mathbf{Q}$ is a reduct generated by a sample data related to a set of objects $\mathbf{X}$ (all objects distinguishable by $\mathbf{Q}$), then for each set $\mathbf{P'} \subset \mathbf{P}$ there exist an equivalence class of **IND(P')** consisting of more than one object.

Proof. Since the objects were all different (i.e., the equivalence classes of the relation **IND(P)** are one-element sets) then for each set $\mathbf{P'} \subset \mathbf{P}$, which is not a reduct, there must be an equivalence class of the relation **IND(P')** consisting of more than one element.

The consequence of Lemma is that to ensure that a set **P** is a reduct, it is sufficient to check if the lemma is satisfied for all subsets **P'** of **P** containing all but one element of **P**. One can easily see that if the reduct is comprised of p attributes then the sample used to generate it must consist of more than p objects.

The problem of evaluating the size of a sample generating a specific configuration of reducts is a difficult one and is somewhat impractical, but it is interesting to know how big it could be in some *border* cases. That knowledge can relieve a practitioner from the fear that her task of analysing data is intractable in the sense of the ratio of the size of output (number of reducts) to the size of input (the number of object; the number of attributes can be considered fixed).

The interesting border case is where the number of reducts is maximal, i.e., $\binom{n}{\lfloor n/2 \rfloor}$. The only way for these reducts to materialise is in the form of all ($\lfloor n/2 \rfloor$)-elements subsets of the sets of attributes. Using Lemma, one can infer that for each ($\lfloor n/2 \rfloor$-1)-elements subset of **Q** there must be two objects undistinguishable on exactly that subset of attributes. Therefore there must be a sufficient number of objects such that there is a distinct pair of them corresponding to each of the ($\lfloor n/2 \rfloor$-1)-elements subsets of **Q**.

Assume that there are m objects, they form $(m-1) * m/2$ different pairs. Therefore

$$(m-1) * m/2 \geq \binom{n}{\lfloor n/2 \rfloor -1}$$

It should be obvious that this is a rough estimation only as the number of object must be much larger. To give a justification for this claim, consider the system of linear equations depicting that *partial* equality of the pairs of objects and compare the number of equations and the number of variables. The latter is much smaller and therefore the solution to that system will consist of all the same objects. This is a contradiction to the assumption that all the objects are different.

An interesting problem is to estimate the maximum possible number of reducts determined by a sample, as a function of the number of attributes and the number of objects.

3. Algorithms for determination of reducts

In this section the practical problem of finding reducts is addressed.

Whatever the sizes of the set of attributes and the number of objects in the sample, the problem of finding reducts is a large one and the possession of an efficient algorithm for their determination is required.

It will be convenient to have a short notation for the two families $\mathbf{P}_\bullet$ and $\mathbf{P}^\bullet$ of sets defined for a set **P**. $\mathbf{P}_\bullet$ will denote the family of all subsets of **P** containing all but one element of **P**, and $\mathbf{P}^\bullet$ will denote the family of all supersets of **P** (in **Q**) containing exactly one more element than **P**.

There are two standard ways to approach the task of finding reducts: *top-down* and *bottom-up*. We outline both of them:

(3.1) Let **J** and **K** be two families of subsets of the set of attributes **Q**. The role of **J** is auxiliary, the returned family **K** consists of all reducts. The initial state of **J** is {**Q**} and the family **K** starts empty.
For each set $\mathbf{P} \in \mathbf{J}$ construct the family $\mathbf{P}_\bullet$ and for each set $\mathbf{S} \in \mathbf{P}_\bullet$ check if all the tuples from **X(S)** are unique.
Each subset $\mathbf{S} \in \mathbf{P}_\bullet$ such that all the tuples from **X(S)** are unique is added to **J**; if at least one such set from $\mathbf{P}_\bullet$ was added to **J** then **P** must be removed from **J**.
If for all subsets $\mathbf{S} \in \mathbf{P}_\bullet$ the tuples from **X(S)** are not unique then move **P** from **J** to **K**.
That procedure terminates when **J** becomes an empty family.

The *bottom-up* strategy is reflected in the following procedure.

(3.2) Let **H** and **K** be two families of subsets of the set of attributes **Q**. The initial state of **H** is the set of attributes **CORE(Q)** (possibly an empty set). The role of **H** is auxiliary, the returned family **K** consists, as in (3.3) of all reducts. The initial state of **K** is empty family.
For each $\mathbf{P} \in \mathbf{Q}$ check **X(P)** for the uniqueness of tuples. If this check is positive, i.e. all tuples are unique then transfer **P** to **K**, otherwise replace **P** in **H** with the family $\mathbf{P}^\bullet$. In this replacement the sets containing a reduct from the current family **K** should be ignored (they are already weak reducts, but not reducts).
The procedure terminates when **H** becomes an empty family.

The above algorithms are conceptually simple, but they carry a heavy load of computations. Moreover the inconvenience of both approaches becomes apparent when one considers the changing decision environment. A good example illustrating the dynamics is the medical diagnostics, where a new case is detected or a new treatment is developed. The next decision tuple must be incorporated and it can result in several changes to the decision algorithm. Other possible changes could be triggered by changes in diagnosis; for instance an old diagnosis could be wrong or an old case may appear to be inconsistent due to an omission of some important decision attribute. One of the ways to update knowledge is to rework the whole decision table. In a small universe such a process does not inflict a big computational effort in absolute terms. If the universe or, more precisely, the decision table is large, then the process of reworking the decision table can be costly. This argument justifies the need for having a procedure which will cut costs by effecting only necessary changes to the decision algorithm which result from the gradual change to the decision table (by one decision at a time).

The procedure for such maintenance of the decision algorithm can be developed in a similar way to another procedure for the dynamic computation of reducts. This is presented here. The general idea of the latter is based on an obvious observation.

If a partition of the set of objects generated by the indiscernibility relation **IND(P)**, $\mathbf{P} \subset \mathbf{Q}$, is given, then the addition of a new object results in one of the following actions:

– The classification of this object to one of the equivalence classes in the existing partition (this object is nondistinguishable from some other objects); or alternatively,

– This object will form another set in the altered partition. In the first case, one can try to distinguish the object by means of its value for other attribute(s) from the set $\mathbf{Q}-\mathbf{P}$. This leads to the conclusion that the set **P** is not sufficient to distinguish the objects and that perhaps one of its supersets can satisfy that condition.

The above observation is essential for the construction of an algorithm for the maintenance of reducts. It will be apparent that the algorithm resembles the earlier described *bottom-up* approach. The algorithm uses the comparison operation **COMP** (x,**X**,**P**), described below.

The notation used is following. x is an object, **X** is a set of objects, $x \notin \mathbf{X}$, $\mathbf{P} \subset \mathbf{Q}$. The objects x and all $y \in \mathbf{X}$ are given by means of their values of the attributes from **Q**.

The result of the comparison procedure **COMP** (x,**X**,**P**) is the set of objects from **X** which, restricted to attributes from the set **P**, have the same values as x restricted to **P**. In other words **COMP** (x,**X**,**P**) is the set of objects complementing x in the equivalence class (containing x) of indiscernibility relation **IND(P)** defined over $\mathbf{X} \cup \{x\}$.

It is obvious that if the set **P** of attributes is a reduct then **COMP**(x,**X**,**P**) is the empty set.

The algorithm for the maintenance of reducts works as follows:

(3.3) ALGORITHM REDUCT MAINTENANCE

Input: Set of objects **X** described using their values of **Q**-attributes.

Output: Set **K** of reducts for the set **X**. Initialy $\mathbf{K} = \emptyset$.

1. Start from one-object set **X** (first object); all single attribute sets are reducts. $\mathbf{K} = \{\{q\}; q \in \mathbf{Q}\}$.

For every new object x

2. Perform procedure **COMP** (x,**X**,**P**) for all $\mathbf{P} \in \mathbf{K}$. If **COMP** (x,**X**,**P**) is empty then retain **P** in **K**. If **COMP** (x,**X**,**P**) is nonempty then remove the set **P** from **K** and add $\mathbf{P}^{\bullet}$ to the family **K'**.

3. For each $\mathbf{P} \in \mathbf{K'}$ check if it is (proper) superset for a set from **K**. If it is the case remove it from **K'**. Otherwise perform **COMP** (x,**X**,**P**) and if it is empty move **P** to **K**. If **COMP** (x,**X**,**P**) is nonempty then remove **P** from **K'** and add $\mathbf{P}_{\bullet}$ to **K'**.

Perform the step 3 for each set from **K'** until **K'** becomes an empty family.

We shall illustrate the algorithm above on an example which will cover all essential possible configurations of data, thus exploiting all possible loops in our algorithm.

EXAMPLE 3. Assume that a system consists of six attributes {a,b,c,d,e,f}. The following twelve objects described in the table below arrive in the given order:

	a	b	c	d	e	f
1)	1	1	2	1	2	1
2)	2	2	2	1	1	2
3)	3	3	2	3	1	1
4)	1	1	2	2	3	1
5)	3	1	3	2	2	2
6)	1	2	2	3	2	2
7)	2	2	2	2	1	3
8)	1	2	1	1	2	2
9)	3	2	3	2	2	2
10)	3	3	3	3	3	1
11)	1	3	1	1	3	1
12)	2	1	3	1	2	1

The beginning is simple; all single attribute sets {a}, {b}, {c}, {d}, {e} and {f} are reducts.

The arrival of the second object, with the same c- and d- values results in the family **K** reduced to {a}, {b}, {e} and {f}. **K'** := {{cd}} and since **COMP**(x_2,{x_1},{cd}) = {x_1} and $\{cd\}^{\bullet} = \varnothing$ (other attributes are in single-attribute reducts), the family **K'** := ∅. The third object can be processed.

The third object does not violate the reducts {a} and {b}. The operation **COMP** performed on the third object and reducts {e} and {f} returns nonempty sets and therefore they are not reducts any longer but their supersets have a chance. The supersets, which should be considered in the first place are $\{e\}^{\bullet}$ and $\{f\}^{\bullet}$, so **K'** = {{ce},{de},{cf},{df},{ef}}. The operation **COMP** performed on sets from **K'** returns nonempty results only for {ce} and {cf}. According to step 3 of the algorithm, **K** := {{a},{b},{de},{df},{ef}} and **K'** := {{cde},{cef},{cdf}}. The new sets from **K'** are containing some sets from **K**, so they should be removed from **K'** and thus **K'** := ∅.

The fourth object violates the reducts {a} and {b} only. The new set **K** := {{de},{df},{ef}} and **K'** := {{ab},{ac},{ad},{ae},{af},{bc},{bd},{be},{bf}}. After performing operation **COMP** on these sets for the fourth object the new intermediate set of reducts

K = {{ad},{ae},{bd},{be},{de},{df},{ef}}

and updated set

K' = {{abc},{abd},{abe},{abf},{acd},{ace},{acf},
{bcd},{bce},{bcf},{bde},{bdf},{bef}}.

The only sets from **K'** which are not supersets to some sets from **K** are {abc}, {abf}, {acf} and {bcf}. Operation **COMP** returns nonempty set for all of them. So far, there is no change to **K**, but

K' := {{abcd},{abce},{abcf},{abdf},{abef},{acdf},{acef},{bcdf},{bcef}}

Except {abcf}, each of above sets contains a reduct from **K**. There is no change to **K**, but **K'** in the next form consists of {abcdf} and {abcef}, which obviously contains some reducts from **K**. In consequence {abcdf} and {abcef} must be removed from **K'**. Thus **K'** becomes an empty family and one can process the next object.

The first few objects usually generate a substantial amount of work (if performed manualy). Let us break the monitoring the algorithm for a few objects.

The processing of the seventh object results in the following **K**.

K = {{ad},{aef},{bcd},{bef},{cef},{de},{df}}

After the operation **COMP** applied for the eight object and reducts from **K**,

K := {{bcd},{cef}}

and

K' := {{abd},{acd},{ade},{adf},{bde},{cde},{def},{bdf}, {cdf},{abef},{acef},{adef},{bcef},{cdef}}

After removing supersets for current reducts

K' := {{abd},{acd},{ade},{adf},{bde},{cde},{def},{bdf},{cdf},{abef},{adef}}

COMP(x_7,**X**,**P**) for all **P** $\in$ **K'**, results in the updated set of reducts:

K = {abd},{acd},{adf},{bcd},{bde},{cde},{cdf},{cef},{def}

and the next updated **K'**

K' = {{abde},{acde},{adef},{abdf},{bcdf},{bdef},{abcef},{abdef},

All of the above sets are supersets for current reducts and can be removed from **K'**. **K'** becomes empty and the processing of the eight objects is complete.

The arrival of the ninth object results in the violation of the reducts {acd},{adf},{cde},{cdf},{cef} and {def}

Hence the updated intermediate set of reducts is reduced to

K := {{abd},{bcd},{bde}}

and new

K' := {{abcd},{acde},{acdf},{abdf},{adef},{bcde}, {cdef},{bcdf},{acef},{bcef},{bdef}}

After checking the sets from **K'** for the containment of reducts

K' := {{acde},{acdf},{adef},{cdef},{acef},{bcef}}

Operation **COMP** returns empty set only for {bcef}. In the new **K'**

K' := {abcde},{abcdf},{abcef},{abdef},{acdef},{bcdef}}

all sets but one, {acdef}, contain a reduct from current **K** and after their removal and update, **K'** = {{abcdef}}. Obviously the result is positive when checking this set from **K'** for the containment of reducts and, as a consequence, **K'** becomes empty. Before the arrival of a new object the set **K** is following:

K := {{abd},{bcd},{bde},{bcef}}

On the arrival of the tenth object the reduct {abd} is violated. **K** := {{bcd},{bde},{bcef}} and **K'** = {{abcd},{abde},{abdf}}. Only {abdf} does not contain a reduct. Operation **COMP** performed on {abdf} disqualifies it as a reduct and consequently the updated **K'** is comprised of {abcdf} and {abdef}. These sets contain reducts from the last set **K** and therefore **K'** becomes empty.

The operation **COMP** applied to the last sets of reducts on the next object (eleventh) returns empty sets and therefore all the reducts **K** := {{bcd},{bde},{bcef}} are waiting for the twelfth object.

The operation **COMP** applied on the twelfth object to the last sets of reducts returns nonempty set only for {bde}. Thus **K** := {{bcd},{bcef}} and **K'** := {{abde},{bcde},{bdef}}. {bcde} contains the reduct {bcd}, operation **COMP** on {abde} has empty result, so it becomes a new reduct. Operation **COMP** on {bdef} gives a nonempty result, and as an effect **K'** := {{abdef},{bcdef}}. Each of these sets contains a reduct so they have to be removed from **K'**. The processing of the twelfth object is complete and the set of reducts is **K** := {{abde},{bcd},{bcef}}

Demonstration of the performance of the algorithm (especially for the last few objects) should convince the reader about the lower complexity of computations than those of standard methods. The worst case of complexity will remain in the same range (non-polynomial with respect to the number of attributes) but there are definitely savings in the expected real-time performance. The implementations of the *bottom-up* and the *maintenance* algorithm were compared through the random samples of 100 objects and 10 attributes. When the number of reducts were small (3-5 reducts during most of the exercise), the computation time of the maintenance algorithm was approximately 11.5 times better than that of the other one. When the numbers used in a sample were from a larger domain and as a consequence the number of reducts was large (10-25), the time was still better but the ratio had dropped to approximately 3.

Obviously, this data should not be treated as imperative, as the preparation of the benchmark probably requires more attention than it had been given.

4. Reducts and decision tables

The Reduct Maintenance Algorithm can be used directly in the reduction of decision algorithms (both consistent and inconsistent) and, with minor alterations, in further reduction of decision algorithms to take an advantage of specific values of condition attributes of a decision or a group of decisions. We will focus rather on the direct application of the Maintenance algorithm, showing the relevant notions associated with the notion of reduct.

First, consider the reduction of a consistent decision algorithm. This notion can be expressed in an equivalent form, using the notion of functional dependency or weak reduct. Consider a decision table with condition and decision attributes, **C** and **D**, respectively.

(4.1) The (**C**,**D**) algorithm is consistent if one of the following conditions is satisfied.
(a) $\mathbf{C} \in \mathbf{WRED}(\mathbf{C} \cup \mathbf{D})$ or
(b) There is a functional dependency $\mathbf{C} \rightarrow \mathbf{D}$.

The reduction of the decision algorithms is related to the notion of left-reduced functional dependency:

(4.2) The subset **C'** of **C** is a reduct of **C** in the algorithm (**C**,**D**) if the functional dependency $\mathbf{C'} \rightarrow \mathbf{D}$ is left-reduced.

The notions of a reduct of **C**, and of a reduct of **C** in the algorithm (**C**,**D**) are different:

(4.3) REMARKS. (a) If the subset **C'** of **C** is a reduct of **C** in the algorithm (**C**,**D**), then **C'** is not necessarily a reduct of **C**.
(b) If **C'** is a reduct of **C** then **C'** is not necessarily a reduct of **C** in the algorithm (**C**,**D**)

An illustration to the Remarks (4.3) is provided on an example (Example 4).

EXAMPLE 4. Consider the decision table:

	C				D	
	a	b	c	d	e	f
1)	0	1	1	1	2	3
2)	1	0	3	1	2	2
3)	0	0	0	2	3	2
4)	1	0	2	1	2	2
5)	0	1	0	2	3	2

The sets {a,d}, {b,d} and {c} are reducts of {a,b,c,d} in the decision algorithm ({a,b,c,d},{e,f}) but none of them is a reduct of {a,b,c,d}. On other hand the set {b,c} is the only reduct of {a,b,c,d}, but it is not a reduct of {a,b,c,d} in the decision algorithm ({a,b,c,d},{e,f}). Notice that {a,d} is not a subset of a reduct of {a,b,c,d} either.

The subsequent conclusion can be drawn from above example.

(4.4) COROLLARY. If **C'** is a reduct of **C**, then **C'** contains a subset **C"** which is a reduct of **C** in the decision algorithm **(C,D)**.
If the decision algorithm **(C",D)** is a reduct of the decision algorithm **(C,D)** then **C"** is not necessarily a subset of a reduct of **C**.

The concept of the reduction of inconsistent decision algorithm can be expressed in terms of functional dependencies.

(4.5) The decision algorithm **(C',D)** is a reduct of the inconsistent decision algorithm **(C,D)** if **C'** $\subseteq$ **C** and there exists a left-reduced functional dependency **C'** $\rightarrow$ **C**.

The concept of reducability of attributes can be applied to the set of decision attributes. If the number of decision attributes is bigger than one, then there is a reason to check if there are dependencies between them. The existence of such dependecies can be interpreted as different levels of importance of attributes or it could serve simply as a means of two-level decision process.

(4.6) The decision algorithm **(C,D')** is a reduct of the decision algorithm **(C,D)** if **D'** $\subseteq$ **D** and there exists a left-reduced functional dependency **D'** $\rightarrow$ **D**.

From the definition one can see that if **(C,D)** is a consistent decision algorithm than the existing functional dependency **C** $\rightarrow$ **D** can be split into the composition of the two: **C'** $\rightarrow$ **D'** and **D'** $\rightarrow$ **D**. The first functional dependency corresponds to the lower level and the second one correponds to the higher level of decision making process.
The definition (4.6) remains in force for inconsistent decision tables as well.

From the characterisations presented for the most vital notions in the Pawlak' knowledge analysis one can see the importance of reducts. The practical computation process of the above notions requires the computation of reducts and this justifies the need for an efficient algorithm for their computation.

The essential question is whether and to which extent the decision tables should be the subject of reduction. On one hand the reduced set of rules represents the minimum knowledge necessary to arrive at appropriate decisions. Therefore the benefit of the knowledge reduction is a concise and condensed set of rules, not containing redundancies. Here, the word minimum means that a proper subset of that reduced knowledge is insufficient to make some decision(s). However, another important observation is that there could exist another reduced knowledge (another reduct) sufficient for that particular decision process. Depending on circumstances, a different minimum of knowledge could be more useful than another one, since it can employ different information, which is available or easily attainable.

The final, fully reduced decision algorithm is not necessarily the best means for the decision making process. The reason is clearly visible on the decision algorithm from Example 4 and explained in the corollary (4.4). The example indicates that there is a danger of losing information in the process of the reduction of the decision algorithm. In Example 4, since the attributes a and d depend on other attributes, they are superfluous and, as such, they can be removed because they are unnecessary for decision making. Consequence these attributes will not appear in the decision rules. One can see however, that they could be sufficient in decision making as the following rules are present:

$d_2 \rightarrow e_2\ f_2$

$a_1 \rightarrow e_2\ f_2$

$a_0\ b_0 \rightarrow e_3\ f_2$

These rules will be ignored if an elimination of dispensable attributes will be done in the first place. If the attributes c_4 or c_1 are easier accessible to the decision maker than c_3, the lack of the above rules imposes some limitations on her. If a decision maker wants to take an advantage of all possible decision rules, and thus increase the possibility of utilisation of the data, which is already on hand, the elimination of dispensable attributes should be performed with great care.

5. Conclusions

The process of reducts determination is an important part of the information analysis in Pawlak's Knowledge Representation Systems. There are straightforward, conceptualy simple algorithms for the determination of reducts. Because of the *quantitative* nature of the problem they are NP-hard (in respect of the number of attributes) and one cannot expect any algorithm of improved assymptotic worst case computational complexity. The only way to solve the problem more efficiently is to propose an algorithm which could be better by a polynomial factor. The proposed here algorithm for reducts determination gives some improvement over traditional approaches (in the real execution time) and moreover, due to its incremental reducts construction principle, it can be used as a tool for the maintenance of reducts in the dynamic environment. That principle can be used successfuly in the construction of an algorithm for the determination of the decision rules (decision algorithm) from the decision table. The construction of such an algorithm and its implementation is underway [6].

The NP completeness of the general problem of reducts determination was contrasted with the problem of the determination of reducts from sample data related to a given set of objects. There is an evidence that in the worst case (maximum possible number of reducts) the ratio of the number of reducts to number of tuples (objects) has a quadratic upper bound (in respect of the number of objects). This suggest that from the practical point of view the problem can be tractable as the size of the output is polynomialy comparable with the size of the output.

Other approach to the reducts determination emerge from other concept – functional dependency – which was defined and investigated in a number of papers. This approach is not data dependent and it is of little value for an application to the determination of value based reduced decision algorithm. The concept of functional dependency and other associated notions are useful however as media for an explanation of the notions of consistency of the decision algorithm, reduct of the decision algorithm, etc.. The notion of functional dependency is also instrumental in constructing examples even if its role is not displayed explicitly.

A brief discussion was provided about the necessity of the elimination of dependent attributes from the decision table and the influence that elimination can have on the quality of the decision making process. There is a need for a real-life example of a knowledge representation system where the elimination of superfluous attributes does not affect that quality.

References

[1] Buszkowski W, Orlowska E (1986), On the Logic of Database Dependencies, Bull. Polish Acad. SCi. Math., 34, pp 345–354.

[2] Cendrowska J (1987), PPRISM: An Algorithm for Inducing Modular Rules, Int. Journal of Man-Machine Studies, 27, pp 349-370.

[3] Codd E F (1970), A Relational Model of Data for Large Shared Data Banks, Communications of ACM, 13, pp 377-387.

[4] Maier D (1983), The Theory of Relational Databases, Computer Science Press, Rockville.

[5] Orlowska E (1983), Dependencies of Attributes in Pawlak's Information Systems, Fundamenta Informaticae, 6, pp 247–256.

[6] Orlowski M (1992), Maintenance of Decision Algorithms, School of Information Systems Report, 2/92, QUT Brisbane

[7] Pawlak Z (1981), Information Systems – Theoretical Foundations, Information Systems, 6, pp 205-218.

[8] Pawlak Z (1985), On Rough Dependency of Attributes in Information Systems, Bull. Polish Acad. SCi. Math., 33, pp 551–559.

[9] Pawlak Z (1990), Rough Sets; Theoretical Aspects of Reasoning about Data, (manuscript; monograph in submission)

[10] Pawlak Z, Slowinski K, Slowinski R (1986), Rough Classification of Patients after Highly Selective Vagotomy for Duodenal Ulcer Int. Journal of Man-Machine Studies, 24, pp 413-433.

[11] Rauszer C M (1985), Dependency of Attributes in Information Systems, Bull. Polish Acad. SCi. Math., 33, pp 551–559.

[12] Rauszer C M (1985), An Equivalence between Theory of Functional Dependencies and Fragment of Intuitionistic Logic, Bull. Polish Acad. SCi. Math., 33, pp 571–579.

[13] Rauszer C M (1988), Algebraic Properties of Functional Dependencies, Bull. Polish Acad. SCi. Math., 38, pp 561–569.

Part III
Chapter 2

THE DISCERNIBILITY MATRICES AND FUNCTIONS IN INFORMATION SYSTEMS

Andrzej SKOWRON Cecylia RAUSZER
Institute of Mathematics
University of Warsaw
02-097 Warsaw, Poland

Abstract. We introduce two notions related to any information system, namely the discernibility matrix and discernibility function. We present some properties of these notions and as corollaries we obtain several algorithms for solving problems related among other things to the rough definability, reducts, core and dependencies generation.

1. Introduction

Information systems (sometimes called data tables, attribute-value systems, condition-action tables etc.) are used for representing knowledge. The information system notion presented here is due to Pawlak and was investigated by several researchers (see e.g. [Paw81a,b, 85, 91], [Nov88a,b], [Grz86], [Che88], [Orl84a,b], [Pag87], [RaS86], [Rau84,85a,b,87], [Sko91], [Vak87,89], [Zia88]). Generalizations of the idea for the case when some data are missing from the table (so called incomplete information systems, or incomplete data bases) are discussed in [Lip79, 81], and others. This approach to information systems overlaps with various others [Cod70], [Sal68], [Wil82] however certain substantial differences should be noted.

An information system may be also viewed as a description of a knowledge base [Paw91]. Among the research topics related to information systems are: rough set theory, problems of knowledge representation, problems of knowledge reduction, dependencies in knowledge bases etc. The rough sets have been introduced by Pawlak [Paw82] as a tool to deal with inexact, uncertain or vague knowledge in AI applications e.g. knowledge based systems in medicine, natural language processing, pattern recognition, decision systems, approximate reasoning. Since 1982 the rough sets have been intensively studied and by now many practical applications based on the developed theory of rough sets have been implemented [Paw91].

A basic problem related to many practical applications of information systems is whether the whole set of attributes is always necessary to define a given partition of a universe. This problem is referred to as knowledge reduction [Grz86], [Los79], [Nov88a,b, 89], [Rau85a,b, 88, 90], [Paw91]. The knowledge reduction problem is related to the general concept of independence [Mar58], [Gla79]. Knowledge reduction consists in removing superfluous attributes from the information system in such a way that the set of remaining attributes preserves only this part of knowledge which is really useful.

Dependencies in an information system ([Paw85,85a], [Nov88a,b,89], [Bus86] and [Rau84,85a,b,87,88,90] are basic tools for drawing conclusions from basic knowledge. They express some of the relationships between the basic categories described in the information system. A process of building theories is based on discovering inference rules of the form "if ... then ... " i.e. how new knowledge can be derived from a given one.

The paper is an extension of the ideas presented in [Sk91], where two basic notions, namely discernibility matrix and discernibility function of an information system, have been introduced. Applying these notions we will show several efficient methods for computing certain objects related to information systems such as reducts, core, dependencies etc. We give an upper bound for the time complexity of proposed solutions. In most cases the upper bound is of order n^2(where n is the number of objects in a given information system) because of the time necessary to process the discernibility matrix or function description. The procedures with polynomial complexity presented here and based on discernibility matrix and function properties seem to be more efficient than those presented before e.g. in [Paw91].

Among the considered problems the one with the highest complexity is the problem of generating the reduct set for a given information system. We show that the problem is polynomially (with respect to deterministic time reducibility) equivalent to the problem of transforming the conjunctive form of any monotonic boolean function to the (reduced) disjunctive form (by applying multiplication and absorption laws whenever possible). We also present some remarks about practical implementations of described procedures [Sk91]. We show that the problems of generating minimal (relative) reducts and of generating minimal dependencies are NP-hard.

2. Information Systems and rough sets

In this section we present basic notions related to information systems and rough sets.

An *information system* is a pair $\mathbf{A} = (U, A)$, where

U - a nonempty, finite set called the *universe* and

A - a nonempty, finite set of attributes i.e.

$a : U \to V_a$ for $a \in A$,

where V_a is called the *value set* of a.

If $\mathbf{A} = (U, A)$ is an information system and $B \subseteq A$, then by $\mathbf{A}|B$ we denote the information system (U, B), called the *restriction* of $\mathbf{A}$ to B.

With every subset of attributes $B \subseteq A$ we associate a binary relation $IND(B)$, called *B-indiscernibility relation*, and defined as follows:

$$IND(B) = \{(x, y) \in U : \text{ for every } a \in B,\ a(x) = a(y)\}.$$

Then $IND(B)$ is an equivalence relation and

$$IND(B) = \bigcap_{a \in B} IND(a).$$

The objects x, y satisfying the relation $IND(B)$ are indiscernible with respect to attributes from B. In other words, we cannot distinguish x from y in terms of attributes in B. By $[x]_A$ we denote the equivalence class of $IND(A)$ including x, i.e. the set $\{y \in U : x\, IND(A)\, y\}$.

By $DIS(B)$, for $B \subseteq A$, we denote the relation $U^2 - IND(B)$ called the *B-discernibility relation.*

For any $B' \subseteq B \subseteq A$ we have $DIS(B') \subseteq DIS(B)$.

The value $a(x)$ assigned to an object x by an attribute a can be viewed as a name (or a description) of the primitive category of a to which x belongs i.e. $a(x)$ is the name of $[x]_{IND(a)}$. The name (description) of a category of $B = \{a_{i_1}, \ldots, a_{i_k}\} \subseteq A$ containing an object x is the sequence

$$(a_{i_1}, a_{i_1}(x)), \ldots, (a_{i_k}, a_{i_k}(x)).$$

Some subsets (categories) of objects in an information system cannot be expressed exactly in terms of the available attributes, they can be only roughly defined.

If $\mathbf{A} = (U, A)$ is an information system, $B \subseteq A$ and $X \subseteq U$ then the sets

$$\{x \in U : [x]_B \subseteq X\} \text{ and } \{x \in X : [x]_B \cap X \neq \emptyset\}$$

are called the *B-lower* and the *B-upper approximation* of X in $\mathbf{A}$, respectively, and they are denoted by $\underline{B}X$ and $\overline{B}X$, respectively. One can interpret $\underline{B}X$ and $\overline{B}X$ as the interior and the closure of a set X in the topology generated by the equivalence classes of the indiscernibility relation $IND(B)$ [Sk88]. The set $BN_B(X) = \overline{B}X - \underline{B}X$ will be called the *B-boundary region* of X.

Clearly, $\underline{B}X$ is the set of all elements of U, which can be with certainty classified as elements of X with respect to the knowledge represented by the attributes in B; and $\overline{B}X$ is the set of those elements of U which can be possibly classified as elements of X with respect of the knowledge represented by the attributes in B; finally, $BN_B(X)$ is the set of elements which can be classified neither in X nor in $-X$ basing on the knowledge B.

A set X is said to be *B-definable* if $\overline{B}X = \underline{B}X$. It is easy to observe that $\underline{B}X$ is the greatest B-definable set contained in X, whereas $\overline{B}X$ is the smallest B-definable set containing X.

Impreciseness of a set (category) is due to the existence of a boundary region. The following *accuracy measure* was introduced in [Paw82]:

$$\alpha_B(X) = \frac{card\, \underline{B}X}{card\, \overline{B}X}$$

where $X \neq \emptyset$.

The accuracy measure $\alpha_B(X)$ is intended to capture the degree of completeness of our knowledge about the set X. We have $0 \leq \alpha_B(X) \leq 1$, for any B and $X \subseteq U$; if $\alpha_B(X) = 1$ then the B-boundary region of X is empty and the set X is B-definable; if $\alpha_B(X) < 1$, then the set X has a non-empty B-boundary region and consequently is B-undefinable.

Example 2.1. Let us consider a simple example of an information system represented in Table 1.

The example is taken from [Paw91]; it is related to the example given in [Cas89] in connection with the situation in the Middle East.

In the example $\mathbf{A} = (U, A)$, where $U = \{1, \ldots, 6\}$, $A = \{a, c, d, e, o\}$ and the values of the attributes are defined as in Table 1.

U	a	c	d	e	o
1	0	1	1	1	1
2	1	1	0	1	0
3	1	0	0	1	1
4	1	0	0	1	0
5	1	0	0	0	0
6	1	1	0	1	1

Table 1

The objects in U correspond to the following six parties:

1 - Israel 4 - Jordan

2 - Egypt 5 - Syria

3 - Palestinians 6 - Saudi Arabia

The relations between those parties are determined by the following issues

a - autonomous Palestinian state on the West Bank and Gaza

c - Israeli military outpost along the Jordan River

d - Israel retains East Jerusalem

e - Israeli military outposts on the Golan Heights

o - Arab countries grant citizenship to Palestinians who choose to remain within their borders.

The table represents the way the participants of the Middle East region interact with the above issues; 0 means that the participant is against and 1 - neutral or favorable towards the issue.

The task is to find essential differences between the participants of the debate, so that we can clearly distinguish their attitudes to the issues being discussed.

Here we will not go into the details of that task. The reader can find more details in [Paw91]. We only use the example to illustrate our method.

3. The discernibility matrix and the discernibility function

In this section we introduce two basic notions, namely those of a discernibility matrix and a discernibility function, which will help us to prove several properties and to construct efficient algorithms related to information systems, e.g. generation of reducts, core, dependencies. We prove also that these two notions are strongly related.

Let $\mathbf{A} = (U, A)$ be an information system and let us assume that $A = \{a_1, \ldots, a_m\}$, $U = \{x_1, \ldots, x_n\}$.

By $M(\mathbb{A})$ we denote an $n \times n$ matrix (c_{ij}), called the *discernibility matrix* of $\mathbb{A}$, such that

$$c_{ij} = \{a \in A : a(x_i) \neq a(x_j)\} \text{ for } i,j = 1,\ldots,n\,.$$

Since $M(\mathbb{A})$ is symmetric and $c_{ii} = \emptyset$ for $i = 1,\ldots,n$, we represent $M(\mathbb{A})$ only by elements in the lower triangle of $M(\mathbb{A})$, i.e. the c_{ij}'s with $1 \leq j < i \leq n$.

A *discernibility function* $f_{\mathbb{A}}$ for an information system $\mathbb{A}$ is a boolean function of m boolean variables $\overline{a}_1,\ldots,\overline{a}_m$ corresponding to the attributes $a_1,\ldots,a_m$, respectively, and defined as follows:

$$f_{\mathbb{A}}(\overline{a}_1,\ldots,\overline{a}_m) = \wedge\{\vee(c_{ij}) : 1 \leq j < i \leq n, c_{ij} \neq \emptyset\}$$

where $\vee(c_{ij})$ is the disjunction of all variables $\overline{a}$ such that $a \in c_{ij}$.

In the sequel we shall write a_i instead of $\overline{a}_i$ when no confusion can arise.

From the definition of the discernibility matrix $M(\mathbb{A})$ we have the following:

Proposition 3.1. *Let $\mathbb{A} = (U, A)$ and $B \subseteq A$. If for some i,j we have $B \cap c_{ij} \neq \emptyset$, then $x_i DIS(B) x_j$. In particular, if $\emptyset \neq B \subseteq c_{ij}$ for some i,j, then $x_i DIS(B) x_j$. Moreover $M(\mathbb{A}|B) = (c_{ij} \cap B)$.* □

Example 3.1. For the information system $\mathbb{A}$ from Example 2.1 we obtain the discernibility matrix presented in Table 2, and the following discernibility function:

$$f_{\mathbb{A}}(a,c,d,e,o) = c \wedge e \wedge o \wedge$$
$$\wedge(a \vee d) \wedge (c \vee e) \wedge (c \vee o) \wedge (e \vee o) \wedge$$
$$\wedge(a \vee c \vee d) \wedge (a \vee d \vee o) \wedge (c \vee e \vee o \wedge$$
$$\wedge(a \vee c \vee d \vee o) \wedge$$
$$\wedge(a \vee c \vee d \vee e \vee o)\,.$$

	1	2	3	4	5	6
1						
2	*ado*					
3	*acd*	*co*				
4	*acdo*	*c*	*o*			
5	*acdeo*	*ce*	*eo*	*e*		
6	*ad*	*o*	*c*	*co*	*ceo*	

Table 2
The discernibility matrix $M(\mathbb{A})$
$c_{ii} = \emptyset$ and $c_{ij} = c_{ji}$ for $i,j = 1,\ldots,6$

□

We assume that any of the tasks listed below can be performed in a number of steps proportional to the number m of attributes in A:

1. for given **A** and i, j, compute c_{ij};
2. for given variables x and y with values $val(x)$ and $val(y)$ being subsets of a given attribute set A, compute $val(x) \cap val(y)$, $val(x) \cup val(y)$, $A - val(x)$ or test whether $val(x)$ is empty.

Moreover, unless otherwise stated, we assume that the number of attributes m is fixed. In consequence, the complexity bounds, related to that case whem m is fixed, are given in terms of the number of objects n in the universe U of the information system **A**.

Proposition 3.2. *The problem of computing the discernibility matrix* $M(\mathbf{A})$ *from* **A** *is in* $DTIME(n^2)$.

Proof. $M(\mathbf{A})$ has n^2 elements of the form c_{ij} and the number of steps for computing of any c_{ij} is bound by a constant. □

Below we present some examples of problems related to information systems. Their solutions are based on the properties of the discernibility matrix and function.

MP. (The Membership Problem)

Input: $M(\mathbf{A})$ for a given information system $\mathbf{A} = (U, A)$ with $U = \{x_1, \ldots, x_n\}$,

$x \in X \subseteq U$ and $B \subseteq A$.

Output: 1: if $x \in \underline{B}X$

0: if $x \notin \underline{B}X$.

We assume that the elements of X are represented by marking in $M(\mathbf{A})$ all rows (columns) corresponding to them. The element x is represented in a similar way.

Proposition 3.3. **MP**$\in$ $DTIME(n)$.

Proof. We choose in $M(\mathbf{A})$ the column j corresponding to x and mark in it all positions which are in rows corresponding to the objects indiscernible from x_j in terms of the attributes from B, i.e. all places (i, j) in that column with the property $c_{ij} \cap B = \emptyset$. The number of performed steps is bounded by n.

We check in a number of steps proportional to n if

$$\{x_i \in U : c_{ij} \cap B = \emptyset \text{ and } 1 \leq i \leq n\} \subseteq X$$

verifying if all previously marked places are in the rows corresponding to the elements of X. If the answer is "yes" the output is 1, otherwise 0. □

Example 3.2. We check if $2 \in \underline{\{c,e\}}\{1,2,3\}$ for the data table from Example 2.1. We choose column 2 in the discernibility matrix $M(\mathbf{A})$ from Example 3.1 and mark rows 1,2,6. Since $6 \notin \{1,2,3\}$, we have $2 \notin \underline{\{c,e\}}\{1,2,3\}$. □

D. (Definability Problems)

D.1 (Definability of Sets)

Input: $M(\mathbf{A})$ for a given information system $\mathbf{A} = (U, A)$

with $U = \{x_1, \ldots, x_n\}$, and $X \subseteq U$, $B \subseteq A$.

Output: 1: if $\underline{B}X = X$

0: if $\underline{B}X \neq X$.

D.2. (Rough Definability of Sets)

Input: $M(\mathbb{A})$ for a given information system $\mathbb{A} = (U, A)$

with $U = \{x_1, \ldots, x_n\}$, and $X \subseteq U$, $B \subseteq A$.

Output: 1: if $\underline{B}X \neq \emptyset$ and $\overline{B}X \neq U$

0: otherwise.

D.3. (Internal Undefinability of Sets)

Input: $M(\mathbb{A})$ for a given information system $\mathbb{A} = (U, A)$

with $U = \{x_1, \ldots, x_n\}$, and $X \subseteq U$, $B \subseteq A$.

Output: 1: if $\underline{B}X = \emptyset$ and $\overline{B}X \neq U$

0: otherwise.

D.4. (External Undefinability of Sets)

Input: $M(\mathbb{A})$ for a given information system $\mathbb{A} = (U, A)$

with $U = \{x_1, \ldots, x_n\}$, and $X \subseteq U$, $B \subseteq A$.

Output: 1: if $\underline{B}X \neq \emptyset$ and $\overline{B}X = U$

0: otherwise.

D.5. (Total Undefinability of Sets)

Input: $M(\mathbb{A})$ for a given information system $\mathbb{A} = (U, A)$

with $U = \{x_1, \ldots, x_n\}$, and $X \subseteq U$, $B \subseteq A$.

Output: 1: if $\underline{B}X = \emptyset$ and $\overline{B}X = U$

0: otherwise.

Proposition 3.4. D.1, D.2, D.3, D.4, D.5 $\in DTIME(n^2)$.

Proof. We prove only the first two cases. The proof of all the other cases is similar and is left to the reader.

In the first case we check successively for every element $x_i \in X$ whether $x_i \in \underline{B}X$. To do this it is sufficient to apply Proposition 3.3.

In the case of **D.2** we check successively for every x in X applying Proposition 3.3, a condition which at the beginning is a pair: $(x \in \underline{B}X,\, x \in \underline{B}(U - X))$. As long as both conditions are false (i.e. $x \in BN_B(X)$) we choose a new element from X (if it exists), and again check the same condition. If one condition is true, then in the next steps we only check the second condition. The procedure halts if it was possible to match both conditions (in different steps), or all elements of X were checked, but it was not possible to match both conditions. In the first case the output is 1, in the second — 0. More precisely, we can describe our algorithm in the following way:

Algorithm D2

(var $M(\mathbb{A})$: **array** $[1 \ldots n(n-1)/2]$ **of ATTRIBUTE_SET**;

```
    var B : ATTRIBUTE_SET;
    var X : OBJECT_SET;
    var ANSWER : BOOLEAN)
begin
    var LOW, UP : BOOLEAN;
    LOW := UP := false;
    TEMP := X;
    while TEMP ≠ ∅ and (LOW and UP =false) do begin
    x := CHOOSE(TEMP); //x is an element of TEMP//
    if LOW =false then
          if x ∈ B̲TEMP then LOW :=true end_if
    //Checking the if condition we apply Proposition 3.3.//
    else if UP =false then
          if x ∈ B̲(U − TEMP) then UP :=true end_if
    //Checking the if condition we apply Proposition 3.3.//
    end_if;
    TEMP := TEMP − [x]; //[x] is the eqivalence class of IND(B) including x//
    end_do;
ANSWER := LOW and UP
end
```

In any computation of the above algorithm the **while** loop is performed at most n times. By Proposition 3.3 the number of steps in the body of that **while** loop is of order $O(n)$. Hence **D.2**$\in DTIME(n^2)$. □

One can extend our considerations to problems related to rough sets with accuracy measure application.

4. Core and reducts in an information system

Two fundamental concepts — reduct and core — are considered in connection with knowledge reduction. A reduct of knowledge is its essential part, which suffices to define all basic concepts occurring in the considered knowledge, whereas the core is the common part of all reducts.

Let $\mathbb{A} = (U, A)$ be an information system.

An attribute $a \in B \subseteq A$ is *dispensable* in B if

$$IND(B) = IND(B - \{b\}),$$

otherwise b is *indispensable* in B.

The set of all indispensable attributes in A is called the *core* of $\mathbb{A}$, and is denoted by $CORE(\mathbb{A})$.

One can verify that $CORE(\mathbb{A})$ can be characterized by $M(\mathbb{A})$ in the following way:

Proposition 4.1. $CORE(\mathbb{A}) = \{a \in A : c_{ij} = \{a\} \text{ for some } i, j\}$.

Proof. Let $B = \{a \in A : c_{ij} = \{a\} \text{ for some } i, j\}$. We will show $CORE(\mathbb{A}) = B$.
($\subseteq$) Let $a \in CORE(\mathbb{A})$. Then $IND(A) \subsetneq IND(A - \{a\})$, so there exist x_i and x_j which are indiscernible with respect to $A - \{a\}$ but discernible by a. Hence $c_{ij} = \{a\}$.

($\supseteq$) If $a \in B$ then for some i and j we have $c_{ij} = \{a\}$. Hence a is indispensable in A. □

CORE (Core Problem).

Input: $M(\mathbf{A})$ for a given information system $\mathbf{A} = (U, A)$
with $U = \{x_1, \ldots, x_n\}$.

Output: $CORE(\mathbf{A})$.

Proposition 4.2. CORE *problem is in* $DTIME(n^2)$.

Proof. It follows from Proposition 4.1 that it is enough to find the union of all singletons in $M(\mathbf{A})$. This will require the number of steps proportional to the number of places in $M(\mathbf{A})$. □

We say that $B \subseteq A$ is *independent* in $\mathbf{A}$ if every attribute from B is indispensable in B. Otherwise the set B is said to be *dependent* in $\mathbf{A}$.

From the last definition and the definition of $M(\mathbf{A})$ we obtain:

Proposition 4.3. *A set* $B \subseteq A$ *is independent in* $\mathbf{A}$ *iff* $B = CORE(\mathbf{A}|B)$.

Proof. Observe that B is independent in $\mathbf{A}$ iff for every $a \in B$ there exist $i, j (1 \leq j < i \leq n)$ such that $c_{ij} \cap B = \{a\}$. Now it is sufficient to apply Propositions 3.1 and 4.1. □

I. (The Independece Problem)

Input: $M(\mathbf{A})$ for a given information system $\mathbf{A} = (U, A)$
with $U = \{x_1, \ldots, x_n\}$, and $B \subseteq A$.

Output: 1: if B is independent in $\mathbf{A}$
0: otherwise.

Proposition 4.4. I$\in$ $DTIME(n^2)$.

Proof. Let $M(\mathbf{A}) = (c_{ij})$. We apply Proposition 4.3. First we construct $M(\mathbf{A}|B)$ computing all intersections $c_{ij} \cap B$ marking all singletons and computing their union W. This will require at most a number of steps of order $O(n^2)$. Next we check, if B is equal to W. If the result of that test is true the output is 1 otherwise 0. □

A set $B \subseteq A$ is called a *reduct* in $\mathbf{A}$ if B is independent in $\mathbf{A}$ and $IND(B) = IND(A)$. The set of all reducts in $\mathbf{A}$ is denoted by $RED(\mathbf{A})$.

It follows from the definition that $B \in RED(\mathbf{A})$ iff B is a minimal (with respect to set theoretical inclusion) subset of A such that $IND(B) \subseteq IND(A)$ (or equivalently $DIS(A) \subseteq DIS(B)$).

Theorem 4.1 establishes some connections between reducts and the properties of the discernibility matrix and function.

By v_B we denote the characteristic function of the set $B \subseteq A = \{a_1, \ldots, a_m\}$, i.e. the function $v_B : \{0,1\}^m \to \{0,1\}$ such that $a \in B$ iff $v_B(a) = 1$.

Theorem 4.1. *Let* $\mathbf{A} = (U, A)$ *be an information system with* $U = \{x_1, \ldots, x_n\}$, $A = \{a_1, \ldots, a_m\}$ *and let* $\emptyset \neq B \subseteq A$. *The following conditions are equivalent:*

(i) B contains a reduct from $RED(\mathbb{A})$ i.e. $IND(A) = IND(B)$;

(ii) $f_{\mathbb{A}}(v_B(a_1), \ldots, v_B(a_m)) = 1$;

(iii) for all i and j such that $c_{ij} \neq \emptyset$ and $1 \leq j < i \leq n$

$$c_{ij} \cap B \neq \emptyset .$$

Proof. The equivalence (ii) $\Leftrightarrow$ (iii) follows from the constructions of the discernibility function $f_{\mathbb{A}}$ and the discernibility matrix $M(\mathbb{A})$.
(iii) $\rightarrow$ (i) By our assumption we have $c_{ij} \cap B \neq \emptyset$ for any i, j such that $1 \leq j < i \leq n$. It means that in B we have enough attributes to discern between all these objects from U, which are discernible with respect to all attributes in A, i.e. B contains a reduct from $RED(\mathbb{A})$.
(i) $\rightarrow$ (iii) If B contains a reduct X from $RED(\mathbb{A})$, then any two objects discernible with respect to some attributes from A are also discernible with respect to some attributes from $B \supseteq X$. Hence if $c_{ij} \neq \emptyset$ then $c_{ij} \cap B \neq \emptyset$ for any i, j. □

Now we prove the following theorem:

Theorem 4.2. *Let $\mathbb{A} = (U, A)$ be an information system, and let $\emptyset \neq B \subseteq A$. Then the set B is minimal with respect to one of the conditions from Theorem 4.1 iff it is minimal with respect to the remaining ones.*

Proof. Follows from Lemmas 4.1 and 4.2 proved below. □

Lemma 4.1. *Let $\mathbb{A} = (U, A)$ be an information system and let $\emptyset \neq B \subseteq A$. Then the set B is minimal with respect to condition (i) of Theorem 4.1 iff it is minimal with respect to condition (iii) of that Theorem.*

Proof. Let us denote by $MIN(M(\mathbb{A}))$ the family of all minimal (with respect to set inclusion) subsets B of A satisfying condition (iii). It suffices to prove that the following equality holds: $RED(\mathbb{A}) = MIN(M(\mathbb{A}))$.

($\Leftarrow$) Let $B \in MIN(M(\mathbb{A}))$. We will show that B is independent in $\mathbb{A}$ and $IND(B) = IND(A)$. Suppose B is dependent. Hence there exists an independent set of attributes $Y \subset B$ such that $IND(Y) = IND(B)$. Due to the minimality assumption about B, we have $Y \cap c_{ij} = \emptyset$ for some nonempty element c_{ij} of the discernibility matrix $M(\mathbb{A})$. Hence $x_i IND(Y) x_j$ and $x_i DIS(B) x_j$, a contradiction. For the rest of the proof it suffices to show that $IND(B) \subseteq IND(A)$. Assume $x_i DIS(A) x_j$. Hence $c_{ij} \neq \emptyset$. Then by the minimality of B we obtain $c_{ij} \cap B \neq \emptyset$, so $x_i DIS(B) x_j$.

($\Rightarrow$) Assume now that $B \in RED(\mathbb{A})$. We will show that

$$B \in MIN(M(\mathbb{A})) .$$

Suppose first that $c_{ij} \cap B = \emptyset$ for some nonempty element c_{ij} of $M(\mathbb{A})$. Since $c_{ij} \neq \emptyset$ and x_i and x_j are discernible by c_{ij}, we would have $x_i DIS(c_{ij}) x_j$ and $x_i DIS(A) x_j$ (because $IND(A) \subseteq IND(c_{ij})$). By the definition of $M(\mathbb{A})$ we would also have

$x_i IND(A - c_{ij}) x_j$. As B is a reduct, $IND(B) = IND(A)$. By our assumption $B \subseteq A - c_{ij} \subseteq A$, so $IND(A - c_{ij}) = IND(A)$. Hence $x_i IND(A) x_j$, a contradiction. Suppose now that B is not minimal. Then for some $B' \subset B$ and for all nonempty elements c_{ij} of $M(\mathbb{A})$ we would have $c_{ij} \cap B' \neq \emptyset$. Since B is a reduct, there exist x_k and x_l discernible with respect to B and indiscernible with respect to B' i.e. $x_k DIS(B) x_l$ and $x_k IND(B') x_l$. Otherwise, if for all objects x, y from U $xDIS(B)y$ implied $xDIS(B')y$, we would have $IND(B') \subseteq IND(B)$ which contradicts the assumption that B is a reduct. However if $x_k IND(B') x_l$ then by definition of $M(\mathbb{A})$ $B' \cap c_{kl} = \emptyset$, a contradiction. So B is minimal. □

Lemma 4.2. *Let $\mathbb{A} = (U, A)$ be an information system and let $\emptyset \neq B \subseteq A$. Then the set B is minimal with respect to the condition (i) of Theorem 4.1 iff it is minimal with respect to condition (ii) of that Theorem.*

Proof. Let us recall that we agred to use the same notation $a_1, \ldots a_m$ for attributes and for boolean variables corresponding to them.

If f is a boolean function of m variables $a_1, \ldots, a_m$ and $v : \{a_1, \ldots, a_m\} \to \{0, 1\}$ then by $f \circ v$ we denote the value $f(v(a_1), \ldots, v(a_m))$. Let $\mathbb{A} = (U, A)$. By $MIN(f)$ we denote the set:

$$\{B \subseteq A : f(v_B(a_1), \ldots v_B(a_m)) = 1 \text{ and } f(v_{B'}(a_1), \ldots v_{B'}(a_m)) = 0 \text{ for any } B' \subset B\}.$$

It is sufficient to prove the following equality:

$$RED(\mathbb{A}) = MIN(f_{\mathbb{A}}).$$

Let g be the reduced disjunctive form of $f_{\mathbb{A}}$ obtained from $f_{\mathbb{A}}$ by applying the multiplication and absorption laws as many times as possible. Then there exist l and $X_i \subseteq \{a_1, \ldots, a_m\}$ for $i = 1, \ldots, l$ such that

$$g = \bigwedge X_1 \vee \ldots \vee \bigwedge X_l.$$

It is sufficient to prove that $RED(\mathbb{A}) = \{X_1, \ldots, X_l\}$.

First we prove that $X_k \in RED(\mathbb{A})$ for $k = 1, \ldots, l$.

Let us observe that if $c_{ij} \neq \emptyset$, then $X_k \cap c_{ij} \neq \emptyset$ for any i, j. In fact, suppose $X_k \cap c_{i_0 j_0} = \emptyset$ and $c_{i_0 j_0} \neq \emptyset$ for some i_0, j_0. From the definition of the discernibility function $f_{\mathbb{A}}$ we have $f_{\mathbb{A}} = (\bigvee c_{i_0 j_0}) \wedge h$ for some h. Hence for the valuation $v(a) = 1$ iff $a \in X_k$ we could have $f_{\mathbb{A}} \circ v = 0$ and $g \circ v = 1$, a contradiction.

X_k is also a minimal set (with respect to set theoretical inclusion) in the family of all subsets X of A satisfying the condition: if $c_{ij} \neq \emptyset$, then $X \cap c_{ij} \neq \emptyset$, for any i, j. Indeed, in the opposite case one could find $Y \subset X$ such that $c_{ij} \neq \emptyset$ implies $Y \cap c_{ij} \neq \emptyset$ for any i, j. Hence for the valuation $v(a) = 1$ iff $a \in Y$ we would have $f_{\mathbb{A}} \circ v = 1$ and $g \circ v = 0$.

We proved that $X_k \in MIN(M(\mathbb{A}))$. Hence by Lemma 4.1 $X_k \in RED(\mathbb{A})$.

Now, assume that $X \in RED(\mathbb{A})$. First we will show $X_i \subseteq X$ for some i. Suppose that $X_i - X \neq \emptyset$ for any i. Hence for every i one could find $a_i \in X_i - X$. Let Y be the set of all those a_i's. Then $Y \cap X = \emptyset$. Hence for the valuation $v(a) = 0$ iff $a \in Y$ we would have $g \circ v = 0$ and $f_{\mathbb{A}} \circ v = 1$ (because from Lemma 4.1, $X \cap c_{ij} \neq \emptyset$ for any i, j, if $c_{ij} \neq \emptyset$, and by definition $v(a) = 1$ for $a \in X$). This contradiction proves that $X_{i_0} \subseteq X$ for some i_0.

Now it is sufficient to prove that $X_{i_0} = X$. Suppose $a' \in X - X_{i_0}$ for some a'. Then, taking the valuation $v(a) = 1$ iff $a \in X_{i_0}$, we would have $g \circ v = 1$ and $f_{\mathbb{A}} \circ v = 0$ (which follows from the minimality of X proved by Lemma 4.1). The contradiction proves that $X = X_{i_0}$. □

Corollary 4.1. *Let $\mathbb{A} = (U, A)$ be an information system. If f' is the result of applying finitely many times the absorption or multiplication laws to the initial conjunctive form of the discernibility function $f_{\mathbb{A}}$ then $MIN(f_{\mathbb{A}}) = MIN(f')$. If M' is the result of substituting the empty set $\emptyset$ for some elements c_{ij} of $M(\mathbb{A})$ such that $\emptyset \neq c_{kl} \subseteq c_{ij}$ for some k, l $(k, l) \neq (i, j)$, $1 \leq j < i \leq n$, $1 \leq l < k \leq n$ then $MIN(M(\mathbb{A})) = MIN(M')$.*

Proof. The corollary follows from the definitions of the discernibility matrix and function, and from the proofs of Lemmas 4.1 and 4.2. □

Remark. It is easy to observe a strong connection between the notions of a reduct in an information system $\mathbb{A}$ and a *prime implicant* [Weg87,Kau68]] of the monotonic boolean function $f_{\mathbb{A}}$, namely we have the following equivalence:

$$\{a_{i_1}, \ldots, a_{i_p}\} \in RED(\mathbb{A}) \text{ iff } a_{i_1} \wedge \ldots \wedge a_{i_p} \text{ is a prime implicant of } f_{\mathbb{A}}.$$

It follows from our remark that for computing the reduct set of a given information system $\mathbb{A}$ one can apply known algorithms for computing all prime implicants of a given monotonic boolean function (in our case, $f_{\mathbb{A}}$).

Using the properties presented above, one can show the correctness of decision procedures related to reducts which we are going to develop in the sequel.

RM. (Reduct Membership Problem)

Input: $M(\mathbb{A})$ for a given information system $\mathbb{A} = (U, A)$
with $U = \{x_1 \ldots, x_n\}$, and $\emptyset \neq B \subseteq A$.

Output: 1: if $B \in RED(\mathbb{A})$
0: if $B \notin RED(\mathbb{A})$.

Proposition 4.5. **RM**$\in DTIME(n^2)$.

Proof. To solve **RM** it suffices to check whether the intersection $c_{ij} \cap B$ is non-empty for any i, j $(1 \leq j < i \leq n)$ such that $c_{ij} \neq \emptyset$. If $c_{ij} \cap B = \emptyset$ and $c_{ij} \neq \emptyset$ for some i, j then the output is 0, because by Theorem 4.1 the set B cannot contain any reduct. Otherwise it suffices to apply Proposition 4.4 and check whether B is independent. If B is independent in $\mathbb{A}$, then the output is 1, otherwise 0. The time complexity of the procedure is $O(n^2)$. □

RS. (Reduct Set Problem)

Input: $M(\mathbb{A})$ for a given information system $\mathbb{A} = (U, A)$
with $U = \{x_1 \ldots, x_n\}$.

Output: RED($\mathbb{A}$).

Theorem 4.3. *The reduct set problem* **RS** *is polynomially equivalent (with respect to polynomial time reducibility) to the problem of transforming any conjunctive form of a monotonic boolean function (different from $f_0 \equiv 0$) to a reduced disjunctive form.*

Proof. In time $O(n^2)$ one construct the function $f_{\mathbb{A}}$ in a conjunctive form. After transforming $f_{\mathbb{A}}$ to a disjunctive form by applying multiplication and absorption laws whenever possible, we get the reduced disjunctive form of $f_{\mathbb{A}}$, which possesses the following property: each disjunct is a conjunction of the variables corresponding to the attributes forming the reduct, and for each reduct there exists a disjunct obtained by taking the conjunction of all variables corresponding to the attributes from that reduct.

This property follows from our last theorem and corollary.

On the other hand, for any monotonic boolean function f (different from $f_0 \equiv 0$) one can construct an information system $\mathbb{A}$ such that $f_{\mathbb{A}} = f$. The time complexity of the construction is bounded by $p(n)$, where n is the length of the standard code of f, and p is a polynomial. □

Example 3.3. Applying the absorption law to the discernibility function $f_{\mathbb{A}}$ from Example 3.2, and performing the multiplications we obtain the following expression:

$$c \wedge e \wedge o \wedge (a \vee d).$$

Hence the reduct set $RED(\mathbb{A})$ is $\{\{a, c, e, o\}, \{c, d, e, o\}\}$. By Proposition 4.1, we have $CORE(\mathbb{A}) = \{c, e, o\}$. □

For practical implementations the conclusion of the last Theorem is not optimistic, at least in the worst case. Fortunately, in applications it is possible to compute the reduct set, or at least to obtain certain description of a set containing the reduct set with an upper bound on the cardinality of the reduct set.

In practical implementation, we apply several simple rules to improve the efficiency of the procedures generating the reduct set (or its description). Let us consider two of them.

In many applications, it is possible to discern different objects using a few attributes only. In the first step we apply the absorption law to eliminate all elements of $M(\mathbb{A})$ (more precisely, we replace them by the empty set $\emptyset$), whose subsets are some other elements of $M(\mathbb{A})$ containing less attributes than a fixed threshold, e.g. 2 or 3. After that we often obtain a matrix with a small number of elements and direct calculation can be applied to obtain the reduct set. Correctness of this rule follows from the properties presented before.

For information systems with a huge number of reducts, we construct the discernibility function and perform reduction applying multiplication and absorption rules. We try to keep in each disjunct at least one conjunct with as small a number of elements as possible. If the time reserved for the reduct set computation is over (or there is no more memory) and the set has not been generated yet, the procedure halts. Then the output contains a description of the received conjunction of expressions together with the upper bounds on the possible number of reducts and the minimal reduct length.

MR. (Minimal Reduct Problem)

Input: information system $\mathbf{A} = (U, A)$

with $A = \{a_1, \ldots, a_m\}$.

Output: $Z \subseteq A$ such that $card(Z) = min\{card(Y) : Y \in RED(\mathbb{A})\}$.

Proposition 4.6. **MR**$\in NP-HARD$.

Proof. Let us consider the so called HITTING SET PROBLEM [GJ79]: For a given collection C of subsets of a finite set S and a positive integer $K \leq card(C)$ determine whether there is a subset $S' \subseteq S$ with $card(S') \leq K$ such that S' contains at least one element from each subset in C. One can easily show that the HITTING SET PROBLEM is polynomial-time reducible to the following HITTING REDUCT PROBLEM : For a given discernibility matrix $M(\mathbb{A})$ of an information system $\mathbb{A}$ and a natural number K determine whether there is a reduct $B \in RED(\mathbb{A})$ with $card(B) \leq K$. In consequence the HITTING REDUCT PROBLEM is NP-COMPLETE. Hence using the standard techniques [GJ79] we obtain that our **MR** (optimization) problem is NP_HARD. □

Despite the above result, one can construct some heuristics for generating solutions "close" to optimal ones. Let us consider one of them.

First we construct the discernibility function $f_{\mathbb{A}}$. Then we look for a pair of conjuncts such that after multiplying them and applying the absorption law, at least one of the obtained disjuncts will have length one. As long as it is possible to find such a pair, we multiply its elements, apply the absorption law and continue the procedure. If such a pair cannot be found, we obtain the output by simply taking all single variables appearing as disjuncts in conjuncts of the function obtained from $f_{\mathbb{A}}$.

5. Relative core and reducts in a decision table

We define in a natural way the notion of an extension of an information system.

Let $\mathbb{A} = (U, A)$ be an information system, and let $a^* \notin A$ be a function from U into a finite set V^*. The information system $\mathbb{A}^* = (U, A \cup \{a^*\})$, is called an *extension* of $\mathbb{A}$ by a^* (or an *a^*-extension* of $\mathbb{A}$). This is a special case of a so called *decision table.* The decision tables [Paw91] are information systems with the set of attributes divided into two disjoint sets C and D, called respectively condition and decision attributes. One can represent the set of attributes $D = \{d_1, \ldots, d_k\}$ by a single attribute a^*, assuming that $a^*(x) = (d_1(x), \ldots, d_k(x))$ for $x \in U$ (or one can take a code of that vector as the value $a^*(x)$). In this way we obtain an information system $(U, C \cup \{a^*\})$ called the representation of the decision table $(U, C \cup D)$, where C, D are nonempty disjoint sets of attributes. We define several notions, e.g. the relative reducts and core related to a decision table $(U, C \cup D)$ using the corresponding representation $(U, C \cup \{a^*\})$ i.e. an extension of (U, C) by a^*.

Example 5.1. The decision table presented here represents the control kiln algorithm for a rotary clinker kiln [Mro87]. The computer control algorithm, which is able to mimic the stoker's performance as a kiln controller, was derived from an observation of the decisions of a stoker who controls the kiln. We will not go into the details (concerning, e.g. the meaning of attributes), which can be found in the literature ([Mro87], [Paw91]). The actions of the stoker can be described by a decision table, where a, b, c and d are condition attributes, and e and f are decision attributes. Each combination of the condition attributes values corresponds to a specific kiln state, and in each state of the kiln appropriate action must be taken in order to obtain the required quality of cement. The objects correspond to different situations, in which the stoker is taking decisions.

Table 3 describes knowledge of an experienced human operator of the kiln obtained by observing the stoker's actions.

U	a	b	c	d	e	f
1	3	3	2	2	2	4
2	3	2	2	2	2	4
3	3	2	2	1	2	4
4	2	2	2	1	1	4
5	2	2	2	2	1	4
6	3	2	2	3	2	3
7	3	3	2	3	2	3
8	4	3	2	3	2	3
9	4	3	3	3	2	2
10	4	4	3	3	2	2
11	4	4	3	2	2	2
12	4	3	3	2	2	2
13	4	2	3	2	2	2

Table 3

U	a	b	c	d	a^*
1	3	3	2	2	1
2	3	2	2	2	1
3	3	2	2	1	1
4	2	2	2	1	2
5	2	2	2	2	2
6	3	2	2	3	3
7	3	3	2	3	3
8	4	3	2	3	3
9	4	3	3	3	4
10	4	4	3	3	4
11	4	4	3	2	4
12	4	3	3	2	4
13	4	2	3	2	4

Table 4

Let $\mathbf{A} = (U, A)$ where $U = \{1, \ldots, 13\}$, $A = \{a, b, c, d\}$ and let the condition attributes in A be defined as in Table 3.

An extension $\mathbf{A}^*$ of the information system $\mathbf{A} = (U, A)$ (with decision attributes e, f) is obtained by coding the two decision attributes e, f as one attribute a^* with possible values 1,2,3,4 (coding the corresponding pairs of values). The extension $\mathbf{A}^*$ has the form presented in Table 4.

The discernibility matrix for $\mathbf{A}^*$ is presented in Table 5.

If $\mathbf{A}^*$ is an a^*-extension of an information system $\mathbf{A} = (U, A)$ and $B \subseteq A$, then the set

$$\bigcup\{\underline{B}X : X - \text{equivalence class of the relation } IND(a^*)\}$$

is called the *B-positive* region of a^* (in $\mathbf{A}^*$) and is denoted by $POS_B(a^*)$.

Let us observe that in the case when $IND(B) \subseteq IND(a^*)$, we have $\underline{B}X = X$ for all $X \in IND(a^*)$, so the B-positive region of a^* (in $\mathbf{A}^*$) is equal to U.

The B-positive region of a^* (in $\mathbf{A}^*$) is the set of all objects of the universe U which can be properly classified into classes of $IND(a^*)$ on the basis of knowledge expressed by the classification defined by $IND(B)$.

PR. (Positive Region Problem)

Input: $\mathbf{A}^*$ — a^*-extension of $\mathbf{A} = (U, A)$

with $U = \{x_1, \ldots, x_n\}$, $B \subseteq A$ and $M(\mathbf{A}^*)$.

Output: $POS_B(a^*)$.

	1	2	3	4	5	6	7	8	9	10	11	12
1												
2	b											
3	b d	d										
4	a b a^* d	a a^* d	a a^*									
5	a a^* b	a a^*	a a^* d	d								
6	b a^* d	d a^*	d a^*	a a^* d	a a^* d							
7	d a^*	b a^* d	b a^* d	a b a^* d	a b a^* d	b						
8	a a^* d	a b a^* d	a b a^* d	a b a^* d	a b a^* d	a b	a					
9	a c a^* d	a d b c a^*	a d b c a^*	a d b c a^*	a b a^* c d	a b a^* c	a a^* c	c a^*				
10	a d b c a^*	a d b c a^*	a d b c a^*	a d b c a^*	a d b c a^*	a b a^* c	a b a^* c	b a^* c	b			
11	a b a^* c	a b a^* c	a d b c a^*	a d b c a^*	a b a^* c	a d b c a^*	a d b c a^*	b c a^* d	b d	d		
12	a a^* c	a b a^* c	a d b c a^*	a d b c a^*	a b a^* c	a d b c a^*	a c a^* d	c a^* d	d	b d	b	
13	a b a^* c	a a^* c	a c a^* d	a c a^* d	a a a^* c	a c a^* d	a d b c a^*	b c a^* d	b d	b d	b	b

Table 5

Proposition 5.1. **PR**$\in DTIME(n^2)$.

Proof. For every equivalence class $[x]$ of $IND(a^*)$ generated by $x \in U$ we construct the set $\underline{B}[x]$. More precisely, for a given x it is easy to find $[x]$ by simply marking in

$M(\mathbf{A})$ all rows of the column corresponding to x with value $a^*(x)$ of the attribute a^* (on objects corresponding to those rows). Next, for each object $y \in [x]$ we construct the equivalence class $[y]_B$ of the indiscernibility relation $IND(B)$ as follows: we find the column corresponding to y in $M(\mathbf{A})$ and mark all its rows corresponding to objects indiscernible from y with respect to the attributes in B, i.e. rows with c_{ij} such that $c_{ij} \cap B = \emptyset$. Then we check, whether $[y]_B$ is included in $[x]$: if all objects corresponding to marked rows are included in $[x]$ then y is included in $POS_B(a^*)$, otherwise not. Then we erase the dispensable rows and columns corresponding to objects in $[y]_B$. We continue the process for the elements of $[x]$ in the remaining part of the matrix (if such exist), or we repeat the process for another equivalence class, if all classes have not been processed yet. □

Example 5.2. We will compute $POS_B(a^*)$ for the $\mathbf{A}^*$ from Example 5.1 and $B = \{b, c\}$. We obtain the following classes of the indiscernibility relation $IND(a^*)$: $\{1,2,3\}$, $\{4,5\}$, $\{6,7,8\}$, $\{9,10,11,12,13\}$. In the next step we compute the classes of the indiscernibility relation $IND(B)$: $\{1,7,8\}$, $\{2,3,4,5,6\}$, $\{9,12\}$, $\{10,11\}$ and $\{13\}$. Finally, we obtain :
$POS_B(a^*) = \underline{B}\{1,2,3\} \cup \underline{B}\{4,5\} \cup \underline{B}\{6,7,8\} \cup \underline{B}\{9,10,11,12,13\} = \{9,10,11,12,13\}$.
One can check that $POS_B(a^*) = POS_{\{c\}}(a^*)$.

Let us take now $B' = \{a, c, d\}$. The equivalence classes of the indiscernibility relation $IND(B')$ are the following: $\{1,2\}$, $\{3\}$, $\{4\}$, $\{5\}$, $\{6,7\}$, $\{8\}$, $\{9,10\}$, $\{11,12,13\}$. Hence $POS_{B'}(a^*) = U$. □

Let $\mathbf{A}^* = (U, A \cup \{a^*\})$ be an extension of $\mathbf{A} = (U, A)$, and $\emptyset \neq B \subseteq A$.

We say that an attribute $b \in B$ is *relatively dispensable* in B, if

$$POS_B(a^*) = POS_{B-\{b\}}(a^*);$$

otherwise b is said to be *relatively indispensable* in B.

If every attribute from B is relatively indispensable in B, we say that B is *relatively independent* in $\mathbf{A}^*$ (or B is *independent with respect* to a^* in $\mathbf{A}^*$).

The concept of independence defined above is a special case of the independence introduced by Marczewski [Mar58], see also an overview paper [Gla75].

RI. (Relative Independence Problem)

Input: $\mathbf{A}^*$ — a^*-extension of $\mathbf{A} = (U, A)$

with $U = \{x_1, \ldots, x_n\}$, $\emptyset \neq B \subseteq A$ and $M(\mathbf{A}^*)$.

Output: 1 if B is relatively independent in $\mathbf{A}^*$

0 otherwise.

Proposition 5.2. RI$\in DTIME(n^2)$.

Proof. First let us define an extension $\mathbf{A}' = (U, A^* \cup \{pos\})$ of the system $\mathbf{A}^*$, where $pos : U \longrightarrow \{0,1\}$ and $pos(x) = 1$ iff $x \in POS_B(a^*)$. It follows from Proposition 5.1 that $M(\mathbf{A}')$ can be constructed in $O(n^2)$ steps. It is easy to observe that $M(\mathbf{A}') = (c'_{ij})$ has the following property:

(*) B is relatively independent in $\mathbf{A}^*$ iff for every $b \in B$ there exist i, j $(1 \leq j < i \leq n)$ such that $c'_{ij} \cap B = \{b\}$ and $[pos \in c'_{ij}$ or $(a^* \in c'_{ij}$ and $x_i, x_j \in POS_B(a^*))]$.

To prove this equivalence let us first assume that B is relatively independent in $\mathbf{A}^*$. Then for arbitrary $b \in B$ we have

$$POS_B(a^*) \supset POS_{B-\{b\}}(a^*).$$

which means that $x_i \in POS_B(a^*) - POS_{B-\{b\}}(a^*)$ for some i. Hence $[x_i]_B \subseteq POS_B(a^*)$ and $[x_i]_{B-\{b\}} \subseteq U - POS_{B-\{b\}}(a^*)$. From the first condition we have that $[x_i]_B \subseteq \mathbf{x}$ for some equivalence class $\mathbf{x}$ of $IND(a^*)$, and by the second $[x_i]_{B-\{b\}}$ has non-empty intersections with at least two different equivalence classes $\mathbf{y}$ and $\mathbf{z}$ of $IND(a^*)$. We can assume that $x_i \in \mathbf{y}$, so $x_i \in \mathbf{y} \cap [x_i]_{B-\{b\}}$ and $x_j \in \mathbf{z} \cap [x_i]_{B-\{b\}}$ for some $x_j \neq x_i$. The objects x_i and x_j are discernible by means of b and not discernible by means of the attributes in $B - \{b\}$. Hence $c'_{ij} \cap B = \{b\}$. If $pos \in c_{ij}$, then the proof is finished. Assume that $pos \notin c_{ij}$. Then both x_i and x_j belong either to $POS_B(a^*)$ or to $U - POS_B(a^*)$. Since $x_i \in POS_B(a^*)$, we have also $x_j \in POS_B(a^*)$.

Let us now assume that the condition (*) is true, and suppose that for some $b_0 \in B$ we have $POS_B(a^*) = POS_{B-\{b_0\}}(a^*)$. By assumption there exist objects x_i and x_j discernible by b_0 and indiscernible by $B - \{b_0\}$ such that either exactly one of them belongs to $POS_B(a^*)$ or they are both in $POS_B(a^*)$, but belong to different classes $\mathbf{y}$ and $\mathbf{z}$ of $IND(a^*)$. In both cases we obtain a contradiction with the assumption $POS_B(a^*) = POS_{B-\{b_0\}}(a)$. In fact, if $x_i \in POS_B(a^*)$ and $x_j \notin POS_B(a^*)$, then $x_i \in POS_{B-\{b_0\}}(a^*)$ and $x_j \notin POS_{B-\{b_0\}}(a^*)$. But this is a contradiction, since $x_i IND(B - \{b_0\}) x_j$ implies that x_i and x_j either both belong or both do not belong to $POS_B(a^*)$. If $x_i, x_j \in POS_B(a^*) = POS_{B-\{b_0\}}(a^*)$, $x_i \in \mathbf{y}$ and $x_j \in \mathbf{z}$, where $\mathbf{y}$,$\mathbf{z}$ are different classes of $IND(a^*)$, then we obtain a contradiction because by definition $POS_{B-\{b_0\}}(a^*)$ cannot include the equivalence class $[x_i]_B$, which has non-empty intersections with two different equivalence classes of $IND(a^*)$.

To check the condition (*) formulated above it suffices to perform one path through the matrix $M(\mathbf{A}')$ which can be realized in a number of steps of order $O(n^2)$. □

Example 5.3. Let $\mathbf{A}^*$ be as in Example 5.1. We check whether $B = \{b, c\}$ is relatively independent in $\mathbf{A}^*$. From the previous example we have

$$pos(a) = 1 \text{ iff } a \in \{9, 10, 11, 12, 13\}.$$

Hence a new matrix for $\mathbf{A}'$ has the form presented in Table 6.

In Table 6 the decision pos is denoted by p. Let us observe that $c'_{12,1} \cap \{b, c\} = \{c\}$ and $pos \in c'_{12,1}$, but there are no i, j satisfying
$c'_{ij} \cap \{b, c\} = \{b\}$ and $[pos \in c'_{ij}$ or $(a^* \in c'_{ij}$ and $x_i, x_j \in POS_B(a^*)]$.

As a conclusion we obtain that the set B is relatively dependent in $\mathbf{A}$. The elimination of b from B will not change the positive region (see also the previous example) i.e.

$$POS_{\{b,c\}}(a^*) = POS_{\{c\}}(a^*).$$

For the set $B' = \{a, c, d\}$ from the previous example we have $POS_{B'}(a^*) = U$, so the characteristic function pos' of the set $POS_{B'}(a^*)$ is always equal to 1. In consequence, no element in the corresponding matrix $M(\mathbf{A}')$ contains *pos*, and it suffices to consider $M(\mathbf{A}^*)$ instead of $M(\mathbf{A}')$.

We have

$c_{4,3} \cap \{a,c,d\} = \{a\}$, $4,3 \in POS_{B'}(a^*) = U$ and $a^* \in c_{4,3}$;
$c_{9,8} \cap \{a,c,d\} = \{c\}$, $9,8 \in POS_{B'}(a^*) = U$ and $a^* \in c_{9,8}$;
$c_{7,1} \cap \{a,c,d\} = \{d\}$, $7,1 \in POS_{B'}(a^*) = U$ and $a^* \in c_{7,1}$.
Hence the set B' is relatively independent in $\mathbf{A}$. □

A subset $B \subseteq A$ is called a *relative reduct* in $\mathbf{A}^*$ if B is relatively independent in $\mathbf{A}^*$ and $POS_A(a^*) = POS_B(a^*)$. We denote by $RED(\mathbf{A}^*)$ the set of all relative reducts in $\mathbf{A}^*$.

The relative reduct in $\mathbf{A}^*$ is a minimal subset of A providing the same classification of objects into elementary categories of $IND(a^*)$ as the whole set A. Let us observe that A can have more than one relative reduct.

RR. (Relative Reduct Problem)

Input: $\mathbf{A}^*$ — a^*-extension of $\mathbf{A} = (U, A)$

with $U = \{x_1, \ldots, x_n\}$, $M(\mathbf{A}^*)$ and $B \subseteq A$.

Output: 1 if B is a relative reduct in $\mathbf{A}^*$

0 otherwise.

Proposition 5.3. **RR**$\in DTIME(n^2)$.

Proof. Follows from Propositions 5.1 and 5.2. □

Example 5.4. For $\mathbf{A}^*$ of Example 5.1 and $B' = \{a,c,d\}$ we know from Example 5.3 that B' is relatively independent in $\mathbf{A}^*$. We also have $POS_{B'}(a^*) = U$. Hence $B \in RED(\mathbf{A}^*)$.□

Let $\mathbf{A}^* = (U, A \cup \{a^*\})$. The set of all relatively indispensable attributes in A is called the *relative core* of $\mathbf{A}$, and is denoted by $CORE(\mathbf{A}^*)$.

Proposition 5.4. [Paw91] $CORE(\mathbf{A}^*) = \bigcap RED(\mathbf{A}^*)$.

The concept of a core can be used, in the case when the latter is nonempty, as a basis for computing all reducts, since the core is included in every reduct, and its computation is straightforward.

RC. (Relative Core Problem)

Input: $M(\mathbf{A}^*)$ for an a^*-extension of $\mathbf{A} = (U, A)$

with $U = \{x_1, \ldots, x_n\}$.

Output: $CORE(\mathbf{A}^*)$.

Proposition 5.5. **RC**$\in DTIME(n^2)$.

	1	2	3	4	5	6	7	8	9	10	11	12
1												
2	b											
3	b d	d										
4	a b a* d	a a* d	a a*									
5	a a* b	a a*	a a* d	d								
6	b a* d	d a*	d a*	a a* d	a a* d							
7	d a*	b a* d	b a* d	a b a* d	a b a* d	b						
8	a a* d	a b a* d	a b a* d	a b a* d	a b a* d	a b	a					
9	a p c a* d	a d b p c a*	a d b p c a*	a d b p c a*	a d b p c a*	a p b c a*	a p c a*	c p a*				
10	a d b p c a*	a d b p c a*	a d b p c a*	a d b p c a*	a d b p c a*	a p b c a*	a p b c a*	b p c a*	b			
11	a p b c a*	a p b c a*	a d b p c a*	a d b p c a*	a p b c a*	a d b p c a*	a d b p c a*	b p c d a*	b d	d		
12	a p c a*	a p b c a*	a d b p c a*	a d b p c a*	a p b c a*	a d b p c a*	a p c d a*	c p d a*	d	b d	b	
13	a p b c a*	a p c a*	a p c d a*	a p c d a*	a p c a*	a p c d a*	a d b p c a*	b p c d a*	b d	b d	b	b

Table 6

Proof. Let $M(\mathbb{A}') = (c'_{ij})$ be defined as in Proposition 5.2. It suffices to conclude as in Proposition 5.2 that the following equivalence holds:

$b \in CORE(\mathbb{A}^*)$ iff there exist i, j $(1 \leq j < i \leq n)$ such that

$$c'_{ij} \cap A = \{b\} \text{ and } [pos \in c'_{ij} \text{ or } (a^* \in c'_{ij} \text{ and } x_i, x_j \in POS_B(a^*)].$$ □

Example 5.5. Let us consider again $\mathbf{A}^*$ of Example 5.1. We have $POS_A(a^*) = U$ (see Example 5.3). Hence our condition equivalent to $b \in CORE(\mathbf{A}^*)$ has the following form: there exist i, j $(1 \leq j < i \leq n)$ such that

$$c_{ij} - \{a^*\} = \{b\} \text{ and } a^*(x_i) \neq a^*(x_j),$$

where $M(\mathbf{A}^*) = (c_{ij})$.

The attribute b is relatively dispensable in A, as no c_{ij} contains only b together with a^*. The attributes a, c, d are relatively indispensable in A, since:

$$c_{4,3} = \{a, a^*\},\ c_{9,8} = \{c, a^*\},\ c_{7,1} = \{d, a^*\}.$$

Hence $CORE(\mathbf{A}^*) = \{a, c, d\}$. □

Now we show that reducts and cores in information systems are special cases of relative reducts and cores.

An extension $\mathbf{A}^* = (U, A \cup \{a^*\})$ of an information system $\mathbf{A} = (U, A)$ is called *standard* if $IND(a^*) = IND(A)$.

Proposition 5.6. *If $\mathbf{A}^*$ is a standard extension of $\mathbf{A}$ and $B \subseteq A$ then the following conditions are equivalent:*

(i) $POS_B(a^*) = POS_A(a^*)$

(ii) $IND(B) = IND(A)$.

Proof. The implication *(ii)* ⇒ *(i)* is obvious. Assume that *(i)* holds. Since $\mathbf{A}^*$ is a standard extension of $\mathbf{A}$, we have $IND(a^*) = IND(A)$, so $POS_A(a^*) = U$. Hence by our assumption $POS_B(a^*) = U$. Therefore every equivalence class of the indiscernibility relation $IND(a^*)$ is definable by B, i.e. is a union of some equivalence classes of $IND(B)$. But every equivalence class of $IND(a^*)$ is also an equivalence class of $IND(A)$. As $B \subseteq A$, we obtain $IND(a^*) = IND(B)$ and finally $IND(A) = IND(B)$. □

Proposition 5.7. *Let $\mathbf{A}^*$ be a standard extension of $\mathbf{A}$. Then*

(i) $CORE(\mathbf{A}) = CORE(\mathbf{A}^*)$ *and*

(ii) $RED(\mathbf{A}) = RED(\mathbf{A}^*)$.

Proof. (i) Assume that $b \in CORE(\mathbf{A})$, i.e. b is indispensable in $\mathbf{A}$, so $IND(A) \neq IND(A - \{b\})$ for $b \in B$. Since $\mathbf{A}^*$ is a standard extension of $\mathbf{A}$, this condition is equivalent to $POS_{A-\{b\}}(a^*) \neq POS_A(a^*)$ by the previous proposition, which means that b is relatively indispensable in $\mathbf{A}^*$ i.e. $b \in CORE(\mathbf{A}^*)$.

(ii) Assume that $B \in RED(\mathbf{A})$ i.e. B is independent in $\mathbf{A}$ and $IND(B) = IND(A)$. Since $\mathbf{A}^*$ is a standard extension of $\mathbf{A}$, by the last proposition the second condition is equivalent to $POS_B(a^*) = POS_A(a^*)$. For every $b \in B$, we have $IND(B) \neq IND(B - \{b\})$, and again by the last proposition $POS_{B-\{b\}}(a^*) \neq POS_A(a^*)$. Hence $POS_{B-\{b\}}(a^*) \neq$

$POS_B(a^*)$, i.e. b is relatively indispensable in B, so B is relatively independent in $\mathbf{A}^*$. Thus we have proved that $B \in RED(\mathbf{A}^*)$.

Assume now that $B \in RED(\mathbf{A}^*)$ i.e. B is relatively independent in $\mathbf{A}^*$ and $POS_B(a^*) = POS_A(a^*)$. The latter condition is equivalent to $IND(B) = IND(A)$. Since B is relatively independent in $\mathbf{A}^*$, for every $b \in B$ we have $POS_{B-\{b\}}(a^*) \neq POS_B(a^*) = POS_A(a^*)$, and by the last proposition $IND(A) \neq IND(B - \{b\})$ i.e. b is indispensable in B. Thus we have that B is independent in $\mathbf{A}$ and $IND(B) = IND(A)$, so $B \in RED(\mathbf{A})$. □

Theorem 4.1 can be generalized to extensions of information systems and relative reducts. To show this, we first prove the following lemma:

Lemma 5.1. *Let $\mathbf{A}^* = (U, A \cup \{a^*\})$ be an extension of an information system $\mathbf{A}$ with $U = \{x_1, \ldots, x_n\}$, and let $\emptyset \neq B \subseteq A$. Let $pos : U \to \{0,1\}$ be the characteristic function of the positive region $POS(a^*)$. The following conditions are equivalent:*

(i) $POS_A(a^*) = POS_B(a^*)$;

(ii) for all i and $j (1 \leq j < i \leq n)$:
if $[(a^* \in c_{ij}$ and $x_i, x_j \in POS_A(a^*))$ or $pos(x_i) \neq pos(x_j)]$ then $B \cap c_{ij} \neq \emptyset$.

Proof. $(i) \to (ii)$ Let $POS_A(a^*) = POS_B(a^*)$ and suppose that for some i, j we have also $[(a^* \in c_{ij}$ and $x_i, x_j \in POS_A(a^*))$ or $pos(x_i) \neq pos(x_j)]$ and $B \cap c_{ij} = \emptyset$. Hence $x_i IND(B) x_j$ and one of the following two cases holds:

1. x_i, x_j are discernible by pos;

2. x_i, x_j are discernible by a^* and $x_i, x_j \in POS_A(a^*)$.

In both cases we obtain a contradiction. In fact, assuming the first condition we have e.g. $x_i \in POS_A(a^*) = POS_B(a^*)$ and $x_j \notin POS_A(a^*) = POS_B(a^*)$. By definition of $POS_B(a^*)$, this contradicts the fact that $x_i IND(B) x_j$. Assuming the second condition, we would obtain $x_i, x_j \in POS_B(a^*)$, which is impossible, because by the definition of $POS_B(a^*)$ two objects discernible by a^* cannot both belong to $POS_B(a^*)$ if they are indiscernible by B.

$(ii) \to (i)$ Assume now that condition (ii) holds and $POS_A(a^*) \supset POS_B(a^*)$. Then $x_i \in POS_A(a^*) - POS_B(a^*)$ for some i. Hence $[x_i]_A \subseteq POS_A(a^*)$ and $[x_i]_B \cap POS_B(a^*) = \emptyset$. From the last condition it follows that there exist two different equivalence classes $\mathbf{y}$ and $\mathbf{z}$ of $IND(a^*)$ and $x_j \neq x_i$ such that $x_i IND(B) x_j$ and $x_i \in \mathbf{y}$, $x_j \in \mathbf{z}$. We consider two possible cases:

1. $x_j \in POS_A(a^*)$;

2. $x_j \notin POS_A(a^*)$.

In the first case $x_i, x_j \in POS_A(a^*)$ and x_i, x_j are discernible by a^*, so by (ii) $B \cap c_{ij} \neq \emptyset$. In the second case $pos(x_i) = 1$ and $pos(x_j) = 0$, and $B \cap c_{ij} \neq \emptyset$ by (ii). Hence we have an attribute $b \in B$ discerning between x_i and x_j but this contradicts $x_i IND(B) x_j$. □

Let $\mathbb{A}^* = (U, A \cup \{a^*\})$, $B \subseteq A$ and $M(\mathbb{A}^*) = (c_{ij})$. We define a matrix $\mathbf{M}(\mathbb{A}^*) = (\mathbf{c}_{ij})$ in the following way:

$$\mathbf{c}_{ij} = \begin{cases} c_{ij} - \{a^*\} & \text{if } (a^* \in c_{ij} \text{ and } x_i, x_j \in POS_A(a^*)) \\ & \text{or } pos(x_i) \neq pos(x_j) \\ \emptyset & \text{otherwise} \end{cases}$$

The relative discernibility function $\mathbf{f}_{\mathbb{A}^*}$ for $\mathbb{A}^*$ is constructed from $\mathbf{M}(\mathbb{A}^*)$ in an analogous way as $f_{\mathbb{A}}$ from $M(\mathbb{A})$.

We have the following theorem:

Theorem 5.1. *Let $\mathbb{A}^* = (U, A \cup \{a^*\})$ be an extension of an information system $\mathbb{A}$, $\emptyset \neq B \subseteq A$ and let $v_B : A \to \{0,1\}$ be the characteristic function of B. The following conditions are equivalent:*

(i) $POS_A(a^*) = POS_B(a^*)$ i.e. B contains a relative reduct from $RED(\mathbb{A}^*)$;

(ii) $\mathbf{f}_{\mathbb{A}^*}(v_B(a_1), \ldots, v_B(a_m)) = 1$

(iii) for all i and j such that $\mathbf{c}_{ij} \neq \emptyset$ and $1 \leq j < i \leq n$:

$$\mathbf{c}_{ij} \cap B \neq \emptyset .$$

Proof. The equivalence of (*ii*) and (*iii*) follows directly from the definitions of $\mathbf{M}(\mathbb{A}^*)$ and $\mathbf{f}_{\mathbb{A}^*}$. The equivalence of (*i*) and (*ii*) [and (*iii*)] follows from Lemma 5.1. □

We also have the following theorem:

Theorem 5.2. *Let $\mathbb{A}^* = (U, A \cup \{a\})$ be an extension of $\mathbb{A}$ and let $\emptyset \neq B \subseteq A$. The set B is minimal with respect to one of the conditions stated in Theorem 5.1 iff it is minimal with respect to the remaining ones.*

Proof. First we prove that the set B is minimal with respect to the condition (*ii*) iff it is minimal with respect to condition (*iii*).

Let us assume that B is minimal with respect to condition (*ii*). It means that $\mathbf{f}_{\mathbb{A}^*}(v_B(a_1), \ldots, v_B(a_m)) = 1$ and $\mathbf{f}_{\mathbb{A}^*}(v_{B'}(a_1), \ldots, v_{B'}(a_m)) = 0$ for any $B' \subset B$. Hence, from the definition of $\mathbf{f}_{\mathbb{A}^*}$ it follows that every non-empty set $c_{ij} - \{a^*\}$ contains an element from B and for any $B' \subset B$ there exists a non-empty set $c_{ij} - \{a^*\}$ disjoint with B'. Assume that B is not minimal with respect to (*iii*). Thus for some $B_0 \subset B$ we would have $\mathbf{c}_{ij} \cap B_0 \neq \emptyset$ for any i, j such that $\mathbf{c}_{ij} \neq \emptyset$. But then

$$\mathbf{f}_{\mathbb{A}^*}(v_{B_0}(a_1), \ldots, v_{B_0}(a_m)) = 1 ,$$

which contradicts our assumption about the minimality of B with respect to (*ii*).

Now assume that B is minimal with respect to condition (*iii*) but it is not minimal with respect to (*ii*), i.e. there exists $B_0 \subset B$ such that

$$\mathbf{f}_{\mathbb{A}^*}(v_{B_0}(a_1), \ldots, v_{B_0}(a_m)) = 1.$$

By the definition of $\mathbf{f}_{\mathbb{A}^*}$, this means that any non-empty set $c_{ij} - \{a^*\}$ has a non-empty intersection with B_0, which contradicts our assumption about the minimality of B with respect to (iii).

Now we will prove that any set $B \subseteq A$ is minimal with respect to condition (i) iff it is minimal with respect to (iii) $[(ii)]$.

Let us assume that B is minimal with respect to (i). Then, for any $B' \subset B$, we have $POS_A(a^*) \supset POS_{B'}(a^*)$. Suppose that B is not minimal with respect to (iii). Then (iii) would hold for some $B_0 \subset B$. It means that for all i and j such that $\mathbf{c}_{ij} \neq \emptyset$ and $1 \leq j < i \leq n$: $\mathbf{c}_{ij} \cap B_0 \neq \emptyset$.

Observe that by definition $\mathbf{c}_{ij} \neq \emptyset$ iff

(**) $\quad [(a^* \in c_{ij}$ and $x_i, x_j \in POS_A(a^*))$ or $pos(x_i) \neq pos(x_j)]$.

Of course, if $\mathbf{c}_{ij} \neq \emptyset$, then condition (**) holds by the definition of $\mathbf{M}(\mathbf{A}^*)$. If $pos(x_i) \neq pos(x_j)$, then $\mathbf{c}_{ij} = c_{ij} - \{a^*\} \neq \emptyset$, because x_i, x_j are in different classes of $IND(B)$, so they are discernible by some attributes from $B \subseteq A$. If $a^* \in c_{ij}$ and $x_i, x_j \in POS_A(a^*)$, then x_i, x_j are discernible by a^*. Hence by the definition of $POS_A(a^*)$ $(= POS_B(a^*))$ the above elements belong to different classes of $IND(B)$, so they are discernible by attributes from B. This proves the equivalence stated above.

Hence condition (ii) of Lemma 5.1 would hold and in consequence $POS_A(a^*) = POS_{B_0}(a^*)$, which contradicts our assumption about the minimality of B with respect to (i).

Assume now that B is minimal with respect to (iii), and suppose that B is not minimal with respect to (i), i.e. $POS_A(a^*) = POS_{B_0}(a^*)$ for some $B_0 \subset B$. Reasoning in an analogous way as before (and applying Lemma 5.1), we would have that B_0 is not minimal with respect to (iii), which contradicts our assumption about the minimality of B. □

Remark. Now it is easy to observe a strong connection between the notion of a relative reduct in $\mathbf{A}^*$ and that of a *prime implicant* [Weg87,Kau68] of the monotonic boolean function $\mathbf{f}_{\mathbb{A}^*}$, namely we have the following equivalence:

$\{a_{i_1}, \ldots, a_{i_p}\} \in RED(\mathbf{A}^*)$ iff $a_{i_1} \wedge \ldots \wedge a_{i_p}$ is a prime implicant of $\mathbf{f}_{\mathbb{A}^*}$.

Example 5.5. The construction of the relative discernibility function $\mathbf{f}_{\mathbb{A}^*}$ for $\mathbf{A}^*$ from Example 5.1 and the application of the absorption law to $\mathbf{f}_{\mathbb{A}^*}$ as many times as possible after multiplication lead to the expression

$$a \wedge c \wedge d.$$

Hence $RED(\mathbf{A}^*) = \{\{a, c, d\}\}$. □

6. Dependencies in an information system

In this section we show how to apply our method based on discernibility matrices and discernibility function to check or generate dependencies in information systems.

A subset C of the set of attributes in an information system $\mathbf{A}$ depends on a subset B of attributes in $\mathbf{A}$, in symbols $B \rightarrow_{\mathbb{A}} C$, if $IND(B) \subseteq IND(C)$.

From the definition $C \rightarrow_{\mathbb{A}} c$ for any $c \in C$ and $B \rightarrow_{\mathbb{A}} C$ iff $B \rightarrow_{\mathbb{A}} c$ for every $c \in C$.

The sets B, C of attributes in $\mathbf{A}$ are equivalent iff $B \rightarrow_{\mathbb{A}} C$ and $C \rightarrow_{\mathbb{A}} B$.

Proposition 6.1. [Paw91] *The following conditions are equivalent:*

1) $B \rightarrow_{\mathbb{A}} C$*;*

2) $IND(B \cup C) = IND(B)$*;*

3) $POS_B(C) = U$*;*

4) $\underline{B}X = X$ *for all* $X \in U/C$*.*

□

The Proposition demonstrates that if C depends on B, then the knowledge connected with C is redundant within the knowledge base in the sense that the knowledge B provides the same characterization of objects.

Proposition 6.2. [Paw91]

(i) If $B \in RED(\mathbb{A}))$ *and* $B \subseteq C$*, then* $B \rightarrow_{\mathbb{A}} C - B$*.*

(ii) If B *is dependent in* $\mathbb{A}$ *and* $IND(B) = IND(A)$*, then there exists a subset* $C \subseteq B$ *such that* $B \rightarrow_{\mathbb{A}} C$ *and* $C \in RED(\mathbb{A})$*.*

The dependence can also be partial in the sense that only part of information represented by C is derivable from B in $\mathbb{A}$.

Let $\mathbb{A} = (U, A)$ be an information system and $B, C \subseteq A$. We say that the set C depends on B in degree k $(0 \leq k \leq 1)$, symbolically

$$B \rightarrow_{\mathbb{A},k} C, \text{ if } k = \gamma_B(C) = \frac{cardPOS_B(C)}{card(U)}\,.$$

If $k = 1$ we will say that C is *totally dependent* on B in $\mathbb{A}$; if $O < k < 1$, we say that C is *roughly dependent* on B in $\mathbb{A}$; and if $k = 0$ we say that C is *totally independent* of B in $\mathbb{A}$.

It follows from the definition of dependence that, if

$$B \rightarrow_{k,\mathbb{A}} C\,,$$

then the positive region of the partition of U defined by the indiscernibility relation $IND(C)$ covers $k \times 100$ percent of all objects in the information system. On the other hand, only the objects belonging to the positive region of that partition can be uniquely classified. This means that $k \times 100$ percent of objects can be classified into the blocks of the partition of U defined by $IND(C)$ on the basis of the attributes in B.

Thus the coefficient $\gamma_B(C)$ can be understood as a degree of dependence between C and B. In other words, if we restrict the set of objects in the knowledge base to the set $POS_B(a^*)$, where a^* represents C, we obtain a knowledge base, in which $B \rightarrow C$ is a total dependence.

The measure k of dependence $B \rightarrow_{k,\mathbb{A}} C$ does not capture the actual distribution of this partial dependence over the classes of $IND(C)$. For example, some decision classes can be fully characterized by B, whereas others may be characterized only partially. The coefficient

$$\gamma_B(X) = card\underline{B}X/cardX \text{ where } X \in U/IND(C)$$

says how precisely the elements of each class of the relation $IND(C)$ can be classified on the basis of the attributes from B.

Thus the two numbers $\gamma_B(C)$ and $\gamma_B(X)$, $X \in U/IND(C)$ give us a better information about the "classification power" of the set B of attributes with respect to the classification defined by $IND(C)$.

As a measure of significance of the subset of attributes $B' \subseteq B$ with respect to the classification induced by a set of attributes C one can consider the difference $\gamma_B(C) - \gamma_{B-B'}(C)$, which expresses how the positive region of the classification $U/IND(C)$, when classifying objects by means of attributes B, will be affected if we drop some attributes (subset B') from the set B.

Now we present two examples of problems related to dependencies in information systems, and their solutions using the discernibility matrix.

DA. (Dependency of Attributes)

Input: $M(\mathbb{A}^*)$ for an information system $\mathbb{A} = (U, A)$

with $U = \{x_1, \ldots, x_n\}$, $B \subseteq A$ and $b \in A - B$.

Output: 1 if $B \rightarrow_{\mathbb{A}} b$

0 otherwise.

Proposition 6.3. **DA**$\in DTIME(n^2)$.

Proof. It suffices to prove that the following two conditions are equivalent for $b \notin B$:

1. $B \rightarrow_{\mathbb{A}} b$
2. For all i and j such that $1 \leq j < i \leq n$:

$$b \in c_{ij} \text{ implies } B \cap c_{ij} \neq \emptyset .$$

Let $B \rightarrow_{\mathbb{A}} b$, i.e. if $x_i DIS(\{b\}) x_j$, then $x_i DIS(B) x_j$ for any $x_i, x_j \in U$. Suppose $b \in c_{i_0 j_0}$ and $B \cap c_{i_0 j_0} = \emptyset$ for some i_0, j_0. Hence

$$x_{i_0} DIS(\{b\}) x_{j_0} \text{ and } x_{i_0} IND(B) x_{j_0} ,$$

a contradiction.

Assume now that $b \in c_{ij}$ implies $B \cap c_{ij} \neq \emptyset$ for any i, j. If $x_i DIS(\{b\}) x_j$ for some i, j, then by the definition of $M(\mathbb{A})$, $b \in c_{ij}$. Hence by assumption we have $B \cap c_{ij} \neq \emptyset$, so (by Proposition 3.1) $x_i DIS(B) x_j$. □

Example 6.1. Let us consider again the information system $\mathbb{A}$ of Example 2.1. We want to find all dependencies of the form $B \rightarrow_{\mathbb{A}} c$, where $B \subseteq A$. Assume that $B \rightarrow_{\mathbb{A}} c$. Considering the last proposition and the discernibility matrix of Example 3.1, we have $\emptyset \neq c_{6,3} \cap B = \{c\} \cap B$. Hence $c \in B$. Of course, if $c \in B$, then $B \rightarrow_{\mathbb{A}} c$. In this way we have the following equivalence:

$$B \rightarrow_{\mathbb{A}} c \text{ iff } c \in B .$$

An analogous result holds for e and o. Hence we have

$$B \rightarrow_{\mathbb{A}} x \text{ iff } x \in B, \text{ for } x \in \{c, e, o\} .$$

In the next example we present the characterization of all attribute sets $B \subseteq A$ such that $B \rightarrow_{\mathbb{A}} x$, where $x \in \{a, d\}$. □

The proof of Proposition 6.3 suggests a method for generating for a given $a \in A$ all (or in some sense minimal) dependencies of the form $B \rightarrow_{\mathbb{A}} a$.

Let $\mathbb{A} = (U, A)$ and $a \in A$. Given a discernibility matrix $M(\mathbb{A}) = (c_{ij})$ we construct a new matrix $M(\mathbb{A}, a) = (c[a]_{ij})$ as follows :

$$c[a]_{ij} = \begin{cases} c_{ij} - \{a\} & \text{if } a \in c_{ij} \\ \emptyset & \text{otherwise} \end{cases}$$

Now we construct from $M(\mathbb{A}, a)$ a boolean function $f_{\mathbb{A},a}$ in the same way as the discernibility function $f_{\mathbb{A}}$ was constructed from $M(\mathbb{A})$.

As a corollary from the equivalence on which the proof of Proposition 6.3 is based we obtain:

Corollary 6.1. *Let $\mathbb{A} = (U, A)$ be an information system and let $B \subseteq A$, $a \in A - B$. The following conditions are equivalent:*

1. *$B \rightarrow_{\mathbb{A}} a$;*
2. *there exists an implicant of $f_{\mathbb{A},a}$ containing exactly all the attributes from B.*

Example 6.2. We shall characterize the dependencies of the form $B \rightarrow_{\mathbb{A}} a$ and $B \rightarrow_{\mathbb{A}} d$ for $\mathbb{A}$ from Example 2.1. To do that, we can use the monotonic boolean function $f_{\mathbb{A},x}$ constructed for $x = a$ and $x = d$. The function $f_{\mathbb{A},x}(a, c, d, e, o)$ is a conjunction of some disjuncts. The disjuncts are formed for each position in $M(\mathbb{A})$ containing x by taking all elements in the position instead of x. (The empty disjunction is by assumption equivalent to false). In our case:
$f_{\mathbb{A},a}(a, c, d, e, o) = (d \vee o) \wedge (c \vee d) \wedge (c \vee d \vee o) \wedge (c \vee d \vee e \vee o) \wedge d$
$f_{\mathbb{A},d}(a, c, d, e, o) = (a \vee o) \wedge (a \vee c) \wedge (a \vee c \vee o) \wedge (a \vee c \vee e \vee o) \wedge a$.
Hence after simple applications of the absorption rule we conclude that $f_{\mathbb{A},a}$ is equivalent to d and $f_{\mathbb{A},d}$ to a. Therefore $B \rightarrow_{\mathbb{A}} a$ iff $a \in B$ or $d \in B$ and $B \rightarrow_{\mathbb{A}} d$ iff $d \in B$ or $a \in B$. □

MD. (Minimal Dependence Problem)

Input: an information system $\mathbb{A} = (U, A)$, $a \in A$.

Output: $B \rightarrow_{\mathbb{A}} a \in MD(\mathbb{A}, a)$ (if such dependence exists),

0 (otherwise),

where $MD(\mathbb{A}, a) = \{B \longrightarrow_{\mathbb{A}} a : card(B) = \min\{card(C) : C \subseteq A - \{a\} \text{ and } C \rightarrow a\}\}$.

We can formulate a corollary allowing to generate an output for the **MD** problem.

Corollary 6.2. *Let $\mathbb{A} = (U, A)$ be an information system and let $B \subseteq A$, $a \in A - B$. The following conditions are equivalent:*

1. *$B \rightarrow_{\mathbb{A}} a$ is in $MD(\mathbb{A}, a)$;*

2. *there exists a prime implicant of* $f_{\mathbb{A},a}$ *containing exactly all the attributes from* B.

It is easy to see from the last corollary and Proposition 4.7 that:

Corollary 6.3. MD∈ NP-HARD.

The last corollary means that it is unrealistic to try to design an efficient algorithm solving the **MD** problem. Nevertheless, one can give some useful heuristics for the **MD** problem, or efficient algorithms solving the **MD** problem for some classes of information systems, with data satisfying some additional conditions.

Our last example is related to the process of generating optimal decision rules considered in [Paw91].

Example 6.3. It is possible to generate optimal decision rules with a more specific right hand side, namely one can look for the optimal decision rule for the value 2 of a^* in Example 5.1. To obtain such a decision rule, it is necessary to modify our information system $\mathbf{A}^*$ by taking the binary attribute a_2^* instead of a^*, where $a_2^*(x) = 1$ if $x \in \{4,5\}$ (i.e. $a^*(x) = 2$) and $a_2^*(x) = 0$ if $x \notin \{4,5\}$. The remaining attributes are the same as in Example 5.1. In this way we obtain the information system $\mathbf{A}_2^*$. For the discernibility matrix $M(\mathbf{A}_2^*)$, we have $a_2^* \notin c_{ij}$ if $a_2^*(x_i) = a_2^*(x_j)$. Now applying Corollary 6.1 to $M(\mathbf{A}_2^*)$ one can observe, that in order to generate the dependencies of the form $B \rightarrow a_2^*$, where $B \subseteq A$, it suffices to consider the fragment of the discernibility matrix $M(\mathbf{A}^*)$ corresponding to the "crossing" of $\{4,5\}$ and $U - \{4,5\}$. The above fragment is presented in Table 7.

The boolean function $f_{\mathbf{A}_2^*,a_2^*}$, corresponding to that fragment of the matrix has, after several applications of the multiplication and absorption laws, the form: a . Hence we have the dependence

$$\{a\} \rightarrow_{\mathbf{A}_2^*} a_2^* .$$

From the initial table in Example 5.1 we obtain the following optimal decision rule [Paw91] for the decision 2 i.e. $e = 1$ and $f = 4$:

$$(a = 2) \rightarrow_{\mathbb{A}} (e = 1) \wedge (f = 4) .$$

□

The method presented here can be extended to the case of partial dependencies.

Conclusions

In the paper several solutions of some basic problems related to information systems are presented. The solutions are based on the notions of discernibility matrix and function. The procedures based on the presented method are simpler and seem to be more efficient than those presented before e.g. in [Paw91].

The procedures presented here have a form suitable for parallel implementations. We will discuss this issue in another paper.

Acknowledgement. We are greatly indebted to Professor Zdzislaw Pawlak for suggesting some problems related to dependencies and stimulating discussion. We would like to thank Dr Beata Konikowska , Dr Wiktor Bartol and Professor Miroslav Novotny for the constructive criticism and comments on an earlier version of this paper.

	1	*2*	*3*	*4*	*5*
4	*a* *b* *d*	*a* *d*	*a*		
5	*a* *b*	*a*	*a* *d*		
6				*a* *d*	*a* *d*
7				*a* *b* *d*	*a* *b* *d*
8				*a* *b* *d*	*a* *b* *d*
9				*a d* *b* *c*	*a* *b* *c d*
10				*a d* *b* *c*	*a d* *b* *c*
11				*a d* *b* *c*	*a* *b* *c*
12				*a d* *b* *c*	*a* *b* *c*
13				*a* *c* *d*	*a* *c*

Table 7

References

[Bus86] Buszkowski W. and Orlowska E. (1986): **On the Logic of Database Dependencies**, Bull. Polish Acad. Sci. Math. vol.**34**, 345–354

[Cas89] Casti J. L. (1989): **Alternate Realities - Mathematical Models of Nature and Man**, John Wiley and Sons

[Cod70] Codd, E.F. (1970): **A Relational Model of Data for Large Shared Data Banks**, Comm. ACM. vol.**13**, 377–387

[Che88] Chen K.H., Ras Z. and Skowron A. (1988): **Attributes and Rough Properties in Information Systems**, Internat.J.Approx.Reason. **2**(4), 365–376

[GJ79] Garey M.R. and Johnson D.S., (1979): **Computers and Intractability: A Guide to the Theory of NP-Completeness**, Freeman and Company, San Francisco

[Grz86] Grzymala-Busse J., (1986): **On the Reduction of Knowledge Representation Systems**, Proc. of the 6-th International Workshop on Expert Systems and their Applications, Avignon, France, April 28-30, 463–478

[Gla79] Glazek K. (1979): **Some Old and New Problems in the Independence Theory**, Colloquium Mathematicum vol.**17**, 127–189

[Ha86] Halpern J.Y. ed. (1986): **Theoretical Aspects of Reasoning about Knowledge**, Morgan Kaufmann

[Hur83] Hurley R. B. (1983): **Decision Tables in Software Engineering**, Van Nostrand Reinhold Company, New York

[Kau68] Kautz W.H. (1968): **Fault Testing and Diagnosis in Combinational Digital Circuits**, IEEE Transactions on Computers, April, 352–366

[Lip79] Lipski W. Jr. (1979): **On Semantic Issues Connected with Incomplete Information Databases**, ACM Transactions on Database Systems vol.**4**, 269–296

[Lip81] Lipski W. Jr. (1981): **On Databases with Incomplete Information**, Journal of the ACM. vol.**28**, 41–70

[Los79] Los J. (1979): **Characteristic Sets of a System of Equivalence Relations**, Colloquium Mathematicum XLII, 291–293

[Mar 58] Marczewski E. (1958): **A general Scheme of Independence in Mathematics**, BAPS, 731–736

[Mo83] Moshkov M.Iu., (1982): **On the Conditional Tests (in Russian)**, Dokl. Akad. Nauk SSSR **265**(2), 550–552.

[Mro87] Mrozek A. (1987): **Rough Sets and Some Aspects of Expert Systems Realization**, Proc. 7-th Internat. Workshop on Expert Systems and their Applications, Avignon, France, 597–611

[Nov88a] Novotny M. and Pawlak Z. (1988): **Independence of Attributes**, Bull. Polish Acad. Sci. Math. vol.**36**, 459–465

[Nov88b] Novotny M. and Pawlak Z. (1988): **Partial Dependency of Attributes** Bull. Polish Acad. Sci. Math.vol.**36**, 453–458

[Nov89] Novotny M. and Pawlak Z. (1989): **Algebraic Theory of Independence in Information Systems**, Institute of Mathematics Czechoslovak Academy of Sciences Report **51**

[Orl83] Orlowska E. (1983): **Dependencies of Attributes in Pawlak's Information Systems**, Fundamenta Informaticae vol.**6**, 247–256

[Orl84a] Orlowska E. and Pawlak Z. (1984): **Expressive Power of Knowledge Representation Systems**, International Journal of Man-Machine Studies vol.**20**, 485–500

[Orl84b] Orlowska E. and Pawlak Z. (1984): **Representation of Nondeterministic Information**, Theoretical Computer Science vol.**29**, 27–39

[Pag87] Pagliani P. (1987): **Polystructured Model Spaces as Algorithm Oriented Models for Approximation Spaces and Pawlak's Information Systems**, Facolta di Science dell Informazione, Internal Raport PAG /3

[Paw81a] Pawlak Z. (1981): **Information Systems - Theoretical Foundations**, (The book in Polish), PWN Warsaw

[Paw81b] Pawlak Z. (1981): **Information Systems - Theoretical Foundations**, Information Systems vol.**6**, 205–218

[Paw82] Pawlak Z. (1982): **Rough sets**, International Journal of Information and Computer Science vol.**11**, 344–356

[Paw85] Pawlak Z. (1985): **On Rough Dependency of Attributes in Information Systems**, Bull. Polish Acad. Sci. Tech. vol.**33**, 481–485

[Paw85a] Pawlak Z. and Rauszer C. M. (1985): **Dependency of Attributes in Information Systems**, Bull. Polish Acad. Sci. Math. vol.**33**, 551–559

[Paw91] Pawlak Z. (1991): **Rough Sets: Theoretical Aspects of Reasoning about Data**, Kluwer

[Par90] Parikh R. ed. (1990): **Theoretical Aspects of Reasoning about Knowledge**, Morgan Kaufmann

[RaS86] Rasiowa H. and Skowron A., (1986): **Approximation Logic**, In: Mathematical Methods of Specification and Synthesis of Software Systems '85, Akademie-Verlag Berlin, Band **31**, 123–139

[Rau84] Rauszer C. M. (1984): **An Equivalence Between Indiscernibility Relations in Information Systems and a Fragment of Intuitionistic Logic**, Lecture Notes in Computer Science vol.**208**, Springer Verlag, 298–317

[Rau85a] Rauszer C. M. (1985): **An Equivalence between Theory of Functional Dependencies and Fragment of Intuitionistic Logic**, Bull. Polish Acad. Sci. Math. vol.**33**, 571–679

[Rau85b] Rauszer C. M. (1985): **An Algebraic and Logical Approach to Indiscernibility Relations**, ICS PAS Reports No **559**

[Rau87] Rauszer C. M. (1987): **Algebraic and Logical Description of Functional and Multivalued Dependencies.** Proc of the Sec. Int. Symp. on Methodologies for Intelligent Systems. October 17, Charlotte, North Holland 1987, 145–155

[Rau88] Rauszer C. M. (1988): **Algebraic Properties of Functional Dependencies**, Bull. Polish Acad. Sci. Math. vol.**33**, 561–569

[Rau90] Rauszer C.:**Decision Logic**, to appear in Fundamenta Informaticae

[Sal68] Salton G. (1968): **Automatic Information Organization and Retrieval**, McGraw-Hill, New York

[Sk88] Skowron A. (1988): **On Topology in Information Systems**, Bull. Acad. Polon. Sci. Math. vol.**36**, pp.477–479

[Sk90] Skowron A. (1990): **An Approach to Synthesis of Decision Algorithms in Interaction with Knowledge Bases and Environment. An Example Based on Some Glaucoma Processes (in cooperation with Medical Center in Warsaw)**, manuscript pp.1–173, Warsaw

[Sko91] Skowron A. and Stepaniuk J. (1991): **Towards an Approximation Theory of Discrete Problems: Part I**, Fundamenta Informaticae **15**(2), 187-208

[Sk91] Skowron A. (1991): **The Discernibility Matrix and Function in Information System**, manuscript

[Tr82] Truszczynski M., (1982): **Algorithmic Aspects of the Minimization of the Set of Attribute Problems**, Fundamenta Informaticae 4(4)

[Vak87] Vakarelov D. (1987): **Abstract Characterization of Some Modal Knowledge Representation Systems and the Logic NIL of Nondeterministic Information**, In: Jorraud, Ph. and Squrev, V. (ed.) Artificial Intelligence II, Methodology, Systems, Applications, North Holland

[Vak89] Vakarelov D. (1989): *Modal Logics for Knowledge Representation Systems*, LNCS vol.**363** 257–277

[Weg87] Wegener I. (1987): **The Complexity of Boolean Functions**, John Wiley and Sons

[Wil82] Wille R. (1982): **Restructuring Lattice Theory: An Approach Based on Hierarchies of Concepts**, In: I.Rival (Ed.), Ordered Sets, Reidel, Dordrecht-Boston, 445–470

[Zia87] Ziarko W. (1987): **On Reduction of Knowledge Representation**, Proc. 2-nd International Symp.on Methodologies of for Intelligent Systems, Charlotte USA, North Holland, 99–113

[Zia88] Ziarko W. (1988): **Acquisition of Design Knowledge from Examples**, Math. Comput. Modeling vol.**10**, 551-554

Part III
Chapter 3

SENSITIVITY OF ROUGH CLASSIFICATION TO CHANGES IN NORMS OF ATTRIBUTES

Krzysztof SŁOWIŃSKI
Department of Surgery
F. Raszeja Mem. Hospital
60-833 Poznań, Poland

Roman SŁOWIŃSKI
Institute of Computing Science
Technical University of Poznań
60-965 Poznań, Poland

Abstract. Rough classification of patients after highly selective vagotomy (HSV) for duodenal ulcer is analysed from the viewpoint of sensitivity of previously obtained results to minor changes in the norms of attributes. The norms translate exact values of pre-operating quantitative attributes into 2 to 4 qualitative terms, e.g. "low", "medium" and "high". An extensive computational experiment leads to the general conclusion that original norms following from medical experience were well defined, and that the results of analysis of the considered information system using rough sets theory are robust in the sense of low sensitivity to minor changes in the norms of attributes.

1. Introduction

The paper refers to rough classification of patients after highly selective vagotomy (HSV) for duodenal ulcer investigated by Słowiński (1992) in this volume (cf. Pawlak et al.(1986) and Fibak et al.,(1986)). The information system summarizing our experience in the treatment of duodenal ulcer by HSV was composed of 122 patients described by 11 pre-operating attributes and classified from the viewpoint of long-term results of the operation into 4 classes of the known Visick grading.

First, in the above mentioned studies, the information system has been reduced so as to get a minimum subset of attributes ensuring an acceptable quality of classification. Then, the reduced information system has been identified with a decision table. From this table, a decision algorithm has been derived, composed of 44 deterministic rules.

Moreover, an analysis of the distribution of values adopted by the most significant attributes in particular classes resulted in the definition of the most characteristic values of these attributes for each class. The set of characteristic values for a given class defines a model of patients belonging to this class.

The decision algorithm, together with the models of patients, represent the knowledge acquired by the clinicist on 122 cases. This representation is free of all redundancies, so typical for all real data bases, which cloud the most important factors of clinical experience. The algorithm and the models can support decisions concerning the treatment of duodenal ulcer by HSV.

All the results obtained until now are valid, however, for some norms translating the original domains of attributes into coded domains. The norms follow from medical and laboratory experience but there are neither certain nor sharp. This dependence has been pointed out by Greenburg (1987) who, in addition to his good opinion about the newly developed methodology, asked, what is the sensitivity of the obtained results to minor changes in the norms. In other words, how robust is the model when minor changes are made in the norms.

In Słowiński and Słowiński (1990), we have undertaken the task of answering this questions. In this paper, we resume the results obtained in the course of a computational experiment. In section 2, we examine the sensitivity of the quality of classification to minor changes in the norms. Using the norms ensuring the highest quality of classification, we perform reduction of attributes in section 3, and a derive decision algorithm in section 4. The conclusions about the robustness of the model and the sensitivity to the changes in the norms are drawn in section 5.

2. Sensitivity of rough classification to minor changes in the norms

The information system considered in Słowiński (1992) was composed of 122 patients with duodenal ulcer treated by HSV, described by 11 pre-operating attributes and a long-term result in the Visick grading. Attributes 1 – 4 concerned anamnesis, and the remaining attributes were related to pre-operating gastric secretion examined with the histaminic test of Kay. Originally, all attributes, except 1 and 4, take arbitrary real values from intervals defined by extreme cases (see Appandix of the paper by Słowiński (1992)). In clinical experience, exact values of these attributes are usually translated into qualitative terms, e.g. "low", "medium", "high" and "very high". This translation is due to some empirical norms defining intervals of attribute values corresponding to the qualitative terms. The terms were then coded by numbers $0,1,2,3$ which create a coded domain of the attributes. The norms adopted are shown in Table 1 of the paper by Słowiński (1992).

In previous studies, we came to conclusion that the minimal set R of attributes ensuring a satisfactory quality of classification ($\gamma_R(\mathcal{Y}) = 0.68$) is $R = \{3,4,6,9,10\}$. Of course, this result is valid for the assumed definition of norms. Now, we will check the sensitivity of this result to minor changes in the norms concerning attributes from set R.

In the course of a computational experiment, we analysed one by one attributes $3,6,9,10$ in the following way. We changed the norms by moving bounds of intervals corresponding to the attribute domain to the left or to the right, and we observed the effect of this movement on the quality of classification $\gamma_R(\mathcal{Y})$. The magnitude of the movement was expressed in the percentage of the boundary values. The movement to the left and to the right corresponds to the negative and positive percentage, respectively. Figs. 1–4 summarize the results of the experiment. The dashed area represents an insensitivity zone of the highest quality $\gamma_R(\mathcal{Y})$.

Let us observe that the change of the norm for attribute 3 in the range of 0% to +30% don't change $\gamma_R(\mathcal{Y})$. Changing the norm for attribute 6, one can improve $\gamma_R(\mathcal{Y})$ which rises to 0.72 in the range of +20% to +21%. A similar consequence can be observed in the case of attribute 9 and 10, when $\gamma_R(\mathcal{Y})$ rises to 0.71 and 0.69 in the range of –25% to –10% and –6% to –3%, respectively.

In conclusion, one can state that the changes in the norms of attributes from set R don't cause any significant improvement in the quality of classification which proves that original norms following from medical experience were well defined.

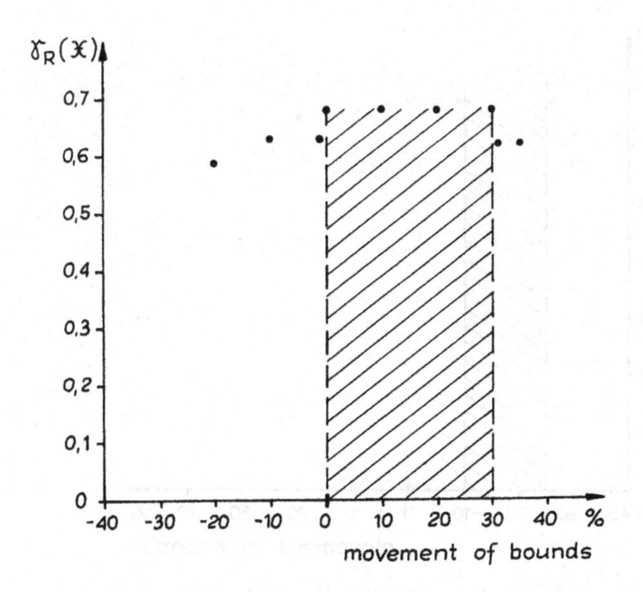

Figure 1. Sensitivity to changes in the norm of attribute 3

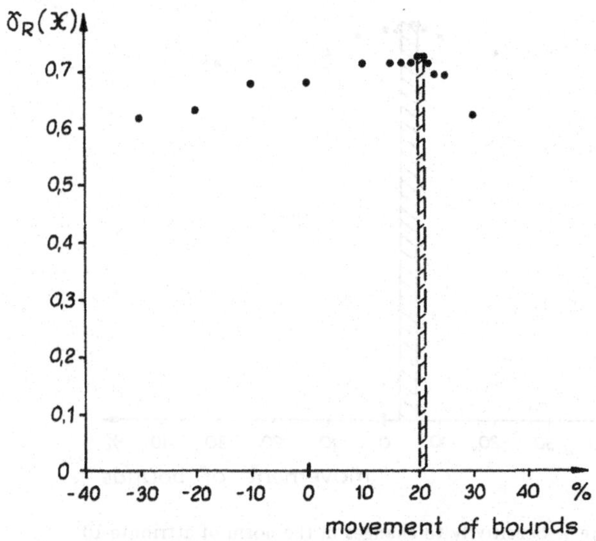

Figure 2. Sensitivity to changes in the norm of attribute 6

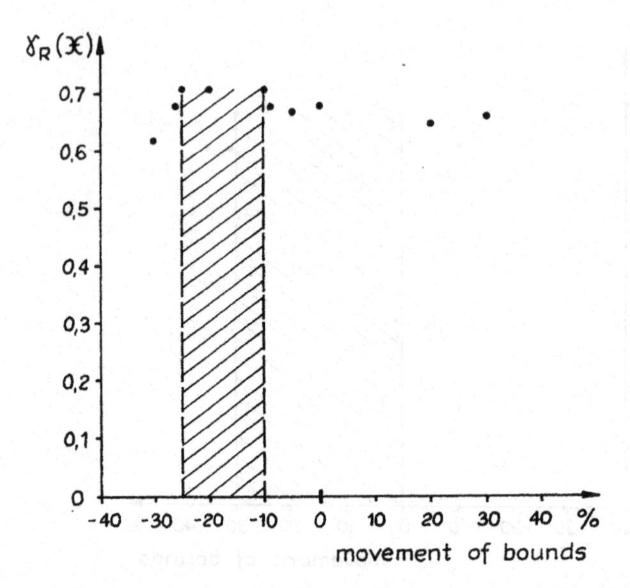

Figure 3. Sensitivity to changes in the norm of attribute 9

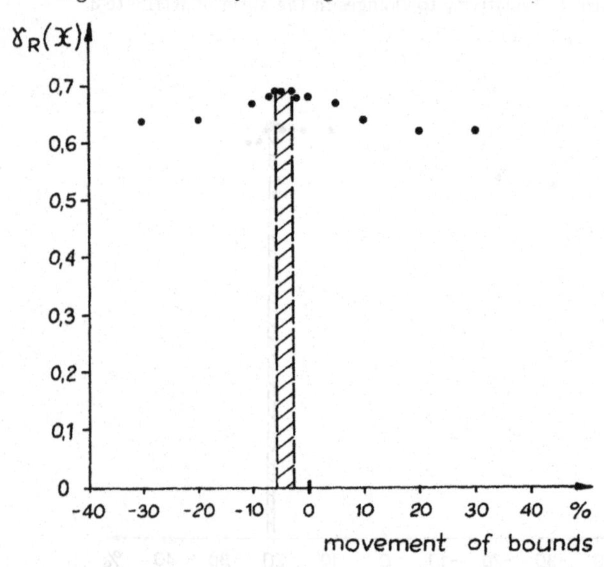

Figure 4. Sensitivity to changes in the norm of attribute 10

The next stage of the sensitivity analysis concerned the set of remaining attributes, i.e. $Q - R$. We changed one by one the norms of attributes 2, 5, 7, 8, 11, in the same way as before, while observing the effect of these changes on the quality of classification $\gamma_Q(\mathcal{Y})$. The results of this experiment are summarized in Table 1.

Table 1. Sensitivity of $\gamma_Q(\mathcal{Y})$ to minor changes in norms

No.of attribute	$\gamma_Q(\mathcal{Y})$ for original norms	The higest $\gamma_Q(\mathcal{Y})$ for new norms	Insensivity zone of the highest $\gamma_Q(\mathcal{Y})$	"Best" norm				Perc. change
				0	1	2	3	
2	0.967	0.967	-20%, +20%	≤ 35	> 35	–	–	0%
5	0.967	0.967	-20%, +30%	≤ 2	2–4	> 4	–	0%
7	0.967	0.967	-30%, +20%	≤ 50	50–100	> 100	–	0%
8	0.967	0.967	-30%, +30%	≤ 2	2–3	> 3	–	0%
11	0.967	0.967	-20%, +5%	≤ 15	15–25	25–40	> 40	0%

The conclusion is clear: the quality of classification $\gamma_Q(\mathcal{Y})$ is not sensitive to minor changes in the norms of attributes 2, 5, 7, 8 and 11.

The experiment showed that the quality of classification is sensitive to changes in the norms concerning the most significant attributes only, i.e. 3, 6, 9, 10. Knowing percentage zones in which these attributes give individually the highest quality of classification, we searched by a trial-and-error procedure the definition of norms which give together the highest quality $\gamma_Q(\mathcal{Y})$. The result is shown in Table 2. The corresponding $\gamma_Q(\mathcal{Y}) = 0.984$.

Table 2. "Best" norms for attributes 3,6,9,10

No. of atribute	Domain			Percentage change
	0	1	2	
3	≤ 0.5	0.5-3	> 3	0%
6	≤ 84.7	84.7-181.5	> 181.5	+21%
9	≤ 7.5	7.5-11.3	> 11.3	-25 %
10	≤ 94	94-235	> 235	-6%

3. Reduction of attributes

In this section, we deal with reduction of attributes subject to the new definition of norms which ensures the highest quality $\gamma_Q(\mathcal{Y})$.

Reduction of attributes is performed in view of answering two different questions: (a) what are the reducts, i.e. the minimal sets of independent attributes which ensure the

same quality of classification as the whole set Q, and (b) what is the smallest subset of attributes which ensure an **acceptable** quality of classification.

Comparison of the above reduction for new and original norms informs about sensitivity of results (a) and (b) to the changes in the norms.

Let us start with question (a). Using the new norms (cf. Tables 1 and 2), we obtained one minimal set composed of 8 attributes:

$$NW = \{2,3,4,5,6,7,9,11\}$$

The result obtained for original norms was the following:

$$OR = \{2,3,4,5,6,7,9,10,11\}.$$

Let us observe that using new norms, we got the minimal set with 8 instead of 9 attributes. The accuracy of classes and the quality of classification for both cases is shown in Table 3.

Table 3. Accuracy of classes and quality of classification for minimal sets NW and OR

Minimal	Quality of	Accuracy of classes			
set	cllassification	1	2	3	4
NW	0.984	1.0	0.9	1.0	0.87
OR	0.967	0.95	0.9	1.0	0.87

In conclusion, one can state that the contents of the minimal set shows a low sensitivity to changes in the norms in the neighbourhood of the best quality of classification.

Let us pass to question (b) of reduction. In previous study (cf. Pawlak et al. (1986), Fibak et al.(1986) and Słowiński (1992), we removed the particular attributes from set Q and observed the decrease in the quality of classification. We did this to find the smallest set of relevant attributes which would give a relatively high and acceptable quality of classification. In the present study, we have repeated this procedure for new and original norms to see the sensitivity of the reduction process.

The reduction of attributes from set Q, according to a given sequence, consists in removing the first attribute, the second and the third, and so on, until all the attributes in the sequence have been removed. Using a trial-and-error procedure we searched for sequences of attributes which are characterized by the least steep descent of the quality of classification in the course of reduction. For original norms, we have found 3 such sequences composed of 6 attributes:

$$E = \{1,8,11,2,7,3\}$$

$$G = \{8,11,2,1,7,10\}$$

$$H = \{8,11,7,1,2,5\}$$

The same sequences were checked with the new norms. The results are summarized in Figs. 5, 6, 7.

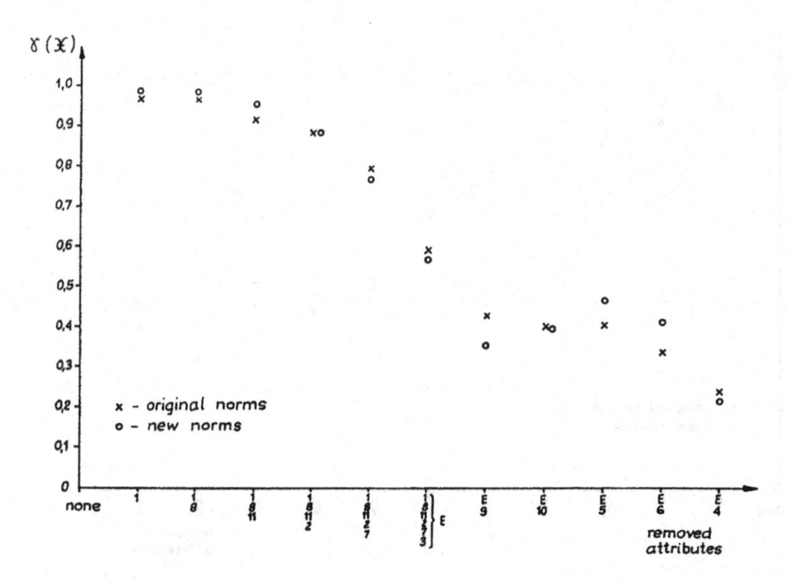

Figure 5. Quality of classification vs. removed attributes for new and original norms

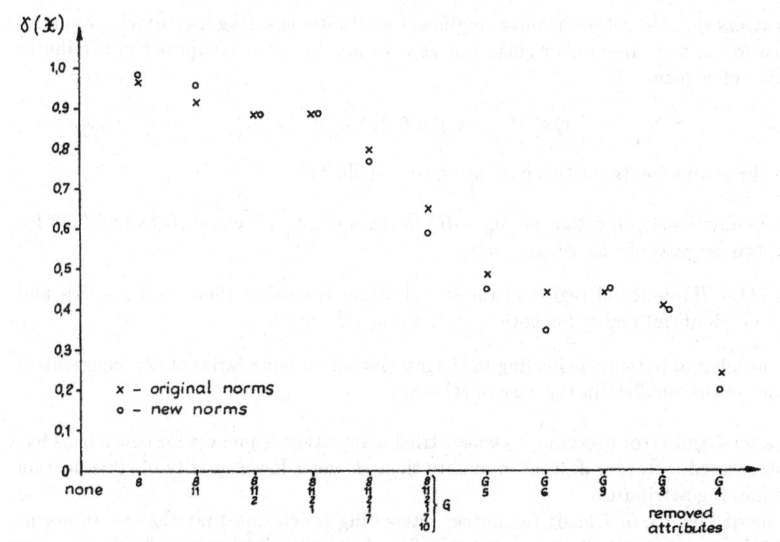

Figure 6. Quality of classification vs. removed attributes for new and original norms

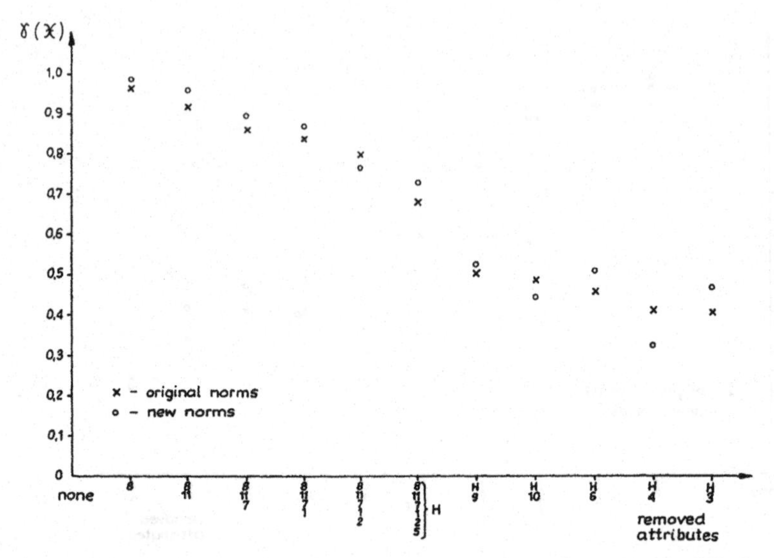

Figure 7. Quality of classification vs. removed attributes for new and original norms

Let us stress that the set of the most significant attributes ensuring a satisfactory quality of classification is the same for original and new norms, i.e. it is composed of attributes from outside of sequence H:

$$Q - H = \{3, 4, 6, 9, 10\} = R$$

There are three arguments for this conclusion (cf. Table 4):

(*i*) The quality of classification for $(Q - H)$ is the highest and equals 0.73 and 0.68 for new and original norms, respectively.

(*ii*) The $(Q - H)$-doubtful region of the classification is smaller then the $(Q - E)$- and $(Q - G)$-doubtful region for both new and original norms.

(*iii*) The number of patients belonging to the intersection of boundaries of non-consecutive classes is the smallest in the case of $(Q - H)$.

Using a trial-and-error procedure, we have tried many other sequences for new norms but all of them showed a steeper descent of quality then H and a lower quality of classification for five remaining attributes.

Analysis of Figs. 5, 6, 7 leads to another interesting conclusion that changes in norms don't perturbe a monotonic character of the quality of classification in the reduction process according to sequences E,G and H.

In view of the above results one can state that the reduction process is not much sensitive to changes of norms from original to new values. Moreover, the number of patients in the intersection of boundaries of non-consecutive classes (i.e. 1 and 4, 2 and 4, 1 and 3) is smaller for original then for new norms in the case of all sequences.

Table 4. Accuracy of classes and quality of classification for 3 sets of attributes

Subsets of attributes	Norms	Quality of classification	Doubtful region of classification (no. of patiens)	No. of patiens in intersection of boundaries of classes		
				1 and 4	2 and 4	1 and 3
$Q - E =$ $\{4,5,6,9,10\}$	new	0.57	49	28	6	8
	original	0.59	50	11	5	20
$Q - G =$ $\{3,4,5,6,9\}$	new	0.58	51	24	6	8
	original	0.65	43	21	3	7
$Q - H =$ $\{3,4,6,9,10\}$	new	0.73	33	16	9	5
	original	0.68	39	13	2	5

Let us also remark that the changes in norms made attribute 4 (qualitative-one) yet more discriminant.

Somebody could be surprised that attribute 10 which was absent in the minimal set for new norms is now in the subset of most significant attributes. This follows from the fact that significance of an attribute, in the sense of its influence on the quality of classification, depends on what other attributes enter the same subset. In other words, the significance of an attribute is not "context-free". We observed such situations quite often in the reduction process.

4. Decision algorithms

We have derived a decision algorithm from the information system identified with the decision table (cf. Słowiński (1992)). Reducing the information system to $(Q - H)$-positive region of the classification, we have received algorithms composed of 44 to 48 deterministic decision rules. Let us recall that the number of rules depends on the order of attributes selected for setting up the decision algorithm (Boryczka and Słowiński (1988)).

Using the new norms (cf. Tables 1 and 2), we obtained algorithms composed of 51 to 57 deterministic decision rules. They are, of course, non-comparable with the rules obtained for original norms because the corresponding coded information systems are different.

In conclusion, one can state that the use of norms which ensure the highest quality of classification $\gamma_Q(\mathcal{Y})$ decision algorithms with a greater number of rules then the original norms, and from this point of view the original norms are better.

5. Conclusions

The sensitivity analysis concerning the norms translating exact values of attributes into several qualitative terms, leads to the following global conclusions.

1. The quality of classification is sensitive to changes in the norms of the most significant quantitative attributes only, i.e.3, 6, 9, 10. The range of this sensitivity wasn't greater then 0.1 for movement of bounds from −30% to +30%.

2. The contents of the minimal set shows a low sensitivity to changes in the norms from the original to the new definition. The new definition is that which ensures the highest quality of classification ($\gamma_Q(\mathcal{Y}) = 0.984$).

3. The set of the most significant attributes ensuring a satisfactory quality of classification is the same for original and new norms, i.e. $R = \{3, 4, 6, 9, 10\}$.

4. Changes in norms don't perturbe a monotonic character of the quality of classification in the reduction process according to sequences of attributes analysed in previous studies for original norms.

5. The number of patients in the intersection of boundaries of non-consecutive classes of the result of operation is in general smaller for original norms then for new ones.

6. The original norms are better then the new ones from the viewpoint of the number of rules in the corresponding decision algorithm.

The above conclusions demonstrate that original norms following from medical experience were well defined. Moreover, the results of analysis of the considered information system using rough sets theory are robust in the sense of low sensitivity to minor changes in the norms of attributes.

References

[1] Boryczka, M., Słowiński, R. (1988). Derivation of optimal decision algorithms from decision tables using rough sets. *Bull. Polish Acad. Sci., Tech. Sci.*, **36** (3-4),251-260.

[2] Fibak, J., Pawlak, Z., Słowiński, K., Słowiński, R. (1986). Rough sets based decision algorithm for treatment of duodenal ulcer by HSV. *Bull. Polish Acad. Sci., Bio. Sci.*, **34** (10-12), 227-246.

[3] Greenburg, A.G. (1987). Commentary on the paper by Z. Pawlak, K. Słowiński and R. Słowiński in Int. J. Man-Machine Studies (1986) **24**, 413-433. *Computing Reviews* **27** (3).

[4] Pawlak, Z. (1984). Rough classification. *Int. J. Man-Machine Studies* **20**, 469-483.

[5] Pawlak, Z., Słowiński, K., Słowiński, R. (1986). Rough classification of patients after highly selective vagotomy for duodenal ulcer. *Int. J. Man-Machine Studies* **24**, 413-433.

[6] Słowiński,K. (1992).Rough classification of *HSV* patients. Chapter I.6 in this volume.

[7] Słowiński,K., Słowiński,R. (1990). Sensivity analysis of rough classification. *Int. J. Man-Machine Studies*, **32**,693-705.

Part III
Chapter 4

DISCRETIZATION OF CONDITION ATTRIBUTES SPACE

Andrzej LENARCIK Zdzisław PIASTA
Kielce University of Technology
25-314 Kielce
Poland

Abstract. Objects in an information system analyzed by the rough sets theory methods are characterized by attributes, which can take on a finite set of values only. In diagnostic experiments, condition attributes are usually treated as continuous variables, taking values from certain intervals. So, to use this theory in such problems, certain discretization (coding) of continuous variables is needed. The optimal classification properties of an information system were taken by the authors as base criteria for selecting discretization. The concepts of a random information system and of an expected value of classification quality were introduced. The method of finding suboptimal discretizations based on these concepts is presented and is illustrated with data from concretes' frost resistance investigations.

1. Introduction

This paper concerns the application of rough sets theory [1] to classification arising in diagnostic problems. The classification is made on the basis of a suitably-gathered set of attributes. As examples, we can consider the evaluation of the frost resistance of porous building materials using absorbability, capillary absorption, apparent gravity density and compressive strength as features [2,3], or the identification of the kinds of inside structure defects of composites with features which are extracted from acoustic emission signals registered during the loading of these composites [4,5,6]. In both cases, the attributes are treated as continuous, and can take on all values from certain intervals. Usually, to investigate the type of problem presented above, the methods of multivariate statistical analysis are used [2-6]. However, the methods of rough sets theory also have been found to be of practical interest in such cases [7]. To use these latter methods, certain discretization (coding) of continuous attributes is needed. In this paper the new approach to this problem is presented. The concept of a random information system is introduced. In a certain sense, it is the converse of the approach presented by Pawlak, Wong and Ziarko [8]. In the paper [8], the information system as a finite set of objects is treated in a deterministic way, but the frequencional solution of the approximation problem is of a probabilistic nature. Our approach is quite the opposite - the information system is random, but the approximation is performed in a deterministic way on the system realizations. A mathematical analysis is made with the assumption of complete probabilistic knowledge about the information system. Qualitative and quantitative conclusions resulting from our considerations can be very helpful in using the widely prevalent deterministic theory, especially in discretization of condition attributes space, and in the way in which the deterministic decision rules are used for classifying objects from outside the system.

2. Random Information System

The domain of our consideration is the probabilistic space (Ω, Σ, P) , in which the elementary events $\omega \in \Omega$ are treated as different variants of the realization of the same experimental scheme. By an experimental scheme $\mathcal{I}$ (or random information system), we understand a sequence of independent experiments:

$$\mathcal{I} = (\mathcal{A}_1, \mathcal{A}_2, \mathcal{A}_3, \ldots). \tag{1}$$

Experiments are represented by the random independent vectors

$$\mathcal{A}_\alpha : \Omega \rightarrow S \times D, \quad \alpha = 1, 2, 3, \ldots \tag{2}$$

with the same distribution. The realizations $A_\alpha = \mathcal{A}_\alpha(\omega)$ are called objects and are the elements of the Cartesian product of the space of states

$$S = \{s_1, s_2, \ldots, s_k\} \tag{3}$$

and of the set of decisions

$$D = \{d_1, \ldots, d_l\}. \tag{4}$$

Comparing sets S and D with concepts from deterministic theory, S is the Cartesian product of condition attributes domains, whereas D is the Cartesian product of decision attributes domains. The object is identified as a pair: (state - decision), so two coordinates of vector $\mathcal{A}_\alpha = (\mathcal{A}_\alpha^1, \mathcal{A}_\alpha^2)$, $\mathcal{A}_\alpha^1(\omega) \in S$, $\mathcal{A}_\alpha^2(\omega) \in D$ can be considered.

By $\mathcal{I}_n$ we denote n-elemental finite random information system, which consists of the first n-elements from system $\mathcal{I}$. The infinite information system $\mathcal{I}$ is only a formalization which permits the investigation of dependencies between $\mathcal{I}_n$ and $\mathcal{I}_{n+1}$. As stated in the introduction, a full probabilistic knowledge about the system is assumed in our considerations. Since vectors (2) are independent and have the same distribution, full knowledge of the system is contained in the common distribution:

$$P(\mathcal{A}_\alpha^1 = s_i, \mathcal{A}_\alpha^2 = d_j) = u_{ij}. \tag{5}$$

It is worthwhile to note that any combinations of condition attributes and the corresponding decisions which are impossible can be removed from considerations simply by introducing null probabilities in formula (5).

Note 1.
The analysis of the random information system will be performed on realizations. This limits our interest to random systems $\mathcal{I}_n$ of a settled number of elements only. Selecting any particular $\omega \in \Omega$, gives the information system:

$$I_n = \mathcal{I}_n(\omega) = (\mathcal{A}_1(\omega), \mathcal{A}_2(\omega) \ldots \mathcal{A}_n(\omega)) = (A_1, A_2 \ldots A_n). \tag{6}$$

The system I_n has a finite set of objects. This makes it possible to use the deterministic rough sets theory to analyze this system. Characteristics of system (6) obtained in this way will be, in fact, random variables, dependent on $\omega \in \Omega$, allowing us to consider means and variances of these quantities.

3. Quality and Correctness of Classification

For any fixed $n > 0$ and for any particular $\omega \in \Omega$, consider a finite n-dimensional realization $I_n = \mathcal{I}_n(\omega)$, which is a finite set of objects:

$$I_n = (A_1, A_2, \ldots, A_n). \tag{7}$$

The number n_{ij} denotes the cardinality of set $\{\alpha : A^1_\alpha = s_i, \quad A^2_\alpha = d_j\}$. So $n_i = \sum_j n_{ij}$ is the number of objects belonging to state i. For further consideration, we will introduce the following differentiation among states, crucial to the underlying structure.

1. If $n_i = 0$, then s_i is an empty state.
2. If j exists, such that $n_{ij} = n_i$, then s_i is a pure state.
3. If the state s_i is neither pure nor empty, it is called a mixed state.

In case 2, if there exists j for state s_i, then it is the only one. Such a state s_i will be called j-pure or d_j-pure. So the natural transformation $K_n : S \rightarrow D$ arises, which is partially defined in the set of states. It maps the pure states to a decision, which realizes this purity. K_n function will be further called empirical, or natural, classification. The domain of K_n function will be called the set of K-pure states. Comparing the proposed conception with Pawlak's rough sets theory one can state that all pure states are the low approximation of the family of sets $D_1, \ldots, D_l$, where D_j is a set of these objects in sequence (7), on which decision d_j is realized. Mixed states are the sums of the borders of these approximations. These two types of states appeared in rough sets theory, while the concept of the empty state seems to be new, and has no analogy in the deterministic approach. We also note that the deterministic quality of classification is the ratio of the number of objects belonging to pure states to the number of all objects. This value will be further named the quality of classification *a posteriori* and denoted by J_1. Thus

$$J_1 = \sum_{s_i\text{-pure}} \frac{n_i}{n}. \tag{8}$$

We can conclude in accordance with note 1, that random variable $\mathcal{J}_1$ has been defined. It is worthwhile to note that the natural empirical classification for system I_n obtained above can be used to classify object A_{n+1}, which may be from outside the system. This classification would be a prediction of the decision value d_j for object A_{n+1}, if state A^1_{n+1} is j-pure. Because not all states are pure, the classification is not always possible. It follows that the probability J_2 of classification of object A_{n+1}, using classification K_n, is equal to the probability that the new object belongs to K_n-pure state:

$$J_2 = \sum_{s_i\text{-pure}} x_i, \tag{9}$$

where $x_i = \sum_j u_{ij}$ is the probability that any object A_{n+1} belongs to state i. Defined in this way, J_2 is further called the quality of classification *a priori* and corresponds to the random variable $\mathcal{J}_2$ in accordance with note 1. Of course, both values J_1 and J_2 ($\mathcal{J}_1$ and $\mathcal{J}_2$) can take only the values from the interval $< 0, 1 >$. We also note, that the possibility of classifying the object A_{n+1} does not ensure the correctness of that classification. This is consequence of the fact that decision $K_n(s_i)$ realizing the purity of state s_i to which the object A_{n+1} $(A_{n+1} = s_i)$ belongs may not coincide with the true classification A^2_{n+1} of this object. Therefore, the degree of correctness of classification P_2 (*a priori*)

$$P_2 = \sum_{s_i\text{-pure}} P(\mathcal{A}^1_{n+1} = s_i; \quad \mathcal{A}^2_{n+1} = K_n(s_i)) \tag{10}$$

is defined as a probability of correct classification of object A_{n+1} using the empirical classification specified by system (7). According to note 1, $\mathcal{P}_2$ is a new random variable. The value $\mathcal{B}_2$ is defined by the equation

$$\mathcal{J}_2 = \mathcal{P}_2 + \mathcal{B}_2 \tag{11}$$

and is called an empirical classification error. There is no reason to define the value P_1 (*a posteriori*) analogical to J_1 , because P_1 is always a unity.

4. Expected Value of the a priori Classification Characteristics

In this section we will deal with obtaining formulas for expected values of random variables: $\mathcal{J}_2, \mathcal{P}_2, \mathcal{B}_2$. First, consider a general situation. Let $I_n = \mathcal{I}_n(\omega)$ be a realization of a random information system. We can consider pure, empty and mixed states as introduced in section 3. The following random variable can be defined:

$$\mathcal{F}(\omega) = \sum_{\substack{s_i\text{-pure} \\ K_n(s_i)=d_i}} a_{ij} + \sum_{s_i\text{-empty}} b_i. \tag{12}$$

When $E(\mathcal{F})$ is known, it is evident that $E(\mathcal{J}_2)$, $E(\mathcal{P}_2)$, $E(\mathcal{B}_2)$ can be considered as known also, because they are special cases of $E(\mathcal{F})$:

$$\begin{aligned} \mathcal{P}_2 &= \mathcal{F}, & &\text{if } a_{ij} = u_{ij}, & b_i &= 0; \\ \mathcal{J}_2 &= \mathcal{F}, & &\text{if } a_{ij} = x_i, & b_i &= 0; \\ \mathcal{B}_2 &= \mathcal{F}, & &\text{if } a_{ij} = x_i - u_{ij}, & b_i &= 0. \end{aligned}$$

Putting $a_i = 0$, $b_i = x_i$ we get the expected value of $\mathcal{E}_2$ which is the probability of object $\mathcal{A}_{n+1}$ belonging to an empty state.

First, the case of one state will be investigated ($k = 1$). Then from the above, $\mathcal{F}$ can take the following values:

$$\mathcal{F} = \begin{cases} a_{1j}, & \text{if } s_1 \text{ j-pure state} \\ b_1, & \text{if } s_1 \text{ empty state} \\ 0, & \text{if } s_1 \text{ mixed state} \end{cases}$$

where $P(s_1 \text{ is j-pure}) = u_{1j}^n$, $P(s_1 \text{ is empty}) = 0$. So

$$E(\mathcal{F}) = \sum_{j=1}^{l} a_{1j} u_{1j}^n.$$

Now consider the case of two states ($k = 2$). We divide the space Ω for $n+1$ mutually exclusive hypotheses:

$$H_r = \{\text{there are } r \text{ objects in } s_2 \text{ state}\}, \quad r = 0, 1, \ldots, n.$$

In determining $E\mathcal{F}$ the Bayes formula can be used:

$$E\mathcal{F} = \sum_{r=0}^{n} E(\mathcal{F}/H_r)P(H_r).$$

The probabilities $P(H_r)$ of H_r hypothesis and $E(\mathcal{F}/H_r)$ are specified below:

$$P(H_0) = x_1^n; \qquad E(\mathcal{F}/H_0) = \sum_{j=1}^{l} a_{1j}\left(\frac{u_{1j}}{x_1}\right)^n + b_2;$$

$$P(H_r) = \binom{n}{r} x_1^{n-r} x_2^r; \qquad E(\mathcal{F}/H_r) = \sum_{j=1}^{l}\left[a_{1j}\left(\frac{u_{1j}}{x_1}\right)^{n-r} + a_{2j}\left(\frac{u_{2j}}{x_2}\right)^r\right]; \qquad r = 1,\ldots,n-1$$

$$P(H_n) = x_2^n; \qquad E(\mathcal{F}/H_n) = \sum_{j=1}^{l} a_{2j}\left(\frac{u_{2j}}{x_2}\right)^n + b_1.$$

So

$$\begin{aligned} E\mathcal{F} &= \sum_{r=0}^{n} E(\mathcal{F}/H_r)P(H_r) = E(\mathcal{F}/H_0) + \sum_{r=1}^{n-1} E(\mathcal{F}/H_r)P(H_r) + E(\mathcal{F}/H_n) = \\ &\left(\sum_{j=1}^{l} a_{1j}u_{1j}^n + b_2 x_1^n\right) + \sum_{i=1}^{k}\sum_{j=1}^{l}\binom{n}{r}(a_{1j}u_{1j}^{n-r}x_2^r + a_{2j}u_{2j}^r x_1^{n-r}) + \left(\sum_{j=1}^{l} a_{2j}u_{2j}^n + b_1 x_2^n\right) = \\ &\left[\sum_{j=1}^{l} a_{1j}(u_{1j}+x_2)^n + (b_1 - \sum_{j=1}^{l} a_{1j})x_2\right] + \left[\sum_{j=1}^{l} a_{2j}(u_{2j}+x_1)^n + (b_2 - \sum_{j=1}^{l} a_{2j})x_2\right]. \end{aligned}$$

Reasoning for $k = 2$ can be easily extended for any k given the following formula:

$$E\mathcal{F} = \sum_{i=1}^{k}\left[\left(\sum_{j=1}^{l} a_{ij}(x_1+\ldots+u_{ij}+\ldots+x_k)^n\right) + (b_i - \sum_{j=1}^{l} a_{ij})(x_1+\ldots+\hat{x}_i+\ldots+x_k)^n\right], \quad (13)$$

where the symbol $\hat{x}_i$ denotes that the number x_i is omitted. We will now show that this extension is true. We present a method of mathematical induction to prove it. The step $(k) \Rightarrow (k+1)$ is seen to be the only one needed, and is only the simple reformulation of the case for $k = 2$. We can define $n+1$ hypotheses

$$H_r = \{\text{there are } r \text{ objects in state } s_{n+1}\} \qquad r = 0, 1, \ldots, n.$$

Then we obtain

$$P(H_r) = \binom{n}{r}(x_1 + \ldots + x_k)^{n-r} x_{k+1}^r \qquad r = 0, 1, \ldots, n.$$

For further conclusions, the following notations will be useful:

$$X_i = x_1 + \ldots + \hat{x}_i + \ldots + x_k; \quad X_{k+1} = x_1 + \ldots + x_k; \quad U_{ij} = X_i + u_{ij}.$$

Then

$$P(H_r) = \binom{n}{r}(X_{k+1}^{n-r} x_{k+1}^r) \qquad r = 0, 1, \ldots, n$$

and by the hypothesis (13)

$$E(\mathcal{F}/H_r) = \sum_{j=1}^{k}\left[\sum_{j=1}^{l} a_{ij}\left(\frac{U_{ij}}{X_{k+1}}\right)^{n-r} + (b_i - \sum_{j=1}^{l} a_{ij})\left(\frac{X_i}{X_{k+1}}\right)^{n-r}\right] + \sum_{j=1}^{l} a_{k+1,j}\left(\frac{u_{k+1,j}}{x_{k+1}}\right)^r$$

for $r = 1, 2, \ldots, n$. So

$$E(\mathcal{F}/H_r)P(H_r) = \sum_{i=1}^{k}\sum_{j=1}^{l}\Big[\binom{n}{r} a_{ij} U_{ij}^{n-r} x_{k+1}^{r} + (b_i - \sum_{j=1}^{l} a_{ij}) \binom{n}{r} X_i^{n-r} x_{k+1}^{r}\Big] + \sum_{j=1}^{l} a_{k+1,j} \binom{n}{r} X_{k+1}^{n-r} u_{k+1,j}^{r}$$

for $r = 1, 2, \ldots, n$. For $r = 0$ we have

$$E(\mathcal{F}/H_0) = \sum_{i=1}^{k}\Big[\sum_{j=1}^{l} a_{ij}\left(\frac{U_{ij}}{X_{k+1}}\right)^n + (b_i - \sum_{j=1}^{l} a_{ij})\left(\frac{X_i}{X_{k+1}}\right)^n\Big] + b_{k+1}.$$

So

$$\begin{aligned} E(\mathcal{F}/H_0)P(H_0) &= \sum_{i=1}^{k}\sum_{j=1}^{l}\Big[\binom{n}{0} a_{ij} U_{ij}^{n} x_{k+1}^{0} + (b_i - \sum_{j=1}^{l} a_{ij})\binom{n}{0} X_i^n x_{k+1}^{0}\Big] + \\ &+ \sum_{j=1}^{l} a_{k+1,j}\binom{n}{0} X_{k+1}^{n} u_{k+1}^{0} + (b_{k+1} - \sum_{j=1}^{l} a_{k+1,j})\binom{n}{0} X_{k+1}^{n} x_{k+1}^{0}. \end{aligned}$$

Hence

$$\begin{aligned} E\mathcal{F} &= \sum_{r=0}^{n} E(\mathcal{F}/H_r)P(H_r) = \\ &= \sum_{i=1}^{k}\Big[\sum_{j=1}^{l} a_{ij}(U_{ij} + x_{k+1})^n + (b_i - \sum_{j=1}^{l} a_{ij})(X_i + x_{k+1})^n\Big] + \\ &+ \sum_{j=1}^{l} a_{k+1,j}(X_{k+1} + u_{k+1,j})^n + (b_{k+1} - \sum_{j=1}^{l} a_{k+1,j})X_{k+1}^{n}, \end{aligned}$$

which upon examination completes the proof. So the following formulas for means have been obtained:

$$E\mathcal{J}_2 = \sum_{i=1}^{k}\Big[x_i \sum_{j=1}^{l}(x_1 + \ldots + u_{ij} + \ldots + x_k)^n - l x_i(x_1 + \ldots + \hat{x}_i + \ldots + x_k)^n\Big], \quad (15)$$

$$E\mathcal{P}_2 = \sum_{i=1}^{k}\Big[\sum_{j=1}^{l} u_{ij}(x_1 + \ldots + u_{ij} + \ldots + x_k)^n - x_i(x_1 + \ldots + \hat{x}_i + \ldots + x_k)^n\Big], \quad (16)$$

$$E\mathcal{B}_2 = \sum_{i=1}^{k}\Big[\sum_{j=1}^{l}(x_i - u_{ij})(x_1 + \ldots + u_{ij} + \ldots + x_k)^n - (l-1)x_i(x_1 + \ldots + \hat{x}_i + \ldots + x_k)^n\Big], \quad (17)$$

$$E\mathcal{E}_2 = \sum_{i=1}^{k} x_i(x_1 + \ldots + \hat{x}_i + \ldots + x_k)^n. \quad (18)$$

5. Expected Value of Quality of Classification a posteriori

The formulas introduced in section 4 do not permit calculation of the quality of classification *a posteriori* directly, but analogous formula for $E\mathcal{J}_1$ can be found. According to (8), $\mathcal{J}_1$ is the number of objects which are in pure states divided by the number of objects. As before, we first examine the case of one state s_1 ($k=1$). In this case, $\mathcal{J}_1$ can be equal only to unity (there can be only one state). Then

$$E\mathcal{J}_1 = \sum_{j=1}^{l} u_{1j}^n.$$

For k=2 we divide Ω for mutually exclusive hypotheses:

$$H_r = \{\text{there are } r \text{ objects in state } s_2\}.$$

It is evident that

$$P(H_r) = \binom{n}{r} x_1^{n-r} x_2^r, \quad \text{for } r = 0, 1, \ldots, n.$$

Thus

$$E(\mathcal{J}_1/H_r) = \frac{n-r}{n} \sum_{j=1}^{l} \left(\frac{u_{1j}}{x_1}\right)^{n-r} + \frac{r}{n} \sum_{j=1}^{l} \left(\frac{u_{2j}}{x_2}\right)^r, \quad \text{for } r = 0, 1, \ldots, n.$$

Hence,

$$E\mathcal{J}_1 = \sum_{r=0}^{n} E(\mathcal{J}_1/H_r)P(H_r) = \sum_{r=0}^{n} \sum_{j=1}^{l} \left[\frac{n-r}{n} \binom{n}{r} u_{1j}^{n-r} x_2^r + \frac{r}{n} \binom{n}{r} u_{2j}^r x_1^{n-r}\right].$$

Using the identity

$$\sum_{i=0}^{n} i \binom{n}{i} a^{n-i} b^i = nb(a+b)^{n-1}$$

we discover that

$$E\mathcal{J}_1 = \sum_{j=1}^{l} \left[u_{1j}(u_{1j} + x_2)^{n-1} + u_{2j}(u_{2j} + x_1)^{n-1}\right].$$

This suggests the following formula for arbitrary k

$$E\mathcal{J}_1 = \sum_{i=1}^{k} \sum_{j=1}^{l} u_{ij}(x_1 + \ldots + u_{ij} + \ldots + x_k)^{n-1}. \tag{19}$$

Concise proof of this formula can be obtained by using the method of mathematical induction. Since formula (19) has been verified for $k=2$, it is enough to show the step $(k) \Rightarrow (k+1)$. All previous notations prove to be valid. Let

$$X_i = x_1 + \ldots + \hat{x}_i + \ldots + x_k \quad \text{for } i = 1, \ldots, k, \quad X_{k+1} = x_1 + \ldots + x_k,$$

and

$$H_r = \{\text{there are } r \text{ objects in state } s_{n+1}\}.$$

Then

$$P(H_r) = \binom{n}{r} X_{k+1}^{n-r} x_{k+1}^r,$$

and by the hypothesis (19)

$$E(\mathcal{J}_1/H_r) = \frac{n-r}{n} \sum_{i=1}^{k} \sum_{j=1}^{l} \frac{u_{ij}}{X_{k+1}} \left(\frac{X_i + u_{ij}}{X_{k+1}} \right)^{n-r-1} + \frac{r}{n} \sum_{j=1}^{l} \left(\frac{u_{k+1,j}}{x_{k+1}} \right)^r \qquad \text{for } r = 0, \ldots, n.$$

So

$$\begin{aligned} E\mathcal{J}_1 &= \sum_{r=0}^{n} E(\mathcal{J}_1/H_r) P(H_r) = \\ &= \sum_{i=1}^{k} \sum_{j=1}^{l} u_{ij} \sum_{r=0}^{n} \frac{n-r}{n} \binom{n}{r} (X_i + u_{ij})^{n-r-1} x_{k+1}^r + \sum_{j=1}^{l} \sum_{r=0}^{n} \frac{r}{n} \binom{n}{r} X_{k+1}^{n-r} u_{k+1,j}^r = \\ &= \sum_{i=1}^{k} \sum_{j=1}^{l} \frac{u_{ij}}{X_i + u_{ij}} \sum_{r=0}^{n} \frac{n-r}{n} \binom{n}{r} (X_i + u_{ij})^{n-r} x_{k+1}^r + \sum_{j=1}^{l} \sum_{r=0}^{n} \frac{r}{n} \binom{n}{r} X_{k+1}^{n-r} u_{k+1,j}^r \end{aligned}$$

and finally we get:

$$\begin{aligned} E\mathcal{J}_1 &= \sum_{i=1}^{k} \sum_{j=1}^{l} u_{ij} (X_i + u_{ij} + x_{k+1})^{n-1} + \sum_{j=1}^{l} u_{k+1,j} (X_{k+1} + u_{k+1,j})^{n-1} = \\ &= \sum_{i=1}^{k+1} \sum_{j=1}^{l} u_{ij} (X_i + u_{ij})^{n-1}, \end{aligned}$$

which succesfully completes the proof of formula (19).

6. Condition Attributes Space Partition

In this section we will find such a discretization of the continous condition attributes space as provides a possibility of a high expected value of the classification quality *a priori* $E\mathcal{J}_2$. We assume that objects in the information system are realizations of the random vector

$$A_\alpha = (A_\alpha^1, A_\alpha^2) \quad : \quad \Omega \to E \times D \tag{20}$$

which takes its values from the Cartesian product of the Euclidian space E, and of the set of decisions (classes) $D = \{d_1, \ldots, d_l\}$. The coordinates of the point $a = (a^{(1)}, \ldots, a^{(m)})$ in the space E are the values of m continuous condition attributes. We assume that the probability that the object will be related to the j-th decision is adequately represented by the probability distribution:

$$\Phi_j(U) = P(A_\alpha^1 \in U | A_\alpha^2 = d_j), \qquad j = 1, \ldots, l. \tag{21}$$

By the partition of the space E, we mean any arbitrary family $R = \{U_1, \ldots, U_k\}$ of the measurable subsets $U_i \in E$, which are such that:

$$U_i \cap U_j = \phi \text{ for } i \neq j \text{ and } U_1 \cup \ldots \cup U_k = E. \tag{22}$$

Consider a sequence of independent experiments $(\mathcal{A}_1, \ldots, \mathcal{A}_n)$ which coincide with a random selection of n-elemental sets of objects. The partition $R = \{U_1, \ldots, U_k\}$ of the space E induces the random information system which was defined in section 2. In this case the family R is a set of states. The formula

$$u_{ij} = P\{\mathcal{A}_\alpha^1 \in U_i, \mathcal{A}_\alpha^2 = d_j\} \qquad i = 1 \ldots k, \;\; j = 1 \ldots l,$$

completes the probability knowledge of the system. Note that

$$u_{ij} = \Phi_j(U_i)\pi_j,$$

where $\pi_j = P(\mathcal{A}_\alpha^2 = d_j)$ is a probability that an object will be from class d_j . The degree of predictivity for the random information system is the quality of classification *a priori*, for which the expected value is defined by formula (15). This quantity depends on the partition of the space; therefore, we will denote it $E\mathcal{J}_2(R)$ or $E\mathcal{J}_2(U_1, \ldots, U_k)$. From formula (15) we have:

$$E\mathcal{J}_2(R) = \sum_{i=1}^{k} x_i \left[\sum_{j=1}^{l} (1 - x_i + u_{ij})^n - l(1 - x_i)^n \right], \tag{23}$$

where $x_i = \sum_j u_{ij}$. Let us define a function of l-variables

$$J(u_1, \ldots, u_l) = x \left[\sum_{j=1}^{l} (1 - x + u_j)^n - l(1 - x)^n \right]. \tag{24}$$

The right side of (24) depends only on $u_1, \ldots, u_l$ because $x = u_1 + \ldots + u_l$. We denote $\bar{u} = (u_1, \ldots, u_l)$. The formula (23) can be rewritten as follows:

$$E\mathcal{J}_2(U_1, \ldots, U_k) = \sum_{i=1}^{k} J(u_{i1}, \ldots, u_{il}) = \sum_{i=1}^{k} J(\bar{u}_i).$$

The quantity $J(\bar{u}_i)$ we will call a contribution to the quality of the classification coming from the set U_i .

Consider the partitions of space E, which are implied by discretization of continuous condition attributes. By discretization of the q-th attribute we understand the sequence of numbers

$$a_1^{(q)} < a_2^{(q)} < \ldots < a_{r_q}^{(q)}, \tag{25}$$

which divides the set of values of this attribute for $r_q + 1$ intervals $(q = 1, \ldots, m)$. Each value of this sequence defines the hiperplane in the d-dimensional space E perpendicular to the q-th attribute axis. As a result of the discretization of all attributes, the space is partitioned into $k = (r_1 + 1) \ldots (r_q + 1)$ m-dimentional intervals (fig.1), which create the partition of space E according to (22). For such a partition, the expected value of the quality of the classification *a priori* is determined.

In the proposed method, we start with an initial discretization (25) of each attribute. The subsequent steps provide an increase in the value of the quality of classification by eliminating values of the sequences (25). In each of these steps, we eliminate such a hiperplane as will result the greatest increase in the value of the quality of classification. This algorithm is continued until the elimination of each element of the remaining values from sequences (25) does not increase the value of $E\mathcal{J}_2$. This concept is similar to the method of elimination of parameters in linear regression (backward method) and, similarly, leads to suboptimal results.

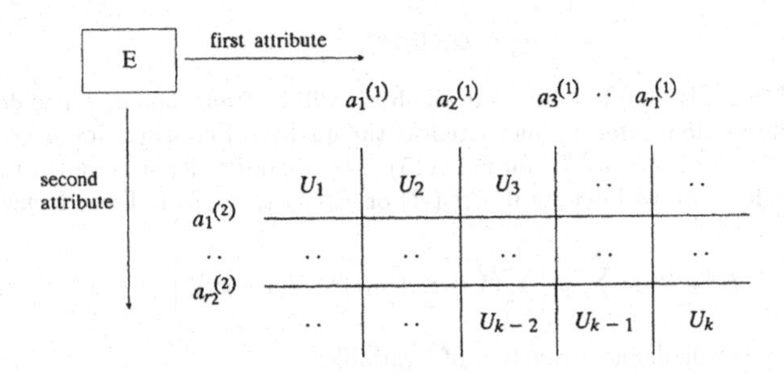

Figure 1. The partition of space E induced by the discretization of continous condition attributes.

A computer implementation of this algorithm requires the creation in the operation memory (RAM) of all the states $\{U_1, \ldots, U_k\}$ which were arrived at by the primary discretization. Therefore, the only limitation to the practical use of this algorithm is the capacity of the RAM. In the example below we consider two classes normally distributed: $N(m_1, \Sigma), N(m_2, \Sigma)$ where $\Sigma = \sigma^2 I$ and I is the identity matrix. We set the values $\sigma = 0.8$ and $m_1 = (1.1, 0.85, 0.5, 0.3)$ and $m_2 = (-1.1, -0.85, -0.5, -0.3)$. The number n of objects is set at 15. The primary discretization is obtained in the interval $< -1.4, 1.4 >$ using 13 equally spaced dividing points (fig.2a).

To obtain the smallest possible 4-dimensional structure for purposes of computation, a primary elimination of dividing points is performed four times, each time along a fixed axis. Fig.2b illustrates the results obtained. Then the elimination of four dimensions is performed. Note that the variables which most weekly separate the classes have been eliminated (fig.2c).

The final number of partitions $k = 6$ does not exceed the number of objects $n = 15$. These qualitative conclusions have been observed in other situations examined by the authors.

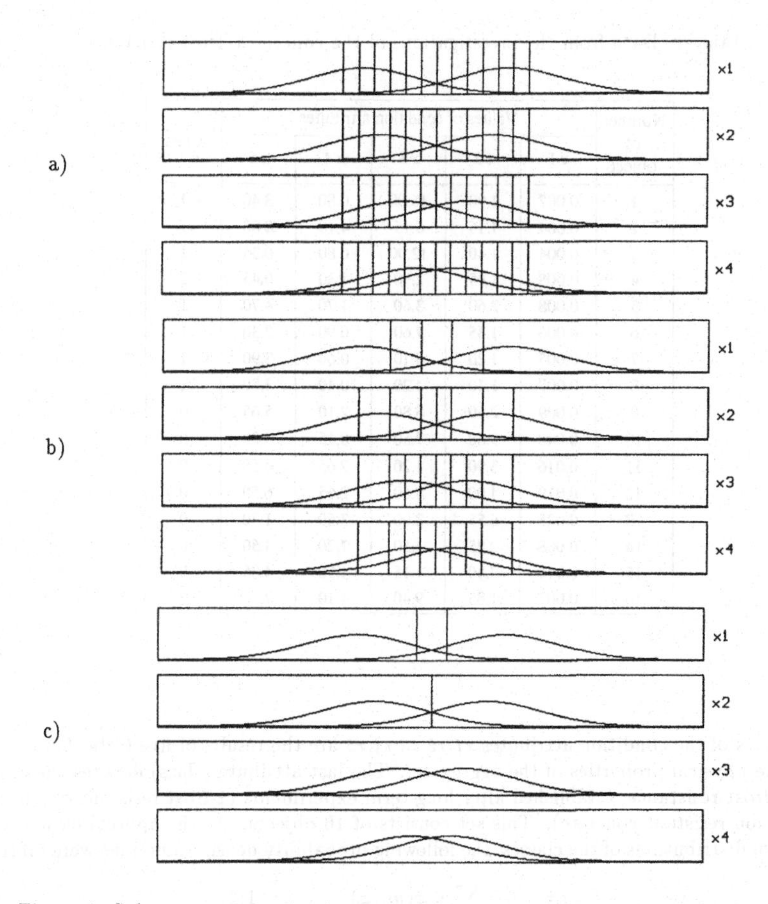

Figure 2. Subsequent stages in the elimination process for two normally distributed classes.

7. Practical Example

The algorithm for eliminating intermediate values of condition attributes, given in the previous section, was applied to the real set of data arising from the investigation of the concretes' frost resistance [9]. This set is shown in Table 1.

Table 1. Data from the investigations of the concretes' frost resistance.

Number of object	Values of condition attributes					Class
	$x1$	$x2$	$x3$	$x4$	$x5$	
1	0.007	1.90	16.00	0.80	3.40	1
2	0.004	1.15	6.15	4.10	8.00	1
3	0.004	1.40	12.00	0.80	0.35	1
4	0.008	1.95	2.90	1.30	0.45	1
5	0.008	2.60	3.60	1.70	4.70	1
6	0.006	1.35	7.60	0.90	2.30	1
7	0.005	1.20	8.10	0.60	3.90	1
8	0.007	1.30	4.20	0.40	3.80	1
9	0.009	2.60	8.80	2.10	5.65	0
10	0.048	9.60	7.50	0.15	0.20	0
11	0.016	3.90	1.20	2.65	6.50	0
12	0.019	4.00	4.60	2.65	6.50	0
13	0.035	6.55	2.00	3.50	1.70	0
14	0.008	2.25	4.50	7.20	4.50	0
15	0.016	2.40	1.15	1.50	4.85	0
16	0.007	1.85	9.40	1.10	7.20	0

The values of the condition attributes $x1, x2, x3, x4, x5$ are the results of five tests characterizing the physical properties of the aggregates. The last attribute (class) denotes the concretes' frost resistance, established after long-term experiments (1-frost resistant concrete, 0-frost non-resistant concrete). This set consists of 16 objects. As the approximations of unknown distributions of the classes, the following probability density functions were taken:

$$f(x) = \frac{1}{n_j} \sum_{class(i)=j} \phi(m_i, x), \qquad j = 1, 2,$$

where $\phi(m_i, x)$ is the density of normal distribution $N(m_i, \Sigma)$. The i-th object position in condition attributes space is determined by the mean vector m_i. The covariance matrix $\Sigma = \sigma^2 I$. The calculations presented below were made for $\sigma = (max_q - min_q)/16$, where max_q and min_q denote the largest and the smallest value of the q-th condition attribute, respectively. Number n_j is the cardinality of the j-th class. Let $x_1^{(q)} < \ldots < x_{\tilde{n}}^{(q)}$, be a sequence of values of the q-th condition attribute, ordered by increase in size ($\tilde{n} \leq n$ because some values may be reiterated). We define $\tilde{n} - 1$ intermediate values of the q-th attribute:

$$a_i^{(q)} = \frac{x_i^{(q)} + x_{i+1}^{(q)}}{2}, \quad i = 1, \ldots, \tilde{n} - 1. \tag{26}$$

The sequence (26) divides space E for $\tilde{n}$ intervals, each of which includes at least one object. If the neighbouring intervals include objects related to the same decision, then we do not consider the intermediate values separating these intervals. The final subsequence of sequence (26) is taken as an initial discretization (25) of the q-th condition attribute. The preliminary discretization (for all of five condition attributes) is shown in fig.3.

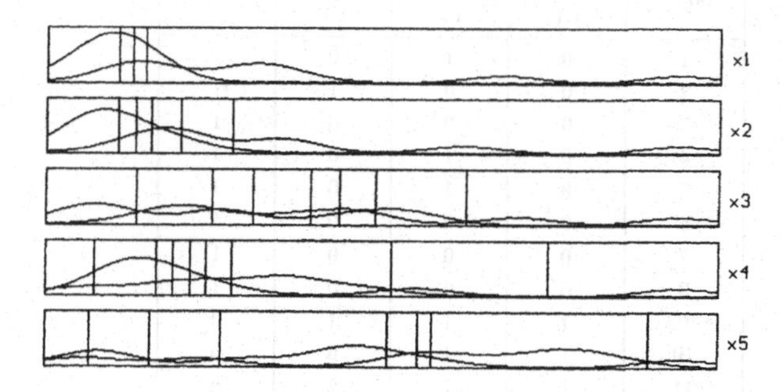

Figure 3. Preliminary discretizations for all condition attributes from Table 1.

The elimination of intermediate values was conducted in four-dimensional spaces, obtained by removing one of five condition attributes. The most valuable results were obtained, when the initial sets were respectively *(x1,x2,x4,x5)* and *(x2,x3,x4,x5)*. The elimination of the intermediate values of the first set of attributes led to the results shown in fig.4.

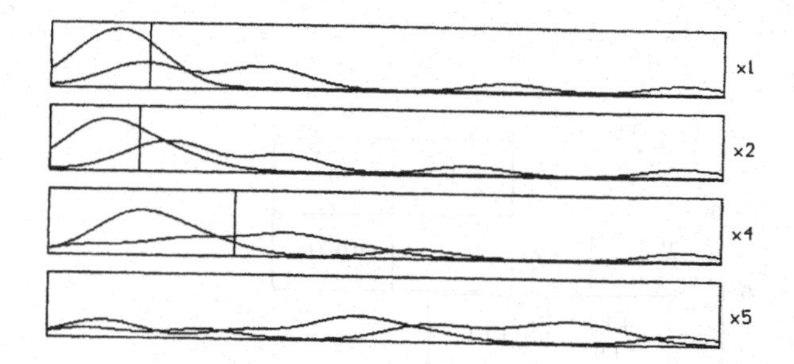

Figure 4. Final result of the elimination process for *x1,x2,x4* and *x5* attributes.

The intermediate points: 0.0085 along *x1* , 1.875 along *x2* and 1.9 along *x4* were obtained. The attribute *x5* has been eliminated. Coding the set from Table 1 on the basis of the intermediate points obtained, we get the information system presented in Table 2 .

Table 2. The coded information system obtained for $x1$,$x2$,$x4$ and $x5$ attributes.

Number of object	Codes of condition attributes			Class
	x 1	x 2	x 4	
1	0	1	0	1
2	0	0	1	1
3	0	0	0	1
4	0	1	0	1
5	0	1	0	1
6	0	0	0	1
7	0	0	0	1
8	0	0	0	1
9	1	1	1	0
10	1	1	0	0
11	1	1	1	0
12	1	1	1	0
13	1	1	1	0
14	0	1	1	0
15	1	1	0	0
16	0	0	0	0

Note that eight states in this system are possible. Five pure states, two empty and one mixed state were obtained (fig.5).

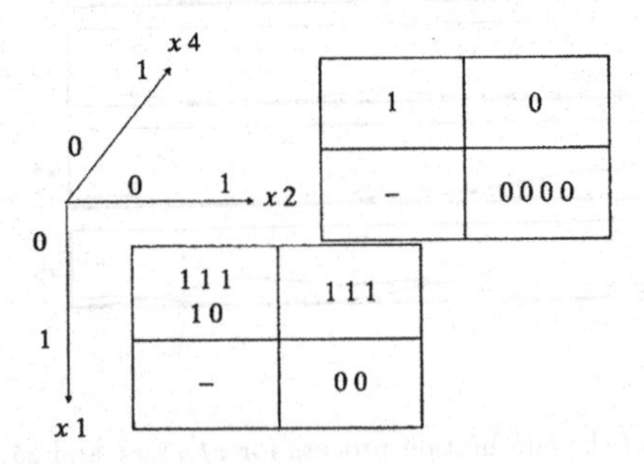

Figure 5. Illustration of states for the coded information system from Table 2.

The elimination of intermediate values for the second set of attribute leads to the result which is illustrated in fig.6.

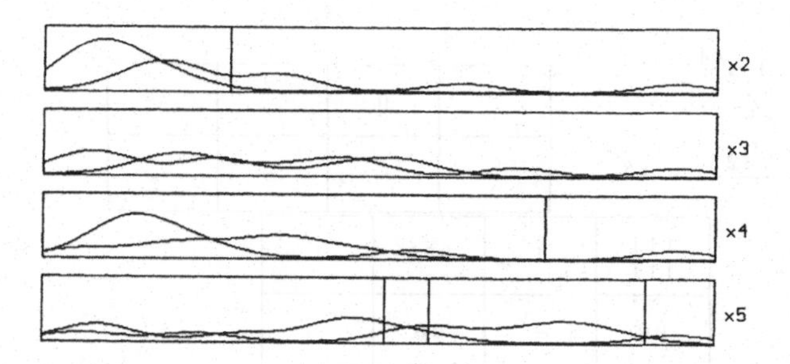

Figure 6. Final result of the elimination process for $x2, x3, x4$ and $x5$ attributes.

The intermediate points: 3.25 along $x2$, 5.65 along $x4$ and 4.2, 4.775, 7.60 along $x5$ were obtained. The attribute $x3$ has been eliminated. After coding the set of data from Table 1 we obtain an information system shown in Table 3.

Table 3. The coded information system obtained for $x2, x3, x4$ and $x5$ attributes.

Number of object	Codes of condition attributes			Class
	x 2	x 4	x 5	
1	0	0	0	1
2	0	0	3	1
3	0	0	0	1
4	0	0	0	1
5	0	0	1	1
6	0	0	0	1
7	0	0	0	1
8	0	0	0	1
9	0	0	2	0
10	1	0	0	0
11	1	0	2	0
12	1	0	2	0
13	1	0	0	0
14	0	1	1	0
15	0	0	2	0
16	0	0	2	0

In this case, seven pure states and nine empty states were obtained (fig.7).

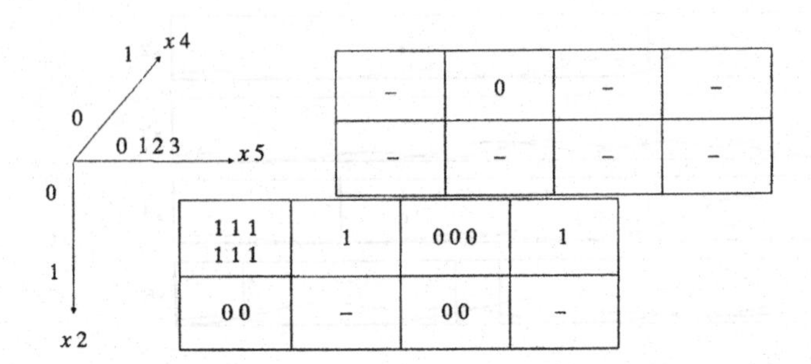

Figure 7. Illustration of states for the coded information system from Table 3.

The classification properties of the coded information systems from Table 2 and Table 3 are suboptimal according to the expected value of the quality of classification *a priori* .

Acknowledgements

We are grateful to Professor Zdzisław Pawlak for inspiration and financial support. This work was also partially supported by the Kielce University of Technology. Thanks are due to our collegues: Robert Kowal, Mateusz Masternak, Maciej Sękalski and Mateusz Wiśniewski for their helpful contribution during the development of the project.

References

[1.] Pawlak, Z. (1991) Rough Sets: Theoretical Aspects of Reasoning about Data, Kluwer Academic Publisher, Dordrecht.

2.] Zygadło, M. and Piasta, Z. (1988) 'Indirect assessment of frost durability of ceramics', Industrial Ceramics 8, 129-133.

[3.] Zygadło, M., Piasta, Z. and Lenarcik, A. (1991) 'A method of diagnostic of brittle materials frost resistance examplified by ceramic bricks', in A.M Brandt and I.H. Marshall (eds.), Brittle Matrix Composites 3, Elsevier Applied Science, London and New York, pp.377-385.

[4.] Roy, C., Maslouhi, A., Gaucher, D. and Piasta, Z. (1988) 'Classification of acoustic emission sources in CFRP assisted by pattern recognition analysis', Canadian Aeronautics and Space Journal 34, 224-232.

[5.] Roy, C., Maslouhi, A., Allard, J. and Piasta, Z. (1991) 'Pattern recognition characterization of microfailures in composites via analytical quantitative acoustic emission', in A.H. Cardon and G. Verchery (eds.), Durability of Polymer Based Composite Systems for Structural Applications, Elsevier Applied Science, London and New York, pp.312-324.

[6.] Feknous, N., Ballivy, G. and Piasta, Z. (1989) 'Monitoring of damaged injected rock with acoustic emission technique' , Proceedings of Conference on Recent Developments on the Facture of Concrete and Rock, Elsevier Applied Science, London.

[7.] Piasta, Z. (1991) 'Diagnostic classification for brittle matrix composites assisted by pattern recognition and rough sets analysis', in A.M Brandt and I.H. Marshall (eds.), Brittle Matrix Composites 3, Elsevier Applied Science, London and New York, pp.258-268.

[8.] Pawlak, Z., Wong, S.K.M. and Ziarko, W., (1988) 'Rough sets: probabilistic versus deterministic approach', International Journal of Man-Machine Studies 29, 81-95.

[9.] Piasta, Z. and Rusin, Z., (1986) 'Evaluation of standard methods for examining usefulness of aggregates for frost resistant concretes', Budownictwo 22, Kielce University of Technology, 119-126 (in Polish).

Part III
Chapter 5

CONSEQUENCE RELATIONS AND INFORMATION SYSTEMS

Dimiter VAKARELOV
Department of Mathematical Logic with Laboratory for Applied Logic
Faculty of Mathematics and Informatics
Sofia University, 1126 Sofia, Bulgaria

Abstract. A representation theory for Scott consequence systems in property systems and in Pawlak's information systems is developed. It is proved that each Scott's information system can be represented in a certain Pawlak's information system.

1. Introduction

The main aim of this paper is to develop a representation theory of Scott consequence systems in some information systems.

Scott consequence system is an abstract set with a binary relation between sets of the type $X \vdash Y$, which satisfies some conditions, coming from logic. Roughly speaking, the relation $X \vdash Y$ between two sets of logical formulas holds iff whenever all members of X are true, some members of Y are also true. However this interpretation of $\vdash$ is not meaningful when we consider abstract sets. So some representation theory for the abstract notion of Scott consequence systems is needed. We treat in an uniform way several types of consequence relations, including Tarski consequence relation and some combinations with the notion of consistency. Consequence relations lay down in some abstract notions of information systems, as Scott's notion of information system, introduced by Scott in [5]. Rasiowa and Epstein applied this notion in [4] for constructing models for approximate information.

As a material in which we will represent consequence systems we take two intuitively very clear notions of information system. The simplest is the notion of Property system, which deals with objects and properties. More detailed is the notion of an information system in the sense of Pawlak [2], [3]. Primitive notions in Pawlak's information systems are objects, attributes (like "color") and values of attributes (like "green", "red").

The technic, which we use in the representation constructions, is a generalization of the Stone representation method for distributive lattices. The main tool is an useful analog of the notion of a prime filter from the theory of distributive lattices. The idea of the presented representation theory came to me during my work on [7].

2. Scott consequence systems and some related notions

Let $W \neq \emptyset$ be a set. We denote by $P(W)$ the set of all subsets of W and by $P_{fin}(W)$ the set of all finite subsets of W. By a Scott consequence relation in W we mean any binary relation $\vdash$ in $P(W)$ satisfying the following conditions for any A, $B,A',B' \in P(W)$ and $x \in W$:

(Refl) If $A \cap B \neq \emptyset$ then $A \vdash B$,
(Mono) If $A \vdash B$, $A \subseteq A'$ and $B \subseteq B'$ then $A' \vdash B'$,
(Cut) If $A \vdash B \cup \{x\}$ and $\{x\} \cup A \vdash B$ then $A \vdash B$,
(Fin) If $A \vdash B$ then there exist finite subsets $X \subseteq A$ and $Y \subseteq B$ such that $X \vdash Y$.

We say that $(W,\vdash)$ is a Scott consequence system, S-system for short, if $W \neq \emptyset$ and $\vdash$ is a Scott consequence relation in W. Another way of treating Scott consequence relations is to consider the relation $\vdash$ only for finite sets, satisfying (Refl), (Mono) and (Cut) and to extend it for arbitrary sets in the following way: $A \vdash B$ iff for some finite subsets $A' \subseteq A$ and $B' \subseteq B$ we have $A' \vdash B'$. Then (Fin) is satisfied by definition.

Normally Scott consequence relations are considered in sets, which elements are formulas and the intended meaning of $A \vdash B$ is the logical one. In this paper our approach is abstract one and we do not assume in general that W is a set of formulas or sentences.

The following more general cut conditions are true in each S-system (see Segerberg [6]):

(Cut0) If $A \vdash B \cup \{x\}$ and $\{x\} \cup A' \vdash B'$ then $A \cup A' \vdash B \cup B'$,
(Cut1) If $A \vdash B \cup C$ and $\{x\} \cup A' \vdash B'$ for every $x \in C$ then $A \cup A' \vdash B \cup B'$,
(Cut2) If $A \vdash B \cup \{x\}$ for every $x \in C$ and $C \cup A' \vdash B'$ then $A \cup A' \vdash B \cup B'$.

Let $S = (W,\vdash)$ be an S-system. We say that $\vdash$ is a Tarski consequence relation in W and that S is a Tarski consequence system (T-system), if it satisfies the following condition for any $A, B \in P(W)$:

(TFin) If $A \vdash B$ then there exist a finite set $X \subseteq A$ and an element $a \in B$ such that $X \vdash \{a\}$.

Instead of $A \vdash \{a\}$ we will write $A \vdash a$. In the literature (for instance Gabbay [1]) Tarski consequence relation is a relation of the type $X \vdash a$. Then $\vdash$ is extended for arbitrary sets as follows: $X \vdash Y$ iff $X \vdash a$ for some $a \in Y$. It is easy to see that the following conditions characterize T-systems:

(TRef) If $a \in A$ then $A \vdash a$,
(TMono) If $A \vdash a$ and $A \subseteq A'$ then $A' \vdash a$,
(TCut) If $A \vdash x$ and $\{x\} \cup A \vdash y$ then $A \vdash y$,
(TFin) If $A \vdash a$ then there exists a finite subset $X \subseteq A$ such that $X \vdash a$.

Let $S = (W,\vdash)$ be an S-system. A set $X \subseteq W$ is called consistent (in S) if $X \not\vdash \emptyset$, otherwise X is called inconsistent. We denote by Con the set of all consistent subsets of W. Note that the notion of consistency of a set can not be defined in T-systems, because Tarski consequence relations do not admit empty set in the second place of the relation $\vdash$. In the next definition T-systems are extended as to admit the notion of consistency.

Let $S = (W,\vdash)$ be an S-system. We say that $\vdash$ is a Tarski-consistency relation and that S is a Tarski-consistency system (TC-system), if the following condition is satisfied for any $A, B \in P(W)$:

(TCFin) If $A \vdash B$ then there exist finite subsets $X \subseteq A$ and $Y \subseteq B$ such that $X \vdash Y$ and Y has at most one element.

The following is another axiomatic characterization of TC-systems. Let $S = (W,\vdash,Con)$ be a system with $W \neq \emptyset$, $\vdash \subseteq P(W) \times W$ and $Con \subseteq P(W)$. The axioms are (TRef),

(Tmono), (TCut), (TFin) and the following for the set Con:

(CMono 1) If $A \in Con$ and $A' \subseteq A$ then $A' \in Con$,

(CMono 2) If $A \not\vdash a$ then $A \in Con$,

(CCut) If $A \vdash x$ and $A \in Con$ then $\{x\} \cup A \in Con$,

(CFin) If each finite subset of A is in Con then $A \in Con$.

Let $S = (W, \vdash, Con)$ be a TC-system. We say tat S is a coherent TC-system if for any $a \in W$ we have $\{a\} \in Con$. We say that 1 is unit of S if for any $A \subseteq W$ we have $A \vdash 1$.

The notion of a coherent TC-system with unit 1 is a modification of the following notion of information system given by Scott [5]. Namely, a system $S = (W, \vdash, Con, 1)$ is called a Scott information system, SI-system for short, if $Con \subseteq P_{fin}(W)$, $\vdash \subseteq Con \times W$, $1 \in W$ and the following axioms are true:

S1. If $a \in A$ and $A \in Con$ then $A \vdash a$,

S2. If $A, A' \in Con$, $A \vdash a$ and $A \subseteq A'$ then $A' \vdash a$

S3. If $A \cup \{a\}$, $A \in Con$, $A \vdash a$ and $A \cup \{a\} \vdash b$ then $A \vdash b$,

S4. If $A \in Con$ then $A \vdash 1$,

S5. If $A \in Con$ and $A' \subseteq A$ then $A' \in Con$,

S6. If $A, A' \in Con$, $A \vdash a$ and $A \subseteq A'$ then $A' \vdash a$

S7. If $a \in W$ then $\{a\} \in Con$.

The following cut-like condition can be derived from the above axioms:

S8. If $A, B \in Con$, $B \vdash a$ and for all $x \in B$ we have $A \vdash x$ then $A \vdash a$.

Let us note that in the original Scott's definition S8 is taken instead of S2 and S3. It is easy to see that S2 and S3 are derivable from S8 and S1. Note also that S1, S2 and S3 are similar to (TRefl), (TMono) and (TCut) but not identical with them. So the relation $\vdash$ in SI-systems is not in general Tarski consequence relation. This is not good for our purposes, because SI-systems in its present formulation could not be considered as S-systems. The next lemma, however, shows that instead of SI-systems we can take coherent TC-systems with unit, which are S-systems.

Lemma 2.1. The notions of SI-system and coherent TC-system with unit 1 are equivalent in a sense that their primitive notions are inter-definable.

Proof. Let $S = (W, \vdash, Con, 1)$ be an SI-system and define $\vdash'$ as follows: for finite $X \subseteq W$ and $a \in W$ let $X \vdash' a$ iff $X \in Con$ & $X \vdash a$ or $X \notin Con$. Define $\vdash^*$ and Con^* for arbitrary subsets of W as follows:

$A \vdash^* a$ iff for some finite subset $A' \subseteq A$: $A' \vdash' a$, and

$A \in Con^*$ iff each finite subset of A belongs to Con.

Then it is easy to show that the system $(W, \vdash^*, Con^*, 1)$ is a coherent TC-system with unit 1. Moreover for each finite subset $X \subseteq W$ and $a \in A$ we have $X \vdash a$ iff $X \notin Con$ or $X \in Con$ & $X \vdash' a$.

Let now $S = (W, \vdash, Con, 1)$ be a coherent TC-system with unit 1. Define for any finite subset $X \subseteq W$ and $a \in W$:$X \vdash' a$ iff $X \in Con$ & $X \vdash a$, $X \in Con'$ iff $X \in Con$ and X is a finite set. Then the system $(W, \vdash', Con', 1)$ is an SI-system. Moreover for finite subsets of W we have the equivalence: $X \vdash a$ iff $X \notin Con$ or $X \vdash' a$. This shows that the notions of SI-system and coherent TC-system with unit are equivalent. ∎

3. Property systems

Property systems are very simple kind of information systems, which will be used to interpret S-systems.

By a property system (see [7]), P-system for short, we mean any triple $P = (Ob, Pr, f)$, where Ob is a nonempty set of objects, Pr is a set, whose elements are called properties, and f is a function which assigns to each object x a subset $f(x) \subseteq Pr$. The elements of $f(x)$ are called properties of the object x. A P-system P is called coherent if for any object x we have $f(x) \neq \emptyset$. This means that each object from Ob has at least one property from Pr. It is not unusual for some P-system to have object x such that $f(x) = \emptyset$. Since in P-systems we are interested not in all possible properties of its objects but only of those which are of some interest, then it is quite possible to have objects which do not possess any property from Pr.

Let $1 \in Ob$, we say that 1 is an universal element of P if $f(1) = Pr$, i.e. 1 possesses all properties from the set Pr.

Now we introduce a special kind of P-systems, called set theoretical ones. Let $S = (W, V, f)$ be a system such that $W \neq \emptyset$, $V \subseteq P(W)$ and for any $x \in W$ $f(x) = \{A \in V | x \in A\}$. We consider S as a P-system with $Ob = W$ and $Pr = V$.

Let $P = (Ob, Pr, f)$ be a P-system. We shall associate with P a system $S(P) = (W_P, \vdash_P)$ in the following way: put $W_P = Ob$ and define $\vdash_P$ first for finite $A, B \subseteq OB$. Let $A = \{a_1, \dots, a_m\}$ and $B = \{b_1, \dots, b_n\}$, then

(*) $A \vdash B$ iff $f(a_1) \cap \dots \cap f(a_m) \subseteq f(b_1) \cup \dots \cup f(b_n)$ iff
$(\forall p \in Pr)(p \in f(a_1) \& \dots \& p \in f(a_m) \longrightarrow p \in f(b_1)$ or $\dots$ or $p \in f(b_n))$

Here we adopt the convention that for $m = 0$ and $n = 0$ we assume $A = \emptyset$ and $B = \emptyset$. When $m = 0$ we put $f(a_1) \cap \dots \cap f(a_m) = Pr$ and when $n = 0$ we put $f(b_1) \cup \dots \cup f(b_n) = \emptyset$.

Now for arbitrary sets $A', B' \subseteq W$ we put:

(**) $A' \vdash_P B'$ iff there exists finite subsets $A \subseteq A'$ and $B \subseteq B'$ such that $A \vdash_P B$.

Lemma 3.1. (i) The system $S(P)$ is an S-system, called the S-system over P.

(ii) If in (*) and (**) B is supposed to be always one element set, then $S(P)$ is a T-system.

(iii) If in (*) and (**) B is supposed to be always one element or empty set, then $S(P)$ is a TC-system.

(iv) If P is a coherent P-system with an universal element 1, and if in (*) and (**) B is supposed to be always one element or empty set, then $S(P)$ is a coherent TC-system with unit 1, so $S(P)$ determines Scott's information system in the sense of lemma 2.1.

Proof – a routine check. Let us, for example, show that in (iv) $S(P)$ is a coherent TC-system with unit 1. Let $a \in Ob$. Since $f(a) \neq \emptyset$ we have $a \not\vdash_P \emptyset$ so $\{a\} \in Con$ and hence $S(P)$ is a coherent $S(P)$ system. Since $f(1) = Pr$ then for every $A \subseteq Ob$ we have $A \vdash_P 1$, hence 1 is unit of $S(P)$. ■

Condition (*) gives the following meaning of $\vdash_P$: $A \vdash_P B$ iff whenever all elements of A possess some property from a given set of properties Pr, the same property is possessed by some elements of B. In the next section we give a representation theorem for S-systems, which shows that the same meaning can be attached to arbitrary Scott consequence relation.

The example connected with P-systems can be generalized in the following way. Let $W \neq \emptyset$ be a set, $\underline{D} = (D, <, 0, 1, \cap, \cup)$ be a distributive lattice with zero 0 and unit 1, and let f be a total function from W into D. Define a binary relation $\vdash_D$ in $P(W)$ as in the definition (*) and (**) where $\cap$ and $\cup$ now are the lattice operations of $\underline{D}$ and instead Pr and $\emptyset$ as subsets of Pr we use 1 and 0. Then as in lemma 3.1 one can verify that the just defined system $S(D) = (W, \vdash_D)$ is an S-system. Taking $W \subseteq D$ and for any $x \in W$ $f(x) = x$, we see that any nonempty subset of D can be considered as an S-system. This example will play a suggestive role in the next section.

4. Representation theorems for Scott consequence systems in property systems

The representation theory, which we are going to develop for S-systems is similar to those for distributive lattices. The main key in the representation construction for distributive lattices is the notion of a prime filter. So our first aim now is to find a good analog of a prime filter in S-systems. The following definition is appropriate.

Let $S = (W, \vdash)$ be an S-system. A subset $x \subseteq W$ is called a prime filter in S if for all finite subsets A and B of W such that $A \vdash B$ and $A \subseteq x$ we have that $x \cap B \neq \emptyset$. Let us note that all prime filters in distributive lattices are also prime filters in the above sense. The next lemma gives other equivalent definitions of a prime filter.

Lemma 4.1. Let $S = (W, \vdash)$ be an S-system.

(1) The following conditions are equivalent for any $x \subseteq W$:

(i) x is a prime filter in S,

(ii) $(\forall A \subseteq W)(x \vdash A \longrightarrow x \cap A \neq \emptyset)$,

(iii) $(\forall A \subseteq W)(x \vdash A \longleftrightarrow x \cap A \neq \emptyset)$,

(iv) $x \not\vdash \overline{x}$, where $\overline{x} = W \setminus \{x\}$.

(2) $\emptyset$ is a prime filter in S iff $(\forall A \subseteq W)(\emptyset \not\vdash A)$,

(3) W is a prime filter in S iff $(\forall A \subseteq W)(A \not\vdash \emptyset)$.

Proof. (1). (i)$\longrightarrow$ (ii). Suppose x is a prime filter in S and let $x \vdash A$ for some $A \subseteq W$. Then by (Fin) there exists a finite $A' \subseteq A$ and finite $x' \subseteq x$ such that $x' \vdash A'$. By (Mono) we get $x \vdash A'$ and by the definition of a prime filter we get $x \cap A' \neq \emptyset$, so $x \cap A \neq \emptyset$.

(ii) $\leftrightarrow$ (iii) — by (Refl).

(ii) $\rightarrow$ (iv). Since $x \cap \overline{x} = \emptyset$ then by (ii) we obtain $x \not\vdash \overline{x}$.

(iv) $\rightarrow$ (i). Let $x \not\vdash \overline{x}$ and suppose for the sake of contradiction that x is not a prime filter. Then for some finite sets A and B we have $A \vdash B$, $A \subseteq x$, but $x \cap B = \emptyset$. Then $B \subseteq \overline{x}$ and by (Mono) we get $x \vdash \overline{x}$ — a contradiction.

(2) ($\rightarrow$) Suppose $\emptyset$ is a prime filter and $A \subseteq W$. Then, since $\emptyset \cap A = \emptyset$ we obtain by (ii) that $\emptyset \not\vdash A$.

($\leftarrow$) Suppose $(\forall A \subseteq W)(\emptyset \not\vdash A)$ and for the sake of contradiction that $\emptyset$ is not a prime filter. Then by (ii) we have some A, such that $\emptyset \vdash A$, which is a contradiction.

(3) ($\rightarrow$) Let W be a prime filter and suppose for the sake of contradiction that for some $A \subseteq W$ we have $A \vdash \emptyset$. Then by (Mono) we obtain $W \vdash \emptyset$ and by (ii) $W \cap \emptyset \neq \emptyset$, which is a contradiction.

($\leftarrow$) Let $(\forall A \subseteq W)(A \not\vdash \emptyset)$. Then $W \not\vdash \emptyset$. Suppose $W \vdash A$. From here we get $A \neq \emptyset$. Then $W \cap A \neq \emptyset$, which shows that W is a prime filter. ∎

Following Segerberg [6] let us call an S-system S left-assertive if it satisfies the condition

(Ass_L) $(\exists A \subseteq W)(A \vdash \emptyset)$,

and right-assertive if it satisfies the condition

(Ass_R) $(\exists A \subseteq W)(\emptyset \vdash A)$.

Then the conditions (3) and (4) of lemma 4.1 can be stated in the following way:

$\emptyset$ is a prime filter in S iff S is not right-assertive,

W is a prime filter in S iff S is not left-assertive.

The following is an analog of the Stone separation theorem in the theory of distributive lattices.

Theorem 4.2. (Separation Lemma for S-systems) Let $S = (W, \vdash)$ be an S-system A, $B \subseteq W$ and $A \not\vdash B$. Then there exists a prime filter x such that $A \subseteq x$ and $x \cap B = \emptyset$.

Proof. Let $M = \{t \subseteq W \mid A \subseteq t \text{ and } t \nvdash B\}$. We shall show that M is inductive set as to apply the Zorn Lemma. Obviously $A \in M$, so $M \neq \emptyset$. Let N be a nonempty linearly ordered by the set inclusion subset of M and let $y = \bigcup\{t \mid t \in N\}$. We shall show that $y \in M$. Since for each $t \in N$ we have $A \subseteq t$, then, obviously $A \subseteq y$. It remains to show that $y \nvdash B$. Suppose for the sake of contradiction that $y \vdash B$. Then, applying (Fin) and (Mono) we obtain a finite subset $X = \{\alpha_1, \ldots, \alpha_n\} \subseteq y$ such that $X \vdash B$. It is not possible for X to be empty, because then by (Mono) we will get $A \vdash B$, which is a contradiction. Then for some $t_1, \ldots, t_n \in N$ we have $\alpha_1 \in t_1, \ldots, \alpha_n \in t_n$. Since N is linearly ordered by the set inclusion, we have for some permutation $i_1, \ldots, i_n$ of $1, \ldots, n$ the following inclusions: $t_{i_1} \subseteq \ldots \subseteq t_{i_n} = z$. Then $X \subseteq z$ and by (Mono) and $X \vdash B$ we get $z \vdash B$ - a contradiction. Thus $y \in M$. Then by the Zorn Lemma M contains a maximal element, say x. We shall show that x is a prime filter in S. Applying lemma 4.1(ii) let $C \subseteq W$ and $x \vdash C$. We will proceed to show that $x \cap C \neq \emptyset$. Suppose for the sake of contradiction that $x \cap C = \emptyset$. From $x \vdash C$ we get $C \neq \emptyset$, because otherwise we will obtain $x \vdash B$, which is impossible. Since x is a maximal element in M, then for any $\chi \in C$ we have $x \cup \{\chi\} \notin M$. Since $A \subseteq x$, then for any $\chi \in C$ we have $x \cup \{\chi\} \vdash B$. Then from $x \vdash C$ and (Cut 1) we obtain $x \vdash B$ — a contradiction. Thus $x \cap C \neq \emptyset$, which shows that x is a prime filter in S. From $x \nvdash B$ we get $x \cap B = \emptyset$. So we have found a prime filter x such that $A \subseteq x$ and $x \cap B = \emptyset$. This ends the proof of the theorem. ∎

Let $S = (W, \vdash)$ be an S-system and let V_S be the set of all prime filters of S. Since $V_S \subseteq P(W)$, then the system (W, V_S, f) with $f(a) = \{x \in V_S \mid a \in x\}$ for $a \in W$, is a P-system of set-theoretical type, called the canonical P-system over S.

Theorem 4.2 says that if $\vdash$ is not the universal relation in $P(W)$ then $V_S \neq \emptyset$. Obviously $\vdash$ is not universal iff $\emptyset \nvdash \emptyset$ iff $\emptyset$ is a consistent set. S-systems for which $\emptyset \nvdash \emptyset$ will be called consistent S-systems.

Corollary 4.3. Let $S = (W, \vdash)$ be an S-system. Then for any $A, B \subseteq W$ the following equivalence holds:

$A \vdash B$ iff $(\forall x \in V_S)(A \subseteq x \longrightarrow x \cap B \neq \emptyset)$

Proof. $(\rightarrow)$ Let $A \vdash B$, $x \in V_S$ and $A \subseteq x$. Then by (Mono) $x \vdash B$ and by lemma 4.1(ii) we get $x \cap B \neq \emptyset$.

$(\rightarrow)$ Let $A \nvdash B$. Then by the Separation Lemma there exists $x \in V_S$ such that $A \subseteq x$ and $x \cap B = \emptyset$. ∎

Theorem 4.4. (Representation theorem for S-systems in P-systems) Let S be an S-system. Then there exists a P-system P such that S coincides with the S-system $S(P)$ over P.

Proof. Let $P = P(S) = (W, V_S, f)$ be the canonical P-system over S. We shall show that S coincides with $S(P)$. By definition $S(P) = (W, \vdash_P)$, where $\vdash_P$ is defined by the equivalence $(*)$ before lemma 3.1. We have to show that for any $A, B \subseteq W$ $A \vdash B$ iff $A \vdash_P B$.

Suppose first that A and B are finite sets and that $A = \{a_1, \ldots, a_m\}$ and that $B = \{b_1, \ldots, b_n\}$. By Corollary 4.3. we have:

$A \vdash B$ iff $(\forall x \in V_S)(A \subseteq x \rightarrow x \cap B \neq \emptyset)$ iff
$(\forall x \in V_S)((\forall a \in A)(a \in x) \rightarrow (\exists b \in B)(b \in x))$ iff
$(\forall x \in V_S)(\forall a \in A)(x \in f(a)) \rightarrow (\exists b \in B)(x \in f(b))$ iff
$(\forall x \in V_S)(x \in f(a_1) \cap \ldots \cap f(a_m) \rightarrow x \in f(b_1) \cup \ldots \cup f(b_n))$ iff
$f(a_1) \cap \ldots \cap f(a_m) \subseteq f(b_1) \cup \ldots \cup f(b_n)$ iff $A \vdash_P B$.

Now, applying (Fin) and (Mono) we obtain that for arbitrary subsets A and B of W we have $A \vdash B$ iff $A \vdash_P B$, which ends the proof of the theorem. ■

In a similar way one can prove the following theorem.

Theorem 4.5. Let S be a T-system (TC-system). Then there exists a P-system P such that S coincides with T-system (TC-system) $S(P)$ over P.

Let us note that the construction of $S(P)$ here is according to the construction described in lemma 3.1 (ii) and (iii).

Theorem 4.6. Let S be a coherent TC-system with unit 1. Then there exists a P-system P such that S coincides with the TC-system over P.

Proof. Let $P = (W, V_S, f)$ be the canonical P-system over S. Since S is coherent then for every $a \in W$ we have that $\{a\} \nvdash \emptyset$. Then by theorem 4.2. there exists a prime filter x such that $\{a\} \subseteq x$, so $a \in x$. This shows that $x \in f(a)$, so $f(a) \neq \emptyset$ for every $a \in W$, which shows that P is a coherent P-system.

We shall show that 1 is an universal element of P. Let x be a prime filter. Since 1 is an unit of S we have $x \vdash 1$. From here we obtain that $x \cap \{1\} \neq \emptyset$, so $1 \in x$. Hence $x \in f(1)$ for every $x \in V_S$. Consequently $f(1) = V_S$, which shows that 1 is an universal element of P. Then by lemma 3.1(iv) $S(P)$ is a coherent TC-system with unit 1. The proof that $S(P)$ coincides with S is similar to those in theorem 4.4. ■

The following theorem shows that the example of S-systems, connected with distributive lattices is also in a sense characteristic.

Theorem 4.7. Let $S = (W, \vdash)$ be an S-system, satisfying the following condition for all $a, b \in W$:

(Antysimm) if $\{a\} \vdash \{b\}$ and $\{b\} \vdash \{a\}$ then $a = b$.

Then there exist a distributive lattice $\underline{D} = (D, \leq, 0, 1, \cup, \cap)$ with zero 0 and unit 1 and an injective function f such that for any two finite sets $A = \{a_1, \ldots, a_m\}$ and $B = \{b_1, \ldots, b_n\}$ we have

$(*)$ $A \vdash B$ iff $f(a_1) \cap \cdots \cap f(a_m) \leq f(b_1) \cup \cdots \cup f(b_n)$.

Proof. By theorem 4.4 there exists a P-system $P = (Ob, Pr, f)$ such that S coincides with the S-system $S(P)$ over P. Take D to be the Boolean algebra of all subsets of Pr. Then $(*)$ is true by the construction of $S(P)$, where $\cap$, $\cup$ are set-theoretical intersection and union and $\leq$ is set-inclusion $\subseteq$. It remains to show that f is an injective function.

Suppose $f(a) = f(b)$. Then we have $f(a) \subseteq f(b)$ and $f(b) \subseteq f(a)$. By $(*)$ we have $\{a\} \vdash \{b\}$ and $\{b\} \vdash \{a\}$. This by (Antysimm) implies $a = b$, which ends the proof.■

This theorem shows that each S-system, satisfying (Antysimm) can be isomorphically embedded in some distributive lattice, even in some Boolean algebra of all subsets of some set.

5. Information systems in the sense of Pawlak

By an information system in the sense of Pawlak, PI-system for short, we will mean any system $\Sigma = (OB, AT, \{VALa \mid a \in AT\}, g)$ where:

- OB is a nonempty set, whose elements are called objects,
- AT is a set, whose elements are called attributes,
- $VALa$ for each $a \in AT$ is a set, whose elements are called values of attribute a,
- g is a total function, called information function, defined in the set $OB \times AT$, which assigns to each $x \in OB$ and $a \in AT$ a set $g(x, a)$ such that $g(x, a) \subseteq VALa$.

This definition is a slight generalization of the notion of many-valued information system given by Pawlak in [3]. When $g(x,a)$ has at most one element for each $a \subseteq AT$, then Σ is called deterministic PS-system. Deterministic systems were introduced by Pawlak in [2]. If all sets of Σ are finite then Σ is called a finite system. Actually Pawlak's systems are always finite.

To obtain a feeling of the meaning of the above notions suppose that the object x is a person, the attribute a is "spoken language" and $VALa$={English, French, German, Russian}. If x speaks only English then we have $g(x,a)$ = {English}, if x speaks only English and Russian then $g(x,a)$ = {English, Russian}. It is possible however for x to speak neither of the four languages from $VALa$. Then $g(x,a) = \emptyset$.

Intuitively the elements of some set $VALa$ can be treated as similar properties collected in one set, the common meaning of which is called attribute . The expression $A \in g(x,a)$ contain two pieces of information: "x has the property A" and "$A \in VALa$". So PI-systems can be considered as property systems, in which the properties are grouped into a groups called attributes. In this way from one property system we can obtain different PI-systems. The following are two special cases.

Let $P = (Ob, Pr, f)$ be a P-system. Let $AT = \{a\}$, $VALa = Pr$ and for any $x \in Ob$ $g(x,a) = f(x)$. Then the system $\Sigma P' = (Ob, AT, \{Pr\}, g)$ is a PI-system with only one attribute a, collecting all properties from Pr. In general, the system $\Sigma P'$ is not deterministic. The second example gives a deterministic system. Let $AT = Pr$ and for each $a \in Pr$ let $VALa = \{a\}$. For each $x \in Ob$ and $a \in Pr$ define $g(x,a) = \emptyset$ if $a \notin f(x)$ and $g(x,a) = \{a\}$ if $a \in f(x)$. Obviously the system $\Sigma P'' = (Ob, AT, \{VALa \mid a \in AT\}, g)$ thus defined is a deterministic PI-system.

Let $\Sigma = (OB, AT, \{VALa \mid a \in AT\}, g)$ be a PI-system. We can associate to Σ a P-system $P_\Sigma = (Ob, Pr, f)$, called the underlying P-system of Σ, in the following way. Take $Ob = OB$, $Pr = \bigcup\{VALa \mid a \in AT\}$ and for $x \in OB$ $f(x) = \bigcup\{g(x,a) \mid a \in AT\}$.

We say that the PI-system S is coherent if the underlying P-system P is coherent, that is, for any $x \in OB$ there exists $a \in AT$ such that $g(x,a) \neq \emptyset$.

We say that 1 is an universal element of Σ if 1 is an universal element of P_Σ, that is, $\bigcup\{g(1,a) \mid a \in AT\} = \bigcup\{VALa \mid a \in AT\}$.

Now, applying the constructions of lemma 3.1 to P we can define a system $S(\Sigma) =_{def} S(P_\Sigma) = (W, \vdash)$ over Σ, such that $W = OB$ and $S(\Sigma)$ has a structure of S-system, T-system, TC-system, and coherent TC-system with unit 1, if S is a coherent PI-system with universal element 1. Then the representation theorems from the preceding section can be reformulated now for PI-systems, even for deterministic PI-systems.

Theorem 5.1. (Representation theorem for S-systems in PI-systems) Let S be an S-system. Then there exists a (deterministic) PI-system Σ such that S coincides with the S-system $S(\Sigma)$ over Σ.

Proof. Let $S = (W, \vdash)$ be an S-system. By theorem 4.4 there exists a P-system $P = (Ob, Pr, f)$ such that S coincides with the S-system $S(P)$ over P. Let Σ be the PI-system $\Sigma P'$ (or $\Sigma P''$) defined above. Then the underlying P-system P_Σ coincides with P. So $S(\Sigma) = S(P_\Sigma) = S(P) = S$. If we choose $\Sigma = \Sigma P''$ then Σ is a deterministic PI-system.
■

In the same way one can prove the following two theorems.

Theorem 5.2. Let S be a T-system (TC-system). Then there exists a (deterministic) PI-system Σ such that S coincides with the T-system (TC-system) $S(\Sigma)$ over Σ.

Theorem 5.4. Let S be a coherent TC-system with unit 1. Then there exists a (deterministic) PI-system Σ such that S coincides with the TC-system $S(\Sigma)$ over Σ.

If we identify coherent TC-systems with unit with Scott information systems, then theorem 5.4 can be considered as a representation theorem for Scott information systems in (even deterministic) Pawlak information systems.

References

[1] Gabbay D. M., Semantical Investigations in Heyting's Intuitionistic Logic, Synthese Library v. **148**, D. Reidel Publishing Company, Holland, 1981.

[2] Pawlak Z., Information Systems — Theoretical Foundations, Information systems, **6**(1981), 205–218.

[3] Pawlak Z.,Systemy Informacyjne, WNT, Warszawa, 1983 (in Polish).

[4] Rasiowa H. and G. Epstein, Approximation Reasoning and Scott's Information Systems. In the proc. of International Symp. on Methodologies for Intelligent Systems'87, Charlotte, October 14–17, 1987, USA, North-Holland, 33–42.

[5] Scott D., Domains for Denotational Semantics. A connected and expanded Version of a Paper Prepared for ICALP 1982, Aarhus, Denmark 1982.

[6] Segerberg K.,Classical Propositional Operators. Clarendon Press, Oxford, 1982.

[7] Vakarelov D.,Logical Analysis of Positive and Negative Similarity Relations in Property Systems, in the proc. of WOCFAI'91, July 1–5, 1991, Paris, ed. Michel De Glas and Dov Gabbay, pp 491–499.

Part III
Chapter 6

ROUGH GRAMMAR FOR HIGH PERFORMANCE MANAGEMENT OF PROCESSES ON A DISTRIBUTED SYSTEM

Zbigniew M. WÓJCIK   Barbara E. WÓJCIK
The Division of Mathematics, Computer Science and Statistics
The University of Texas at San Antonio
San Antonio, TX 78249, U.S.A.

Abstract. A novel methodology for high performance allocation of processors to tasks based on an extension of the rough sets to the novel rough grammar is presented. It combines effectively a global load balancing with a dynamic task scheduling on a multiprocessor machine. Our methodology does not require a *priori* knowledge about run times of tasks and prevents users from manual distribution of job tasks into available processors. The production rules are constructible from a concurrent program by a compiler. In order to control the flow of tasks and data the set of the rough grammar production rules is updated and rolled in a pipeline fashion through the multiprocessor net during job execution together with codes of the processes (tasks). This pipeline fashion of rolling the jobs defines the global job balancing. The rough grammar uses any operators and metrics inside its production rules (not only the concatenation). Therefore, it is active and capable of driving purposeful processing by demanding software and data on a multiprocessor (e.g., MIMD) system. Performance parameters for our dynamic management of tasks are derived and compared to a statically scheduled multiprocessor. Based on these parameters, our decentralized methodology is shown to attain a much higher performance level. For example, the speedup of the order of tenths is easily feasible for an arbitrary algorithm. Moreover, the fault tolerance is highly improved with our decentralized strategy. The production rules instantiated as a result of a distributed computation at a previous rough grammar graph level are kept as permanent checkpoints for the current computation level. More than one layer of permanent checkpoints is definable to increase the level of fault tolerance.

1. Introduction

One of the most serious problems arising on a distributed system is an uneven migration of tasks (e.g., processes) in the net. Some nodes are often overloaded postponing execution of jobs. Some other nodes may sit idle waiting for requests and thus contributing to an inefficiency of a distributed system.

Most sequentially executed processes sit idle waiting for requests and blocking other jobs if full preallocation of nodes to a job for the entire job execution is incorporated. The remedy developed in this paper is a reduction of the idle and wait times of nodes to a minimum by a dynamic sequencing of all the tasks in a processor net in a pipeline fashion. In attempting to construct a pipeline, a problem arises with sending the results back from the server to the caller residing on a proceeding node. Any wait for the results from longer tasks and then resumption may postpone other processes from using the pipeline and thus disrupt current tasks computing in the pipeline even if each node supports multiprogramming. Furthermore, a server passing a distributed computation in the direction opposite to the pipeline may leave some nodes idle in front of the pipeline.

Any caller's wait on a separate node definitely increases the average job wait time on a net. Attempting to keep all nodes busy, we suggest sending the requestor's code to the servers followed by returning the requestor's node to the distributed operating system (and to a process coming next). Our dynamic task management is the most advantageous when the cost of waiting or being blocked is higher than the cost of sending the requestor's code to the server.

Load balancing (as *scattered decomposition, nearest neighbor balancing* [14] and the *simulated annealing* [1,2] does not involve the casual or semantic relationships between tasks (i.e., which tasks are dependents of which predecessors) necessary for preallocation of appropriate codes and data near predecessors to reduce efficiently the communication links, node idleness and task waits to minimum. Load balancing is capable of solving temporal overloading of single nodes (i.e., only locally). Task scheduling methods (as *connectivity matrix* and the *task table [11])* consider fully the program structure, but are static and do not aim specifically towards minimization of task waits and communication links. *Optimal scheduling* or *bin packing* [1,6,12] are centralized, *NP-complete* and assume *a priori* knowledge about execution time of each task which are difficult to accept for many distributed jobs. The *diffusing computation* [4] involves the dynamic program structure, but does not balance the load and does not minimize waits. The total waits of job tasks become significant if the dependents get blocked and may lead to deadlocks. Centralized approaches organize effectively global load balancing but are not acceptable on large distributed systems due to communication problems and low fault tolerance. Fully decentralized approaches are not appropriate on large systems because of unpredictable fluctuations of local loads. A global load balancing should be incorporated as well to put the local loads in a reasonable order.

Minsky's conjecture [8] summarizes inefficiency of traditional approaches to parallel computations using static schedules of tasks: the speedup of a parallel computer is a logarithmic function of the number of processors. Though many particular algorithms (especially embedded on a SIMD architecture) have linear speedup even for 100 processors [10], there has not been any general strategy yet showing linear improvement for any arbitrary parallel algorithm. Our simulation of a MIMD architecture with static schedules shows its speedup degraded significantly even with respect to a uniprocessor system (see Tables 1 - 6).

Our rough architecture [20] obeys Amdahl's law limiting the speedup S on a parallel machine composed of p processors with an algorithm in which a fraction f of tasks is inherently sequential. At any two subsequent stages of a distributed system the time T_p of execution of a MIMD program having p tasks and thus needing and getting p processors equals: $T_p = f * T_s + (1 - f) * T_s/p$, where T_s is the time of sequential execution of all the tasks in the entire MIMD program; $f * T_s$ is the time of execution of the sequential fraction f of tasks of the MIMD job in one of the two stages (i.e., before or after the parallel fraction $(1 - f)$ on another stage). The maximum speedup S on a two-stage parallel machine is: $S = T_s/T_p = 1/(f + (1 - f)/p)$.

We consider a general case of a multiprocessor system involving two or more stages. The speedup in average on our rough architecture is linear (see section 4) as a function of the number of tasks executed at any single stage for an arbitrary algorithm. This is because a full availability of processors for tasks is assumed at each stage of parallel execution and the pipelining assures this. Statistical number of tasks is generated in our simulation in the range between the minimum and maximum number of tasks at each stage giving the average speedup equal to the value S given by Amdahl's law in which p is the average number of tasks at the stage (with full availability of processors) and $f = 0$ (because only one statistically selected stage of processing is considered). The run times of tasks (not requited to be known in advance) at any stage and the run times of tasks have been generated by a pseudo-random number generator in our simulation and thus reflect any possible situation on a distributed system.

Our methodology will provide a convenient way of configuring and scaling on site for any new algorithm or application. Concurrent programs developed elsewhere will become portable. High cost of using parallel systems will be significantly reduced.

Our methodology provides one interesting feature more. In a fault tolerant mode each caller's code is used as a permanent checkpoint [18]. If one of servers fails, the job is resumed from the last permanent checkpoint by calling other servers capable of servicing the request.

After [11] we assume, that a task (a process, or an operator) is a compact code fragment (e.g., a single program statement, a sequence of statements, or a subroutine or function) executable sequentially without any inputs inside it. A schedule of tasks for execution is static, if it is prepared by a compiler and is not updated during execution. Otherwise is called dynamic. In the static schedules the necessary nodes (processors) are fully preallocated to the job tasks before the run time.

The rough grammar with its ROUGH MATRIX is the main tool for the dynamic management of tasks and are presented in section 2. Two novel dynamic task management approaches using the ROUGH and pipelining are presented in section 3. Section 4 establishes a significant improvement in the performance parameters of our dynamic task management over static task schedulers.

2. The Rough Grammar for Distributed Management of Processes

We introduce the rough grammar [16,19,20] as an extension of the rough sets [9] in order to increase drastically the efficiency of distributed processing. Let U denote the entire space of data (operands), and R be the space of operators (i.e., the available software modules). The concept of the approximation space adapted from the rough sets [9] is used here as the couple (U, R). The equivalence relations defined for the rough sets are extended here to R, and the concept of the universe U is adapted here in whole. High performance distributed processing is composed of broadcasts of software modules and data, and then executions, all will be arranged and controlled by the rough grammar.

DEFINITION 1. An elementary demand for software and data is a couple (X, r), $X \in U$, $r \in R$, $(X, r) \in U \times R$, consisting of: a) operands (data) $X = \{x_1, \ldots x_i, \ldots x_N\}$ specified by names, types, meanings and their number with specific metrics $m(X)$ (as addresses on a disk and memory requirements); b) an operator r (a software module) specified by a triple: (i) a name; (ii) a type, meaning and number of formal arguments; (iii) metrics $m(r)$.

A demand basically results in a broadcast of the code of the operator to a destination node, then a request for actual parameters (data), and finally the execution and the outputting of results. A result (called *object_approximation*) $a(X,r)$ of execution of r on X is composed of: a) The lower approximation $\underline{A}(X,r)$ of the data X (i.e., meanings, types and number of elements). More specifically, the lower approximation defined for the rough sets [9,15] must contain the greatest definable set of all possible results returned by r as far as meaning of the results, type and number of the output data is concerned; b) The upper approximation $\overline{A}(X,r)$ (i.e., a specific symbolic (semantic) representation $S(\overline{A}(X,r))$ and numerical values $V(\overline{A}(X,r))$). More specifically, the upper approximation [9,15] must be the least definable set of all possible output data during processing of X in its prespecified domain by appropriate operators (software) from R; c) Metrics $m(a(X,r))$ of data and software modules (i.e., addresses and memory requirements). *DEFINITIONS 2.a,b,c* specified below give semantic, quantitative and rough equalities for two demands (X,r) and (Y,r):

DEFINITION 2.a of the semantic equality $\equiv$ *in* (U,R): Two results of processing by r are equal semantically, if a software module r working on operands X and Y returns the same lower approximations. Then we say, that the demands (X,r) and (Y,r) are the same semantically:

$$(X,r) \equiv (Y,r), \quad if \ \underline{A}(X,r) = \underline{A}(Y,r). \tag{1}$$

DEFINITION 2.b. of the quantitative equality $=$ *in* (U,R): Two evaluations by r are equal quantitatively if (X,r) and (Y,r) return the same upper approximations (i.e., symbolic and numeric values):

$$(X,r) = (Y,r), \quad if \ \overline{A}(X,r) \ is \ identical \ to \ \overline{A}(Y,r). \tag{2}$$

If (2) is satisfied, the two demands (X,r) and (Y,r) are said to be equal quantitatively.

DEFINITION 2.c. of the rough equality $\approx$ *in* (U,R): $(X,r) \approx (Y,r)$ (i.e. (X,r) is roughly equal to (Y,r) in (U,R)), if (X,r) and (Y,r) are equal semantically and if the output data are within prespecified tolerances:

$$(X,r) \approx (Y,r), \quad if \ (X,r) \equiv (Y,r) \ \wedge \tag{3}$$
$$(\mid V(\overline{A}(X,r)) - V(\overline{A}(Y,r)) \mid \leq T, \ if \ \overline{A}(X,r), \ \overline{A}(Y,r) \ contain \ numerical \ values,$$
$$and \ S(\overline{A}(X,r)) \cup S(\overline{A}(Y,r)) \subseteq \underline{A}(X,r), \ if \ \overline{A}(X,r), \ \overline{A}(Y,r) \ involve \ symbols),$$

where T denotes some assumed admissible tolerances for the numerical values returned by r, and $\underline{A}(X,r)$ is a concept (meaning) or a type of the result returned by r. For instance, two software modules may give slightly different numerical results within the same meaning and a tolerance T, so they are roughly the same.

DEFINITION 3. **Rough set** is any collection of approximations $a(X,r)$ in (U,R).

POSTULATE 1. If $(X,r_X) \equiv (Y,r_Y)$, then there exists a tolerance T such that $(X,r_X) \approx (Y,r_Y)$.

This postulate is justified by the fact, that a human being classifies two slightly different results as carrying the same meaning or a similar name.

LEMMA 1. The following obvious relations apply for the above three equalities [15]:

$$(X,r_X) = (Y,r_Y) \ \Rightarrow \ (X,r_X) \equiv (Y,r_Y). \tag{4}$$

$$(X, r_X) = (Y, r_Y) \quad \Rightarrow \quad (X, r_X) \approx (Y, r_Y). \tag{5}$$

$$If \ \ (X, r_X) \approx (Y, r_Y), \ \ then \ \ (X, r_X) \equiv (Y, r_Y). \tag{6}$$

$$\underline{A}\underline{A}(X, R) = \underline{A}(X, R),$$

$$\overline{A}\overline{A}(X, R) = \overline{A}(X, R),$$

$$\overline{A}\underline{A}(X, R) = \underline{A}(X, R),$$

$$\underline{A}\overline{A}(X, R) = \underline{A}(X, R). \tag{7}$$

LEMMA 2. The following properties hold true (compare also [9,15]):

$$If \ \ \overline{A}(X, R) = \overline{A}(Y, R),$$
$$then \ \ \underline{A}(X, R) = \underline{A}(Y, R) \ \ and \ \ (X, R) \approx (Y, R). \tag{8}$$

$$If \ \ \underline{A}(X, R) = O, \ \ then \ \ \overline{A}(X, R) = O \ \ and \quad (X, R) \approx O, \tag{9}$$

where O is the empty set;

$$If \ \ (X, r_X) \equiv O, \ \ then \ \ (X, r_X) \approx O \ \ and \ \ (X, r_X) = O. \tag{10}$$

$$If \ \ (X, R) = O, \ \ then \ \ \underline{A}(X, R) = O. \tag{11}$$

$$If \ \ (X, r_X) \approx (Y, r_Y), \ \ then \ \ (X, r_X) - (Y, r_Y) \approx O, \tag{12}$$

where $-$ is the difference of sets;

$$\overline{A}(X, r_X) - \overline{A}(Y, r_Y) = O, \ \ if \ \ (X, r_X) = (Y, r_Y). \tag{13}$$

$$If \ \ \overline{A}(X, r_X) = \overline{A}(Y, r_Y), \ \ then \ \ (X, r_X) \cap (Y, r_Y) = (X, r_X) = (Y, r_Y). \tag{14}$$

$$If \ \ \underline{A}(X, r_X) = \underline{A}(Y, r_Y), \ \ then \ \ (X, r_X) \cup (Y, r_Y) \equiv (X, r_X) \equiv (Y, r_Y). \tag{15}$$

The proofs result directly from *DEFINITIONS 2.a,b,c, LEMMA 1* and *POSTULATE 1.* The above equations are applicable for finding out which operators can be used in place of other ones, and which intermediate results or their copies are equally useful during a distributed processing.

DEFINITION 3. The rough grammar $RG = \{N, T, P, S\}$ for a distributed computing is a quadruple composed of: a) Start symbols $\underline{S}$, being input data U for a distributed processing; b) Terminals T being approximations $\overline{A}(X, r)$ or $\underline{A}(X, r)$ of some specific data X in U to be returned as a result of some distributed computation in (U, R); c) Nonterminals N, i.e., all the operators R and intermediate results of processing and the start symbols; d) Production rules P (executable on a distributed system) involving both operands U and operators r_i, $\otimes \in R$, (usually different than concatenation). Typical production rules are of the *single form* for a two place operation $\otimes$:

$$a(x_1 \quad \otimes \quad x_2) \longrightarrow object\_approximation, \tag{16}$$

where $a \in \{\underline{A}, \overline{A}\}$; $x_1, x_2 \subset U$. In many cases the operator $\otimes$ may have only one operand x_1, or more than two operands.

For a given input composed of: a) some data (of a rough nature within prespecified types, ranges, domains and/or meanings), b) operation (procedure or function) assumed by an algorithm, the operation (production of the rough grammar) produces some output as a result of execution. The operation is not necessarily a concatenation of input data (or symbols) in S or N. The exact output of the operation is not easily predictable, this is why a computer processing is necessary. The rough grammar production rule predicts only a rough structure of the output as a type, range, domain and/or meaning of the results of processing using an operation. By analogy to simple computer programming concepts, the value or address parameters of the operations are the start and nonterminal symbols. The address parameters are the nonterminals and the terminals, all specified by types in the headings of the modules. The productions of the rough grammar are executions of the modules (procedures or functions).

A more sophisticated elaboration of rough grammar is needed for control of a distributed processing. Values 1 in an entire column of the rough matrix (Fig.1) indicate parallel predecessors of a node labeled by the right side of a rough grammar production rule and being a column pointer. The left sides of the production rules are the values of the matrix column pointer (and are listed in the row above the rough matrix). The predecessors are treated as operands and operators providing computation for a JOIN node. Any branch condition is also an operator (see Fig. 1). The node uses for its execution the operator and the operands received from the predecessors and returns the approximation of the received operands. The possible outcomes (right sides of the production rules) are listed as the column pointers to the rough matrix. Each outcome (if it is not empty) is applied as a predecessor (a row pointer) to the next stage of the parallel processing (compare Fig.1). Synchronization is provided by the rough table (i.e., by the rough matrix extended by metrics of operators and operands): the computation is fired when a full set of the operands arrives for the operations of a production rule specified by '1's of a particular column. A few complete sets of '1's in different columns may fire a few different productions at a time in parallel, e.g., pointed to by column pointers b_1 or b_3 in Fig.1. A sequence of values 1 in one row indicates parallel successors for a FORK node, i.e., a set of destinations of a single operand or an operator. The rough table may be treated not only as a casual schedule of tasks, but also as a parallel program.

Any right side production rule in the rough grammar graph of a lower-level (e.g., an *add_rule* : $(b_2,\ add,\ b_3) \longrightarrow c_1)$ can be demanded by the left-side of a higher-level rough-grammar rule (e.g., by *root_add_rule* : $(c_0,\ add,\ c_1,\ c_1 < 0) \longrightarrow sum)$. Hypotheses can be spread in this way (e.g., as demands for software checking data properties). Some practical results of the implementation of the rough grammar in computer vision are published in [16,19,20].

The following points (a) through (f) describe the differences between the rough grammar and the classical grammars: **a)** Classical Chomsky grammars accept only the operation of concatenation on the symbols S and N. The rough grammar accepts any operation (arithmetical, logical and any modules written by programmers including the operation of concatenation); **b)** Chomsky grammar concatenates symbols accepted by its production rules. The rough grammar executes operations on symbols and data (including concatenation of symbols). The acceptance and execution takes place if the input symbols, appropriate operation and output are specified in a production rule listed as an entry in a rough table

(Fig.1); **c)** Start symbols, nonterminals and terminals are not necessarily known in advance in the rough grammar. Subsequent nonterminals and terminals are computed during distributed or sequential processing. Chomsky grammars predict all results for a given set composed of S, N, T and P without necessarily using any sophisticated processing by a computer system. Meanings (optional), data structures for inputs and for outputs of operations are only known in the rough grammar, as well as operations for processing given by an algorithm and by a programmer. Classical (Chomsky) grammars could be used to model computer processing, but because of a possible large variety of start symbols or data, unpredictable nonterminals and terminals, too large amount of productions would have to be used; **d)** For unpredictable (i.e., not listed in N or S) variables and start symbols the Chomsky grammars fail. The rough grammar does not necessarily use specific symbols, but rather types for the N, S or T and their ranges, domains for data and meanings for symbols. By this unpredictable (too specific) inputs are taken care of; **e)** Roughness (inaccuracy) of symbols S and N within prespecified type, range, domain and meaning makes all the symbols (including variables carrying data) desired for execution of any operation (not only for concatenation) acceptable by the rough grammar. The necessity of processing of highly unpredictable and thus rough data (or symbols) makes the rough grammar very useful for computer science applications, specifically for high performance distributed processing yielding linear improvement and node busyness very close to one what is shown in this paper; **f)** Despite of the above roughness, the rough grammar requires precise specification of the operands for execution by using the lower and upper approximations to select the most appropriate elements among all possible symbolic and numeric values of data for processing.

Any production rule (located in the rough grammar graph at a lower-level) can be demanded indirectly by a higher-level rough-grammar rule through its right side. Recursive demands for computations are spread in this way. All the productions leading nowhere are disregarded by not demanding them. The rough production rule returns NIL, if demands cannot be satisfied (e.g., if a datum is missing). Simply, semantics for the right side of the rough grammar rule are checked (i.e., whether it reflects an expectation, or not). All the rough grammar graph branches continue until one (in the AND graph) or all (in the OR graph) values of the operand are NIL or until a terminal T is reached. All parents getting operands equal to NIL die thereby stopping any further unnecessary derivation. Typical formal context-free grammar terminates at terminal symbols only. Constraining all productions by appropriate operators (e.g., checking some relations on data) permits an explicit planning of the productions in the run time. Explicit dynamic modification of the schedule of tasks is not permitted using the formal context-free grammars.

The rough grammar is active: its rough table (Fig.1) is suitable to control parallel computations by demanding both operators and data and then their executions. The flow of the parallel execution may be affected by explicit conditions listed as operators in the production rules. A classical formal grammar production rule is passive: it waits until an external procedure makes all the possible replacements of symbols. The rough grammar is also capable of spreading in parallel the demands (to achieve goals) from the graph root or updated demands from the grammar graph leaves towards the root in the data-driven execution.

	b_0	b_1	b_2	b_3	c_0	c_1	*sum*	*sub*
a_0	1							
a_1	1							
a_2		1						
a_3		1						
a_4			1					
a_5			1					
a_6				1				
a_7				1				
b_0					1			
b_1					1			
b_2						1		
b_3						1		
c_0							1	1
c_1							1	1
$c_1 < 0$							1	
add	1	1	1	1	1	1	1	
$c_1 \geq 0$								1
sub								1

Fig.1. A rough matrix (ROUGH TABLE) to control distributed computations. Rows represent operators (as add) and operands (as a_2, a_3). Operators with prespecified operands producing expected outcomes (being column pointers) are marked with '1's in the outcome column. Some of the rows (operators) can be the branch conditions (e.g., $c_1 < 0$). Elements filled with blanks are irrelevant for the processing. Root of the entire distributed computation is the entire ROUGH TABLE. The root dependents are the start symbols of the rough grammar. The start symbols are the operands pointing to a row but not pointing to a column (e.g., elements a_i). The terminal symbols are the column pointers not specified as row pointers (e.g., sum, $diff$). The root may start computations from a terminal symbol (as a top-down execution) or from start symbols (in a bottom-up processing). Using the ROUGH TABLE in this Figure the following is done: $c_0 = a_0 + a_1 + a_2 + a_3$; $c_1 = a_4 + a_5 + a_6 + a_7$; IF $c_1 < 0$ THEN $sum = c_0 + c_1$; IF $c_1 \geq 0$ THEN $sub = c_0 - c_1$. Branch (e.g., IF $c_1 < 0$ THEN ...) is represented by an operation in a row entry which needs to return TRUE to fire an entire column of 1s the branch belongs to

The rough grammar can spread distributed computations in a systematic fashion: a) as a data driven execution (requests for software when some data become available); b) as a demand - driven (top-down) broadcast of codes for data; c) as a demand for both data and codes for a required production rule of the rough grammar. This demand broadcast for data and codes in one selected direction of a processor net gains the pipeline-like global oad balancing.

Our rough grammar demanding data and codes is dynamic: its production rules (listing operators (e.g., procedures and functions) and their operands) can be created or deleted at the run time. Each production rule may be seen to correspond to a task and its operands in the Gonzalez' and Ramamoorthy's [11] task table. The main difference is, that production rules of the rough grammar are passed with acknowledged requests for service together with the predecessor's code to the servers in one prespecified direction of the net making the execution similar to a pipeline. The codes and data demanded at a current moment are swapped by servers or by predecessors for direct use making the distributed system relatively free from waits and from idle subspaces of net, by this purposeful and entirely dynamic.

The rough grammar is capable of exploiting directly its knowledge (e.g., metrics and meanings of operands [18]) for a parallel search of the necessary data and for the most appropriate software during a distributed processing. The demands are composed of some global information concerning data to be processed (as data types (formats and size of strings) and their locations, and data meanings), specification of the operators (as names and their locations) to be used for the processing of the data, their compatibilities and memory characteristics. The receiver of a demand will identify the readiness of the requested code and data (e.g., their names, types and addresses) and will schedule them for transfer to the destination address. Every demand may need to wait for a code (if the code does not already exist in a processor memory) and for data. Receiving or having the requested code, a target processor can poll the predecessor for the values of the operands matching formal specification of the required data types and the data meanings (optionally). Failing to get the expected code or data, a new demand can be formed (as a top-down request), modified by recent responses (i.e., by a full metrics of the recently received code or data).

Each demand (call) for software or for data can be made by a predecessor through a message containing a name of the operation $\underline{A}(r)$ or a name of operation together with operand types $\underline{A}(X, r)$, e.g., through a pattern that must match specification of library software or data with types, meanings (optional) and numbers of formal parameters needed for the next stage of distributed execution. So, the rough grammar graph (representing a computation process) can be activated through a mechanism similar to pattern matching. Then, the demanded code and data must be swapped into the allocated processor. Full preallocation of nodes for entire job is not required.

THEOREM 1. The rough grammar is context-sensitive.
PROOF. It will be shown that productions of the rough grammar are not applied merely to one of the symbols at a time (as in the context free grammars) and that it matters what its neighbors are (this is why we call the grammar like this the context sensitive grammar [3]). We will use *modus ponens* method of proof: $\frac{(p, p \rightarrow q)}{q}$, where p is the proposition: *"it matters what are the neighbors of nonterminal symbols of a grammar"* , and q is: *"the grammar is context sensitive"*. By definition, the neighbors of an operator (a nonterminal symbol) are the operands. The production is processed using an operator and returns a specific approximation depending not only on the current operands but also on their values. Hence, the production (as (16)) is applied to more than to one symbol (e.g., operator) at a time: to an operator and its operands. So, it matters what are the neighbors of symbols such as the operators, and therefore the rough grammar is context sensitive. In particular, the derivation can be stopped by a NIL of an expected actual parameter necessary for execution (e.g., by the operand absence or by wrong operand type). Hence, the derivation is sensitive to the context: to the presence, to the values, meanings and type of operands listed on the left side of the production rule. ◇

Any context-free grammar cannot stop derivation of a string of symbols when failing to find one of necessary string components. Context-free grammar productions are not affected by the absence, value or wrong type of a neighbor in a string [3].

In a context - free grammar a unit production is a production $N1 \longrightarrow N2$, where $N1$ and $N2$ are nonterminals. Two productions $N1 \longrightarrow N2$ and $N2 \longrightarrow N3 \mid N4 \mid \ldots \mid Nn$ are reducible to one production: $N1 \longrightarrow N3 \mid N4 \mid \ldots \mid Nn$, where $N2 \longrightarrow N3 \mid N4 \mid \ldots \mid Nn$ is not a unit production. A unit production is eliminated in this way in the context- free grammar [3]. This is because unit productions in a context - free grammar are strictly deterministic. It is purposeful to get rid of any unit productions in a grammar to reduce the number of computations to a minimum by trying to reach immediately a non-unit production.

THEOREM 2. The rough grammar unit productions of the single form:
$a(x_1 \otimes x_2) \longrightarrow object\_approximation$ are not eliminative.
PROOF (by contradiction). Let us assume a chain of a few unit productions. Suppose, depending on a current operand value, one of two different operands of the operation listed on the left side of one of the unit productions (16) is returned. Now let us eliminate this unit - production. The remaining chain is now unable to determine the right value to be passed to the subsequent production rule, because the operator computing the value is dropped, thereby contradicting the possibility of elimination of the unit production rule of the rough grammar. Furthermore, the derivation would not be stopped by the value NIL or by a wrong type of one of the previously used operands (by virtue of *THEOREM 1)* which would lead to meaningless outcomes of processing (e.g., to a false representations of a reality given by data). ◇

ASSUMPTION 1. The operators used for distributed computations are *associative,* i.e.:

$$\forall(x_1)\ \forall(x_2)\ \forall(x_3)\ \{a[(x_1 \otimes_1 x_2) \otimes_2 x_3] = a[x_1 \otimes_1 (x_2 \otimes_2 x_3)]\} \tag{17}$$

where: $x_1 \cup x_2 \cup x_3 \subset U$; x_1, x_2, x_3 are data or rough approximations of the intermediate results of computations, and $\otimes_1$, $\otimes_2$, $\otimes_3$ are operators.

THEOREM 3. Any context-sensitive rough grammar production with the left side represented by the compound form:

$$a(x_1 \otimes_1 x_2 \ldots \otimes_{N-1} x_N) \longrightarrow object\_approximation \tag{18}$$

is representable by a set of single form productions executable sequentially.
PROOF. Let a_k denote a rough approximation of some intermediate results of a computation. Taking advantage of the assumed associativity, the compound form can be evaluated sequentially from the left (top) to the right (bottom):

$$\begin{array}{rcl} a(x_1 \otimes_1 x_2) & \longrightarrow & a_1 \\ a(a_1 \otimes_2 x_3) & \longrightarrow & a_2 \\ \cdot & \cdot & \cdot \\ a(a_{N-2} \otimes_{N-1} x_N) & \longrightarrow & object\_approximation \end{array} \tag{19}$$

◇

THEOREM 4. Any context-sensitive rough grammar production with the left side represented by a compound form (18) is representable by a parallel structure and executable by a parallel algorithm of the complexity $O(log_2 N)$.

PROOF. Taking advantage of the associativity *(ASSUMPTION 1)*, the original compound production (18) involving $N = 2^n$ operands (and $N-1$ operations) can be rewritten as (Fig.7):

$$STEP\ 1:$$

$$a(x_1 \otimes_1 x_2) \longrightarrow a_1$$

$$a(x_3 \otimes_3 x_4) \longrightarrow a_2$$

$$\cdots$$

$$a(x_{N-1} \otimes_{N-1} x_N) \longrightarrow a_{N/2};$$

$$STEP\ 2:$$

$$a(a_1 \otimes_{2\times 1} a_2) \longrightarrow a_1^2$$

$$a(a_1 \otimes_{2\times 2} a_2) \longrightarrow a_2^2$$

$$\cdots$$

$$a(a_{N/2-1} \otimes_{N-2} a_{N/2}) \longrightarrow a_{N/4}^2;$$

$$\cdots$$

$$STEP\ n = log_2 N:$$

$$a(a_1^{n-1} \otimes_{N/2} a_2^{n-1}) \longrightarrow object\_approximation = a^n \tag{20}$$

$\diamond$.

THEOREM 5. Any rough grammar language can be generated by the productions of the single form: $a(x_1 \otimes x_2) \longrightarrow object\_approximation.$

The proof results directly from the last two *THEOREMS* 3 and 4. The rough language over the rough grammar is the set of all the *object approximations* generated from data (or nonterminals) by operators as a result of parallel or sequential computations. *THEOREMS* 4 and 3 define applicability of the rough grammar production rules to a more complex distributed or sequential processing.

THEOREM 6. The **Chomsky Normal Form (CNF)** for the context-free grammars:

1. *nonterminal* $\longrightarrow$ *string of two nonterminals*

2. *nonterminal* $\longrightarrow$ *one terminal*

can be replaced by the single form of the rough grammar rule:
$a(x_1 \otimes x_2) \longrightarrow object\_approximation.$

PROOF. The rough grammar (as a context sensitive grammar) involves the context-free grammar (CFG), so, the rough grammar single form (16) is powerful enough. Replacing the $\otimes$ in the single form by the concatenation, using only nonterminals in the $a(x_1 \otimes x_2) \longrightarrow object\_approximation$ and ignoring any problems with approximation (i.e., discarding a and parentheses on the left side of the single form (16)) we get immediately the reverse of the first statement of the CNF. Having only one terminal to be used on the left side of the rule: $a(x_1 \otimes x_2) \longrightarrow object\_approximation$ without any need to involve the approximation a, having no necessity to incorporate any operation we get the reverse of the second statement of the CNF. ◇.

THEOREM 6 shows formally higher generality of the rough grammar production rule (16) over the CNF. On a distributed architecture the rough grammar production rule is more useful, primarily because the concatenation used in CNF does not demand any operators nor is capable to spread the demands and data along the grammar graph.

3. ROUGH TABLE for Dynamic Task Management

Let us consider now our decentralized control of a distributed execution by rolling the entire code and the rough table. The rough grammar graph is not mapped statically into the processor net. The codes of all processes and the ROUGH TABLE are transferred through the processor net in one prespecified direction. The nodes to receive the codes and the ROUGH TABLE, especially the JOIN nodes, are scheduled by predecessors at each parallel computation stage and are specified in the ROUGH TABLE. Parallel processes update only their ROUGH TABLE fragments. The JOIN node (compare any column with 1s in Fig.1) must merge the updated segments of the ROUGH TABLE received from the parallel branches. The entire code and ROUGH TABLE can be divided at a FORK node according to the processes executed by the parallel branches (e.g., compare a FORK node with c_o having more than one 1 in its row in Fig.1). Only the codes and the ROUGH TABLE fragments pertinent to the branch nodes can be passed through the FORK branches. The major codes (not required currently for some immediate FORK successors) and the ROUGH TABLE fragments corresponding to the processes different from the current (immediate) parallel FORK path can be sent through the shortest path to a destination JOIN node.

The properties of this dynamic distributed task control are as follows: **a)** The organization is very time effective if process (task) execution time is longer in average than the average time of code transfer. If a bus is not used for code transfer (because of a high bus contention), each successor code (and by this the major code) must be passed by direct links sooner or later and then the effect of pipelining is overwhelming because of eliminating waits and idle times; **b)** There is no excessive contention for the ROUGH TABLE, because the ROUGH TABLE is distributed and each parallel process has a copy of the ROUGH TABLE (in case the ROUGH TABLE is not split into parallel branches of a processor net). After completing a current process, updating and transferring the ROUGH TABLE, the node (processor) may be returned to the operating system and marked *free*; **c)** A ROUGH TABLE update code can be kept permanently with each processor node; **d)** The ROUGH TABLE is easily accessible asynchronously and can be updated by predecessor when the successor is starting to execute its process. Assuming, that a process execution is longer than the ROUGH TABLE update, the updated table can be sent by DMA to the successor before the successor process completion.

Advantages of rolling the entire code are as follows: **a)** An extremely high fault-tolerant architecture is achieved if direct predecessors are not set free immediately after their completion but are kept inactive with their codes for some time as checkpoints for quick fault or deadlock recovery. During a recovery the process net is ready to resume immediately, without any additional code transfers, from the processor node preceding the faulted node [18,20]; **b)** Global pipeline-like load balancing during execution is important in preventing migration of codes and the ROUGH TABLE in any direction through the entire multiprocessor net and considerably improves the distributed system performance (see section 4); **c)** The ROUGH TABLE update can be done without any delay, because the ROUGH TABLE is kept together with each currently executing process and the update request does not need to be sent to the host through many links nor through a common bus (usually very busy); **d)** The net is deadlock [17] free, since the processors required for the entire job are only the dependent (immediate) nodes and are fully preallocated by a dynamic pipeline manner of execution.

There are also disadvantages: **a)** Entire jobs are transferred, including ROUGH TABLES. Fortunately, the DMA is advantageous in that problem, and if a bus is not incorporated, all the transfers of codes must take place through direct links between processors sooner or later; **b)** Job size should basically be smaller than the size of the smallest private storage at each node. To remedy this, the job may be split into pages, or predecessor code and its successor code are to be kept on at least two nodes, but then fault tolerance may be affected. Enormous private memory space may be required. However, private memories are becoming inexpensive.

Another possibility is a direct rolling of the ROUGH TABLES only. The root node (or host) gets only the main code. Any subsequent successor gets its own code upon request sent by its predecessor (based on the ROUGH TABLE) either at the time the predecessor terminates its process in a *lazy swapper mode* or at the time the predecessor starts to run its process in the *preallocation mode* since all the successors become known to the predecessor immediately. Any code is received from the main memory or from the disk (e.g., through a bus). Each predecessor node handles: a) the entire ROUGH TABLE (if the predecessor does not execute its process in parallel with other nodes); b) or only a ROUGH TABLE fragment pertinent to its current process (if another process executes in parallel in a shorter branch).

Advantages of rolling the ROUGH TABLE only are as follows: **a)** Only the codes scheduled for current or for the following execution are transferred through direct links; **b)** A global dynamic load balancing guiding selection of successor processors is accomplished by rolling all the ROUGH TABLES in one selected direction in the net in a pipeline manner, improving thus significantly the performance parameters. Predecessors may be freed immediately after allocating a new nodes to the successor processes (the successor's codes are then transferred immediately), after own process completion and updating the ROUGH TABLE and after passing the ROUGH TABLE to the successor, thereby reducing idle times of nodes and waits of tasks; **c)** There is highly increased fault-tolerance by keeping the updated codes as checkpoints in private memories over a limited time or by keeping the checkpoints in a main storage; **d)** The approach is effective for the code transfer by a bus. Direct links between processors require several intermediate transfers from a disk; **e)** Deadlock freeness.

Disadvantages are as follows: **a)** Smaller reliability when compared with the rolling of the entire jobs, as seen, for example, in the case of a fault of central storage of the job code or a fault of the bus; **b)** There is deadlock proneness if the predecessor nodes wait a very short time for execution limited to an assumed extent of diffusing computation. However, after receiving partial results the predecessor codes are moved forward to the dependent nodes reducing significantly the possibility of a deadlock. A remedy is a deadlock detection and recovery [17].

4. Simulation and Results

The performance of MIMD computations on our distributed rough architecture has been estimated by a simulation. The performance parameters such as mean and average allocation time, node busyness, speedup and improvement have been evaluated. The program sets the number MaxStageCard of tasks to its smallest initial value at the beginning of the simulation. A pseudo-number generator generates the number of tasks for subsequent stages of the simulation between 2 and MaxStageCard. Simulation iterations are performed for various values of RoughTableOverhead associated with handling the rough tables by every task. The duration of each stage is set to time in which the desired QuantumFraction fraction of tasks complete their execution. A few different fractions QuantumFraction are set for each rough table overhead. Run time of tasks generated for each number of tasks CardOfStage[stage] at a current stage are collected in the arrays StageTasks and StageTaskRunTime. StageTasks collects durations of all tasks originally generated, and StageTaskRunTime includes also all tasks suspended by the QuantumFraction at the previous stage and passed forward to the next stage of distributed execution in the pipeline.

A procedure *Process_Optimal* sorts all tasks coming for a current stage according the their durations. The number of tasks completed within assumed QuantumFraction are computed. The number of tasks completed is set to one if the number TasksCompleted (within the QuantumFraction) is smaller than one (this is to consider stages with very small number of parallel tasks). The sum of run times of all tasks completed within the QuantumFraction is determined and all the tasks whose run time exceed the Quantum are collected in the StageTaskRunTime and passed to the next stage.

The performance parameters: allocation time, node busyness and speedup have been computed (see Tables 1 - 6) for: **a)** A classical static schedule (marked with 'stat') when the nodes must be allocated for all the tasks before the run time; **b)** Our decentralized task scheduling (marked with 'diff') in which static preallocation is not needed but *wait_factor* of nodes wait until their requests are serviced [20]; **c)** The entirely dynamic control of execution (marked with 'dyn') in which all the tasks are rolled in one direction of the net of processors and the shortest tasks must wait until the longest task complete execution at each stage [20]. Overhead associated with handling the rough table is neglected; **d)** The entirely dynamic execution (marked with 'opt') in which all tasks are rolled in one direction in the pipeline manner. The tasks are moved to the next stage when the Quantum set by the QuantumFraction expires. Overhead associated with the rough table is considered (see Tables 1 - 6).

5. Conclusions

All the performance parameters have been shown to be at least a few orders higher compared to classical static schedules. The improvement is linear at any stage of any distributed algorithm, both on a SIMD or MIMD architecture. Allocation time is increased when rough control overhead is involved, however, the busyness is also highly increased. The linear improvement at subsequent stages is achieved because of keeping busyness all the time very close to 1.

Linearity of the improvement makes the rough architecture relatively predictable: for a given deadline it becomes basically necessary to divide the sequential average run time T_s of all tasks at an average stage by the speedup and multiply the result by the number

TABLE 1: Simulation of mean performance parameters. Rough Table Overhead = 208. Quantum is set to time in which the fraction 0.1 of tasks complete.

S.	W.	Opt	Task	Allocation time:				Node busyness:				S p e e d u p:			
no	f.	S c	card	stat	diff	dyn	opt	stat	diff	dyn	opt	stat	diff	dyn	opt
1	0.0	177	52.2	687186	1027	521	2600	0.0008	0.51	0.51	0.99	0.040	26.84	26.8	77.3
2	0.0	176	45.2	594525	1027	530	2599	0.0009	0.52	0.52	0.99	0.041	23.92	23.9	68.5
1	0.2	176	46.1	607264	1027	522	2595	0.0009	0.51	0.51	0.99	0.040	23.72	23.7	68.9
2	0.2	176	50.7	668216	1028	525	2601	0.0008	0.51	0.51	0.99	0.040	26.24	26.2	76.2

Task duration is between 1 and 1048.
Preallocation Failure Slope for static scheduling = 0.1.
S.no means simulation number; W.f. is the wait factor;
S c is the count of stages in optimal simulation.

TABLE 2. Average values of mean performance for 2 simulations per each value of the wait factor. Quantum is set to time in which fraction 0.1 of tasks complete.

W.	Task	Allocation time:				Node busyness:				S p e e d u p :			
f.	card	stat	diff	dyn	opt	stat	diff	dyn	opt	stat	diff	dyn	opt
0.0	48.7	640855	1027	526	2599	0.001	0.51	0.51	0.99	0.04	25.38	25.38	72.90
0.2	48.4	637740	1028	523	2598	0.001	0.51	0.51	0.99	0.04	24.98	24.98	72.56

TABLE 3: Simulation of mean performance parameters. Rough Table Overhead = 208. Quantum is set to time in which the fraction 0.4 of tasks complete.

S.	W.	Opt	Task	Allocation time:				Node busyness:				S p e e d u p:			
no	f.	S c	card	stat	diff	dyn	opt	stat	diff	dyn	opt	stat	diff	dyn	opt
1	0.0	138	46.2	608945	1028	529	1052	0.0009	0.51	0.51	0.89	0.041	24.13	24.1	50.1
2	0.0	138	45.9	604678	1027	525	1048	0.0009	0.51	0.51	0.89	0.040	23.81	23.8	49.6
1	0.2	139	50.9	672650	1030	524	1046	0.0008	0.51	0.51	0.89	0.040	26.36	26.4	54.4
2	0.2	136	50.7	666302	1025	522	1043	0.0008	0.51	0.51	0.89	0.040	26.23	26.2	55.1

TABLE 4. Average values of mean performance for 2 simulations per each value of the wait factor. Quantum is set to time in which fraction 0.4 of tasks complete.

W.	Task	Allocation time:				Node busyness:				S p e e d u p :			
f.	card	stat	diff	dyn	opt	stat	diff	dyn	opt	stat	diff	dyn	opt
0.0	46.1	606812	1027	527	1050	0.001	0.51	0.51	0.89	0.04	23.97	23.97	49.84
0.2	50.8	669476	1028	523	1045	0.001	0.51	0.51	0.89	0.04	26.29	26.29	54.74

TABLE 5: Simulation of mean performance parameters. Rough Table Overhead = 208. Quantum is set to time in which the fraction 1.0 of tasks complete.

S.	W.	Opt	Task	Allocation time:				Node busyness:				S p e e d u p:			
no	f.	S c	card	stat	diff	dyn	opt	stat	diff	dyn	opt	stat	diff	dyn	opt
1	0.0	128	50.1	660408	1028	532	740	0.0008	0.52	0.52	0.60	0.041	26.62	26.6	22.0
2	0.0	128	49.7	655751	1030	523	731	0.0008	0.51	0.51	0.59	0.040	25.51	25.5	21.2
1	0.2	128	47.4	622656	1024	520	728	0.0008	0.51	0.51	0.59	0.041	24.63	24.6	20.4
2	0.2	128	46.6	613892	1027	520	728	0.0008	0.51	0.51	0.59	0.041	24.28	24.3	20.1

TABLE 6. Average values of mean performance for 2 simulations per each value of the wait factor. Quantum is set to time in which fraction 1.0 of tasks complete.

W.	Task	Allocation time:				Node busyness:				S p e e d u p :			
f.	card	stat	diff	dyn	opt	stat	diff	dyn	opt	stat	diff	dyn	opt
0.0	49.9	658079	1029	527	735	0.001	0.51	0.51	0.59	0.04	26.06	26.06	21.61
0.2	47.0	618274	1025	520	728	0.001	0.51	0.51	0.59	0.04	24.45	24.45	20.24

of parallel stages (assuming an unlimited amount of processors at each stage of parallel execution). To the above result the difference between the run time of the longest tasks moved to the last stage and the quantum of the last stage must be added. If the deadline is not met, the run times must be reduced by using faster processors, because it is not possible to run the tasks faster.

The above results are significant because linear improvement has been obtained up to now only for some specific algorithms. Static schedules are still in use making parallel computers cost ineffective. Static schedules for arbitrary algorithms makes the MIMD architectures much less cost effective than on a sequential machine (speedup is degraded much below 1, see Tables 1 and 2).

Fig. 2 shows improvement of our rough computer which is linear independently of the overhead associated with handling of the rough tables (for higher overheads the improvement is smaller but still linear).

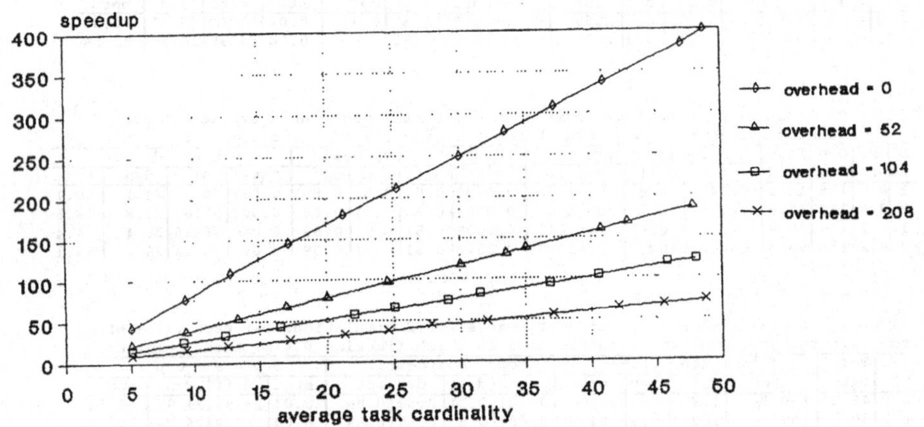

Fig. 2. Linear improvement at an arbitrary stage of a distributed system achieved for different overheads associated with handling of the rough table

Our decentralized task scheduling eliminates the idle times of nodes by allocation of the processors in one selected direction of the net to currently executing parallel processes or to processes expected to run immediately (i.e., by pipelining the distributed MIMD computation in the net). It also eliminates the wait times of processes by passing the predecessor's code to the dependents and returning the predecessor's node to the next jobs. By this it improves significantly the multiprocessor system speedup and other performance parameters over the existing approaches (as *scattered decomposition, nearest neighbor balancing, simulated annealing, connectivity matrix* and *task tables).*

In both approaches to the control of distributed execution high contention for the host is eliminated: the host only dispatches the new jobs. All jobs migrate in only one direction of the net and the global load balancing is the pipelining which is manageable dynamically by the ROUGH TABLES. Each processor memory can be split into a number of frames to better fit each process locality in a case of having a process code longer than processor private memory capacity. Nodes are allocated only in the direction of the rolling of the jobs (e.g., using scattered decomposition for the nearest nodes). A simulation shows that our approach gains a considerable improvement in typical performance parameters over a static multiprocessor task scheduling. Higher linear improvement is achieved by allocating the next nodes for a smaller quantum (e.g., equal to the time when 10% of tasks complete their execution) and passing all the tasks forward when the quantum expires. Jobs may be rolled around a multiple interconnected ring net more than once to complete.

The ROUGH TABLE provides coordination between tasks during execution. Branches of the rough grammar can be traced back using the production rules if a re-schedule is needed. Parallel tasks will meet at some JOIN node specified in the ROUGH TABLE. Each predecessor code is basically sent with the request and the ROUGH TABLE fragment. Pipelining and synchronization between tasks are enabled in this way.

Not only performance parameters are highly improved in our approach. Fault tolerance of our decentralized task scheduling is shown to be much better than for static schedules [18].

References

[1] R. Bettati, J.W. Liu, "Algorithms for End-to-End Scheduling to Meet Deadlines", *Proc. 2nd IEEE Symp. on Parallel and Distributed Processing,* Dallas, Texas, Dec. 1990, pp62-67.

[2] H.A. Choi, B. Narahari, "Scheduling Precedence Graphs to Minimize Total System Time in Partitionable Parallel Architectures", *Proc. 2nd IEEE Symp. on Parallel and Distributed Processing,* Dallas, Texas, Dec. 1990, pp402-406.

[3] D.I.A. Cohen, *Introduction to Computer Theory,* Wiley, 1988.

[4] E.W. Dijkstra, C.S. Scholten, "Termination detection for Diffusing Computation", *Inf. Process. Letters,* 11, 1, Aug. 1980, pp1-4.

[5] M.J. Gonzalez, C.V. Ramamoorthy, "Parallel Process Execution in a Decentralized System", *IEEE Trans. Comp.,* vol.c-21, No.12, 1972, pp1310-1322.

[6] J.Ji, K. Jeng, "Bin Packing Adjustable Rectangles and Applications to Task Scheduling on Partitionable Parallel Computers", *Proc. 2nd IEEE Symp. on Parallel and Distributed Processing,* Dallas, Texas, Dec. 1990, pp312-315.

[7] M.L. Mathews, "Hypercube software performance metrics", in: *Hypercube Multiprocessors,* ed. M. T. Heath, SIAM, 1986, pp155-160.

[8] M. Minsky, S. Papert, "On some Associative Parallel and Analog Computations", in: *Associative Information Techniques,* E.J.Jacks, ed., American Elsevier, 1971.

[9] Z. Pawlak, "Rough Classification", *International Journal of Man-Machine Studies,* 20, 1984, pp469-483.

[10] M.J. Quinn, *Designing Efficient Algorithms for Parallel Computers,* McGraw-Hill, 1987.

[11] C.V. Ramamoorthy, M. J. Gonzalez, "A survey techniques for recognizing parallel processable streams in computer programs", in: *1969 Fall Joint Comput. Conf., AFIPS Conf. Proc.,* v.35, Montvale, N.J.: AFIPS Press, 1969, pp1-17.

[12] G.C. Sih, E.A. Lee, "Dynamic-Level Scheduling for Heterogeneous Processor Networks", *Proc. 2nd IEEE Symp. on Parallel and Distributed Processing,* Dallas, Texas, Dec. 1990, pp42-49.

[13] H. Sulivan, T.R. Bashkov, "A large scale homogeneous, fully distributed parallel machine, I", *Proc. Fourth Symp. on Comput. Arch.,* March 1977, pp 105-117.

[14] W.I. Williams, "Load Balancing and Hypercubes: a preliminary look", in: *Hypercube Multiprocessors 1987,* eds. M.T. Heath, SIAM, Philadelphia, 1987, pp108-113.

[15] Z.M. Wojcik, "Rough Approximation of Shapes in Pattern Recognition", *CVGIP,* 40, 228-249, 1987.

[16] Z.M. Wojcik, "The Rough Grammar for the Parallel Shape Coding", *Proc. ACM South Central Regional Conference,* Tulsa, Nov. 1989, pp84-90.

[17] B.E. Wojcik, Z.M. Wojcik, "Sufficient Condition for a Communication Deadlock and Distributed Deadlock Detection", *IEEE Trans. on Software Eng.,* vol.15, no.12, Dec. 1989, pp1587-1595.

[18] Z.M. Wojcik, B.E. Wojcik, "Fault Tolerant Distributed Computing Using Atomic Send-Receive Checkpoints", *Proc. 2nd IEEE Symp. on Parallel and Distributed Processing,* Dallas, Texas, Dec. 1990, pp212-222.

[19] Z.M. Wojcik, B.E. Wojcik, "A Rough Grammar for a Linguistic Recognition of Image Patches", *Signal Processing,* 19, 1990, pp119-138.

[20] Z.M. Wojcik, "Rough Grammar for Efficient and Fault Tolerant Computing on a Distributed System", *IEEE Trans. Software Engineering,* , July 1991.

Part III
Chapter 7

LEARNING CLASSIFICATION RULES FROM DATABASE IN THE CONTEXT OF KNOWLEDGE–ACQUISITION AND –REPRESENTATION *

Ramin YASDI
Hochschule Bremerhaven
An der Karlstadt 8
2850 Bremerhaven, Germany

Abstract. The bottle-neck of manual knowledge-acquisition is a major obstacle for knowledge representation. The research described in this paper addresses this problem by suggesting a method for learning knowledge from a database. We attempt to improve representation with the assistance of experts and from computer resident knowledge. We describe the knowledge representation in the framework of a conceptual schema consisting of a semantic model and an event model. In these models a concept classifies a domain into different sub-domains. As a method of knowledge acquisition, we apply inductive learning techniques for rule generation. The theory of Rough Sets is used in designing the learning algorithm. Examples of certain concepts are used to induce general specifications of the concepts called classification rules. The basic approach here is to partition the information into equivalence classes, and derive conclusions based on equivalence relations. In a sense we are involved in a data-reduction process, where we want to reduce a large database of information to a small number of rules describing the domain. This is a completely integrated approach that includes user interface, semantics, constraints, representation of temporal events, induction, etc.

1. Introduction

1.1 Motivation

The main subject of this research is knowledge-acquisition and -representation. In essence, the objective is to develop the theoretical and practical background to be able to manage the problem of capturing and representing knowledge about a particular subject domain. This is a very practical problem faced by the designer of every Information System(IS). Questions related to this problem are:

---

1. How to represent the experience and background of an expert as closely as possible to the real world by means of system attributes.

2. How to acquire knowledge in a systematic way, translate it into rules that can be utilized by the system, and aggregate these rules with the existing knowledge in a manner that is consistent with the existing knowledge base.

We acknowledge that not all of the difficulties of developing an information system will be eliminated by knowledge acquisition or representation. Further, some of the difficulties will be shifted from knowledge acquisition to knowledge representation and vice versa. With the growing complexity of today's knowledge based systems this task is becoming more difficult even for the designers themselves particularly because of the lack of knowledge of specific application. For instance the size of facts in a relational database can increase rapidly after a short time. A lot of information can be stored in duplicated form or even as irrelevant. As a consequence the system performance will decrease. We have also realized that most knowledge representation techniques are not suited for knowledge acquisition in machine learning and vice versa.

Learning is one of the most important characteristics of human and machine intelligence. Learning techniques have improved significantly in the last years [15]. Clearly, it is advantageous to apply those results in the design of database systems. By learning from databases, knowledge rules can be extracted from the large amount of data and replace substantial portions of data.

Therefore the main subject of this research is the integration of the knowledge representation and acquisition processes into a unique framework. In particular we want to enhance the knowledge acquisition capabilities by adding a learning component.

Generally, the solution to these problems can be provided by the following:

1. Representing the information needs of an enterprise in a conceptual schema that covers both static and dynamic aspects of the information. A data model like semantic net represents the former whereas an event model represents the latter.

2. Acquiring knowledge through a learning component that enables the system to extract knowledge from experiments or sample expert decisions. Inductive learning techniques like probability and Rough Sets theory can be used for this purpose. These techniques adopt the machine learning paradigm "Learning By Examples" and apply an attribute-oriented representation in the learning-process.

We consider these closely related disciplines when we design a (new) conceptual schema. In the case the system already exists; for example a relational database model, and we want to enhance the knowledge acquisition capability by extending it with a learning component, the additional task will be mapping the conceptual model to the relational database model [16,11].

1.2 Goals

The goal of this research project is to design an intelligent tool for assisting the expert to transfer his knowledge to a system. Once the information system has been generated by the expert, it can be used by an end-user. Here, we define a knowledge engineer as an expert in the field of information system design. He is knowledgeable in computer science. On the other hand, the expert is a specialist in the application domain and not necessarily familiar with computer science. Once the system is operating, the end user can use the information by quering the system or adding new facts into the knowledge base.

This research is dictated by the following specific objectives.

- Specifying the information needs of an enterprise in a Conceptual Schema comprising a semantic model and an event model.

  The semantic model provides a set of semantic constructs. By means of these constructs the objects of the real world can be described by a set of attributes, and classified into several mutually exclusive and exhaustive classes. We construct the classes in the form of a semantic network with related integrity constraints. The defined schema has the capability to serve as knowledge representation. Here, an *example* is a description of a possibly unclassified member of the class. A *classification rule* predicts to which class an example might belong. The descriptions are in the form of logical expressions or production rules.

  The event model describes the behavior of the system. The vast majority of processes in nature exhibit dynamical behavior, i.e. they change over time. We propose a method of learning the behavior of such systems from examples. This model can be used to control or predict dynamical systems.

- Incorporating a learning component that enables the system to extract knowledge automatically or interactively. The source of knowledge acquisition is human experts or an information base of an existing system. We use the inductive learning techniques based on the Rough Sets theory for rule generation. This method like, any other empirical learning system, induces a rule from pre-classified examples. In general, many examples are required to generate rule(s). Providing a large amount of test examples to the system is not always easy. To overcome this drawback we connect the learning component to an existing database to gather the necessary information.

- In contrast to the conventional machine learning systems which generate rules during the design time of the knowledge base system, we propose generation of rules during the operation of the system so-called "accumulative learning".

In this paper we refrain from convincing that the conceptual schema presented is a useful tool for knowledge representation. We also refrain from the justification of Rough Set theory as a computationally good learning algorithm. Here we extend these techniques and apply them in knowledge base design. The major challenge for this research is to modify or redefine the above methods in such a way that they fit to each other and show how these techniques can be integrated in one unique system.

This paper is organized as follows. Having defined the terms used in this section we present the conceptual schema in section 2. Thereafter, we introduce the learning algorithm in section 3. In section 4, we illustrate the purposed system architecture. In section 5, we provide a summary of the features of this design and briefly touch upon some related work.

1.3. Accumulative Learning

By our definition, an accumulative learning system is one which improves performance while it is actually operating. The system must contain some initial information to begin. The system exhibits four main characteristics.

1. The capability of modifying its behavior on the basis either of its mistakes or its successes.
2. The capability of evaluating its results on the basis of its previous solutions.
3. The capability of identifying the action that leads to success and assigning credit to it.
4. The capability of modifying its knowledge itself or suggesting to the expert the possible modification.

1.4 EXIS

EXIS derives from an expert system for an information system design. We use the general term information system which refers to a knowledge base system, an expert system, database system, or a deductive database system. All these systems maintain rules which usually can be regarded as knowledge, expertise or integrity constraints. They are specified at the conceptual level by a model such as a relational or semantic model. EXIS generates a conceptual schema consisting of a semantic and event models for covering the static and dynamic aspects of the application. EXIS assists the knowledge engineer by design such a schema and provides an efficient way of extracting knowledge rules from the existing base. Therefore we refer to EXIS as a type of expert system that simulates the function of a Knowledge Engineer.

This approach is characterized by the architecture illustrated in Figure 7. In this paper we are mainly concerned with the learning aspects of the system.

2. Knowledge Representation

2.1 The Conceptual Schema

What is a conceptual schema?
The starting-point for design of a knowledge base system is some abstract and general

description of the part of the universe which is to be represented by information in the knowledge base. This part of the universe is often called the Universe of Discourse (U of D). The abstract description is called a conceptual schema. It also serves as a specification of knowledge.

2.2 Semantic Model

The semantic network presented hereafter contains most structure like classification, generalization, and aggregation. There are several integrity constraints related to these constructs. These constraints contribute to a more precise definition of the concepts. We distinguish the following constraints:

Schema constraints
are those constraints that maintain integrity of the schema regardless of the particular application. For example, the following constraints force the principle of attribute inheritance.

" A sub-class and its super-class may not both own an attribute with the same name. Otherwise, one cannot distinguish whether it is an attribute of a super class or sub-class".

These constraints are specified in a "Meta Schema " and are not visible to the expert or end-user (for more detail please refer to [21]).

Application constraints
are those constraints that express the integrity of the application. They are regarded as user constraints for example:

" All employees must be older than 18 ".

This kind of constraint is specified or modified by the expert and will be taken into account when the end user queries the system.

Useful constraints
These are rules that may be true in the current state of the database. An example of such a derived rule is:

" All students take CS100 course ".

This constraint may not be true for an other day. However, it may be useful to take advantage of it as long as it is true.

For each type of constraint we use the **with** operator to express a set of constraints associated with each concept. The **with** operator is an option and can be omitted if there are no constraints. We use the predicate calculus as a specification language to introduce the constructs of the schema.

2.2.1. Constructs of semantic model

Classification of entities
An entity is a thing which can be distinctly identified. Objects of the real world with similar qualities can be grouped into entity classes. Entities may have properties to which we refer as attributes. We specify the entity-class in predicate logic as a predicate and the attributes of the entity-class as arguments to the predicate in the form:

$$ec(x, a_1, \cdots, a_n) \equiv_{def} \exists x[a_1(x) = v_1, a_2(x) = v_2, \cdots, a_n(x) = v_n]$$

We define the entity-class by using the following notation:

class-name(Unique-identifier, Attribute-list) with constraint-list

In general form: E_c *with* ϕ

$e_1, \cdots, e_n \in E_c$ are n entities that belong to entity class E_c with a set of $\phi_1, \cdots, \phi_m \in \phi$ constraints.

For example, the entity class employee can be defined by the six- argument predicate:

employee(Ssn, Employ#, Status, Office, Category, Salary) with Salary $\geq$ 22T

Where the first argument is used as a key of the relation.

$\exists$ x [Employ#(x)=2910, Status(x)=single, Office(x)=ER52, Category(x) = prof, Salary(x) = 35T].

Declaration
The arguments of attributes are associated with their types or with a domain by the predicate dom. For example, if we consider the attribute Age and Birthplace. The domain can be defined as

dom(Age, [1...100]),
dom(Birthplace, [FRG, Canada, USA]),
dom(Ontario, [Ottawa, Toronto]),
dom(Hessen, [Frankfurt, Darmstadt]), dom(Michigan, [Detroit, Dearborn]).

The predicate represents a taxonomy of values in an attribute domain, which can be organized in a hierarchical form as:

```
Birthplace    FRG       Hessen      Frankfurt
                                    Darmstadt

              Canada    Ontario     Toronto
                                    Ottawa

              USA       Michigan    Detroit
                                    Dearborn
```

Such information can be represented by an implication

dom(Birthplace = frg) ⇒ dom(frg = [hessen,..]) ⇒ dom(hessen = [frankfurt, darmstadt,..])

The transivity of the above relation can be defined as:

$$\forall x, y, z(dom(x,y) \& dom(y,z)) \rightarrow dom(x,z)$$

Association
Association accounts for the binary relationship between two entity-classes. We specify the relationship by:

$$r(E_{c1}, E_{c2}) \equiv_{def} \forall x, y[r(x,y) \rightarrow \exists x\ E_{c1}(x)\ \&\ \exists y\ E_{c2}(y)]$$

We define a relationship by the notation:

$$\text{relation-name}(E_{c1}, E_{c2})\ \textit{with}\ \phi$$

For example, works-for(person(x), company(y)) indicates that entities from class person are working for entities from class company. Note that relationships are recursive and linked bi-directional, usually one direction is shown.

Generalization
Generalization provides a concise way to express constraints that define some class as a more general class of other ones. This construct is also called specialization when viewed the other way around, i.e. going from more general to the more specific definition. We use *is-a* to show this hierarchy

$$\text{is-a}(E_{c1}, E_{c0}) \equiv_{def} \forall x, y\ \text{is-a}(x,y) \rightarrow \exists x E_{c1}(x)\ \&\ \exists y\ E_{c0}(y)],\ \text{where}\ E_{c0}(x) \subset E_{c1}(y)$$

In general: $(E_1, \cdots, E_n)$ *is-a* E_0 *with* ϕ.

For example, is-a(student(x), person(y)) indicates that entities from class student are subset of the class person.

We use the notation *student is-a person* equivalent to *is-a(student, person)*.

Aggregation
Grouping classes into higher level classes is called aggregation. In a mathematical sense aggregation corresponds to the cartesian product.

$$\textit{has-a}(E_{c0}(x), E_{c1}(y), E_{c2}(z)) \equiv_{def} \forall x E_{c0}(x) \rightarrow \exists y E_{c1}(y)\ \&\ \exists z E_{c2}(z))$$

In general: E_0 *has-a* $(E_1, E_2, \cdots, E_n)$ *with* ϕ

For example, the fact that author, publisher, and title are components of the entity-class book is represented as

has-a(book, author, publisher, title) with publisher = springer

Concept
A subset of attribute values $< v_{i1}, v_{i2}, \cdots, v_{in} >$ is referred as a concept. For example CS100 is an entity from entity-class *course* that is associated with a relationship *student-take-course*. A subset of this relationship is a concept. We can infer the rule:

"All undergraduate students have taken the CS100 course ".

Taxonomy
A taxonomy is a finite tree whose nodes are labelled by sub-concepts. Any node other than a leaf node has two or more successors. The successor concept of each is subsumed by its parent concepts. The union of all successor concepts of any non-leaf node is equal to the parent concept. Moreover all sibling concepts are mutually exclusive.

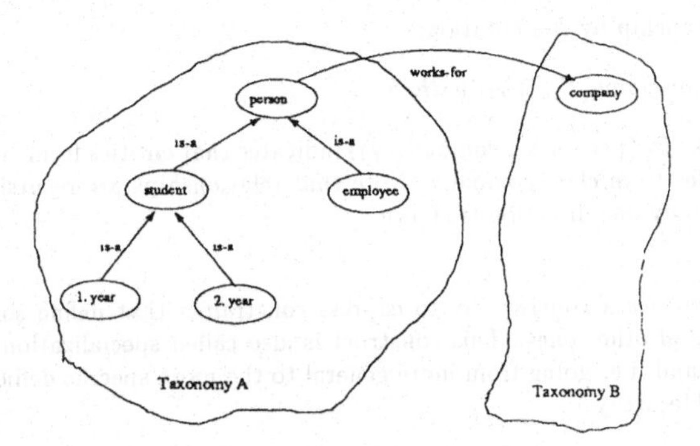

Figure 1. Example of a Taxonomy

The schema below illustrates an extract from a conceptual schema applied to an university information system. Figure 2 shows the schema in a graphical form.

person(Ssn, Name, Address, Birth-date)
employee(Id, Office, Status, Category, Salary) with φ_1, φ_2
professor(Ssn, Rank, Publication) with φ_3
teacher(Working-area)
student(Student#, Program, Year)
course(Course#, Dept, Class-room)

teacher teaches course

employee is-a person
student is-a person

dom(rank, [full prof, assit-prof, asso-prof])

φ_1:
Age $\geq$ 18

φ_2:
Category = worker, Salary $\geq$ 22T
Category = librarian, Salary $\geq$ 29T
Category = Faculty, Salary $\geq$ 32T

φ_3:
Rank=professor, paper published $\geq$ 20
Rank=associate, paper published $\geq$ 10
Rank=assistant, paper published $\geq$ 5

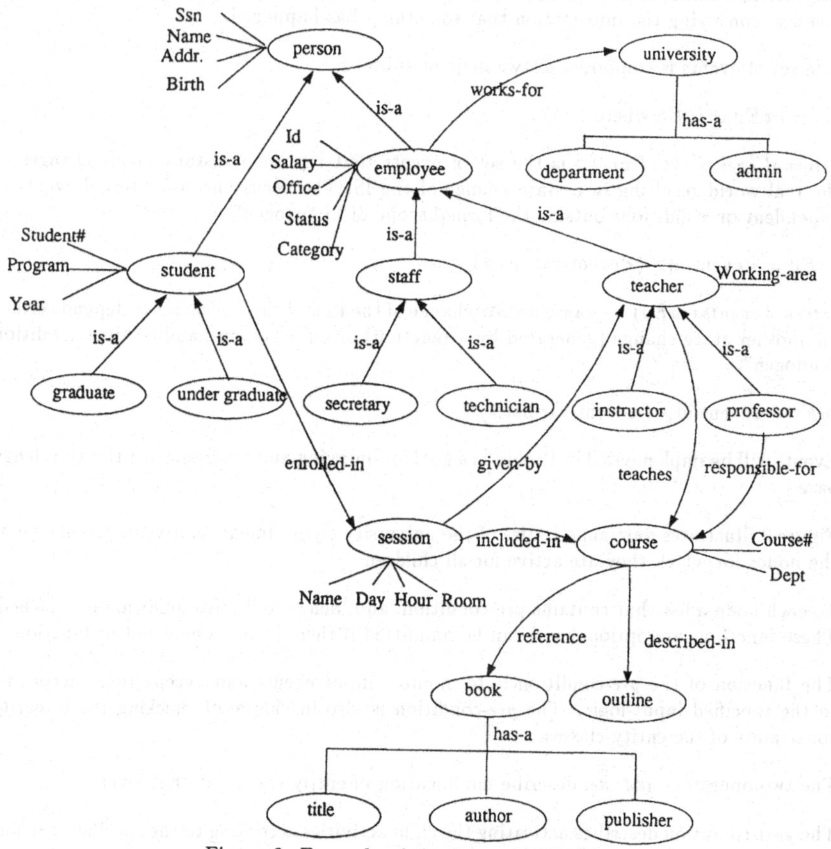

Figure 2. Example of the Semantic Model

2.3 Event Model

An event is constituted by the fact that something has happened either in the UoD or in the Information System (IS). By our definition an event is a "signal" , "message" or a "sensor" which arrives from outside the system or is generated within the system (compare this with placing a token in a Petri net, and activating a goal in Prolog). For example, suppose the light goes on in a dark room ; or the world simply changes for some mysterious reason. We use events for synchronization of the behavior of the activities within the system. We also use events to transmit control information into or within the system. An event happens between two points in time, called "start" and "end" times. At all times between them the event is said to be active. Messages are signals containing some data. They allow us to pass specific information between the activities. In this kind of setting a signal is a primitive message conveying the information that something has happened.

The set of events is composed of two disjoint subsets:

$Ev = exEv \cup inEv$ where

external events $(exEv)$ is the set of events that represent instantaneous changes of the real world resulting in a state change of the IS. The occurrence of external events is dependent on conditions outside the formal scope of IS "exogen".

$exEv$ = <event-id> (<event-var-list>)

internal events $(inEv)$ cause a state change of the IS and their occurrence depends either on another state change (generated by a function) or on a system-handled time condition "endogen".

$inEv$ =<event-id> (<event-var-list>)

Events will be implemented in Prolog as a goal for querying and manipulating the knowledge base.

Figure 3 illustrates partial activities of the university environment. Activities are shown at the nodes for which they are active for all children.

To each node a list that contains pre-condition, add, delete and post-condition is attached. These functions are optional and can be ommitted if there is only one event or function.

The function of the *pre*-condition is to monitor input events and accept them according to the specified input-logic. The *pre*-condition is also in charge of checking the integrity constraints of the entity-classes.

The two operators *add, del* describe modification of entity-classes at that level.

The *post*-condition describes activating the child activities according to the specified out-put logic.

The in-logic, out-logic are: *and, xor, sequence, or.*

These functions are optional and can be omitted if there is only one event or no function. For instance:

pre: $and(ev_1, ev_2)$
add: $student(Student = hans)$
del: nil
post: $seq(ac_{11}, ac_{12}, ev_3)$

states that if events ev_1 and ev_2 occur (in any order) then add student hans into the entity-class student, trigger the actions ac_{11}, ac_{12} and the internal event ev_3.

In the diagram an arc without an arrow-head specifies flow of information and control. This information is transferred into the system via event parameters. Indication of the event parameters as information or control is determined by the specification of the events. Usually, if the value of parameters will be stored in the system, then they will be regarded as information. If the value is used for testing the integrity constraints and will disappear after use, then they will be regarded as control information.

An arc with one arrow-head is used to model a read or write access. It is clear that only active elements can read or write. An arc with a double arrow specifies read and write, i.e. an entity taken from the class is put back into the same class again. Triangles specify the events, and circles the activities.

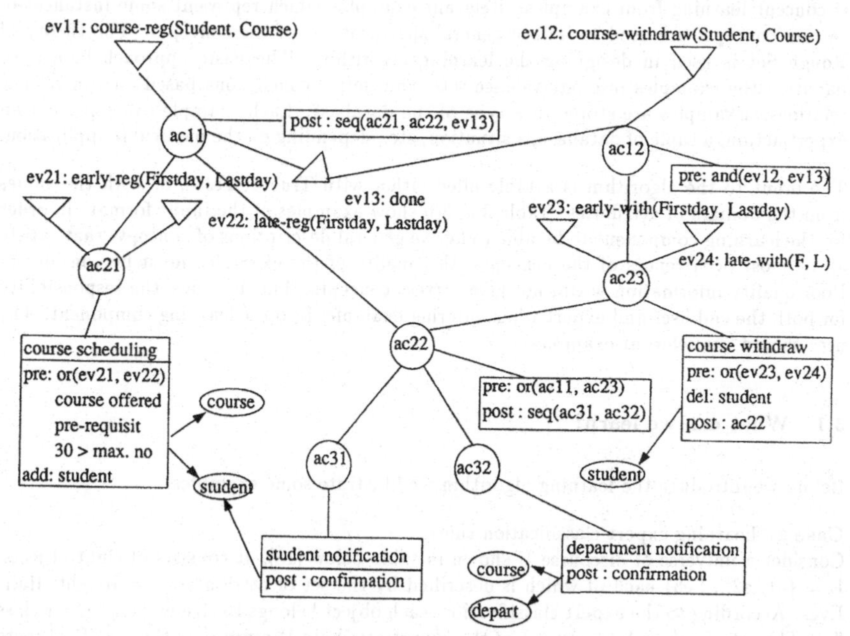

Figure 3. Example of the Event Model

3. Knowledge Acquisition

There are several techniques in the literature for acquiring knowledge from the real world. Since man is known for his capability to learn from experience, it is a challenge to see if he can create other entities that can think and learn just as man can. This research is a small step in this direction. Here we apply inductive learning as a method for knowledge acquisition.

Inductive learning is concerned with algorithms for generating generalizations from experience. These algorithms are viewed as examples of the general concepts of a hypothesis discovery system which, in its turn is placed in a framework in which it is seen as one component in a multi-stage process which includes stages of data gathering, forming hypothesis by generalization, hypothesis criticism or justification and adding hypotheses to the existing theory.

Learning from examples is one form of inductive learning. Concepts are derived from the characteristic description of a set of objects or entities. The quality of information plays an important role in the generation of rules. Poor quality information may give rise to erroneous or contradictory conclusions.

The type of inductive learning that we use for the design of the EXIS learning component is concept learning from examples. Relevant examples which represent some instances of certain concepts are used to induce general specifications of the concepts. The theory of Rough Set is used in designing the learning algorithm. The basic approach here is to partition the examples into equivalence sets, and derive conclusions based on equivalence relations. Examples are stored in a type of the database. Each example can represent an expert action, a tuple of database, a situation, etc., depending on the particular application.

The input to the algorithm is a table filled either with True or False, or with the values from the domain of attributes. Table 3.a, 3.b show examples of the input format. In order for the learning component to be able to derive general descriptions of concepts that satisfy all or most occurrences of the concepts; the quality of the examples must be considered. Poor quality information would not give correct concepts. This becomes the responsibility for both the end user and expert when entering examples into the learning component. The user should use relevant examples.

3.1 What can we learn?

Before we introduce the learning algorithm we illustrate some examples.

Case 1: Learning expert classification rules.
Consider a universe of discourse U shown in table below [18]. It consists of eight objects: E = {e1, e2,..., e8} each of which is described by the set of attributes C = Height, Hair, Eyes. According to the expert classification, each object belongs to either class "+" or class "-". The set of values V4 = {+, -} of the expert attribute D represents the set of concepts which the system wants to learn based on attribute values of C.

At	C			D
E	Height	Hair	Eyes	Expert classification
e1	short	dark	blue	-
e2	tall	dark	blue	-
e3	tall	dark	brown	-
e4	tall	red	blue	+
e5	short	blond	blue	+
e6	tall	blond	brown	-
e7	tall	blond	blue	+
e8	short	blond	brown	-

Table 1.

The generated rules are:

r1: hair(red) → decision(+)

r2: hair(blond), eyes(blue) → decision(+)

It can be shown that rules r1 is deterministic whereas r2 is non-deterministic.

Case 2: Learning schema rules.
Given the structure below, we can learn a rule such as:

"if a person owns a red car then his wife is young with certainty 0.8. "

The rule may be written as:

owns(man, car), car(colour = red), man-wife(man, woman) → woman(age = young) with cf = 0.8

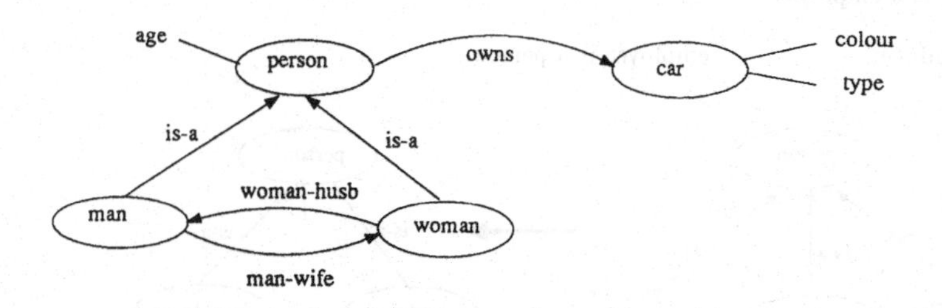

Figure 4. Case 2

Case 3: Introduction of a super-class relationship.
If sub-classes of a generalization are associated to an entity-class then the entity-class can be moved to the super-class; i.e. from

professor works-for university
secretary works-for university

professor is-a person
secretary is-a person

we can induce: person works-for university.

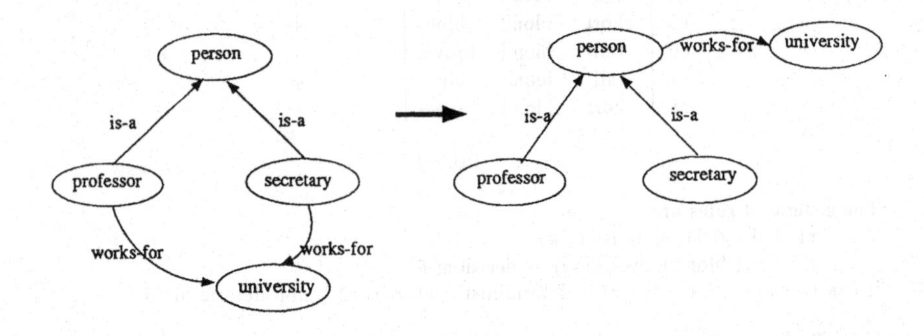

Figure 5. Case 3

Case 4: Introduction of a new entity-class
If two sub-classes share the same properties then it is possible to introduce a new super-class as a generalization of the sub-classes. In other words; from

student is-a person
professor is-a person
secretary is-a person
professor is-a employee
secretary is-a employee

we can induce: employee is a person.

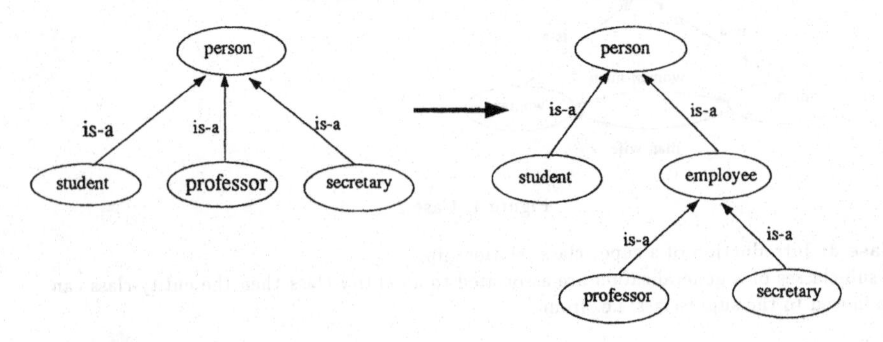

Figure 6. Case 4

Case 5: Discovering cause-effect relationship
Consider a collection of cars characterized by their attributes shown in the table below [22]. We are interested in discovering the cause-effect relationship between a set of condition attribute C and a conclusion attribute D. The relative contribution or significance of an individual attribute is represented by a *significance factor* that reflects the degree of decrease of dependency between C and D due to the removal of an attribute *a* from C.

In practice, the stronger the influence of an attribute *a* on a relationship between C and D the higher the value of its significance. From the table we conclude that there is a strong interaction between engine and max-speed with respect to affecting the acceleration, and both factors seem to contribute almost equally to the outcome.

Cars	Size	Engine	Color	Max-speed	Acceleration
1	full	diesel	black	high	good
2	full	propone	black	high	good
3	compact	gas	white	high	excellent
4	full	gas	white	high	excellent
5	medium	diesel	red	low	good
6	compact	gas	white	low	good
7	medium	gas	black	average	excellent
8	medium	diesel	red	average	poor

Attribute	Engine	Max-speed
Significence	0.75	0.875

Table 2.

3.2 The Rough Set learning method

We specify the theory by mean a tuple

S = (U, At, V, f, R) where:

- U Is a non-empty, finite set called the Universe. The elements of U are interpreted as "objects". We will assign different meanings to the objects of the universe depending on the particular application and the purpose for which we are employing the system. For instance, events, entities, situations, books, human beings, entity-classes, predicates or expert actions can be considered as objects.

- At The knowledge about these objects is expressed through the assignment of some features (attributes) to the objects. At is a non-empty set of attributes, of which the disjoint subsets of attributes C and D are referred to as condition and conclusion (expert decision) respectively. If the objects are characterized in the system by predicates we will refer to them as "attribute predicates".

- V Is a set of attribute values. V =∪ Va, where Va is the domain of attribute a ∈ At.
- f Is a function which assigns attribute values to each object in U. It is defined as: $f = U \times At \rightarrow V$ such that f(e, a) ∈ Va for every e ∈ U and a ∈ At. The set

  Des(S) = (a,v): a ∈ At, v ∈ Va called a set of descriptors of S.
- R Is a set of rules. The basic structure of a rule is a predicate in the form

  rule (no, rule-body, cf)

  The first argument is a rule number that uniquely identifies a rule, the second argument consists of head and body (conclusion and condition). A rule with a null condition expresses a fact. The degree of certainty with which the rule holds or, equivalently, the possibility for a rule to be true is shown by the third argument. The value of cf is a real number in the interval [0, 1].

Consider S = (U, At, V, f, R) for U = e1, e2,..., e10, At = a1, a2, a3, V = v0, v1, v2 and f is given by the Table 3.a.

	At		
	C		D
U	a1	a2	a3
e1	v1	v0	v2
e2	v0	v1	v0
e3	v2	v0	v0
e4	v1	v1	v0
e5	v1	v0	v2
e6	v2	v0	v0
e7	v0	v1	v1
e8	v1	v1	v1
e9	v1	v0	v2
e10	v0	v1	v1

U	w	x	y	z	a1(x, y)	a2(x, y)	a3(x, z)	a4(w, x)	a5(y, w)	a6(z, w)
e1	w1	x1	y1	z1	T	F	F	F	F	AC1
e2	w2	x2	y2	z2	T	F	F	F	T	AC3
e3	w3	x3	y3	z3	T	F	F	T	F	AC4
e4	w4	x4	y4	z4	T	F	F	T	T	AC3
e5	w5	x5	y5	z5	T	F	T	F	F	AC1
e6	w6	x6	y6	z6	T	F	T	F	T	AC3
e7	w7	x7	y7	z7	T	F	T	T	F	AC4
e8	w8	x8	y8	z8	T	F	T	T	T	AC3

Table 3.a, 3.b

Table 3.b shows switching from propositional calculus to predicate calculus. The problem that arises is that attributes must be generated dynamically. For the sake of notational simplicity and generality we use letters for the attribute value. The variables x, y, z are arguments of the attribute predicates $a_1, a_2, a_3, \cdots$ The predicates are satisfied with the attribute value being true or false V = {T, F} thereafter, the actions $AC_1, AC_2, \cdots$ take place.

Indiscernibility relation and partitions
Usually it is the case that information about objects represented by attributes and attribute values is not sufficient for distinguishing all the objects of a system. In Table 3.a, objects e1, e4, e5, e8, e9 cannot be distinguished by attribute a1, objects e1, e5, and e9 cannot be distinguished by the attribute a1 and a2. To deal with such cases the "equivalence relation" $\{\tilde{A} : A \subseteq At\}$ on the set U is defined as follows:

$$(e_i, e_j) \in \tilde{A} \text{ if and only if } \forall a \in A\ (f(e_i, a) = f(e_j, a)).$$

The equivalence classes of the relation A are called "A-elementary sets" in S. If a ∈ A, the equivalence class determined by a will be denoted by $\tilde{a}$.
Thus every subset A ⊆ At defines a certain "classification (partition)" of U denoted by $\tilde{A}$ and the equivalence classes of the relation $\tilde{A}$ are called "blocks of the classification". Some elementary sets in S defined in table 2.a are as follows:

{a1} - elementary sets
E1 = {e2, e7, e10}
E2 = {e1, e4, e5, e8, e9}
E3 = {e3, e6}.

{a3} - elementary sets
E1 = {e2, e3, e4, e6}
E2 = {e7, e8, e10}
E3 = {e1, e5, e9}

{a1, a2} - elementary sets
E1 = {e1, e5, e9}
E2 = {e2, e7, e10}
E3 = {e3, e6}
E4 = {e4, e8}.

Any finite union of A-elementary sets will be called an "A-definable set" in S. An empty set is A-definable for every A ⊆ At in every S.

3.3 Approximate Classification of a Set

Let U consist of a set of objects, and let $\tilde{A}$ be an equivalence relation induced by the partition
E = {E1, E2,..., En} on U. The pair (U, $\tilde{A}$) or (U, A) is called an "approximation space" defined by attributes A. Let Des(Ei) denote the description of an equivalence class (elementary set) Ei of the relation $\tilde{A}$ on E. (The terms "equivalence class" and "elementary set" are interchangeable). Any two objects e1, e2 ∈ Ei, have the same description: Des(e1) = Des(e2) = Des(Ei), if they both belong to the same elementary set Ei (i.e. e1, e2,∈ E).

Let us assume that attribute a3 generates a classification Y = {Y1, Y2, Y3}. We want to classify the set Y based on the values of attribute a1 (i.e. A = {a1}). In order to measure how well the set of description {Des (Ei) | Ei ∈ a1} can specify the membership functions of objects in classes of Y, we use the following notions [17]:

1. The A-lower approximation of a class Yj of Y in {U, A} is defined as
   $\underline{A}Yj = \cup\{Ei \in \tilde{A} \mid Ei \subseteq Yj\}$,
   i.e. $\underline{A}Yj$ is the union of all those elementary sets each of which is contained in Yj.

2. The A-upper approximation of a class Yj of Y in (U, A) is defined as
   $\bar{A}Yj = \cup\{Ei \in \tilde{A} \mid Ei \cap Yj \neq 0\}$
   i.e. $\bar{A}Yj$ the union of all those elementary sets each of which has a non-empty intersection with Yj.

3. The set $\bar{A}Yj - \underline{A}Yj$ is called the A-doubtful region of Ei in (U, A). It is not possible to decide whether or not any object in $\bar{A}Yj - \underline{A}Yj$ belongs to Ei based only on the descriptions of elementary sets of $\tilde{A}$. The ratio

$$\alpha_A(Yj) = \frac{\mid U \mid - \mid \bar{A}Yj - \underline{A}Yj \mid}{\mid U \mid},$$

   provides a measure of the degree of certainty in determining whether or not objects in U are member of the subset Y.

It is obvious that $1 \geq \alpha \geq 0$, and we say that:

- Set Yj is totally A-definable if $\alpha = 1$. This means that in this case the membership of objects in Y can be uniquely specified by the description of the elementary sets of A;

- Set Yj is partially A-definable if $1 > \alpha > 0$.
  $1 > \alpha > 0$ iff $\bar{A}Y \neq \underline{A}Y$ and ($\bar{A}Y \neq U$ or $\underline{A}Y \neq 0$). That is, not all objects in U can be classified with certainty into two disjoint subsets Y and U - Y;

- Set Yj is totally A-nondefinable if $\alpha = 0$.
  $\alpha = 0$ iff $\bar{A}Yj =$ U and $\underline{A}Yj = 0$. In this case we are unable to specify the membership of objects in Yj at all.

The lower approximation $\underline{A}$Y of the partition Y is defined as a collection of lower approximations of all its classes. Similarily, the upper approximations of $\bar{A}Y$ of the partition Y is a set of all upper approximation of all its classes. Obviously all blocks of $\underline{A}$Y are disjoint, although, in general, they do not form a partition of U.

Example 1: Consider the collection of objects as shown in table 2.a.
Let Y = {Y1 = (e2, e3, e4, e6), Y2 = (e7, e8, e10), Y3 = (e1, e5, e9)} be a partition of U defined by a3 as conclusion and let A = a1, a2 be conditions which generate the following classification on U:

E1 = (e2, e7, e10) Des (E1) = (a1: = v0, a2: = v1);
E2 = (e1, e5, e9) Des (E2) = (a1: = v1, a2: = v0);
E3 = (e3, e6) Des (E3) = (a1: = v2, a2: = v0);
E4 = (e4, e8) Des (E4) = (a1: = v1, a2: = v1).

From Ei ⊆ Y we obtain $\underline{A}Y$ = {{e3, e6}, { }, { e1, e5, e9 }}.

From
E1 ∩ Y : {{e2, e7, e10}, {e2, e7, e10}, { }}
E2 ∩ Y : {{ }, { }, {e1, e5, e9 }}
E3 ∩ Y : {{e3, e6} , { } , { }}
E4 ∩ Y : {{e4, e8}, {e4, e8}, { }}
By taking the union of these it follows:
$\bar{A}Y$ = {{e2, e3, e4, e6, e7, e8, e10} {e2, e4, e7, e8, e10}, {e1, e5, e9}}.

Hence the A-doubtful region Y is given by

$\bar{A}Y - \underline{A}Y$ = {{e2, e4, e7, e8, e10}, {e2, e4, e7, e8, e10}, { }} and

$\alpha_{A(Y1)} = \frac{10-5}{10} = 0.5;\qquad \alpha_{A(Y2)} = \frac{10-5}{10} = 0.5;\qquad \alpha_{A(Y3)} = \frac{10-0}{10} = 1.$

Extracting Decision Rules From Examples
A rule may represent, for example, actions (decisions) of an expert. Rules are built up as follows:

$$rij : Des(Ei) \Longrightarrow Des(Yj) \mid Ei \cap Yj \neq \{\}$$

where Des(Ei) and Des(Yj) are the unique descriptions of the condition class Ei and conclusion class Yj, respectively. A rule rij is deterministic if Ei ∩ Yj = Ei, and rij is non-deterministic if Ei ∩ Yj ≠ Ei. In other words, if Des(Ei) uniquely " implies " Des(Yj), then rij is deterministic; otherwise rij is non-deterministic.

Example 1 continued: Consider the collection of objects given in Table 3.a. The partition corresponding to equivalence relation $(a1,a2)$ is E = {E1, E2, E3, E4} where:

Des (E1) = (a1 := v0, a2 := v1),
Des (E2) = (a1 := v1, a2 := v0),
Des (E3) = (a1 := v2, a2 := v0),
Des (E4) = (a1 := v1, a2 := v1).

The portion corresponding to equivalence relation Y is Y = {Y1, Y2, Y3} where:

Des (Y1) = (a3 := v0)
Des (Y2) = (a3 := v1)
Des (Y3) = (a3 := v2)

By definition the rules for class Y1 are:
r11: Des (E1) → Des (Y1) ; a1(v0) & a2(v1) → a3(v0)
r21: not exist because $E2 \cap Y1 = 0$
r31: Des (E3) → Des (Y1) ; a1(v2) & a2(v0) → a3(v0)

r41: Des (E4) → Des (Y1) ; a1(v1) & a2(v1) → a3(v0)

The rules for class Y2 are:

r12: Des (E1) → Des (Y2); a1(v0) & a2(v1) → a3(v1)

r22: not exist because E2 ∩ Y2 = 0.

r32: not exist because E3 ∩ Y2 = 0.

r42: Des (E4) → Des (Y2); a1(v1) & a2(v1) → a3(v1)

The rules for class Y3 are:

r13: not exist because E1 ∩ Y3 = 0

r23: Des (E2) → Des(Y3); a1(v1) & a2(v0) → a3(v2)

r33: not exist because E3 ∩ Y3 = 0

r43: not exist because E4 ∩ Y4 = 0

3.4 Learning Algorithm

Based on the concepts presented, we give the learning algorithm:

Input:

1. Type of rule to be generated.
2. Number of conditions and conclusion attribute list.
3. A sequence of examples with the corresponding set of attribute values.
4. A *flag* indicating whether the conclusion attribute is given by the user or not.

The user can input the above information interactively or through an input file.

Procedure Learning Algorithm;

Initialize i = 1, i denotes the ith conclusion attribute;

Step0: Initialize j = 1, j denotes the partition of the values in the conclusion attribute;

Step1: Initialize:

ep = list of all examples numbers;

cp = list of all condition attributes;

b = empty set represents a subset of condition attributes;

y = list of situation numbers in which the jth condition attribute is equal to 't' (true);

Step2: Compute the set of certainty values $\alpha(y)$ for each bp such that bp = b ∪ {c}, for all {c} ∈ {cp}. Select the set of attributes bp = b ∈ {c} with the highest certainty value $\alpha(y)$, set b = bp;

Step3: If b-lower approximation is an empty set then go to Step 5;
For those equivalence classes in b-lower approximation, output the deterministic decision rule;

Step4: Set ep = ep - [(ep - b-upper) ∪ (b-lower)]; y = y - b-lower;
If ep is an empty set then go to Step 6;
Set cp = cp - b;
If cp is non-empty then go to Step 2;

Step5: For those equivalence classes in b-upper approximation, output the non-deterministic decision rule, then go to Step 4;

Step6: Set j = j + 1;
if j $\leq$ no. partition sets of conclusion attributes then goto Step 1; else go to Step 7.

Step7: if *flag* = n then goto Step 9;

Step8: retract all dynamic variables and terminate;

Step9: Set i = i+1;
if i $\leq$ number of conclusion attributes then initialize the variables and goto Step 0; else retract all dynamic variables and terminate.

End Procedure Learning Algorithm; Table

4. System Architecture

The EXIS methodology is as follows:

First the knowledge engineer configures the knowledge base and defines the meta-schema by declaring the constructs as relations, generalizations, aggregations, etc. with their associated schema integrity constraints. After the system has been created the domain expert performs an initial schema generation step, in which he defines his application by using the constructs provided to him by the designer. This conceptual schema will be stored in the knowledge base.

Finally, the user enters data into the database or applies queries to the system. In this step, the learning component extracts a set of rules from the knowledge base or database, which can be added to the existing knowledge base.

To discover knowledge, the system needs to:

- Gather information by making some queries from the expert or database;
- Form hypothesis, such that all positive examples and no negative examples are covered;

- Evaluate the result by consulting with the expert and perhaps modifying or discarding the hypotheses;
- Continue this cycle until a correct rule emerges and is added to knowledge base.

Both theoretical and practical considerations led us to the architecture which is illustrated in Figure 7. The architecture contains the following main components:

Knowledge Base We distinguish the term knowledge from data. Knowledge is information about general concepts and data are information about entities or relationship. For example:

"all students have taken CS100"	is knowledge
"Jim has taken CS100"	is data.

The knowledge base employed by the learner contains two different kinds of predicates:

meta predicates
Representing the meta schema and defined by the designer and assumed to be correct. They are specified by arbitrary Horn-clauses.

application predicates
Representing the constraints and are defined by the expert or derived from the database.

Learner The learning component generates rules by using the knowledge base or explicit examples.

Evaluator The generated rules can be added to the knowledge base after expert approval.

Schema generator The expert has the possibility of defining the conceptual schema interactively. The declarative language that is used, is very close to Prolog, for example:

entity class and its attributes: *person(Name:string, ID#: integer, colour = red)*
generalization: *is-a(person, employee, student)*

Update processor The generated rule can be added to the knowledge base after expert approval. In order to generate the consistency of the knowledge base after an update, we need an integrity maintenance algorithm for preventing updates from violating integrity constraints. Such algorithms already exist and have been successfully implemented and tested in a real application. We make use of this subsystem [3].

DB access DB contains the set of data relevant to the specific learning task. This data are obtained by performing operations, such as retrieval, selection, projection, and join to collect the necessary data for learning. We will adopt in our future work a similar system as has been presented in [20].

The architecture is scalable. The extended layer resides on one workstation, while the existing system may reside at another end relay on special purpose hardware.

Many of EXIS's modules have already been implemented, as a stand-alone system. The revised version of EXIS incorporating the new aspects is under development. The system runs on the Sun workstation using Prolog.

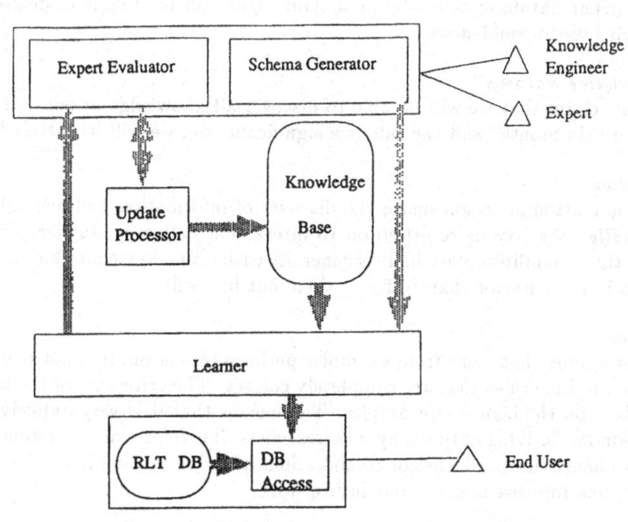

Figure 7. Architecture of EXIS

5. Conclusion

We have described a system that integrates knowledge-representation and -acquisition in a unique framework. The approach ranges across the spectrum, from transfer of expertise from experts to the creation of equivalent expertise through empirical induction.

In the following the main features of the system are summarized.

More expressive knowledge representation
The representation of knowledge in a semantic model and event model facilitates knowledge acquisition. It has more expressive facilities such as generalization, specialization, associations, aggregation. The model can be formed using the constructs. They relieve the expert by defining explicitly all domain knowledge.

Approximate rules
The rules we find capture general knowledge, while recognizing that there will be uncertainty to those rules. We do not expect what we discover are perfectly consistent with the data, or to account for all of it. Our algorithms have the same flexibility.

Efficiency
The method employed can be applied to realistic databases, i.e. $\geq 10^5$ records. This implies that the algorithm is efficient and scales well.

Relational database
Most current databases are relational. Our algorithm is capable of dealing with relations and their inter-dependencies.

Accumulative learning
We don't claim that we will be able to discover all knowledge at once. If we generate one rule every six months and the rule is a significant one, we will be satisfied.

Interaction
We do not attempt to automate the discover of information and generate knowledge *automatically*. We pay more attention to interaction between designer, expert and user to exploit the capabilities of each. It is generally easier for a system to be *told* about necessary changes in its behavior than to figure them out by itself.

Correctness
Since programs that learn from examples perform induction, it is not reasonable to expect them to produce rules that are completely correct. The error rate of the learner is strongly dependent on the choice of examples. This means that it is very unlikely that an efficient algorithm for deriving optional hypotheses exists. Therefore, many systems in the literature resort to heuristic approaches or combinations of deductive and inductive learning. We will look at such improvements in our future work.

5.1 Related work

Machine learning is commonly recognized as a fundamental way for overcoming the knowledge acquisition bottle-neck. As a result knowledge based systems, especially expert systems, are now showinging extraordinary growth. There have been increasing efforts to induce knowledge from existing data. As a result, there have been improvements in the technology. In the following we mention some of them.

INDUCT is a subcomponent of a knowledge acquisition tool KSSO [8]. It uses entity attribute grid techniques to elicit knowledge from the expert and uses the empirical induction approach to build a knowledge base in terms of classes, objects, properties, values and methods(rules). The tool also accepts rules entered by experts and entity-attribute data from database. INDUCT deals nicely with noisy data and missing values.

The ASSISTANT professional program [1] was developed at the Jozef Stefan Institute of Ljubljana, Yugoslavia. It uses a version of the ID3 algorithm by Quinllan [4], and is similar to the work of Leo Breiman et all [2]. The primary purpose of the algorithm is to extract rules from sets of events preclassified into a small number of classes. The division of the set of events may be described either by finding for each cluster a *description* that covers all events in the cluster, or a *discrimination* that will specify features that makes a particular cluster (and hence the contained events) unique from the other clusters. An event is a description of some phenomena, articulated as a vector of attributes. Each attribute takes only one value at a time. The function of the discrimination algorithm is to capture and display the relation between the attribute values of an event and its classification. The Assistant algorithm in its present incarnation is not capable of doing what is proposed here on larger data sets than 30 events, at least not on the IBM-PC version. By assigning each event its own class, a discriminant description of each event can be generated. If the learning set is constrained to contain unique events only, the *Information-Theoretic* measure and the associated computation can be simplified.

KATE [13] is a tool that learns from examples using the induction technique ID3. It includes an object-based language to represent training examples and background domain knowledge. The language is used to represent hierarchies of clauses and descriptions(taxonomies), constraints, and procedural calls, which enables the deduction of descriptors from other descriptors. KATE's designers claim that the frame-oriented hierarchy representation conveys more information than taxonomies of descriptors or semantic nets. KATE differs also from EXIS in the same way as functional programming does from logic programming.

The INLEN system [12] combines a database expert system and machine learning capabilities in order to provide a user with a powerful tool for manipulating both data and knowledge, and for extracting knowledge from data and/or knowledge. INLEN provides three sets of operations: data management, knowledge management operators and knowledge generation operators for communication to the system in the tree area database, knowledge base and machine learning respectively.

In summary, we have described a completly integrated approach that includes several concepts. We have defined and redefined these concepts in such way that they fit to each other and form one unique system. We hope, we have made a useful contribution to overcoming the difficulties of design such a systems.

References

[1] I.Bratko and N.Lavrac. Progress in machine learning. In *Proceedings of EWSL87: 2nd european working session on lerning.* Sigma press, England, 1987.

[2] L.Breiman, J.H.Friedman, R.A.Olshen, and C.J.Stone. *Classification and Regression trees.* Wadsworth International Group, California, 1984.

[3] F.Bry, H.Decker, and R.Manthey. A uniform approach to constraint satisfaction and constraint satisfability in deductive databases. In *Proceedings of the Conference Extending Data Base Technology,* Venice, 1988. Springer Verlag.

[4] J.G.Carbonell, R.S.Michalski, and T.M.Mitchell. An overview of machine learning. In R.S.Michalski, J.G.Carbonell, and T.M.Mitchell, editors, *Machine Learning: An Artificial Intelligence Approach,* volume 1. Morgan Kaufmann, Los Altos, California, 1983.

[5] E.F.Codd. Further normalization of the database relational model. In R.Rustin, editor, *Current Computer Science Symposia 6: Data Base Systems.* Prentice Hall, 1972.

[6] R.Fagin. Multivalued dependencies and a new normal form for relational databases. *ACMTODS*, 2:262-278, 1977.

[7] R.Fagin and M.Y.Vardi. The theory of database dependencies - a survay. In *Proceedings of the ACM, ahort course on the Mathematics of Information Processing*, Louisville, Kentucky, 1984.

[8] B.R.Gainies. Rapid prototyping for expert systems. In *Proceedings of the International Conference on Expert Systems and Leading Edge in Production Planning and Control*, University of South Carolina, 1987.

9] M.R.Genesererth and N.J.Nilson. *Logical Foundations of Artificial Intelligence*. Morgan Kaufmann, Los Altos, California, 1987.

[10] J.J.Hopfield. Neural networks and physical systems with collective computional abilities. In *Proceedings of the National Academy of Science*, 1982.

[11] P.Johannesson and K.Kalman. A method for translating relational schemas into conceptual schemas. Technical Report 69, SYSLAB, University of Stockholm Sweden, 1989.

[12] K.Kaufmann, R.Michalski, and L.Kershberg. Mining for knowledge in db, goals and general discription of the inlen system. In Piatetsky-Shapiro, editor, *Workshop's Proceedings of the 11.th IJCAI, Detroit*, 1989.

[13] Y.Kodratoff. Kate, the knowledge acquisition tool for expert systems. Technical report, IntelliSoft, Orsay, 1988.

[14] D.Maier. *The theory of relational databases*. Computer Science Press, Rockville, 1983.

[15] R.Michalski, J.G.Carbonell, and T.M.Mitchell. *Machine Learning, An Artificial Intelligence Approach*, volume 1,2,3. Springer Verlag, Berlin, 1983, 86, 90.

[16] S.Navathe and A.Wong. Abstracting relational and hierarchical data with a semantic data model. In *Proceedings of the Seventh International Conference on Entity-Relationship Approach*, 1987.

[17] Z.Pawlak. Rough classification. *International Journal of Man-Machine Studies*, 20:469-493, 1984.

[18] J.R.Quinlan. Learning efficient classification procedure and their application to chess and games. In J.G.Carebonell, R.Michalski, and T.M.Mitchell, editors, *Machine Learning: An Artificial Intelligence Approach*. Tioga Press, Palo Alto, CA, 1983.

[19] C.E.Shannon. A mathematical theory of communication. *Bell System Technical Journal*, 4:379-423, 1984.

[20] Siegel, Sciore, and Salveter. Rule discovery for query optimization. In *Workshop Proceedings of the IJCAI*, Detroit, 1989. Morgan Kaufmann.

[21] R.Yasdi and W.Ziarko. An expert system for conceptual schema design: A machine learning approach. *International Journal of Man-Machine Studies*, 29(4), October 1988.

[22] W.Ziarko. A techniques for discovery and analysis of cause-effect relationships in empirical data. In *Workshop's Proceedings of the 11.th IJCAI-89*, Detroit, 1989.

Part III
Chapter 8

'ROUGHDAS' AND 'ROUGHCLASS' SOFTWARE IMPLEMENTATIONS OF THE ROUGH SETS APPROACH

Roman SLOWIŃSKI Jerzy STEFANOWSKI
Institute of Computing Science
Technical University of Poznań
60-965 Poznań, Poland

Abstract. Two software systems based on the rough sets theory are described in this chapter. The first one, called 'RoughDAS', performs main steps of the analysis of data in an information system. The other one, called 'RoughClass', is intended to support classification of new objects. Some problems of sensitivity analysis referring to the handling of quantitative attributes in the rough sets analysis are also discussed.

1. Introduction

The rough sets theory is a tool for analysis of data in an information system. In this analysis, the most essential problems refer to: evaluation of an importance of attributes for classification of objects, reduction of all superfluous objects and attributes, discovering of most significant relationships between condition attributes and objects' assignments to decision classes, and representation of these relationships, e.g. in a form of decision rules.

Once data analysis accomplished, other problems refer to an application of these results to classification of new objects basing on previous experience.

From a practical point of view, the rough sets analysis of real data sets should be performed by a computer software, except of very small data sets. In this chapter an interactive computer program, called 'RoughDAS' [4], is described. It was created for the rough sets based analysis of information systems and enables: approximation of each class of objects, calculation of accuracies of approximations and quality of classification, searching for reducts of attributes, reduction of non-significant attributes, derivation of the decision algorithm from reduced information system. This program has been successfully applied in several practical problems, among others, analysis of treatment of duodenal ulcer by highly selective vagotomy (cf. [3],[10],[12]), analysis of data from peritoneal lavage in acute pancreatitis [13], analysis of the structure of compounds in pharmacy [5], analysis of the technical state of technical objects like reducers or rolling bearings [7],[8].

In practical applications, information systems often contain data of discrete and continuous character. It is known that the rough sets analysis of information systems gives satisfactory results when domains of attributes are finite sets of rather low cardinality. This requirement is often met naturally when attributes have a qualitative (discrete) character. However, attributes taking arbitrary values from given intervals, i.e. having a quantitative character, can be also handled in the analysis after translation of their values into some qualitative terms, e.g. low, medium or high levels. This translation involves a division of the original domain into some subintervals and an assignment of qualitative codes to these subintervals. The definition of boundary values of the subintervals should take into account experience, knowledge, habits and conventions used by the experts and, possibly, an error of measurement of some attributes. Such an approach to handling quantitative attributes was used in the analysis of the above mentioned problems and was implemented in the considered software. The definition of boundary values of the subintervals can influence, however, the quality of classification, so the problem of checking the sensitivity of results to changes in the definition of these values is becoming important and will be briefly discussed in the paper. A computer program for sensitivity analysis will also be presented.

The results obtained using the rough sets analysis, in particular, the reduction of attributes and the decision algorithm, have a great practical importance. For instance, in medicine, reduction may cause elimination of superfluous clinical tests harmful for patients; in technical diagnostics it may decrease the cost and time of diagnosis. The decision algorithm shows all important relationships using a minimum number of decision rules and/or a minimum number of attributes appearing in all decision rules. So, the decision algorithm is more readable for the user than the original information system.

On the other hand, these results represent a knowledge gained by an expert (decision maker) on all cases from his experience recorded in the information system. It is interesting and desirable, in fact, to use this knowledge for supporting decisions concerning classification of new objects (cf. [14]). By new objects one understands objects unseen in the expert's experience (i.e. information system) which are described by values of condition attributes only (all or from a reduced set). The assignment of these objects to classes is unknown. The aim of the expert is to predict this assignment on the basis of his knowledge coming from previous experience.

The idea of such supporting was presented in [14]. It was based on looking through the decision algorithm performed to find a rule whose condition part is satisfied by the description of the new object. When such a rule is found, it is used to support classification. In some cases, when the proper rule does not exist in the algorithm, it is necessary to find, so called, 'nearest rules'. The nearest rules are rules which are close to the description of the classified objects according to a chosen distance measure. These ideas were implemented in the software system called 'RoughClass'.

In the next section, a general description of the 'RoughDAS' is given. It is followed by a short description of the program enabling sensitivity analysis of the roughs sets results. Finally, the 'RoughClass' software for classification support is briefly presented.

2. 'RoughDAS' - Rough Sets based Data Analysis System

2.1. Basic features

'RoughDAS' (i.e. **Rough** Sets based **D**ata **A**nalysis **S**ystem) is a microcomputer program for the analysis of information systems by means of the rough sets theory. It performs all main steps of the rough sets approach, i.e. approximation of classes, calculation of accuracies of approximations and quality of classification, searching for reducts of attributes, reduction of the information system, derivation of the decision algorithm from the reduced information system. It is implemented as an interactive software and runs on IBM PC microcomputers under PC-DOS. All options from the menu are presented in a tree form in Fig. 2.1. Next paragraphs give more information about main options.

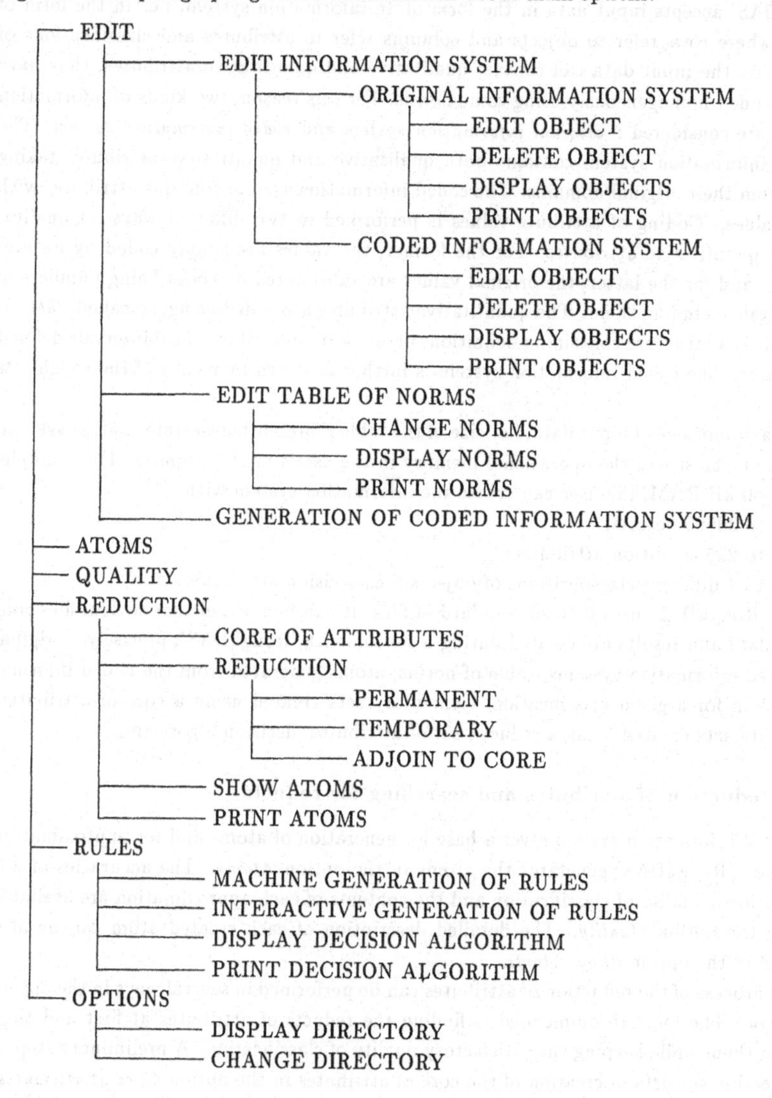

Figure 2.1. Options available in 'RoughDAS'

2.2. Edition of input data

'RoughDAS' accepts input data in the form of an information system, i.e. in the form of a table where rows refer to objects and columns refer to attributes and classifications of objects. As the input data can contain qualitative and quantitative attributes, they have to be encoded homogeneously using some norms. For this reason, two kinds of information systems are considered : *original information system* and *coded information system*. The original information system contains both qualitative and quantitative attributes taking values from their original domains. The coded information system contains attributes with coded values. Coding of attribute values is performed in two different ways for qualitative and quantitative attributes. For the former, the values are simply coded by natural numbers, and for the latter, the original values are substituted by codes being numbers of subintervals including them. The quantitative attributes are coded using, so called, *table of norms*. It is a structure containing definitions of all boundary values of subintervals defined by the user. The coded information system is further analysed by means of the rough sets theory.

A maximum size of input data, i.e. size of the table representing the information system, depends on the size of the operational memory in the user's microcomputer. For example, having 640 kB RAM, the user can create the information system with:

- up to 30000 objects,
- up to 255 condition attributes,
- up to 4 different classifications of objects, i.e. decision attributes.

The 'RoughDAS' uses its own standard of files. It can be noticed that several files containing data and results are created during a session with 'RoughDAS', precisely : original and coded information systems, table of norms, atoms generated from the coded information system for a given classification, elementary sets created using a core of attributes, elementary sets created using a reduced set of attributes, decision algorithm.

2.3. Reduction of attributes and searching for reducts

The coded information system gives a base for generation of atoms and for approximation of classes. 'RoughDAS' generates the atoms in the option *Atoms*. The accuracies of approximations, quality of classification and the contents of each approximation are available through the option *Quality*. The detailed description of each created atom can be also obtained in the option *Show Atoms*.

The process of the reduction of attributes can be performed in several ways in the option *Reduction*. The most recommended is finding the reducts of attributes at first and then reducing them while keeping the satisfactory quality of classification. A preliminary step in this direction consists in creation of the core of attributes in the option *Core of attributes*. The algorithm for finding the reducts of attributes is implemented using the idea of the "depth-first-like" algorithm proposed by Romański [11]. The implemented algorithm finds all existing reducts of attributes. It is much more efficient than previously used procedures (see [1]) because it avoids superfluous checkings of unnecessary subsets owing to the concept of, so called, fences and cones [11] restricting the area of searching. It is particularly useful in situations when the core of attributes is empty and the number of attributes is relatively high.

For a chosen reduct, the user can perform further reduction of attributes in the option *Permanent reduction.* If he has any doubts concerning the choice of the set of attributes, he can experiment with reduction in the option *Temporary reduction.* This option allows to delete some attributes only temporarily and check the result of approximation of the classification by the reduced set of attributes. The temporarily deleted attributes are restored after this test, so the user can make another trial immediately after. It is possible to make up to 5 trials at once (on one screen).

If the user gets too many reducts of attributes and is unable to choose one of them, he should use option *Adjoin to core.* In this option, when the core is not empty , the user can built a reduct by adding attributes to the core. In each step of this procedure he is informed about a possible increase of the quality of classification after addition of one of remaining attributes. In this way the user can guide the construction of one reduct according to his preferences.

2.4 Derivation of a decision algorithm

The decision algorithm can be derived from every (reduced) information system . Generation of decision rules in 'RoughDAS' can be performed in two ways : *automatically* or *interactively.* In both cases this generation is performed using the procedure based on the discriminatory coefficient $ALPHA_P(Y)$, where

$$ALPHA_P(Y) = (card(U) - card(Bn_P(Y))/card(U)$$

(see [2],[3] and [14]). The scheme of the procedure (in pseudoPascal) is presented below. Let $\mathcal{Y} = \{Y_1, \ldots, Y_n\}$ be a classification of U, C and B subsets of attributes, D is a decision attribute referring to the classification of objects.

Considered procedure,

```
begin
  for j:=1 to n do {set the class number}
  begin {for each class}
    U' := U; { U' denotes remaining objects in class Y_j}
    C' := C; { C' denotes remaining condition attributes }
    B  := []; { the subset of condition attributes used to built decision rules, B is
                  empty in the first step }
    Y := Y_j ; { choose the current class}
    repeat
      for i := 1 to card(C') do
      begin
        B' := B ∪ [c_i]; {c_i ∈ C'}
```

```
      calculate discriminant coefficient ALPHA_B'(Y);
    end;
    select c_k with the highest value of ALPHA_B'(Y);
    B := B ∪ [c_k];
    if B̲Y <> [] then
    begin
      identify B-elementary sets {X_1,...,X_m} contained in B̲Y;
      generate deterministic decision rules:
      {Des_B(X_i) ⇒ Des_D(Y)};
    end
    U' := B̄Y − B̲Y; {remove objects from outside B̄Y and from inside B̲Y}
    Y := Y − B̲Y; {remove objects belonging to B̲Y which are already described
                  by created decision rules}
    C' := C' − B';
  until (U'=[]) or (C'=[]);
  if U' <> [] then
  begin
    identify C-elementary {X_1,...,X_m} on U' and ;
    generate non-deterministic decision rules:
    {Des_C(X_i) ⇒ Des_D(Y)} if they have not been generated for previously considered
    classes;
  end
 end
end {procedure}
```

Let us notice that $ALPHA_{B'}(Y_j)$ may attain the highest values for several subsets B' - it is typical for early stages of building decision rules, when subset $B' = B + \{c_i\}$ is of relatively low cardinality. In this case, depending on selection of attribute c_i appended to subset B, one can get slightly different decision algorithms. So, the choice of this attribute should be done by the user. In general, the user should check all possibilities in order to find the decision algorithm which best fits his preferences. In the *interactive generation*, the user controls the process of building rules by selecting attributes appended to B. In the *machine generation*, he cannot influence the selection of attributes. The attributes are selected automatically according to the criterion of the highest value of $ALPHA$. In the case of ties, the most significant attribute (according to the definition of significance given in [9]) is chosen. If this criterion is still not discriminatory, the first attribute from the list is chosen. The obtained decision algorithm is presented to the user and stored in a file.

3. Sensitivity analyser

In the rough sets approach quantitative attributes are handled by transforming their original values into some coded qualitative terms. This involves a division of the original domain of the quantitative attributes into some subintervals, as it was mentioned in the introduction. The definition of boundary values of these subintervals is a key point at the beginning of the sensitivity analysis. It is usually done by experts according to their experience, knowledge, habits or conventions. The definition of boundary values can influence results of the analysis, in particular the quality of classification. Until now, there is no general way to define the optimal boundary values. Sometimes, heuristics resulting from practical experience can be used, as it was in the case of methods defining symptom limit values in vibroacoustic diagnostics of rolling bearings (see [7],[8]). It is thus interesting to check how sensitive are the obtained results to minor changes in the definition of boundary values. In other words, what is the robustness of the model to minor changes in the input data.

In order to get answer to above questions, one can use the program called Sensitivity Analyser [6]. Changing the definition of boundary values for several attributes at once is rather not appropriate, so in the Sensitivity Analyser this possibility is restricted to one attribute at a moment. The program allows to move bounds of subintervals of a chosen attribute to the left and to right of the original bounds and to observe the influence of this movement on the quality of classification. The magnitude of the movement is expressed in the percentage of the original boundary values. The movement to the left and to the right corresponds to the negative and positive percentage respectively. The Sensitivity Analyser uses the same file standard as 'RoughDAS'. It is possible to select the set of attributes to be considered, e.g. the reduced set chosen as a result of analysis with 'RoughDAS'. The maximum range of the movement (in percents) to the left and to the right has to be given by the user. The definition of the range is accompanied by the number of control points in this range. The program builds new coded information systems and computes the quality of classification for all control points. The process of this examination can be repeated for each attribute leading to a better definition of the boundary values.

4. 'RoughClass' software for classification support

4.1 Basic concepts

'RoughClass' is a software system created to support classification of new objects, i.e. objects described by values of condition attributes only.

The system performs an analysis of the decision algorithm in order to find a rule whose condition part satisfies the description of the new object. If the system succeeds, the rule is presented to the decision maker. The decision rule may be either deterministic or non-deterministic. If the system fails to find an exactly matched decision rule, i.e. the description of the new object does not match any rule, it means that there is no such case in experience represented in the information system. The latter case is a consequence of the fact that the rough sets analysis of experience represented in the information system is performed under the *'closed world assumption'* [9], whereas the classification of new objects is often performed in practice under the *'open world assumption'* (cf. p 4.2.).

If an exactly matched decision rule does not exist, the 'RoughClass' helps the decision maker by the analysis of 'nearest rules' [14]. The nearest rules are rules which are close to the description of the classified object using a chosen distance measure. The procedure implemented in 'RoughClass' finds a set of nearest rules and presents them to the decision maker. Rules are characterized by a distance from the new object.

It should be noticed that matching rules and finding nearest ones is connected with two other points: a kind of inconsistency in a decision algorithm and different reliability of rules. The first one is another consequence of the 'closed world assumption'. Application of a decision algorithm obtained under this assumption to classification of objects which come from an 'open world' may lead to a special kind of inconsistency which consists in a non-univocal classification of an object by deterministic rules matching exactly its description. The second point is connected with the fact that decision rules are not equally important for the decision maker. Some rules can be based on many objects while others can be based on very few or single objects. They have then a different strength. Both problems can be examined by 'RoughClass' and will be discussed in the next paragraph.

'RoughClass', similarly to 'RoughDAS', is a program implemented for IBM PC compatible micros running under PC-DOS. It uses the window-menu technique. The program accepts the 'RoughDAS' standard of files. So, the user can analyse his problem using both programs for the same data. The input data for the 'RoughClass' are files taken from 'RoughDAS' containing the decision algorithm and the coded information system. Some main options concerning the process of classification are presented in Fig. 4.1. They will be described briefly in the next points.

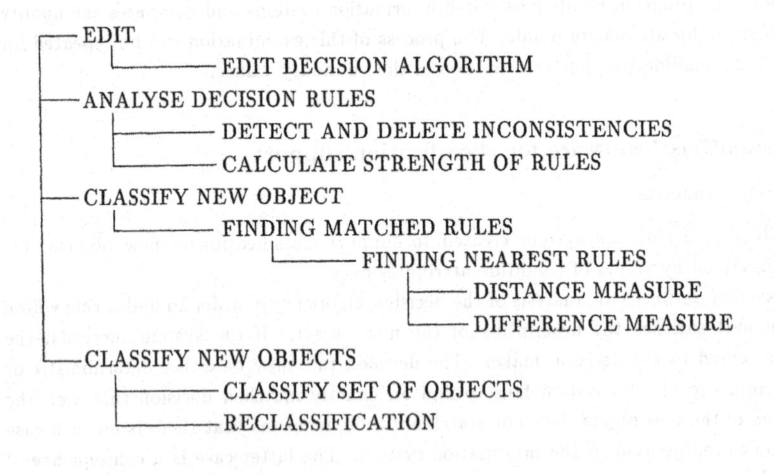

Figure 4.1. Options available in 'RoughClass'

4.2. A special kind of inconsistency in the decision algorithm

As it was mentioned in p 4.1. the procedures for derivation of decision algorithms by means of the rough sets theory (cf. [2],[9],[15]) work under the 'closed world assumption' (i.e. it is assumed that experience contained in the analysed information system is complete). In consequence the algorithms which work correctly for objects having description consistent with the previous experience, may lead to non-univocal classifications for some possible objects having descriptions not represented in the information system. This case is typical for the 'open world assumption' where experience represented in the information system is not complete.

For example, let the analysed decision algorithm contain two following rules (these rules where derived from the information system presented in [14]):

$\#1 \; : \; if \; (a1 = 1) \; and \; (a2 = 2) \; then \; (d = 1)$

$\#2 \; : \; if \; (a2 = 2) \; and \; (a3 = 2) \; then \; (d = 2)$

If it is assumed that the experience represented in the information system is complete, both rules work correctly. If the experience is not complete, a new object may appear with the following description : $(a1 = 1)$ and $(a2 = 2)$ and $(a3 = 2)$. Such an object cannot be classified univocally using rules #1 and #2. This kind of inconsistency in the decision algorithm results from the fact that procedures for the derivation of the decision algorithm analyse only objects represented in the given information system. They try to build the shortest decision rules by eliminating all redundancies perceived in the objects' description (cf. p.2.4). For instance, in rule #1, it is sufficient to use conditions $(a1 = 1)$ and $(a2 = 2)$ only because in the considered information system only objects from class 1 have this combination of the values of $a1$ and $a2$. In other words, only these objects can create the a1,a2-elementary set which belongs to the lower approximation of class 1. Hence conditions for $a3$ and $a4$ are not added.

For this reason, it is necessary to check the consistency of the algorithms before using them to classify new objects.

In 'RoughClass' the option called *Detect and Delete Inconsistencies* detects the above described inconsistencies in the analysed information system and removes them by adding to one of two inconsistent rules additional conditions for the attributes occurring only in the other rule (cf. [14]).

In the presented example, the inconsistency can be removed by adding to rule #1 the condition for attribute $a3$. The procedure must check what is the value of a3 for objects used to built rule #1 in the coded information system. As a result, the following rules are obtained :

$\#1' \; : \; if \; (a1 = 1) \; and \; (a2 = 2) \; and \; (a3 = 3) \; then \; (d = 1)$

$\#2 \; : \; if \; (a2 = 2) \; and \; (a3 = 2) \; then \; (d = 2)$

which are not inconsistent.

4.3. Reliability of decision rules

Let us notice, that decision rules in the decision algorithm are not equally important or reliable. They can be build on a base of different numbers of objects from the information system. One rule may be supported by a single object while another by many objects. So, the first one will be 'weaker' then the latter, i.e. less reliable for the decision support.

For this reason, in 'RoughClass', the description of each decision rule is extended by information about the number of objects in the information system which support the given rule. This number is called *strength* of the rule.

The introduction of this coefficient seems to be useful for analysis of non-deterministic rules. In the non-deterministic rules, decisions are not univocally determined by conditions. So, the strength is defined for each possible class. If the strength of one class is much greater than the strength of other classes indicated by this coefficient in the non-deterministic rule, one can conclude, that according to this rule, the classified object most likely belongs to the strongest class.

4.4. Classification of new objects

Using 'RoughClass' it is possible to classify single objects introduced by the user or sets of objects, i.e. from the files. It is also possible to perform reclassification of the coded information system using the *'leaving-one-out'* method.

Firstly, 'RoughClass' looks over the set of decision rules to find the rule matching the description of the new object. If such a rule exists, the user gets the following information : possible assignments of the object on the basis of this rule, its strength and particular objects used to create it.

Otherwise the user is informed by 'RoughClass' about the rules nearest to the description of the new object. 'RoughClass' finds the nearest rules using the distance measure [14] understood in the way described below.

Let A denote the set of condition (coded) attributes $(a_1, a_2, \ldots, a_n)$. These attributes can have different nature : ordinal or nominal. The ordinal attributes are denoted by O and nominal by N, so $A = O \cup N$. When comparing two description from the viewpoint of a condition attribute, one has to check whether its value are identical or not. In the latter case, if the attribute is ordinal, one has to measure the magnitude of the difference.

As a decision rule may contain less condition attributes than the description of a classified object, the distance between the decision rule and the classified object is computed for attributes existing in the condition part of the rule only. For other attributes it is assumed that there is no difference.

Let a given new object x be described by values $x_1, x_2, \ldots, x_m$ $(m < n)$ of the attributes occurring in the condition part of rule y described by values $y_1, y_2, \ldots, y_m$ of the same attributes. The distance of object x from the rule y can be defined as follows :

$$D = \frac{1}{m}\left(\sum_{i \in N} k_i\, d_i^p + \sum_{j \in O} k_j\, d_j^p\right)^{1/p}$$

where :

m – number of attributes occurring in the rule,
$p = 1,2, \ldots$ - natural number to be chosen by the analyst,
d_i – partial distance for nominal attribute a_i ,
$d_i = 0$ if $x_i - y_i = 0$ and $d_i = 1$ otherwise,
d_j – partial distance for ordinal attribute a ,
$d_j = \mid x_j - y_j \mid / (v_{jmax} - v_{jmin})$,
v_{jmax} and v_{jmin} – maximal and minimal value of a_j , respectively,
k_i and k_j are importance coefficients of the attributes.

Let us notice that depending on the value of p, the distance measure is more or less compensatory. For $p = 1$ the distance measure is perfectly compensatory and for $p = \infty$, it is perfectly non-compensatory. The greater is the value of p the greater is the importance of the largest partial distance in D. Typically used values of p are 1,2, and ∞ .

The importance coefficients can be determined either subjectively or taking into account the classificatory ability of the attributes (cf. [9], where the classificatory ability of the attribute a_i is expressed by the difference between the quality of classification for the set of all considered attributes and the quality of classification for the same set of attributes but without attribute a_i)

For a given object the distance from each rule is calculated and rules are ranked according to the value of D. The rules which are not more distant from the given object than a threshold τ defined by the user are displayed to him. Other restrictions for the distance between the given object and the rule can also be introduced by the user, e.g. the maximal number of differences on corresponding positions of attributes.

The presented nearest rules are described by the following information : possible assignments of the object on the basis of a given rule, its distance from the object, its strength, objects creating it. In addition, other parameters can be displayed, e.g. the number of differences between the object and the rule.

5. Conclusions

Three different microcomputer programs useful in the rough sets analysis were shortly described in this paper. The first one – 'RoughDAS' is intended to be used for to the analysis of information systems by means of the rough sets theory. The second one – 'Sensitivity Analyser', is a useful tool for the sensitivity analysis of the rough sets to changes in the definition of boundary values for quantitative attributes. The third program –'RoughClass', allows to apply the results, obtained by 'RoughDAS', for supporting classification of new objects.

References

[1] Boryczka M.: The computer algorithm to the rough analysis of data [in Polish], *Podstawy Sterowania* **25** (1986) 249-254.

[2] Boryczka M., Słowiński R.: Derivation of optimal decision algorithm from decision tables using rough sets, *Bull. Polish Acad. Sci. Techn. Ser.* **34** (3-4) (1988) 251-260.

[3] Fibak J., Pawlak Z., Słowiński K., Słowiński R. : Rough sets based decision algorithm for treatment of duodenal ulcer by HSV, *Bull. Polish Acad. Sci., Biol. Ser.* **34** (10-12) (1986) 227-246.

[4] Gruszecki G., Słowiński R., Stefanowski J.: RoughDAS - Data and Knowledge Analysis Software based on the Rough Sets Theory. User's Manual. APRO SA, Warsaw 1990, 49 pp.

[5] Krysiński J.: Rough Sets Approach to the Analysis of the Structure Activity Relationship of Quaternary Imidazolium Compounds, *Arzneimittel-Forschung/ Drug Research* **40** (II), (1990) 795-799.

[6] Matusiak D, Stefanowski J.: Sensitivity Analyser. User's Manual, [in Polish], Technical University of Pozna , 1991 (manuscript 34 pp).

[7] Nowicki R., Słowiński R., Stefanowski J. : Rough sets analysis of diagnostic capacity of vibroacoustic symptoms, *Computers and Mathematics with Applications* (1991) (to appear).

[8] Nowicki R., Słowiński R., Stefanowski J. : Evaluation of vibroacoustic symptoms by means of the rough sets theory, *Computers in Industry*, (1991) (to appear).

[9] Pawlak Z.: *Rough Sets. Some Aspects of Reasoning About Data*, Kluwer Academic Publishers, Dordrecht 1991.

[10] Pawlak Z., Słowiński K., Słowiński R.: Rough classification of patients after highly selective vagotomy for duodenal ulcer. *International Journal of Man-Machine Studies* **24** (1986) 413-433.

[11] Romański S.: Operations on Families of Sets for Exhaustive Search, Given a Monotonic Function. In Beeri C., J.W. Schmidt, Dayal W. (eds.) *Proc. 3rd Int. Conf. on Data and Knowledge Bases*, Jerusalem, Israel, June 28-30, 1988, 310-322.

[12] Słowiński K., Słowiński R.: Sensitivity analysis of rough classification. *International Journal of Man-Machine Studies* **32** (1990), 693-705.

[13] Słowiński K., Słowiński R., Stefanowski J.: Rough sets approach to analysis of data from peritoneal lavage in acute pancreatitis.*Medical Informatics*, **13** (3) (1989), 143-159.

[14] Stefanowski J. : Classification Support Based on the rough sets theory.*Proceedings of the IIASA Workshop on User-Oriented Methodology and Techniques of Decision Analysis and Support*, Serock, Sept.9-13, 1991, (to appear).

[15] Wong.S.K.M.,Ye Li, Ziarko R.: Comparison of rough sets and statistical methods in inductive learning. *International Journal of Man-Machine Studies* **24** (1986) 53-72.

APPENDIX

GLOSSARY OF BASIC CONCEPS

1. Information system

An **information system** is the 4-tuple $S = < U, Q, V, f >$, where

U is a finite set of **objects** (observations, states, cases, events,...),

Q is a finite set of **attributes** (features, variables, characteristics,...),

$V = \bigcup_{q \in Q} V_q$ and V_q is a **domain** of the attribute q and

$f: U \times Q \rightarrow V$ is a total function (alternative denotation ρ) such that $f(x,q) \in V_q$ for every $q \in Q$, $x \in U$, called **information function**. Any pair (q,v), $q \in Q$, $v \in V_q$ is called **descriptor** in S.

The information system is in fact a finite data table, columns of which are labelled by attributes, rows are labelled by objects and the entry in columns q and row x has the value $f(x,q)$. Each row in the table represents information about an object in S. The information system is also called **knowledge representation system**

2. Indiscernibility relation

Let $S = < U, Q, V, f >$ be an information system and let $P \subseteq Q$ and $x, y \in U$. We say that x and y are **indiscernible** by the set of attributes P in S iff $f(x,q) = f(y,q)$ for every $q \in P$. Thus, every $P \subseteq Q$ generates a binary relation on U, which is called **indiscernibility relation**, denoted by $IND(P)$ (alternative denotation $\tilde{P}$). $IND(P)$ is an equivalence relation for any P and the family of all equivalence classes of relation $IND(P)$ is denoted by $U \mid IND(P)$ or, in short, by $U \mid P$ (alternative denotation P^*). The family is also called **P-basic knowledge** about U in S.

Eqivalence classes of relation $IND(P)$ are called **P-elementary sets (concepts or categories of P-basic knowlege)** in S. $[x]_{IND(P)}$ denotes a P-elementary set including object x. Q-elementary sets in S are called **atoms.**

$Des_P(X)$ denotes a **description** of P- elementary set $X \in U \mid IND(P)$ in terms of values of attributes from P, i.e.

$$Des_P(X) = \left\{(q, v) : \ f(x, q) = v, \ \forall x \in X, \ \forall q \in P\right\}$$

3. Approximation of sets

The idea of the **rough set** consists of the approximation of a set by a pair of sets, called the lower and the upper approximation of this set. Suppose we are given information system S. With each subset $Y \subseteq U$ and an equivalence relation $IND(P)$ we associate two subsets:

$$\underline{P}Y = \bigcup\{X \in U \mid IND(P) : \ X \subseteq Y\}$$

$$\overline{P}Y = \bigcup\{X \in U \mid IND(P) : \ X \cap Y \neq \emptyset\}$$

called the **P-lower** and the **P- upper approximation** of Y, respectively.
The **P-boundary** of set Y is defined as

$$BN_P(Y) = \overline{P}Y - \underline{P}Y$$

Set $\underline{P}Y$ is the set of all elements of U, which can be **certainly** classified as elements of Y, employing the set of attributes P. Set $\overline{P}Y$ is the set of elements of U which can be **possibly** classified as elements of Y, using the set of attributes P. The set $BN_P(Y)$ is the set of elements which **cannot be certainly** classified to Y using the set of attributes P.

The following denotations are also employed:

$POS_P(X) = \underline{P}X$, P– **positive region** of X

$NEG_P(X) = U - \overline{P}X$, P– **negative region** of X

$BN_P(X)$, P– **borderline (or doubtful) region** of X

(alternative denotation $Pos_P(X)$, $Neg_P(X)$ and $Bn_P(X)$).

An **accuracy of approximation** of set $Y \subseteq U$ by P in S or, in short, accuracy of Y, is defined as:

$$\alpha_P(Y) = \frac{card(\underline{P}Y)}{card(\overline{P}Y)}$$

(alternative denotation $\mu_P(Y)$ or $\beta_P(Y)$) .

4. Approximation of classifications

Let $\mathcal{Y} = \{Y_1, Y_2, \ldots, Y_n\}$ be a **classification** (partition) of U in S, and $P \subseteq Q$. Subsets Y_j $(j = 1, \ldots, n)$ are classes of $\mathcal{Y}$. Then, $\underline{P}\mathcal{Y} = \{\underline{P}Y_1, \underline{P}Y_2, \ldots, \underline{P}Y_n\}$ and $\overline{P}\mathcal{Y} = \{\overline{P}Y_1, \overline{P}Y_2, \ldots, \overline{P}Y_n\}$ are called the ***P*-lower** and the ***P*-upper approximation** of classification $\mathcal{Y}$, respectively. $\underline{P}\mathcal{Y}$ is also called ***P*– positive region** of $\mathcal{Y}$. $BN_P(\mathcal{Y}) = \overline{P}\mathcal{Y} - \underline{P}\mathcal{Y}$ is called ***P*–boundary** of $\mathcal{Y}$ or ***P*– doubtful region** of $\mathcal{Y}$.

The **accuracy of classification** is defined as

$$\alpha_P(\mathcal{Y}) = \frac{\sum_{j=1}^{n} card(\underline{P}Y_j)}{\sum_{j=1}^{n} card(\overline{P}Y_j)}$$

(alternative denotation $\mu_P(\mathcal{Y})$ or $\beta_P(\mathcal{Y})$).

The coefficient

$$\gamma_P(\mathcal{Y}) = \frac{\sum_{j=1}^{n} card(\underline{P}Y_j)}{card(U)}$$

is called the **quality of approximation** of $\mathcal{Y}$ by P or, in short, **quality of classification.** It expresses the ratio of all P-correctly classified objects to all objects in the system.

5. Reduction of attributes

The set of attributes $R \subseteq Q$ **depends** on the set of attributes $P \subseteq Q$ in S (denotation $P \to R$) iff $IND(P) \subseteq IND(R)$. Discovering dependencies between attributes is of primary importance in the rough sets approach to data analysis.

Another important issue is that of **attribute reduction**, in such a way that the reduced set of attributes provides the same quality of classification as the original set of attributes. The minimal subset $R \subseteq P \subseteq Q$ such that $\gamma_P(\mathcal{Y}) = \gamma_R(\mathcal{Y})$ is called **$\mathcal{Y}$- reduct** of P (or, simply, **reduct** if there is no ambiguity in the understanding of $\mathcal{Y}$). $\mathcal{Y}$-reduct of Q is also called **minimal set (or subset)** in S.

Let us notice that an information system may have more than one $\mathcal{Y}$-reduct. $RED_{\mathcal{Y}}(P)$ is the family of all $\mathcal{Y}$-reducts of P. Intersection of all P-reducts of P is called **$\mathcal{Y}$-core** of P, i.e. $CORE_{\mathcal{Y}}(P) = \bigcap RED_{\mathcal{Y}}(P)$. The $\mathcal{Y}$-core of Q is the set of the most characteristic attributes which cannot be eliminated from S without decreasing the quality of approximation of classification $\mathcal{Y}$.

6. Decision tables

An information system can be seen as decision table assuming that $Q = C \cup D$ and $C \cap D = \emptyset$, where C is a set of **condition attributes**, and D, a set of **decision attributes.**

Decision table $DT = < U, C \cup D, V, f >$ is **deterministic (consistent)** iff $C \to D$ and, otherwise, it is **non-deterministic (inconsistent)**. The deterministic decision table **uniquely** describes the **decisions (actions)** to be made when some **conditions** are satisfied. In the case of a non-deterministic table, decisions are **not uniquely** determined by the conditions. Instead, a subset of decisions is defined which could be taken under circumstances determined by conditions.

From the decision table a set of decision rules can be derived. Let $U \mid IND(C)$ be a family of all C-elementary sets called **condition classes** in DT, denoted by X_i ($i = 1, \ldots, k$). Let, moreover, $U \mid IND(D)$ be a family of all D-elementary sets called **decision classes** in DT, denoted by Y_j ($j = 1, \ldots, n$).

$Des_C(X_i) \Rightarrow Des_D(Y_j)$ is called **(C, D)-decision rule.** The rules are logical statements (if...then...) relating descriptions of condition and decision classes.

The set of decision rules for each decision class Y_j ($j = 1, \ldots, n$) is denoted by $\{r_{ij}\}$. Precisely,

$$\{r_{ij}\} = \{Des_C(X_i) \Rightarrow Des_D(Y_j) : \ X_i \cap Y_j \neq \emptyset, \ i = 1, \ldots, k\}$$

Rule r_{ij} is **deterministic (consistent or certain)** in DT iff $X_i \subseteq Y_j$, and r_{ij} is **non-deterministic (inconsistent or possible)** in DT, otherwise.

The set of decision rules for all decision classes is called **decision algorithm.**

Reference

Pawlak, Z. (1991). *Rough Sets. Theoretical Aspects of Reasoning about Data.* Kluwer Academic Publishers, Dordrecht.

SUBJECT INDEX

THEORY AND DECISION LIBRARY

SERIES D: SYSTEM THEORY, KNOWLEDGE ENGINEERING AND PROBLEM SOLVING

1. E.R. Caianiello and M.A. Aizerman (eds.): *Topics in the General Theory of Structures*. 1987 ISBN 90-277-2451-2
2. M.E. Carvallo (ed.): *Nature, Cognition and System I*. Current Systems-Scientific Research on Natural and Cognitive Systems. With a Foreword by G.J. Klir. 1988 ISBN 90-277-2740-6
3. A. Di Nola, S. Sessa, W. Pedrycz and E. Sanchez: *Fuzzy Relation Equations and Their Applications to Knowledge Engineering*. With a Foreword by L.A. Zadeh. 1989 ISBN 0-7923-0307-5
4. S. Miyamoto: *Fuzzy Sets in Information Retrieval and Cluster Analysis*. 1990 ISBN 0-7923-0721-6
5. W.H. Janko, M. Roubens and H.-J. Zimmermann (eds.): *Progress in Fuzzy Sets and Systems*. 1990 ISBN 0-7923-0730-5
6. R. Slowinski and J. Teghem (eds.): *Stochastic versus Fuzzy Approaches to Multiobjective Mathematical Programming under Uncertainty*. 1990 ISBN 0-7923-0887-5
7. P.L. Dann, S.H. Irvine and J.M. Collis (eds.): *Advances in Computer-Based Human Assessment*. 1991 ISBN 0-7923-1071-3
8. V. Novák, J. Ramík, M. Mareš, M. Černý and J. Nekola (eds.): *Fuzzy Approach to Reasoning and Decision-Making*. 1992 ISBN 0-7923-1358-5
9. Z. Pawlak: *Rough Sets*. Theoretical Aspects of Reasoning about Data. 1991 ISBN 0-7923-1472-7
10. M.E. Carvallo (ed.): *Nature, Cognition and System II*. Current Systems-Scientific Research on Natural and Cognitive Systems. Vol. 2: On Complementarity and Beyond. 1992 ISBN 0-7923-1788-2
11. R. Słowiński (ed.): *Intelligent Decision Support*. Handbook of Applications and Advances of the Rough Sets Theory. 1992 ISBN 0-7923-1923-0

KLUWER ACADEMIC PUBLISHERS – DORDRECHT / BOSTON / LONDON